Therapeutic Effects of Natural Products on Human Diseases

Therapeutic Effects of Natural Products on Human Diseases

Seung Ho Lee

Basel • Beijing • Wuhan • Barcelona • Belgrade • Novi Sad • Cluj • Manchester

Editor
Seung Ho Lee
Department of
Nano-Bioengineering
Incheon National University
Incheon
Korea, South

Editorial Office
MDPI AG
Grosspeteranlage 5
4052 Basel, Switzerland

This is a reprint of articles from the Special Issue published online in the open access journal *Life* (ISSN 2075-1729) (available at: www.mdpi.com/journal/life/special_issues/1I6AUEYCF3).

For citation purposes, cite each article independently as indicated on the article page online and as indicated below:

Lastname, A.A.; Lastname, B.B. Article Title. *Journal Name* **Year**, *Volume Number*, Page Range.

ISBN 978-3-7258-2284-3 (Hbk)
ISBN 978-3-7258-2283-6 (PDF)
doi.org/10.3390/books978-3-7258-2283-6

© 2024 by the authors. Articles in this book are Open Access and distributed under the Creative Commons Attribution (CC BY) license. The book as a whole is distributed by MDPI under the terms and conditions of the Creative Commons Attribution-NonCommercial-NoDerivs (CC BY-NC-ND) license.

Contents

About the Editor vii

Seung-Ho Lee
Therapeutic Effects of Natural Products on Human Diseases
Reprinted from: *Life* **2024**, *14*, 1166, doi:10.3390/life14091166 1

Irshad Wani, Sushruta Koppula, Aayushi Balda, Dithu Thekkekkara, Ankush Jamadagni and Prathamesh Walse et al.
An Update on the Potential of Tangeretin in the Management of Neuroinflammation-Mediated Neurodegenerative Disorders
Reprinted from: *Life* **2024**, *14*, 504, doi:10.3390/life14040504 5

So-Young Han and Dong-Soon Im
Evodiamine Alleviates 2,4-Dinitro-1-Chloro-Benzene-Induced Atopic Dermatitis-like Symptoms in BALB/c Mice
Reprinted from: *Life* **2024**, *14*, 494, doi:10.3390/life14040494 27

Ju-Hyun Lee and Dong-Soon Im
Magnolol Reduces Atopic Dermatitis-like Symptoms in BALB/c Mice
Reprinted from: *Life* **2024**, *14*, 339, doi:10.3390/life14030339 38

Yeong-Geun Lee, Tae Hyun Kim, Jeong Eun Kwon, Hyunggun Kim and Se Chan Kang
Cytotoxic Effects of Ardisiacrispin A from *Labisia pumila* on A549 Human Lung Cancer Cells
Reprinted from: *Life* **2024**, *14*, 276, doi:10.3390/life14020276 51

Carmen Valadez-Vega, Olivia Lugo-Magaña, Lorenzo Mendoza-Guzmán, José Roberto Villagómez-Ibarra, Raul Velasco-Azorsa and Mirandeli Bautista et al.
Antioxidant Activity and Anticarcinogenic Effect of Extracts from *Bouvardia ternifolia* (Cav.) Schltdl.
Reprinted from: *Life* **2023**, *13*, 2319, doi:10.3390/life13122319 61

Chan-Ok Son, Mi-Hyeon Hong, Hye-Yoom Kim, Byung-Hyuk Han, Chang-Seob Seo and Ho-Sub Lee et al.
Sibjotang Protects against Cardiac Hypertrophy In Vitro and In Vivo
Reprinted from: *Life* **2023**, *13*, 2307, doi:10.3390/life13122307 74

Keun Tae Park, Yong Jae Jeon, Hyo In Kim and Woojin Kim
Antinociceptive Effect of *Dendrobii caulis* in Paclitaxel-Induced Neuropathic Pain in Mice
Reprinted from: *Life* **2023**, *13*, 2289, doi:10.3390/life13122289 92

Giulia Minniti, Lucas Fornari Laurindo, Nathalia Mendes Machado, Lidiane Gonsalves Duarte, Elen Landgraf Guiguer and Adriano Cressoni Araujo et al.
Mangifera indica L., By-Products, and Mangiferin on Cardio-Metabolic and Other Health Conditions: A Systematic Review
Reprinted from: *Life* **2023**, *13*, 2270, doi:10.3390/life13122270 109

Sibhghatulla Shaikh, Shahid Ali, Jeong Ho Lim, Khurshid Ahmad, Ki Soo Han and Eun Ju Lee et al.
Virtual Insights into Natural Compounds as Potential 5-Reductase Type II Inhibitors: A Structure-Based Screening and Molecular Dynamics Simulation Study
Reprinted from: *Life* **2023**, *13*, 2152, doi:10.3390/life13112152 127

Hlalanathi Gwanya, Sizwe Cawe, Ifeanyi Egbichi, Nomagugu Gxaba, Afika-Amazizi Mbuyiswa and Samkele Zonyane et al.
Bowiea volubilis: From "Climbing Onion" to Therapeutic Treasure—Exploring Human Health Applications
Reprinted from: *Life* **2023**, *13*, 2081, doi:10.3390/life13102081 . . . . **140**

Sulaiman A. Alsalamah, Mohammed Ibrahim Alghonaim, Mohammed Jusstaniah and Tarek M. Abdelghany
Anti-Yeasts, Antioxidant and Healing Properties of Henna Pre-Treated by Moist Heat and Molecular Docking of Its Major Constituents, Chlorogenic and Ellagic Acids, with *Candida albicans* and *Geotrichum candidum* Proteins
Reprinted from: *Life* **2023**, *13*, 1839, doi:10.3390/life13091839 . . . . **159**

Wendy N. Phoswa and Kabelo Mokgalaboni
Comprehensive Overview of the Effects of *Amaranthus* and *Abelmoschus esculentus* on Markers of Oxidative Stress in Diabetes Mellitus
Reprinted from: *Life* **2023**, *13*, 1830, doi:10.3390/life13091830 . . . . **178**

Neli Milenova Vilhelmova-Ilieva, Ivanka Nikolova Nikolova, Nadya Yordanova Nikolova, Zdravka Dimitrova Petrova, Madlena Stephanova Trepechova and Dora Ilieva Holechek et al.
Antiviral Potential of Specially Selected Bulgarian Propolis Extracts: In Vitro Activity against Structurally Different Viruses
Reprinted from: *Life* **2023**, *13*, 1611, doi:10.3390/life13071611 . . . . **198**

Aisha M. H. Al-Rajhi, Husam Qanash, Majed N. Almashjary, Mohannad S. Hazzazi, Hashim R. Felemban and Tarek M. Abdelghany
Anti-*Helicobacter pylori*, Antioxidant, Antidiabetic, and Anti-Alzheimer's Activities of Laurel Leaf Extract Treated by Moist Heat and Molecular Docking of Its Flavonoid Constituent, Naringenin, against Acetylcholinesterase and Butyrylcholinesterase
Reprinted from: *Life* **2023**, *13*, 1512, doi:10.3390/life13071512 . . . . **212**

Kyung-Hyun Cho, Ji-Eun Kim, Tomohiro Komatsu and Yoshinari Uehara
Protection of Liver Functions and Improvement of Kidney Functions by Twelve Weeks Consumption of Cuban Policosanol (Raydel®) with a Decrease of Glycated Hemoglobin and Blood Pressure from a Randomized, Placebo-Controlled, and Double-Blinded Study with Healthy and Middle-Aged Japanese Participants
Reprinted from: *Life* **2023**, *13*, 1319, doi:10.3390/life13061319 . . . . **232**

About the Editor

Seung Ho Lee

Seung Ho Lee is a professor in the Department of Nano-Bioengineering at Incheon National University, South Korea. He earned his Ph.D. in biochemistry and cellular biology from the Medical School of Osaka University, Japan, in 2006. Following this, he conducted postdoctoral research in glycobiology and cancer at the Sanford Burnham Prebys Institute in San Diego, USA. Since 2012, he has been a faculty member at Incheon National University, where his research focuses on developing therapeutic candidates from natural products and investigating their mechanisms of action in the progression of various diseases, including Alzheimer's disease, atopic dermatitis, cancer metastasis, and photoaging.

Editorial

Therapeutic Effects of Natural Products on Human Diseases

Seung-Ho Lee

Department of Nano-Bioengineering, Incheon National University, 119 Academy-ro, Incheon 22012, Republic of Korea; seungho@inu.ac.kr; Tel.: +82-31-835-8269

1. Introduction

Natural products have long served as potential sources of therapeutic drugs. Numerous pathophysiological conditions, such as inflammation, cancer, viral infection, immunological disorders, and metabolic diseases, have been treated with concoctions or concentrated extracts of natural products [1]. Although numerous drugs have been developed to treat the aforementioned conditions, many diseases still cannot be effectively cured with existing drugs. In addition, discovering new drugs that cause little or no side effects poses significant challenges. Natural products, which are compounds or substances derived from living organisms, could provide answers to these problems because they contain various components that can exhibit unexpected biological properties. Furthermore, individual ingredients derived from natural sources are expected to have synergistic effects when used in traditional medicine [2]. In the last 20 years, approximately one-third of the drugs approved by the Food and Drug Administration have been developed from medicinal plants and their derivates [3]. In addition, the use of various medicinal plant extracts as prescription drugs has gradually expanded in developed countries [4–6]. Considering the complex nature of disease progression, drugs containing a single active compound may be ineffective. Natural products for which historical records of therapeutic efficacy are available afford significant advantages in terms of the development of innovative medicines, and recent advances in scientific technologies that can be utilized in profiling analyses of active components, computational prediction, and in vitro/in vivo disease models have reduced the time required to develop innovative drugs based on natural products.

In this Special Issue of *Life* titled, "Therapeutic Effects of Natural Products on Human Diseases", we have gathered 11 research articles and four reviews that aim to enhance our understanding of the therapeutic effects of various natural products. This Special Issue highlights the previously undiscovered therapeutic effects of natural products against various diseases.

2. Highlights from This Special Issue

Vilhelmova-Ilieva et al. evaluated the antiviral potential of Bulgarian propolis extracts. In their study, propolis extracts from Bulgaria were found to be capable of attenuating the attachment and entry of enveloped viruses into host cells, suggesting the potential of Bulgarian propolis extracts for use as antiviral agents [7]. Al-Rajhi et al. found that the chemical composition of laurel leaf extracts (LLEs) can be altered by moist heating. The moist-heated LLEs exhibited enhanced antibiofilm activity. Additionally, the moist-heated LLEs showed higher antioxidant activity and greater inhibitory effects against α-glucosidase and butyrylcholinesterase than LLEs prepared without moist heating. Furthermore, naringenin, a constituent of LLEs, was suggested to serve as a functional component, as the feasibility of molecular docking between naringenin and acetylcholinesterase 1E66 was predicted using computational software, molecular environment 2019 (ver. MOE 2019.0102). These findings suggest that moist-heated LLEs have potential as antibacterial, antioxidant, antidiabetic, and anti-Alzheimer's disease agents [8]. Interestingly, the chemical constituents of *Lawsonia inermis* extracts decreased with moist heating, resulting in diminished antiyeast and antioxidant properties. These results indicate that the effects of moist heating vary depending on the natural product used [9].

Valadez-Vega et al. conducted an initial exploration of the therapeutic activity of *Bouvardia ternifolia* (Cav.) Schltdl., a plant traditionally used to treat inflammation in Mexico. Different parts of *B. ternifolia* (flowers, leaves, and stems) were extracted using hexane, ethyl acetate, and methanol, respectively. Each extract exhibited varying biological properties in terms of their antioxidant and anticancer activities. These findings suggest that different parts of *B. ternifolia* contain distinct constituents that can be developed as potential therapeutic agents [10].

Park et al. evaluated the antinociceptive effects of *Dendrobii caulis*, which is used as a traditional tonic in China. They found that the oral administration of *D. caulis* effectively attenuated paclitaxel-induced neuropathic pain. Additionally, vicenin-2, which has been identified as a key component of *D. caulis*, can alleviate paclitaxel-induced neuropathic pain by regulating the expression of the transient potential vanilloid 1 (TRPV1) receptor in the spinal cord. These results suggest that *D. caulis* and vicenin-2 could be promising candidates for the development of novel antinociceptive agents [11].

Son et al. investigated the protective effects of Sibjotang, a traditional Korean medicinal formula, against cardiac hypertrophy. They found that administering Sibjotang to an animal model of isoproterenol-induced cardiac hypertrophy effectively reduced the ratio of left ventricular weight to body weight as well as the expression of cardiac hypertrophy markers such as atrial natriuretic peptide and brain natriuretic peptide. These findings suggest the potential of Sibjotang in managing cardiac hypertrophy and its associated heart failure [12].

Lee et al. identified the active compound in *Labisia pumila* leaves through spectroscopic analyses, including nuclear magnetic resonance, infrared spectroscopy (IR), and mass spectrometry. In this study, for the first time, ardisiacrispin A was identified as a functional compound of *L. pumila* leaves that demonstrates anti-lung cancer efficacy [13].

The study by Shaikh et al. showed the inhibitory properties of two flavonoids, eriocitrin and silymarin, on 5α-reductase type II (5αR2) activity. Molecular dynamic simulations between the two flavonoids and 5αR2 provided evidence that both flavonoids have strong interactions with 5αR2. These results suggest the potential of these natural products as strong anti-androgenic alopecia agents [14].

In this Special Issue, two natural products with anti-atopic dermatitis properties are highlighted. Han et al. demonstrated the anti-atopic dermatitis activity of evodiamine, an alkaloid found in *Evodia* fruits [15]. Additionally, magnolol, a major component of *Magnolia officinalis*, was shown to have anti-atopic dermatitis effects by Lee et al. [16]. Both natural products were able to improve atopic dermatitis-like symptoms in 1-chloro-2,4-dinitrobenzene (DNCB)-treated BALB/c mice, suggesting their potential as anti-atopic dermatitis agents.

Cho et al. have presented clinical evidence supporting the reliability of Cuban policosanol. A randomized, placebo-controlled, double-blind clinical study demonstrated that consumption of policosanol (20 mg/day for 12 weeks) effectively protected liver function and improved kidney function. These findings provide valuable insights for the potential use of policosanol in the development of drugs aimed at reducing the toxicity of innovative therapeutics [17].

The review articles cover the biological properties of several natural products and the mechanisms underlying their therapeutic efficacy. Minniti et al. summarized the effects of *Mangifera indica* L. (mango fruit), its by-products, and mangiferin on human health, providing useful information regarding the results of clinical trials on *M. indica* L. or its derivatives [18]. Gwanya et al. highlighted the therapeutic potential of *Bowiea volubilis* subsp. *Volubilis*, commonly known as climbing or sea onion. The authors reviewed the traditional uses of *B. volubilis* and provided insights into its phytochemical composition, bioactive constituents, and therapeutic potential [19]. Phoswa et al. reviewed the effects and underlying mechanisms of *Amaranthus* and *Abelmoschus esculentus* on oxidative stress in the management of diabetes mellitus. The antioxidant properties and therapeutic potential of *Abelmoschus esculentus* summarized in this review may serve as key information for the development of antidiabetic treatments [20].

Finally, Wani et al. reported the therapeutic potential of tangeretin, a polymethoxy flavonoid, against neuroinflammation-induced neurodegenerative disorders. They summarized its effects in various neurodegenerative disease models and provided an overview of its physicochemical properties, pharmacokinetic profile, safety, and toxicity. This review demonstrates the strong potential of using tangeretin as a natural agent to treat neurodegenerative diseases [21].

3. Final Reflections

This Special Issue summarizes the biological properties of several natural products and elucidates their therapeutic potential. These findings enhance our understanding of the potential of natural products in meeting the growing demand for safe and effective therapeutics against human diseases.

Funding: This work was supported by the National Research Foundation of Korea (NRF) grant funded by the Korea government (MSIP) (No. NRF-2022R1A2C1004358).

Acknowledgments: I would like to thank all the authors involved in this publication.

Conflicts of Interest: The author declares no conflicts of interest.

References

1. Yuan, H.; Ma, Q.; Ye, L.; Piao, G. The Traditional Medicine and Modern Medicine from Natural Products. *Molecules* **2016**, *21*, 559. [CrossRef] [PubMed]
2. Kiyohara, H.; Matsumoto, T.; Yamada, H. Combination Effects of Herbs in a Multi-Herbal Formula: Expression of Juzen-Taiho-To's Immuno-Modulatory Activity on the Intestinal Immune System. *Evid. Based Complement. Alternat Med.* **2004**, *1*, 83–91. [CrossRef]
3. Newman, D.J.; Cragg, G.M. Natural Products as Sources of New Drugs over the 30 Years from 1981 to 2010. *J. Nat. Prod.* **2012**, *75*, 311–335. [CrossRef]
4. Ruhsam, M.; Hollingsworth, P.M. Authentication of Eleutherococcus and Rhodiola Herbal Supplement Products in the United Kingdom. *J. Pharm. Biomed. Anal.* **2018**, *149*, 403–409. [CrossRef]
5. Thomford, N.E.; Dzobo, K.; Chopera, D.; Wonkam, A.; Skelton, M.; Blackhurst, D.; Chirikure, S.; Dandara, C. Pharmacogenomics Implications of Using Herbal Medicinal Plants on African Populations in Health Transition. *Pharmaceuticals* **2015**, *8*, 637–663. [CrossRef] [PubMed]
6. Yatoo, M.I.; Dimri, U.; Gopalakrishnan, A.; Karthik, K.; Gopi, M.; Khandia, R.; Saminathan, M.; Saxena, A.; Alagawany, M.; Farag, M.R.; et al. Beneficial Health Applications and Medicinal Values of Pedicularis Plants: A Review. *Biomed. Pharmacother.* **2017**, *95*, 1301–1313. [CrossRef]
7. Vilhelmova-Ilieva, N.M.; Nikolova, I.N.; Nikolova, N.Y.; Petrova, Z.D.; Trepechova, M.S.; Holechek, D.I.; Todorova, M.M.; Topuzova, M.G.; Ivanov, I.G.; Tumbarski, Y.D. Antiviral Potential of Specially Selected Bulgarian Propolis Extracts: In Vitro Activity against Structurally Different Viruses. *Life* **2023**, *13*, 1611. [CrossRef] [PubMed]
8. Al-Rajhi, A.M.H.; Qanash, H.; Almashjary, M.N.; Hazzazi, M.S.; Felemban, H.R.; Abdelghany, T.M. Anti-*Helicobacter pylori*, Antioxidant, Antidiabetic, and Anti-Alzheimer's Activities of Laurel Leaf Extract Treated by Moist Heat and Molecular Docking of Its Flavonoid Constituent, Naringenin, against Acetylcholinesterase and Butyrylcholinesterase. *Life* **2023**, *13*, 1512. [CrossRef]
9. Alsalamah, S.A.; Alghonaim, M.I.; Jusstaniah, M.; Abdelghany, T.M. Anti-Yeasts, Antioxidant and Healing Properties of Henna Pre-Treated by Moist Heat and Molecular Docking of Its Major Constituents, Chlorogenic and Ellagic Acids, with *Candida albicans* and *Geotrichum candidum* Proteins. *Life* **2023**, *13*, 1839. [CrossRef]
10. Valadez-Vega, C.; Lugo-Magana, O.; Mendoza-Guzman, L.; Villagomez-Ibarra, J.R.; Velasco-Azorsa, R.; Bautista, M.; Betanzos-Cabrera, G.; Morales-Gonzalez, J.A.; Madrigal-Santillan, E.O. Antioxidant Activity and Anticarcinogenic Effect of Extracts from *Bouvardia ternifolia* (Cav.) Schltdl. *Life* **2023**, *13*, 2319. [CrossRef]
11. Park, K.T.; Jeon, Y.J.; Kim, H.I.; Kim, W. Antinociceptive Effect of *Dendrobii caulis* in Paclitaxel-Induced Neuropathic Pain in Mice. *Life* **2023**, *13*, 2289. [CrossRef] [PubMed]
12. Son, C.O.; Hong, M.H.; Kim, H.Y.; Han, B.H.; Seo, C.S.; Lee, H.S.; Yoon, J.J.; Kang, D.G. Sibjotang Protects against Cardiac Hypertrophy in Vitro and in Vivo. *Life* **2023**, *13*, 2307. [CrossRef] [PubMed]
13. Lee, Y.G.; Kim, T.H.; Kwon, J.E.; Kim, H.; Kang, S.C. Cytotoxic Effects of Ardisiacrispin a from *Labisia pumila* on A549 Human Lung Cancer Cells. *Life* **2024**, *14*, 276. [CrossRef] [PubMed]
14. Shaikh, S.; Ali, S.; Lim, J.H.; Ahmad, K.; Han, K.S.; Lee, E.J.; Choi, I. Virtual Insights into Natural Compounds as Potential 5alpha-Reductase Type Ii Inhibitors: A Structure-Based Screening and Molecular Dynamics Simulation Study. *Life* **2023**, *13*, 2152. [CrossRef] [PubMed]

15. Han, S.Y.; Im, D.S. Evodiamine Alleviates 2,4-Dinitro-1-Chloro-Benzene-Induced Atopic Dermatitis-like Symptoms in Balb/C Mice. *Life* **2024**, *14*, 494. [CrossRef]
16. Lee, J.H.; Im, D.S. Magnolol Reduces Atopic Dermatitis-Like Symptoms in Balb/C Mice. *Life* **2024**, *14*, 339. [CrossRef]
17. Cho, K.H.; Kim, J.E.; Komatsu, T.; Uehara, Y. Protection of Liver Functions and Improvement of Kidney Functions by Twelve Weeks Consumption of Cuban Policosanol (Raydel®) with a Decrease of Glycated Hemoglobin and Blood Pressure from a Randomized, Placebo-Controlled, and Double-Blinded Study with Healthy and Middle-Aged Japanese Participants. *Life* **2023**, *13*, 1319. [CrossRef]
18. Minniti, G.; Laurindo, L.F.; Machado, N.M.; Duarte, L.G.; Guiguer, E.L.; Araujo, A.C.; Dias, J.A.; Lamas, C.B.; Nunes, Y.C.; Bechara, M.D.; et al. *Mangifera indica* L., by-Products, and Mangiferin on Cardio-Metabolic and Other Health Conditions: A Systematic Review. *Life* **2023**, *13*, 2270. [CrossRef]
19. Gwanya, H.; Cawe, S.; Egbichi, I.; Gxaba, N.; Mbuyiswa, A.A.; Zonyane, S.; Mbolekwa, B.; Manganyi, M.C. *Bowiea volubilis*: From "Climbing Onion" to Therapeutic Treasure-Exploring Human Health Applications. *Life* **2023**, *13*, 2081. [CrossRef]
20. Phoswa, W.N.; Mokgalaboni, K. Comprehensive Overview of the Effects of *Amaranthus* and *Abelmoschus esculentus* on Markers of Oxidative Stress in Diabetes Mellitus. *Life* **2023**, *13*, 1830. [CrossRef]
21. Wani, I.; Koppula, S.; Balda, A.; Thekkekkara, D.; Jamadagni, A.; Walse, P.; Manjula, S.N.; Kopalli, S.R. An Update on the Potential of Tangeretin in the Management of Neuroinflammation-Mediated Neurodegenerative Disorders. *Life* **2024**, *14*, 504. [CrossRef] [PubMed]

Disclaimer/Publisher's Note: The statements, opinions and data contained in all publications are solely those of the individual author(s) and contributor(s) and not of MDPI and/or the editor(s). MDPI and/or the editor(s) disclaim responsibility for any injury to people or property resulting from any ideas, methods, instructions or products referred to in the content.

Review

An Update on the Potential of Tangeretin in the Management of Neuroinflammation-Mediated Neurodegenerative Disorders

Irshad Wani [1,†], Sushruta Koppula [2,†], Aayushi Balda [1], Dithu Thekkekkara [1], Ankush Jamadagni [3], Prathamesh Walse [3], Santhepete Nanjundaiah Manjula [1] and Spandana Rajendra Kopalli [4,*]

[1] Department of Pharmacology, JSS College of Pharmacy, JSS Academy of Higher Education & Research, Mysuru 570015, India
[2] College of Biomedical and Health Science, Konkuk University, Chungju-si 380-701, Republic of Korea; koppula@kku.ac.kr
[3] Fortem Biosciences Private Limited (Ayurvibes), No. 24, Attur, 4th Cross, Tirumala Nagar, A Block, Bangalore 560064, India
[4] Department of Integrated Bioscience and Biotechnology, Sejong University, Gwangjin-gu, Seoul 05006, Republic of Korea
* Correspondence: spandanak@sejong.ac.kr; Tel.: +82-2-6935-2619
† These authors contributed equally to this work.

Abstract: Neuroinflammation is the major cause of neurodegenerative disorders such as Alzheimer's and Parkinson's disease. Currently available drugs present relatively low efficacy and are not capable of modifying the course of the disease or delaying its progression. Identifying well-tolerated and brain-penetrant agents of plant origin could fulfil the pressing need for novel treatment techniques for neuroinflammation. Attention has been drawn to a large family of flavonoids in citrus fruits, which may function as strong nutraceuticals in slowing down the development and progression of neuroinflammation. This review is aimed at elucidating and summarizing the effects of the flavonoid tangeretin (TAN) in the management of neuroinflammation-mediated neurodegenerative disorders. A literature survey was performed using various resources, including ScienceDirect, PubMed, Google Scholar, Springer, and Web of Science. The data revealed that TAN exhibited immense neuroprotective effects in addition to its anti-oxidant, anti-diabetic, and peroxisome proliferator-activated receptor-γ agonistic effects. The effects of TAN are mainly mediated through the inhibition of oxidative and inflammatory pathways via regulating multiple signaling pathways, including c-Jun N-terminal kinase, phosphoinositide 3-kinase, mitogen-activated protein kinase, nuclear factor erythroid-2-related factor 2, extracellular-signal-regulated kinase, and CRE-dependent transcription. In conclusion, the citrus flavonoid TAN has the potential to prevent neuronal death mediated by neuroinflammatory pathways and can be developed as an auxiliary therapeutic agent in the management of neurodegenerative disorders.

Keywords: *Citrus* fruits; flavonoids; tangeretin; neuroinflammation; neurodegeneration; microglia; anti-inflammatory; antioxidant

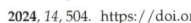

Citation: Wani, I.; Koppula, S.; Balda, A.; Thekkekkara, D.; Jamadagni, A.; Walse, P.; Manjula, S.N.; Kopalli, S.R. An Update on the Potential of Tangeretin in the Management of Neuroinflammation-Mediated Neurodegenerative Disorders. *Life* 2024, *14*, 504. https://doi.org/10.3390/life14040504

Academic Editor: Stefania Lamponi

Received: 21 March 2024
Revised: 10 April 2024
Accepted: 11 April 2024
Published: 14 April 2024

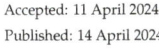

Copyright: © 2024 by the authors. Licensee MDPI, Basel, Switzerland. This article is an open access article distributed under the terms and conditions of the Creative Commons Attribution (CC BY) license (https://creativecommons.org/licenses/by/4.0/).

1. Introduction

The global economic impact of neurodegenerative illnesses is increasing considerably as life expectancy increases [1]. The pathogenic pathways that cause neurodegenerative diseases such as Alzheimer's disease (AD), Parkinson's disease (PD), and Amyotrophic lateral sclerosis (ALS) are not fully understood. Several factors are at work, including genetic, environmental, and endogenous impacts. Common pathogenic pathways include aberrant protein dynamics, oxidative stress with reactive oxygen species (ROS), mitochondrial dysfunction, DNA damage, neurotrophin dysfunction, and neuroinflammatory processes [2]. Neuroinflammation is a defense mechanism that initially protects the brain by removing or reducing different pathogens [3]. This inflammatory reaction may be advantageous, stimulating tissue repair and cleaning away cellular waste. Persistent inflammatory responses, on the other hand, are detrimental and hinder regeneration [4]. Inflammatory activation may persist due to endogenous (such as genetic mutations and protein aggregation) or exogenous (such as infection, trauma, and medicines) reasons [5]. Further, it is well known that microglia and astrocytes are engaged in persistent inflammatory responses that may lead to neurodegenerative diseases [6]. Despite extensive pre-clinical investigations, the outcome of stringent therapeutic strategies is poor.

Flavonoids, a group of natural chemicals with a polyphenolic structure, are present in almost 9000 varieties of plants. Familiar sources of these phytochemicals include fruits, tea, chocolate, vegetables, and many beverages [7,8]. Plant flavonoids are classified as flavanones, flavanols, isoflavonoids, flavones, chalcones, and anthocyanins according of their structural resemblance to two benzene rings joined by a heterocyclic ring and their availability as glycosides, aglycones, or methylated derivatives [9,10]. *Citrus* plants are the primary source of polymethoxyflavones (PMFs), flavones with two or more than 2 -OCH$_3$ groups attached to the benzo-γ-pyrone (C_6-C_3-C_6, 15-carbon ring) structure and in the C4th position, with one carbonyl group attached to the ring. Over 105 million metric tons of *Citrus* are produced annually around the world, with oranges making up more than half of that total.

Traditionally, *Citrus* peel, such as orange peel, is used by several countries in food, beverage, and cosmetic preparations, and to treat conditions like ringworm infections, stomach upsets, coughs, skin inflammation, changes in blood pressure, and muscle discomfort [11–14]. A prominent component, in addition to terpenoids and other volatile oils, is PMFs, which are abundant in *Citrus* peels [14]. *Citrus* flavonoids, especially methoxylated flavones [15,16], have shown immense therapeutic potential [17]. The most popular *Citrus*-plant-derived flavonoids include tangeretin (TAN), naringin, hesperidin, nobiletin, naringenin, and hesperetin. Among these, TAN, found only in *Citrus* fruit peels such as *C. aurantium* L., and *C. reticulata Blanco* species, is known to be used in traditional Chinese medicine, wherein TAN is the principal active component [18]. Ethnopharmacological studies on TAN have revealed numerous biological and functional properties, including antioxidant, anti-carcinogenic, anti-inflammatory, anti-atherogenic, and hepatoprotective properties, as well as actions on the neurological and proliferative fronts [19,20]. Pre-clinical studies have also indicated that TAN possesses neuroprotective and cognitive-enhancing benefits in cellular and in vivo experimental models including AD and PD [21–24]. Mechanistic-based studies indicated that TAN is involved in the regulation of multiple oxidative and inflammatory-related signaling pathways such as the inhibition of ROS, pro-inflammatory mediators, the regulation of NF-κB, MAPKs and JNK, and AKT activation [25]. These investigations indicated TAN's potential role in regulating inflammatory-mediated signaling pathways.

Several studies have investigated the pharmacological importance and favorable biological activities of TAN in various disorders. Many reviews have explored the beneficial effects of TAN in terms of renoprotection, its anti-tumor, antidiabetic, immunomodulatory and antioxidant properties, and neuroprotection [7]. Further, the effects of TAN on the modulation of inflammation-mediated cancer-related pathways including PI3K/AKT/mTOR, Notch, and MAPK signaling, suppressing cell proliferation and inducing autophagy in various types of cancer cells, has also been addressed [10]. Although previous studies indicated the beneficial role of *Citrus*-fruit-derived flavonoids, including TAN, in neuroprotection [26], these studies mainly focused on the overall role of *Citrus* flavonoids in cholinergic regulation, regulating N-methyl-D-aspartate (NMDA) receptor hypofunction, ischemic injury, and oxidative stress signaling in toxin-induced models, with little emphasis on TAN and its pathways in neuroinflammation [26]. In this review, we aimed to understand the recent updates regarding the potential of TAN, its physiochemical characteristics, its safety profile, its pharmacokinetic parameters and the underlying molecular mechanisms involved in targeting neuroinflammation-mediated neurodegenerative disorders. The data for the

2. Chemistry and Sources of TAN

Natural O-methylated flavones, known as PMFs, primarily found in *Citrus* plants, are well-known for their health-improving qualities. In *Citrus* plants and other natural sources of PMFs, the enzyme O-methyltransferases is involved in manufacturing PMFs [27]. At least two methoxy groups are present on the flavone skeleton in PMFs (Figure 1A). Among the approximately 80 PMFs found in plants, TAN (Figure 1B) and nobiletin (Figure 1C) are the two most common [27]. TAN, a polymethoxy-flavonoid (5,6,7,8,4'-pentamethoxy flavone), is a low-molecular-weight (average mass 372.369 Daltons and molecular formula $C_{20}H_{20}O_7$) secondary metabolite that is a member of the flavonoid class of polyphenols. The International Union of Pure and Applied Chemistry's (IUPAC's) name for TAN is 5,6,7,8-tetramethoxy-2-(4-methoxyphenyl)-chromen-4-one. Within the flavone structural subclasses, apigenin (Figure 1D) is TAN's most straightforward structural counterpart. The presence of two aromatic rings (rings A and B), connected by a three-carbon chain that cyclizes to form ring C, a double bond at the C2/C3 position, and the absence of oxygenation at the C-3 position are the distinguishing structural characteristics of flavones. TAN is most likely produced synthetically by methylating all the hydroxyl groups of its nearest structural counterpart, i.e., Nortangeretin (5,6,7,8,4'-pentahydroxyflavone; Figure 1E) [26].

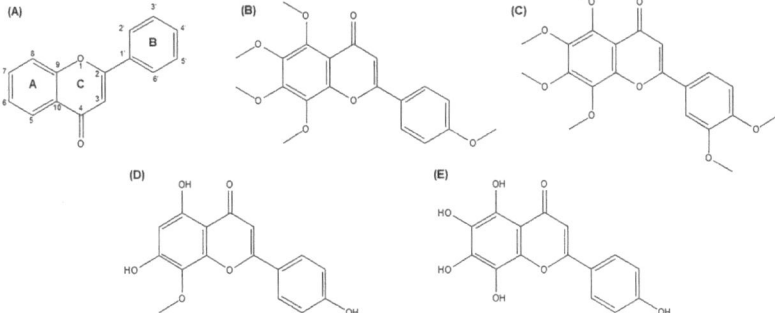

Figure 1. Citrus flavonoid structures. (**A**) Flavone skeleton, (**B**) tangeretin, (**C**) nobiletin, (**D**) apigenin, and (**E**) nortangeretin.

The polymethoxylated flavonoids, including TAN, are widely distributed in citrus fruits and notably rich in fruit peels. Numerous other, smaller-scale, natural sources of TAN have been noted, including Fructus species [28]. TAN and other related polymethoxylated flavonoids are frequently used for quality control of citrus juices due to their abundance in citrus fruits and variable concentrations between species. TAN is currently primarily consumed through the consumption of *Citrus* fruits. *C. tangerina*, *C. sinensis*, and *C. aurantium* fruit peels are highly known for their high levels of TAN, as well as the 4'-methoxylated derivatives of TAN and nobiletin. TAN, hesperidin, and beta-sitosterol have been extracted from *C. jambhiri* fruit peel, TAN, hesperidin, and beta-sitosterol [29]. The *Citrus* fruit bergamot, *C. bergamia*, also contains TAN [30]. TAN has also been found in several different types of *Citrus* species, including *C. unshiu*, *C. reticulata*, *C. tachibana*, *C. depressa*, *C. paradisi* (grapefruit) [31], and *C. poonensis* [32]. The sources of TAN are summarized in Table 1.

Many laboratories have concentrated on developing extraction procedures for TAN and related polymethoxylated derivatives due to the multiple pharmacological actions of TAN and related compounds that have been identified in the last few decades. These molecules have high lipophilicity, making it convenient to extract them using organic solvents and supercritical CO_2 [33]. A comprehensive synthetic strategy has also been developed to create a consistent chemical supply of these compounds [34].

Table 1. Naturally occurring sources of TAN from plants (* study discovered PMFs but not explicitly TAN).

TAN Source	Plant Part	Reference
Citrus poonensis (C. poonensis)	Peel	[32]
C. exocarpium Rubram	Exocarp	[35]
C. reticulata	Peel	[36]
C. unshiu	Peel	[37]
C. depressa Hayata	Peel	[38]
Hura crepitans	Leaves, bark, roots	[39]
Fructus aurantia	Fruit	[28]
C. aurantifolia	Peel, fruit pulp	[40]
C. mitis Blanco	Peel	[41]
C. aurantium	Peel	[42]
C. reticulate Cv. Suavissima	Fruit	[43]
C. grandis Osbeck	Leaves	[44]
C. reticulata *, Citrus paradisi C. clementina C. sinensis C. paradise C. lumia Risso	Fruit	[45]
C. ichangensis Swingle *	Peel	[43]

2.1. Physicochemical Properties of TAN

TAN has a melting point of 154 °C, and a solid, light yellow, needle-like appearance with a 372 M + EI-MS m/z ratio [46]. TAN has hydrophilicity because of the polarity of its pyran ring. All the additional functional groups that extend from the benzene rings are CH_3 groups, which give the cell membrane a high degree of permeability [7]. In bergamot oil, TAN is highly soluble, especially at 60 °C or higher. It is insoluble in water but soluble in methanol or ethyl acetate. The applications of TAN in pharmaceutical or food science are limited by its poor water solubility and unpleasant taste [47,48]. The various physicochemical properties of TAN are summarized in Table 2. The solubility and passive permeability of TAN were found to be 19 µg/mL (low soluble) and 1.62×10^{-6} cm/s (highly permeable) using LYSA and PAMPA methods [49,50].

Table 2. Physicochemical properties of TAN.

Properties	Values
Physical Description	Solid
HBDC	0
HBAC	7
BP	566 °C @ 760 mm Hg
MP	154 °C
Solubility	8.70 mg/L @ 25 °C
LogP	1.78

Abbreviations: HBDC: hydrogen bond donor count; HBAC: hydrogen bond acceptor count; BP: boiling point; MP: melting point.

2.2. Metabolism and Pharmacokinetic Profile of TAN

The biotransformation of PMFs in vivo results in the production of metabolites and has various pharmacological characteristics. The discovery and metabolism of these PMF metabolites has attracted attention. The primary enzyme system involved in the metabolism of PMFs is cytochrome P450 (CYP), which can catalyze the hydroxylation and demethylation processes. It is believed that all *Citrus* species have the same metabolic process for producing PMFs. The number and position of hydroxyl and methoxyl groups significantly affect how PMFs are metabolized [14]. In an in vitro experiment, when TAN was incubated with Aroclor-induced rat liver microsomes, the three primary metabolites of TAN were identified as 4′-dihydroxytangeretin (Figure 2A), 3′,4′-dihydroxylnobiletin (Figure 2B), and 5,4′-dihydroxytangeretin (Figure 2C) [51]. However, four of the five metabolites in a different in vitro metabolism study using human CYP were demethylated at position 4′, with 4′-demethyltangeretin being the principal metabolite. It was interesting to note that 5,6-dihydroxylnobiletin (Figure 2D) was the second most abundant metabolite [52].

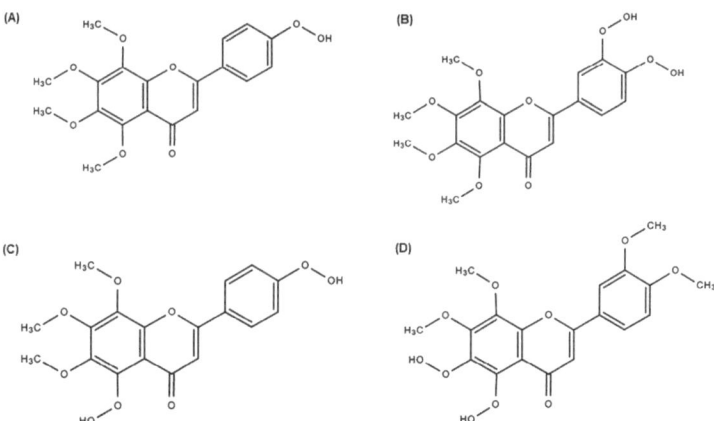

Figure 2. Metabolites of tangeretin: (A) 4′-dihydroxytangeretin, (B) 3′,4′-dihydroxylnobiletin, (C) 5,4′-dihydroxytangeretin, and (D) 5,6-dihydroxylnobiletin.

Ten metabolites were found in an in vivo biotransformation investigation of TAN using repeated gavage feeding on rats. Seven of the ten identified metabolites, including the significant metabolites 4′-demethyltangeretin and 3′,4′-dihydroxylnobiletin, were demethylated at position 4′. The second most typical location for demethylation was position 6 of ring A. Therefore, it may be inferred that TAN position 4′ is the primary site for demethylation, while position 6 is the secondary site. Additionally, the investigation determined that glucuronate–sulfate conjugates make up 38% of the TAN metabolites that are eliminated in the urine [53].

Since TAN demonstrates significant absorption and is consequently bioavailable, it offers a particular advantage over other chemically identical flavones [54]. Furthermore, the oral administration of TAN is safe [55]. Hung et al. [56] found the plasma level of TAN to be 87 ± 0.33 µg/mL (C_{max}) following oral treatment (50 mg/kg b.w.), with T_{max} and $t_{1/2}$ values of 340 ± 48.99 and 342.43 ± 71.27 min, respectively. On the other hand, the plasma concentration of TAN after intravenous (5 mg/kg b.w.) administration was 1.11 ± 0.41 µg/mL, with T_{max} and $t_{1/2}$ values of 1.11 ± 0.41 and 69.87 ± 15.72 min, respectively. After the oral and intravenous administration of TAN, the AUC values were 213.78 ± 80.63 and 78.85 ± 7.39 min µg/mL, and the absolute oral bioavailability was 27.11%. Additionally, only TAN concentrations of 0.0026 and 7.54% were found in the urine and feces after oral dosing, indicating that it was eliminated as a metabolite. The kidney had the highest concentration of TAN, followed by the lung, liver, spleen, and heart. Furthermore, the stomach and small intestine had the highest concentration at 4 h, while the cecum, colon, and rectum had the highest concentration at 12 h in the digestive tract [56].

In another investigation, when TAN was administered intravenously (10 mg/kg b.w.) to rats, the C_{max} was 2470 ± 557 ng/mL with a $t_{1/2}$ value of 166 ± 42 min and was rapidly cleared. When administered as suspension (50 mg/kg b.w.), poor bioavailability was observed. However, an oral dose of 50 mg/kg body weight resulted in a C_{max} of 65.3 ± 20.1 ng/mL and a T_{max} value of 90–120 min. When TAN was prepared with methylated β-cyclodextrin, the oral bioavailability improved (C_{max}: 135 ± 46 ng/mL) and the T_{max} value reduced to 30–90 min. This formulation increased the absolute oral bioavailability twofold (6.02%). Additionally, this research revealed that TAN was stable (recovery rates of 87–100%) under different storage settings [57]. Furthermore, this indicated that the plasma concentration of TAN (50 mg/kg b.w.) was the highest at 30 min, at 4.5 µg/mL, and declined to 0.65 µg/mL after 12 h following oral or i.p. administration [58].

2.3. Safety and Toxicity of TAN

TAN was used as a model compound for safety evaluations to examine the possibility of oral toxicity since it is one of the most common PMFs obtained from natural sources. To established TAN-induced acute oral toxicity in mice, doses of 1000, 2000, and 3000 mg/kg of TAN in an oil solution were administered to mice via gavage. Fourteen days following the injection, no fatalities were noted. However, daily, low-dose TAN administration led to a U-shaped dose–response curve with liver changes. Thus, PMF (a component of the human diet) may be employed safely under various circumstances [59].

In 2005, Vanhoecke et al. supported the results of earlier work by demonstrating the safety of TAN when administered orally to experimental mice. One piece of evidence was the absence of significant organ damage or a decline in function. These findings paved the way for subsequent human safety assessments [55]. In another study, the potential genotoxicity of TAN, determined by adding various concentration of PMF mixture to five distinct bacterial strains in vitro, was evaluated. According to the results, no mutations were found, regardless of whether ribosomal protein S9 was activated. The reported results demonstrated a good safety profile for the PMF combination with no chance of genotoxicity when using in vitro test methods [60]. However, the same group discovered a statistically insignificant positive connection between rising PMF concentrations and mice's spleen weight in a separate investigation where mice were inoculated with sheep's red blood cells (SRBCs). There was no proof that the spleen weight changed in mice without vaccination [61].

3. Neuroinflammation and Therapeutic Potential of TAN

Neuroinflammation-related neurodegenerative disorders such as PD and AD have become a significant burden as the population ages. In addition to being a pathogenic element causing many neurodegenerative illnesses, neuroinflammation is an essential biological response to neuronal tissue dysfunction [62]. The severity and frequency of neuroinflammation are the leading indicators of whether it will progress [63].

Microglia are the principal immune cells in the central nervous system (CNS), and under physiological conditions, these act similarly to macrophages. They serve as the initial line of immune defense for neurons, followed by tissue healing. But excessive microglial activation can result in neuronal death, a hallmark of neurodegenerative disorders, due to the ensuing robust cytokine production. Chronic or traumatic stresses, age-related microglial sensitization, and extended microglial activation are linked to neuroinflammation's pathological and destructive course [62,63]. Excessive cytokine production, which leads to the elevation of various cytokine-regulated signaling systems involved in inflammation, apoptosis, and autophagy, underlies the disturbed communication balance between the brain and the immune system. One of these signaling systems is the nuclear factor-κB (NF-κB) pathway, whose deregulation has already been connected to the pathophysiology of PD and AD [64,65]. Further, it is well understood that the neurophysiological function of the CNS is critically dependent on astrocytes. Astrocytes are a double-edged sword in neurodegenerative disorders. When pathological alterations associated with neurodegenerative diseases increase, inflammatory responses stimulate astrocytes and convert them into an activated state [66,67]. The presence of reactive astrocytes was observed in AD and PD experimental models, wherein they released pro-inflammatory cytokines and neuroinflammatory mediators such as TNF-α, IFN-γ, IL-1, and IL-6, leading to harmful effects and directly contributing to neurodegeneration [67–69].

The mitogen-activated protein kinase (MAPK) family [70] and the phosphatidylinositol 3-kinase (PI3K)/serine/threonine kinase (Akt)/mammalian target of the rapamycin kinase (mTOR) signaling cascade are two additional pathways regulating inflammatory response and apoptosis, whose dysfunction is crucial in transforming neuroinflammation from a biological course to a pathological response [71]. ROS, nitric oxide (NO), and prostaglandins (PGs) are secondary messengers that activate these pathways and also act as proinflammatory mediators.

On the one hand, increased ROS production from hyperactivated microglia induces the release of cytokines [62]. On the other hand, cytokines promote the generation of ROS in the mitochondria [72], which can act as secondary messengers to activate the NF-κB and MAPK family pathways [73]. Oxidative stress may also lead to the expression of an inducible isozyme of nitric oxide synthases (iNOS), resulting in excess NO that damages cell proteins, ultimately leading to neuronal cell death [74]. Currently available and advised pharmacological treatments treat neuroinflammation as a concurrent pathology of neurodegenerative illnesses like AD. The cyclooxygenase (COX) pathway is also a key participant in neuroinflammation. It is interesting to note that the increased production of prostaglandin E2 (PGE2), a highly potent neuroinflammatory mediator, is mediated by both constitutive COX-1 and inducible COX-2 isoforms. The stimulation of stress-responsive pathways, such as the NF-κB pathway, is related to COX-1 upregulation and microglia activation, whereas the overexpression of COX-2 results in direct neuronal injury [62,75].

Given the various crosstalk across signaling cascades and the proinflammatory mediator's interchangeable functions as an upstream activator and downstream effector in inflammatory signal transduction, it is clear that the process of neuroinflammation is exceptionally complex. The pathogenic development of neuroinflammation is schematically depicted in Figure 3.

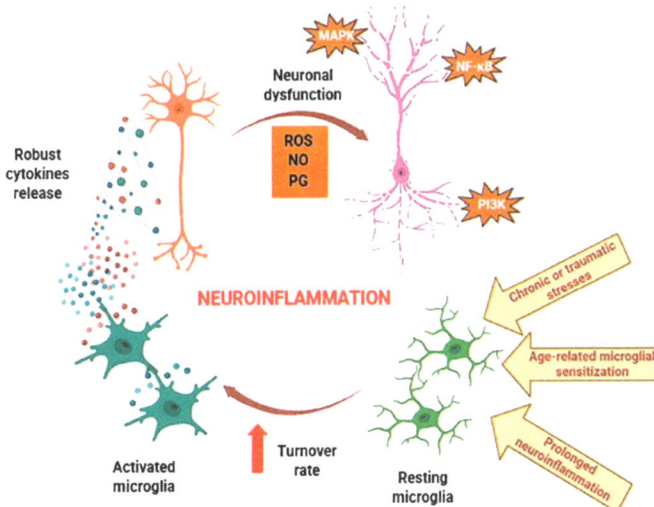

Figure 3. Schematic overview of pathologic neuroinflammation progression. Chronic or traumatic stresses, age-related microglial sensitivity, and the protracted inflammation of neurons, which cause the significant release of cytokines, significantly enhance the overall turnover rate of microglia. Cytokines and secondary messengers such as ROS, PG, and NO interfere with nerve cells' ability to function normally. NF-κB, PI3K, and MAPK pathway failures are the first cause of progressive cell degeneration. ROS: reactive oxygen species; PG: prostaglandins; NO: nitric oxide; NF-κB: nuclear factor-kapa B; PI3K: phosphoinositide 3-kinase; MAPK: mitogen-activated protein kinase.

Currently available treatments for neuroinflammation include traditional anti-inflammatory drugs [76–79], modulators of tumor necrosis factor (TNF-α)/NF-κB signaling [80–83], and plant-based natural therapies [84]. The urgent need for novel treatment approaches in neuroinflammation, exacerbated by the steadily aging global population, has led to the search for well-tolerated and brain-penetrant anti-inflammatory drugs of plant origin. Thus, scientists are interested in a vast family of flavonoids that are present in edible fruits because they have the potential to function as potent nutraceuticals that can delay the onset and progression of illnesses associated with neuroinflammation. There is mounting evidence that *Citrus* PMFs intervene in neurodegeneration and enhance brain functions, which are traits of neuroprotective agents.

PMFs, particularly TAN, have drawn the attention of natural product scientists due to their broad spectrum of bioactivities. Some of the bioactivities of PMFs include anti-inflammatory, anti-cancer, and anti-photoaging properties, the thermogenesis of white and brown adipose tissues, attenuation of the metabolic syndrome, skeletal muscle damage prevention, control of the gut flora, and neuroprotective and anti-atherogenic qualities [85–93].

3.1. Effect of TAN in PD Models

PD, which affects about 1% of people over 60 and over 10 million people globally, is a multisystem neurodegenerative disorder with the progressive loss of midbrain dopamine (DA) neurons and subsequent dopaminergic deafferentation of the basal ganglia, which results in recognizable motor disturbances like the slowing down of movement, rigidity of the muscles, and resting tremors [94,95]. In the following section, we outline the available literature and scientific studies on the potential of TAN against PD.

The first in vivo study to demonstrate that TAN can effectively cross the blood–brain barrier and has a neuroprotective effect on the brain was studied by Dexter et al. at the Department of Neuroinflammation, Imperial College of Science, London [21]. The authors reported that the chronic administration of TAN (10 mg/kg/day for 28 days) led to a marked increase in the levels of TAN in various brain regions of rats, including the hypothalamus, hippocampus and striatum. Further, TAN improved striatal dopamine depletion and reduced the 6-hydroxydopamine (6-OHDA)-induced decline in tyrosine hydroxylase-positive (TH$^+$) cells.

A previous report on an experimental PD model showed that TAN (50, 100, or 200 mg/kg body weight) treatment for 20 days after MPTP intoxication was induced in rats significantly improved motor dysfunction, memory impairments, and cognitive disabilities. Further, the altered levels of proinflammatory cytokines in MPTP-induced rats were reversed by lowering the levels of inflammatory cytokines including interleukin (IL)-1β, IL-6, and IL-2. TAN treatment recovered the dopaminergic neuronal degeneration and hippocampus neuronal loss. The authors indicated that the positive effects of TAN might be achieved through inhibiting the TNF-α, COX-2, and iNOS signaling pathways in an MPTP-induced rat PD model [48].

In addition, after chronic 1-methyl-4-phenyl-1,2,3,6-tetrahydropyridine/probenecid (MPTP/P) injections in mice, TAN increased the messenger ribonucleic acid (mRNA) levels of unfolded protein response (UPR) target genes in dopaminergic neurons and astrocytes and enhanced neuronal protection [96,97]. The data were convincing and in agreement with previous studies suggesting that neuroprotective agents targeting neuroinflammatory pathways in PD pathology were known to preserve the striatonigral integrity partially mediated by activated astrocytes [98,99].

In a transgenic *Drosophila* PD model, abnormal activity patterns were observed in flies, including a reduced climbing ability, an increased concentration of dopamine, and disrupted antioxidative enzyme status, when compared with their respective controls. However, when TAN (5, 10, and 20 µM) was provided in the diet of PD model transgenic flies for twenty-four days, the authors found that the altered climbing ability and dopamine levels were better restored compared with the results obtained using similar doses of L-Dopa, which was used as a standard. Further, the altered antioxidative defense parameters were positively regulated with TAN treatment. These results indicate that TAN is indeed able to reduce the PD symptoms in experimental models and could be further clinically explored [100].

An increased risk of incident epileptic seizures has been reported in individuals with incident Parkinson's, according to retrospective cohort research conducted in the United Kingdom. When compared to individuals without PD who do not have any seizure-provoking comorbidities, researchers discovered that PD patients with other brain disorders or multiple seizure-provoking comorbidities had the highest chance of epileptic seizures [101]. TAN treatment at 50, 100, or 200 mg/kg for 10 days in pilocarpine (30 mg/kg)-induced mice was shown to downregulate the suppression of PI3K/Akt signaling. The authors indicated that TAN alters neuronal apoptosis and ameliorates seizure severity in epilepsy-induced rats by enhancing the PI3K/Akt signaling pathway, reducing seizure-induced matrix metalloproteinases-2 (MMP-2) and matrix metalloproteinases-9 (MMP-9) activation, and lowering the level of apoptosis-inducing factor (AIF) in the nucleus via blocking AIF translocation [102]. These data indirectly suggest that TAN might be helpful in PD patients with epileptic seizure conditions.

3.2. Effect of TAN in AD Models

AD is a multifaceted, intricate CNS disorder that includes both neurodegeneration and persistent neuroinflammation, along with a significant loss of memory and cognitive functions. Although research on the beneficial effects of TAN in relation to AD and AD pathology are limited, some of the published data indicate that it can play a positive role in the management of AD pathology. In recent study by Shu et al. at the Institutes of Biomedical Sciences, School of Medicine, Jianghan University, China, the effects of TAN in an AD model of mice were evaluated [22]. The authors found that TAN treatment (100 mg/kg body weight/day) significantly ameliorated the cognitive dysfunction and reduced amyloid beta (Aβ) aggregation and synaptic loss in an APPswe/PSEN1dE9 transgenic (Tg) mice AD model. A mechanistic study revealed that TAN could inhibit β-secretase in both in vitro and in vivo studies, proving TAN's potential role in the management of AD pathogenesis [22].

In a recent report, Prof. Hsu et al. [103], from the Department of Food Science, National Ilan University, Taiwan, evaluated the effect of TAN and nobiletin against Aβ-induced toxicity in primary rat neurons. TAN and/or nobiletin were used to treat primary cortical neuron toxicity induced by $A\beta_{1-42}$. The authors indicated that TAN (25 µM) exhibited stronger neuroprotective effects when compared with nobiletin by attenuating the free radical damage induced by Aβ, thereby reducing intracellular oxidative damage. Additionally, TAN suppressed the $A\beta_{1-42}$ monomer aggregation, indicating its potential as a neuroprotective agent [103].

3.3. Effect of TAN on Ischemic Brain Injury Models

It is commonly known that ischemic stroke is caused by the brain damage brought on through the blockage of the cerebral arteries after an extended period of ischemia. Further, reports have suggested that citrus flavonoids influence cardiovascular function and cerebral ischemia, making them especially pertinent in relation to neurodegenerative disorders [104,105]. In human hepatocellular carcinoma cells (HepG2) subjected to hypoxic conditions, TAN was shown to improve cell viability and decrease apoptosis. Further, Lee et al. [18] from Kyungpook National University, Daegu, Republic of Korea, studied the natural herb *Aurantii immatri* pericarpium, containing nobiletin and TAN as major constituents, for its brain-protecting effects in a rat model of ischemia-reperfusion. The authors indicated that TAN shielded the brain from damage by inhibiting apoptosis in human hepatocellular carcinoma cells (HepG2) and ameliorating brain injury in an ischemic-reperfusion model of rats [18].

Recently, Zhang et al., from the Department of Neurology, Huaihe Hospital of Henan University, Kaifeng, China, proved that TAN protected human brain microvascular endothelial cells (HBMECs) against oxygen–glucose deprivation (OGD) insult. TAN increased the superoxide dismutase activity and HBMEC survival while decreasing ROS and MDA levels. These effects were produced via suppression of the neuroinflammatory JNK signaling pathway [106].

In a much recent study, the effect of TAN on cognition and memory deficits in a cerebral ischemia rat model was studied [24]. The authors indicated that the decrease in cognitive and motor behavioral functions in bilateral common carotid artery occlusion (BCCAO) and reperfusion injured rats was ameliorated by TAN. Further, TAN improved the altered acetylcholine enzyme activity, oxidative enzyme status, inflammatory mediators, and apoptotic biomarkers in BCCAO rats. The authors concluded that TAN has potential neuroprotective effects against cerebral ischemia.

Further, considering the neuroprotective effects of TAN, Zan et al. [107], from the School of Biomedical and Pharmaceutical Sciences, Guangdong University of Technology, China, investigated the effects of TAN on cerebral ischemia reperfusion-induced neuronal injury in vivo in a middle cerebral artery occlusion/reperfusion (MCAO/R) mice model and oxygen–glucose deprivation and reoxygenation (OGD/R) injury in a hippocampal HT22 cell in vitro model. The authors claimed that TAN significantly attenuated cerebral ischemia and reperfusion (I/R) injury-induced neuronal absent in melanoma 2 (AIM2) inflammasome activation-mediated brain damage and inhibited pyroptosis in mice. Further, TAN regulated NRF2 signaling in hippocampal HT22 cells, indicating that TAN inhibits AIM2 inflammasome activation by the regulating NRF2 pathway. These studies provide strong insights into the therapeutic benefits of TAN in I/R-induced brain damage [107]. A summary of the established neuroprotective effects of TAN in experimental models of neuroinflammation and neurodegeneration is provided in Table 3.

Table 3. Summary of established neuroprotective effects of TAN in experimental models of neurodegeneration.

No.	Model	Experimental Design/Dose	Parameters Tested	Mechanism	Conclusion	Ref.
1	6-OHDA-induced PD rat model	20 mg/kg/day for 4 days; p.o.	TH$^+$ cells and striatal dopamine content	Reduced TH$^+$ cells; increased striatal dopamine content	Neuroprotective agent	[21]
2	MPTP-induced rat PD model	50, 100 or 200 mg/kg body weight for 20 days	Rotarod, working memory, object recognition, inflammatory mediators, cytokines	Enhanced memory and locomotion; decreased COX-2, iNOS, IL-1β, IL-6, and IL-2	Neuroinflammation and dementia associated with PD	[48]
3	Transgenic Drosophila PD model	5, 10 and 20 μM in diet for 24 days	Climbing ability, dopamine levels, antioxidant enzymes	Enhanced climbing ability and dopamine content; decreased oxidative stress	Enhanced behavioral pattern and antioxidant	[100]
4	Pilocarpine-induced mice	50, 100, or 200 mg/kg for 10 days	Neuronal apoptosis and seizure severity	Regulation of PI3K/Akt signalling; decreased seizure-induced MMP-2, MMP-9, and AIF	Recovered PD-associated epileptic seizures	[102]
5	APPswe/PSEN1dE9 transgenic AD mice model	100 mg/kg body weight/day	Cognitive functions, Aβ aggregation	Inhibited β-secretase both in vitro and in vivo	Anti-dementia effect	[22]
6	Aβ-induced rat primary neurons	25 μM	Oxidative damage, Aβ aggregation	Reduced free radical damage and suppressed Aβ neurotoxicity	Neuroprotective effect	[103]

Table 3. Cont.

No.	Model	Experimental Design/Dose	Parameters Tested	Mechanism	Conclusion	Ref.
7	HepG2 cells in vitro and ischemic-reperfusion rat model	100 µg/mL for in vitro and 200 mg/kg in vivo	Apoptosis and cell viability	Inhibited apoptosis, and reduced brain injury	Neuroprotection and ischemic stroke protection	[18]
8	OGD insult in HBMEC cells	2.5, 5 and 10 µM	Cell viability, ROS levels, inflammatory pathways	Reduced ROS levels; ameliorated apoptosis; regulated JNK signaling	Protects brain injury and related neurogenerative diseases	[106]
9	Global cerebral ischemia in rats	5, 10, and 20 mg/kg, oral	Cognition and memory, AchE, Ach levels, ROS levels, inflammation markers	Increased memory and cognition; attenuated AchE and Ach activities; inhibited IL-6 and TNF-α, mitigating apoptosis	Neuroprotection, and antineuroinflammation	[24]
10	In vivo MCAO/R mice model and OGD/R injury in hippocampal HT22 cell in vitro	5, 10 and 20 µM in vitro and 10 µM in vivo	Cell viability, neuronal pyroptosis	Attenuated pyroptosis and regulated Nrf-2 signaling	Neuroprotective effects	[107]
11	LM mice model	5, 10 and 15 mg/kg	Cognitive functions, novel object recognition, inflammatory mediators	Recovered cognitio; decreased ERK 1/2, TNFα; and IL-1β expression; modulated RORα/γ target genes	Cognitive deficiency and related diseases	[23]
12	Potassium dichromate-induced brain injury in rats	50 mg/kg; orally, for 14 days	Behavioral indices, ROS markers, inflammatory markers	Reduced ROS levels; inhibited TNF-α and IL-6; regulated Nrf2 signaling pathway	Neuroprotective effect, anti-neuroinflammation, antioxidant	[108]

Abbreviations: 6-OHDA: 6-Hydroxydopamine, MPTP: 1-methyl-4-phenyl-1,2,3,6-tetrahydropyridine; MMP: matrix metalloproteinases; AIF: apoptosis-inducing factor; Aβ: amyloid beta; HepG2: human hepatocellular carcinoma cells; HBMECs: human brain microvascular endothelial cells; OGD: oxygen–glucose deprivation; AchE: acetylcholine esterase; Ach: acetylcholine; MCAO/R: middle cerebral artery occlusion/reperfusion; OGD/R: oxygen–glucose deprivation and reoxygenation; LPS: lipopolysaccharide; LM: LPS plus midazolam; PD: Parkinson's disease; AD: Alzheimer's disease; ROS: reactive oxygen species; TNF-α: tumor necrosis factor-alpha; IL: interleukin; Nrf-2: nuclear factor erythroid 2-related factor 2; TAN: tangeretin; TH+: tyrosine hydroxylase positive; COX- cyclooxygenase; iNOS: inducible nitric oxide synthase; PI3K: phosphatidylinositol-3-kinase; AKT: protein kinase B; JNK: c-Jun N-terminal kinase; ERK 1/2: extracellular-signal-regulated kinase 1/2; RORα/γ: retinoic-acid-receptor-related orphan receptors α and γ.

3.4. Effect of TAN on Neurogenesis and Cognitive Functions

It is widely known that several neurodegenerative diseases dysregulate cAMP response element (CRE) transcription [109]. Additionally, the cAMP/CREB/ERK/PKA signaling pathway is vital for memory and learning [110]. In PC12D cells [111], and hippocampus neurons [112], TAN increases nerve outgrowth and induces CRE-dependent transcription. These data show that TAN may induce neuroprotection in neuronal cells through CRE-mediated transcription coupled with the upstream cAMP/CREB/ERK/PKA pathway.

In a previous study from Wu et al. at the Institute of Molecular Rhythm and Metabolism, Guangzhou University of Chinese Medicine, China, the authors explored the cognitive enhancing properties of TAN in a delirious mice model by injecting mice with an LPS plus midazolam (LM) model [23]. The impaired cognitive and attentional functions in delirium-induced mice were challenged by the administration of TAN at various doses (5, 10, and

15 mg/kg). Mice with LM-induced delirium exhibited decreased attentional and cognito-behavioral functions and TAN significantly ameliorated these changes. Mechanistic studies revealed that TAN functioned as an RORα/γ agonist, leading to enhanced memory and cognition in LM-delirium mice. Further, the increased expression of neuroinflammatory pathways, including ERK ½, TNF-α, and IL-1β, was reduced in LPS-stimulated microglia. The authors suggest that TAN regulates the expression of RORα/γ-related genes such as *E4bp4* and *Bmal1*, thereby promoting enhanced cognition in mice with LM-induced delirium, indicating the importance of TAN in preventing memory and cognitive deficits.

In a recent study, the effect of TAN on brain injury induced by chromium was evaluated [108]. Potassium dichromate, a xenobiotic, was administered to Wistar rats and the neuroprotective potential of TAN was explored. The altered inflammation and oxidative stress parameters were studied. Potassium-dichromate-induced changes in the brain's MDA and GSH levels, as well as the increased levels of TNF-α and IL-6, were significantly reversed in TAN-treated subjects. Furthermore, this study revealed that TAN significantly reversed the potassium-dichromate-induced alterations in behavior and cholinergic activities. The authors suggest that the mechanism involved in the potential neuroprotective benefits exhibited by TAN might be the regulation of the Nrf2 and caspase-3 signaling pathways [108]. A schematic diagram of the possible signaling pathways involved in the neuroprotective benefits of TAN is shown in Figure 4.

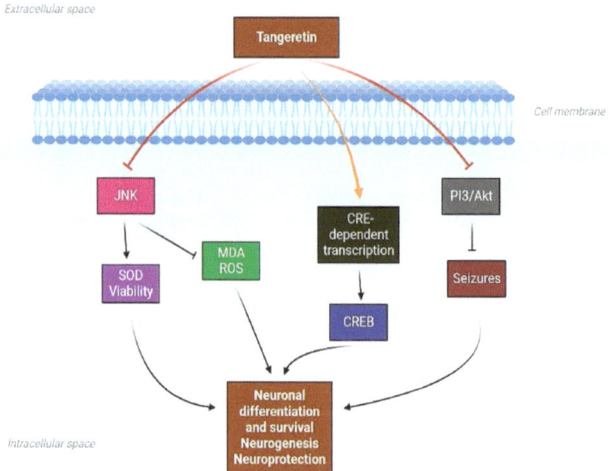

Figure 4. Signaling pathways involved in the neuroprotective benefits of TAN. PI3K: phosphoinositide 3-Kinase; SOD: superoxide dismutase; JNK: c-Jun N-terminal kinase; MDA: malondialdehyde; CREB: cAMP-response element binding protein; ROS: reactive oxygen species.

4. Other Supporting Mechanisms of TAN in Neurodegeneration

4.1. Antioxidant Effect of TAN

Numerous diseases or pathological changes, such as cancer and metabolic problems, are linked to oxidative damage, including neurodegenerative diseases [113–115]. Since the brain is more susceptible to oxidants than other organs, oxidative damage frequently coexists with neurodegenerative disorders. In PD, increased monoamine oxidase B (MAO-B) levels and impaired mitochondrial activity are linked to elevations in ROS generation [116–118] and cellular damage [118,119], which are responsible for neurodegeneration and neuroinflammation [118,120]. Altered glutathione levels in the dopaminergic neurons of transgenic mice provide more evidence for the hypothesis that oxidative stress drives neuronal loss in PD [121].

In LPS-stimulated microglia, TAN exerts an antioxidant impact by reducing ROS production while boosting heme oxygenase-1 (HO-1) expression [121] via the nuclear factor erythroid 2-related factor 2 (Nrf2) signaling pathway, which is one of the therapeutic targets of PD [122]. It has recently been determined that moderate mitochondrial depolarization, which prevents neurotoxic mitochondrial calcium overload during neural insults, represents a unique neuroprotective mechanism. Due to its inhibitory actions in (H_2O_2)-induced mitochondrial calcium ion (Ca^{2+}) accumulation, TAN significantly boosts HT-22 neuron survival of H_2O_2-induced cell death in a dose-dependent manner, thus preventing neuronal apoptosis [123]. Furthermore, TAN was shown to substantially inhibit ultraviolet B (UVB)-induced ROS production in JB6 P+ cells; this ability could be used to restrict endogenous ROS production [124].

In the kidney tissue, TAN significantly decreased lipid peroxides, and inflammatory cytokines, further improving the levels of both enzymatic and nonenzymatic antioxidants [125,126]. TAN effectively normalized the antioxidant enzymes SOD, CAT, GPx, and GR in diabetic rats [127].

4.2. Anti-Inflammatory Effect of TAN

Inflammation is a typical physiologic reaction to tissue damage, microbial pathogen infection, and chemical irritation. However, inflammatory mediators, including free radicals, cytokines, and chemokines, will infiltrate the body and be produced in excessive amounts, damaging cellular and tissue functions. PGE2 and COX-2 are created by the inducible enzyme COX, which also plays a crucial role in the creation of PGE2 in inflammatory areas and the unregulated NO produced by the iNOS. The CNS macrophages, known as microglial cells, maintain the nervous system's balance by removing damaged neurons and preventing the spread of infection. This macrophage population produces cytokines that cause inflammation, creating ROS as the first line of defense against microorganisms [128]. However, a pro-inflammatory and pro-oxidant condition sustained over time by chronic microglia activation is harmful because it may speed up the neurodegenerative process via neuroinflammation, as occurs in the case of AD and PD [129,130].

Chemokine fractalkine (CX3CL1), produced mainly by neurons, has recently been suggested as a potential biomarker for PD [131]. The interaction between neurons and glial cells is mediated by the G-protein-coupled receptor CX3CL1, a regulator of microglial activity [131]. Additionally, older adults with PD have been reported to have peripheral inflammation, which is indicated by an increase in the blood levels of IL-8 and MIP-1 and reduction in the levels of IL-9 and MIP-1 [132].

Evidence suggests that the key signaling pathway for TAN's neuroprotective benefits might be via its anti-inflammatory properties. TAN showed a solid ability to suppress neuroinflammation caused by microglial activation stimulated by lipopolysaccharide [97]. These results indicate that TAN inhibits the production of COX-2 and iNOS at the transcriptional and protein levels in microglial cells. The production of proinflammatory substances, including NO, IL-1, TNF-α, and IL-6, was also inhibited in a dose-dependent fashion. Thus, IκB kinase and MAPKs may function separately or jointly to control TAN's inhibition of proinflammatory mediators [97]. MMPs are vital for many physiological tasks, including proliferation, differentiation, cell motility, apoptosis, and host defense. Blood–brain barrier (BBB) collapse, neuronal cell death, and peripheral immune cell infiltration in neuropathological illnesses, including AD, PD, and MS, are all unfavorable outcomes of abnormal MMP expression. MMPs influence the development of neuroinflammatory diseases by modulating TNF-α activation. TAN reduces MMPs, including MMP-3 and MMP-8, in LPS-stimulated microglia, which aids its anti-inflammatory effects. A schematic diagram of the possible anti-inflammatory mechanism of TAN is presented in Figure 5.

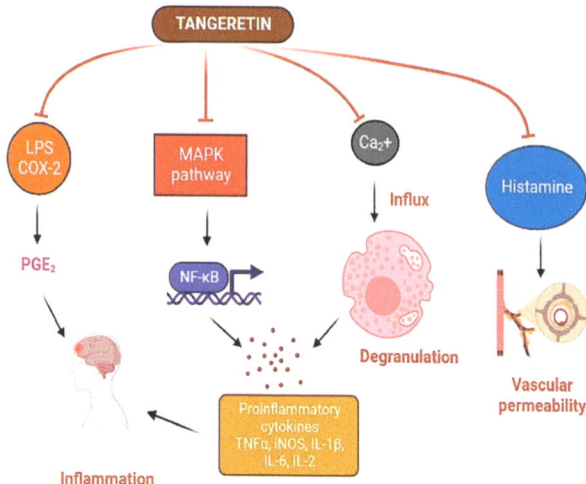

Figure 5. TAN's anti-inflammatory mechanisms. TAN reduces the production of proinflammatory cytokines by controlling certain essential enzymes involved in the MAPK pathway and cell degranulation. TAN also interferes with the formation of PGE2 generated by LPS and COX-2 and reduces vascular permeability caused by histamine to prevent allergies. TAN: tangeretin; MAPK: mitogen-activated protein kinases; PGE2: prostaglandin E2; COX: cyclooxygenase; LPS: lipopolysaccharide; IL: interleukin; TNF-α: tumor necrosis factor-alpha; iNOS: inducible nitric oxide synthase; NF-κB: nuclear factor-kappa B.

4.3. Diabetes-Mediated Neurodegeneration and TAN

The pathophysiology of Diabetes Mellitus (DM) and neurodegeneration is influenced by environmental and genetic variations [133]. Some DM animal models have comparable clinical traits to PD and AD animal models [133,134]. DM and neurodegeneration have the characteristics of impaired glucose metabolism, insulin resistance, and mitochondrial dysfunction, which manifests before clinical diagnosis of neurodegenerative illness. Other causes of the pathophysiology of both disorders include microbial dysbiosis, the aggregation of misfolded proteins, persistent inflammation, and oxidative stress [135]. Dopaminergic neurons and insulin receptors coexist in the substantia nigra, supporting the hypothesis that DM and PD are directly related. The reduced insulin signaling in the basal ganglia correlates with dopamine depletion in the striatum [136]. DM is also associated with a significant decline in cognitive function and the elevated susceptibility to dementia seen in AD. Several common pathways, including apoptosis, neuron aging, lipid peroxidation, tau phosphorylation, and brain shrinkage, are all adversely affected in DM-mediated AD. One possible strategy to treat AD is to repurpose anti-diabetic agents as beneficial therapeutics that may avert or lower the risk of cognitive decline and neurodegeneration [137,138].

Experimental studies in diabetic rats showed a decrease in plasma glucose and glycosylated hemoglobin (HbA1c) levels, although the hemoglobin and insulin levels drastically increased because of TAN. The liver's major enzyme processes for glycolysis, glycogenolysis, the pentose phosphate pathway, and glycogenesis were nearly fully recovered [139]. TAN decreased the capacity of cells to secrete insulin, reduced the amounts of ROS and the messenger RNA in insulin 1 and 2, and enhanced the expression of SOD, CAT, and GPX activities to prevent streptozotocin-induced apoptosis. TAN's protective effects resulted from its blocking the NF-κB signaling pathway [140].

In another study, Chen et al., investigated how TAN inhibits human renal mesangial cells (MC) extracellular matrix production in response to high glucose levels. TAN substantially boosted SOD activity while decreasing MC growth, ROS, and MDA levels. TAN partially inhibited the ERK signaling pathway to provide these protective effects [141]. Based on these reports, we can conclude that TAN can aid in countering DM-mediated neuroinflammatory responses and can provide therapeutic benefits in the management of neurodegenerative disorders. The possible role of TAN in DM-mediated neuroinflammation is depicted in Figure 6.

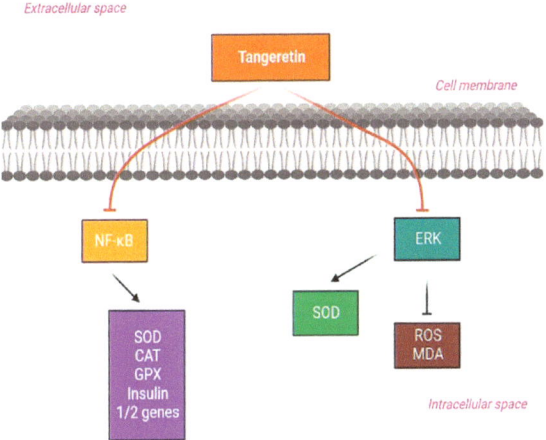

Figure 6. Possible signaling pathways showing the effect of TAN against Diabetes Miletus-mediated neuroinflammation. TAN: tangeretin; MDA: malondialdehyde; NF-κB: nuclear factor kappa B; CAT: catalase; GPX: glutathione peroxidase; ERK: extracellular-signal-regulated kinase.

4.4. Peroxisome Proliferator Receptor-Gamma (PPAR-γ) Agonistic Effects of TAN

The nuclear receptor superfamily, which includes PPAR-γ, controls mitochondrial activity and modifies lipid and glucose metabolism. PPAR-γ agonists, such as pioglitazone, lower inflammation by preventing the production of IL-6 and TNF-α [142]. In the brains of MPTP-treated monkeys, pioglitazone reduced inflammatory responses while maintaining dopaminergic nigrostriatal function [143]. Additionally, pioglitazone treatment reduced the glial activation and arrested the loss of dopaminergic neurons in the substantia nigra (SN) of mice treated with MPTP [144,145]. In the SN of mice treated with MPTP, rosiglitazone, another PPAR-γ agonist, reversed the loss of dopaminergic neurons [146]. These findings suggest the use of PPAR agonists as potential anti-inflammatory treatments to slow the course of neurodegeneration. TAN suppressed UVB-induced COX-2 expression and PGE2 production in human epidermal keratinocytes (HaCaT) cells through PPAR-γ activation [147]. Further, Kim et al., Li et al., and Kurowska et al. have also reported on the PPAR-agonistic activities of TAN in their studies; thus, these effects of TAN can halt the progression of PD [148–150].

5. Limitations and Future Perspectives

One of the main issues with flavonoids is the length of time required for epidemiological investigations due to the extensive exposure times, data collection, and analyses of the presence and absence of flavonoids that are required. Due to their very low yield (mg–gm per kg of plant weight) and the fact that they must be continuously extracted from *Citrus* plants to avoid extinction, TAN's production cost is another significant concern. Such issues might be resolved by a targeted study on the production of flavonoids by microbes or any other natural sources. The purification procedure presents another significant challenge because flavonoids generally tend to form tight complexes with other secondary metabolites, vitamins, minerals, and fibers, making it challenging to separate them.

In order to improve the purity of the separated flavonoids and allow for shorter extraction procedures with a lower cost, it is also necessary to create an effective purification method. TAN's low water solubility, as a result of its lipophilic makeup, is another serious issue. To boost its bioavailability in the studied animal systems, future research should concentrate on various targeted delivery methods, including the novel nano-emulsion-based delivery systems. In addition, a detailed, pharmacokinetic, dosage regimen should be established, as well as research on the safety issues and toxicity profile, and pre-clinical research in various other neuroinflammatory animal models, to better understand the intrinsic mechanisms of TAN. To the best of our knowledge and understanding, no published clinical data are yet available on the impact of the chronic long-term intake of TAN and its neuroprotective benefits. To close these gaps, a systematic research plan of action, extending pre-clinical observations into clinical settings, is necessary to demonstrate TAN's neuroprotective potential. An illustrative diagram of the possible overall benefits of TAN as a potential agent against neuroinflammation-mediated neurodegenerative disorders is presented in Figure 7.

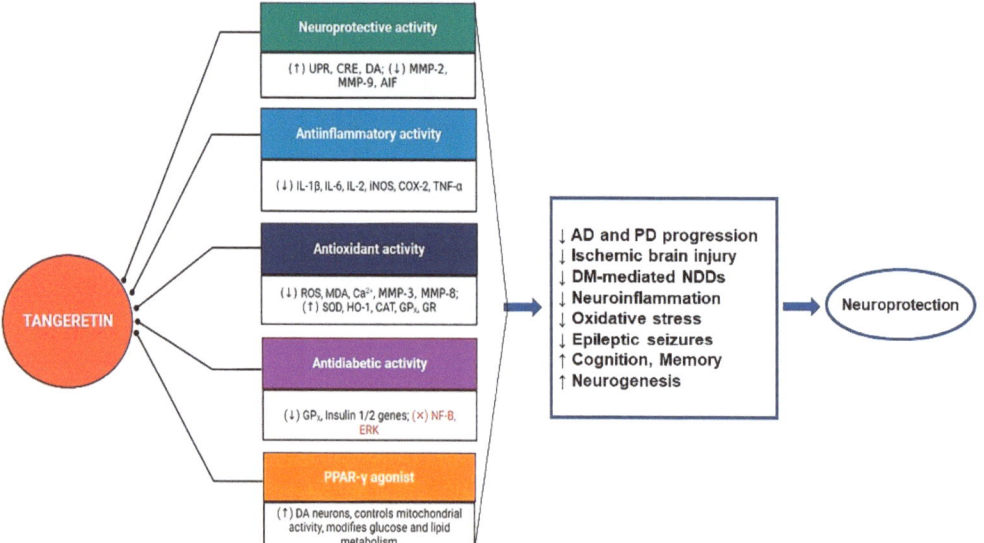

Figure 7. TAN's potential as a candidate for treatments of neuroinflammation-mediated neurodegenerative disorders. AD: Alzheimer's disease; PD: Parkinson's disease; DM: Diabetes Miletus; NDDs: neurodegenerative disorders; UPR: unfolded protein response; CRE: cAMP response element; DA: dopamine; MMP: matrix metalloproteinases; AIF: apoptosis-inducing factor; IL: interleukin; iNOS: inducible nitric oxide synthase; COX: cyclooxygenase; TNF-α: tumor necrosis factor-alpha; ROS: reactive oxygen species; MDA: malondialdehyde; SOD: superoxide dismutase; HO: heme oxygenase; CAT: catalase; GPx: glutathione peroxidase; GR: glutathione reductase; NF-κB: nuclear factor-kappa B; ERK: extracellular-signal-regulated kinase; ↑: increase; ↓: decrease; ×: inhibited.

6. Conclusions

Based on the available literature and reported studies, TAN, one of the active flavonoids present in *Citrus* species of plants, has the potential to prevent neuronal death in various neurodegenerative diseases. Due to TAN's excellent physiochemical and pharmacokinetic profile, TAN is a promising candidate and can be used as an auxiliary therapy to prevent or delay the progression of neuroinflammation-mediated neurodegenerative disorders.

Author Contributions: S.K., I.W., S.N.M. and S.R.K. conceptualized and designed the manuscript; S.K., I.W. and S.R.K. wrote the manuscript; S.K., I.W., A.B., D.T., A.J., S.N.M. and S.R.K. revised the manuscript; S.K., I.W., A.B., D.T., A.J., P.W., S.N.M. and S.R.K. participated in drafting and critically revising the article; S.K. and S.R.K. were involved in funding acquisition. All authors have read and agreed to the published version of the manuscript.

Funding: This research received no external funding.

Institutional Review Board Statement: Not applicable.

Informed Consent Statement: Not applicable.

Data Availability Statement: Not applicable.

Acknowledgments: This work was supported by Sejong University, Seoul, Republic of Korea and partly by Konkuk University, Republic of Korea.

Conflicts of Interest: Authors Ankush Jamadagni and Prathamesh Walse were employed by the company Fortem Biosciences Private Limited (Ayurvibes). The remaining authors declare that the research was conducted in the absence of any commercial or financial relationships that could be construed as a potential conflict of interest.

References

1. Cova, I.; Markova, A.; Campini, I.; Grande, G.; Mariani, C.; Pomati, S. Worldwide Trends in the Prevalence of Dementia. *J. Neurol. Sci.* **2017**, *379*, 259–260. [CrossRef]
2. Jellinger, K.A. Basic Mechanisms of Neurodegeneration: A Critical Update. *J. Cell Mol. Med.* **2010**, *14*, 457. [CrossRef]
3. Wyss-Coray, T.; Mucke, L. Inflammation in Neurodegenerative Disease—A Double-Edged Sword. *Neuron* **2002**, *35*, 419–432. [CrossRef]
4. Kempuraj, D.; Thangavel, R.; Natteru, P.A.; Selvakumar, G.P.; Saeed, D.; Zahoor, H.; Zaheer, S.; Iyer, S.S.; Zaheer, A. Neuroinflammation Induces Neurodegeneration. *J. Neurol. Neurosurg. Spine* **2016**, *1*, 1003.
5. Glass, C.K.; Saijo, K.; Winner, B.; Marchetto, M.C.; Gage, F.H. Mechanisms Underlying Inflammation in Neurodegeneration. *Cell* **2010**, *140*, 918–934. [CrossRef]
6. Zhang, W.; Xiao, D.; Mao, Q.; Xia, H. Role of Neuroinflammation in Neurodegeneration Development. *Signal Transduct. Target. Ther.* **2023**, *8*, 267. [CrossRef]
7. Ashrafizadeh, M.; Ahmadi, Z.; Mohammadinejad, R.; Ghasemipour Afshar, E. Tangeretin: A Mechanistic Review of Its Pharmacological and Therapeutic Effects. *J. Basic Clin. Physiol. Pharmacol.* **2020**, *31*, 20190191. [CrossRef]
8. Panche, A.N.; Diwan, A.D.; Chandra, S.R. Flavonoids: An Overview. *J. Nutr. Sci.* **2016**, *5*, e47. [CrossRef]
9. Kumar, S.; Pandey, A.K. Chemistry and Biological Activities of Flavonoids: An Overview. *Sci. World J.* **2013**, *2013*, 162750. [CrossRef]
10. Raza, W.; Luqman, S.; Meena, A. Prospects of Tangeretin as a Modulator of Cancer Targets/Pathways. *Pharmacol. Res.* **2020**, *161*, 105202. [CrossRef]
11. Lv, X.; Zhao, S.; Ning, Z.; Zeng, H.; Shu, Y.; Tao, O.; Xiao, C.; Lu, C.; Liu, Y. Citrus Fruits as a Treasure Trove of Active Natural Metabolites That Potentially Provide Benefits for Human Health. *Chem. Cent. J.* **2015**, *9*, 68. [CrossRef]
12. Kelebek, H.; Selli, S. Determination of Volatile, Phenolic, Organic Acid and Sugar Components in a Turkish Cv. Dortyol (*Citrus sinensis* L. Osbeck) Orange Juice. *J. Sci. Food Agric.* **2011**, *91*, 1855–1862. [CrossRef]
13. He, D.; Shan, Y.; Wu, Y.; Liu, G.; Chen, B.; Yao, S. Simultaneous Determination of Flavanones, Hydroxycinnamic Acids and Alkaloids in Citrus Fruits by HPLC-DAD–ESI/MS. *Food Chem.* **2011**, *127*, 880–885. [CrossRef]
14. Li, S.; Pan, M.-H.; Lo, C.-Y.; Tan, D.; Wang, Y.; Shahidi, F.; Ho, C.-T. Chemistry and Health Effects of Polymethoxyflavones and Hydroxylated Polymethoxyflavones. *J. Funct. Foods* **2009**, *1*, 2–12. [CrossRef]
15. Arafa, E.S.A.; Shurrab, N.T.; Buabeid, M.A. Therapeutic Implications of a Polymethoxylated Flavone, Tangeretin, in the Management of Cancer via Modulation of Different Molecular Pathways. *Adv. Pharmacol. Pharm. Sci.* **2021**, *2021*, 4709818. [CrossRef]
16. Walle, T. Methoxylated Flavones, a Superior Cancer Chemopreventive Flavonoid Subclass? *Semin. Cancer Biol.* **2007**, *17*, 354–362. [CrossRef]
17. Meiyanto, E.; Hermawan, A. Anindyajati Natural Products for Cancer-Targeted Therapy: Citrus Flavonoids as Potent Chemopreventive Agents. *Asian Pac. J. Cancer Prev.* **2012**, *13*, 427–436. [CrossRef]
18. Yang, E.J.; Lim, S.H.; Song, K.S.; Han, H.S.; Lee, J. Identification of Active Compounds from Aurantii Immatri Pericarpium Attenuating Brain Injury in a Rat Model of Ischemia-Reperfusion. *Food Chem.* **2013**, *138*, 663–670. [CrossRef]
19. Manthey, J.A.; Grohmann, K.; Montanari, A.; Ash, K.; Manthey, C.L. Polymethoxylated Flavones Derived from Citrus Suppress Tumor Necrosis Factor-Alpha Expression by Human Monocytes. *J. Nat. Prod.* **1999**, *62*, 441–444. [CrossRef]
20. Agrawal, K.K.; Murti, Y. Tangeretin: A Biologically Potential Citrus Flavone. *Curr. Tradit. Med.* **2022**, *8*, 31–41. [CrossRef]
21. Datla, K.P.; Christidou, M.; Widmer, W.W.; Rooprai, H.K.; Dexter, D.T. Tissue Distribution and Neuroprotective Effects of Citrus Flavonoid Tangeretin in a Rat Model of Parkinson's Disease. *Neuroreport* **2001**, *12*, 3871–3875. [CrossRef]
22. Bao, J.; Liang, Z.; Gong, X.; Zhao, Y.; Wu, M.; Liu, W.; Tu, C.; Wang, X.; Shu, X. Tangeretin Inhibits BACE1 Activity and Attenuates Cognitive Impairments in AD Model Mice. *J. Agric. Food Chem.* **2022**, *70*, 1536–1546. [CrossRef]

23. Chen, M.; Xiao, Y.; Zhang, F.; Du, J.; Zhang, L.; Li, Y.; Lu, D.; Wang, Z.; Wu, B. Tangeretin Prevents Cognitive Deficit in Delirium through Activating RORα/γ-E4BP4 Axis in Mice. *Biochem. Pharmacol.* **2022**, *205*, 115286. [CrossRef]
24. Alla, N.; Palatheeya, S.; Challa, S.R.; Kakarla, R. Tangeretin Confers Neuroprotection, Cognitive and Memory Enhancement in Global Cerebral Ischemia in Rats. *3 Biotech* **2024**, *14*, 9. [CrossRef]
25. Chen, K.-H.; Weng, M.-S.; Lin, J.-K. Tangeretin Suppresses IL-1β-Induced Cyclooxygenase (COX)-2 Expression through Inhibition of P38 MAPK, JNK, and AKT Activation in Human Lung Carcinoma Cells. *Biochem. Pharmacol.* **2007**, *73*, 215–227. [CrossRef]
26. Braidy, N.; Behzad, S.; Habtemariam, S.; Ahmed, T.; Daglia, M.; Nabavi, S.M.; Sobarzo-Sanchez, E.; Nabavi, S.F. Neuroprotective Effects of Citrus Fruit-Derived Flavonoids, Nobiletin and Tangeretin in Alzheimer's and Parkinson's Disease. *CNS Neurol. Disord. Drug Targets* **2017**, *16*, 387–397. [CrossRef]
27. Feng, S.L.; Yuan, Z.W.; Yao, X.J.; Ma, W.Z.; Liu, L.; Liu, Z.Q.; Xie, Y. Tangeretin, a Citrus Pentamethoxyflavone, Antagonizes ABCB1-Mediated Multidrug Resistance by Inhibiting Its Transport Function. *Pharmacol. Res.* **2016**, *110*, 193–204. [CrossRef]
28. Chen, H.F.; Zhang, W.G.; Yuan, J.; Li, Y.G.; Yang, S.L.; Yang, W.L. Simultaneous Quantification of Polymethoxylated Flavones and Coumarins in Fructus Aurantii and Fructus Aurantii Immaturus Using HPLC–ESI-MS/MS. *J. Pharm. Biomed. Anal.* **2012**, *59*, 90–95. [CrossRef]
29. Chaliha, B.P.; Sastry, G.P.; Rao, P.R. Chemical Examination of the Peel of Citrus Jambhiri Lush: Isolation of a New Flavone. *Tetrahedron* **1965**, *21*, 1441–1443. [CrossRef]
30. Russo, M.; Arigò, A.; Calabrò, M.L.; Farnetti, S.; Mondello, L.; Dugo, P. Bergamot (*Citrus bergamia* Risso) as a Source of Nutraceuticals: Limonoids and Flavonoids. *J. Funct. Foods* **2016**, *20*, 10–19. [CrossRef]
31. Eun, S.H.; te Woo, J.; Kim, D.H. Tangeretin Inhibits IL-12 Expression and NF-ΚB Activation in Dendritic Cells and Attenuates Colitis in Mice. *Planta Med.* **2017**, *234*, 527–533. [CrossRef]
32. Mitani, R.; Tashiro, H.; Arita, E.; Ono, K.; Haraguchi, M.; Tokunaga, S.; Sharmin, T.; Aida, T.M.; Mishima, K. Extraction of Nobiletin and Tangeretin with Antioxidant Activity from Peels of Citrus Poonensis Using Liquid Carbon Dioxide and Ethanol Entrainer. *Sep. Sci. Technol.* **2020**, *56*, 290–300. [CrossRef]
33. Lee, Y.H.; Charles, A.L.; Kung, H.F.; Ho, C.T.; Huang, T.C. Extraction of Nobiletin and Tangeretin from Citrus Depressa Hayata by Supercritical Carbon Dioxide with Ethanol as Modifier. *Ind. Crops Prod.* **2010**, *31*, 59–64. [CrossRef]
34. Qiuan, W.; Zheng, W.; Li, L.; Lianghua, Z.; Ming, L. Synthesis of Citrus Bioactive Polymethoxyflavonoids and Flavonoid Glucosides. *Youji Huaxue* **2010**, *30*, 1682–1688.
35. Zhao, Y.; Kao, C.P.; Liao, C.R.; Wu, K.C.; Zhou, X.; Ho, Y.L.; Chang, Y.S. Chemical Compositions, Chromatographic Fingerprints and Antioxidant Activities of Citri Exocarpium Rubrum (Juhong). *Chin. Med.* **2017**, *12*, 6. [CrossRef]
36. Nakajima, A.; Nemoto, K.; Ohizumi, Y. An Evaluation of the Genotoxicity and Subchronic Toxicity of the Peel Extract of Ponkan Cultivar 'Ohta Ponkan' (*Citrus reticulata* Blanco) That Is Rich in Nobiletin and Tangeretin with Anti-Dementia Activity. *Regul. Toxicol. Pharmacol.* **2020**, *114*, 104670. [CrossRef]
37. Kim, D.S.; Lim, S. bin Semi-Continuous Subcritical Water Extraction of Flavonoids from Citrus Unshiu Peel: Their Antioxidant and Enzyme Inhibitory Activities. *Antioxidants* **2020**, *9*, 360. [CrossRef]
38. Mizuno, H.; Yoshikawa, H.; Usuki, T. Extraction of Nobiletin and Tangeretin from Peels of Shekwasha and Ponkan Using [C_2mim][(MeO)(H)PO_2] and Centrifugation. *Nat. Prod. Commun.* **2019**, *14*, 1934578X19845816. [CrossRef]
39. Adindu, E.A.; Elekwa, I.; Ikedi, O.I. Phytochemical Comparative Screening of Aqueous Extracts of the Leaves, Stem Barks, and Roots of *Hura crepitans* (L) Using GC—FID. *IOSR J. Biotechnol. Biochem.* **2016**, *2*, 11–18.
40. Brito, A.; Ramirez, J.E.; Areche, C.; Sepúlveda, B.; Simirgiotis, M.J. HPLC-UV-MS Profiles of Phenolic Compounds and Antioxidant Activity of Fruits from Three Citrus Species Consumed in Northern Chile. *Molecules* **2014**, *19*, 17400–17421. [CrossRef]
41. Lou, S.N.; Hsu, Y.S.; Ho, C.T. Flavonoid Compositions and Antioxidant Activity of Calamondin Extracts Prepared Using Different Solvents. *J. Food Drug Anal.* **2014**, *22*, 290–295. [CrossRef] [PubMed]
42. Ernawita; Wahyuono, R.A.; Hesse, J.; Hipler, U.C.; Elsner, P.; Böhm, V. In Vitro Lipophilic Antioxidant Capacity, Antidiabetic and Antibacterial Activity of Citrus Fruits Extracts from Aceh, Indonesia. *Antioxidants* **2017**, *6*, 11. [CrossRef] [PubMed]
43. Wang, F.; Chen, L.; Chen, H.; Chen, S.; Liu, Y. Analysis of Flavonoid Metabolites in Citrus Peels (*Citrus reticulata* "Dahongpao") Using UPLC-ESI-MS/MS. *Molecules* **2019**, *24*, 2680. [CrossRef] [PubMed]
44. Kim, H.; Moon, J.Y.; Mosaddik, A.; Cho, S.K. Induction of Apoptosis in Human Cervical Carcinoma HeLa Cells by Polymethoxylated Flavone-Rich Citrus Grandis Osbeck (Dangyuja) Leaf Extract. *Food Chem. Toxicol.* **2010**, *48*, 2435–2442. [CrossRef] [PubMed]
45. Barreca, D.; Mandalari, G.; Calderaro, A.; Smeriglio, A.; Trombetta, D.; Felice, M.R.; Gattuso, G. Citrus Flavones: An Update on Sources, Biological Functions, and Health Promoting Properties. *Plants* **2020**, *9*, 288. [CrossRef] [PubMed]
46. Jang, S.-E.; Ryu, K.-R.; Park, S.-H.; Chung, S.; Teruya, Y.; Han, M.J.; Woo, J.-T.; Kim, D.-H. Nobiletin and Tangeretin Ameliorate Scratching Behavior in Mice by Inhibiting the Action of Histamine and the Activation of NF-ΚB, AP-1 and P38. *Int. Immunopharmacol.* **2013**, *17*, 502–507. [CrossRef]
47. Batenburg, A.M.; de Joode, T.; Gouka, R.J. Characterization and Modulation of the Bitterness of Polymethoxyflavones Using Sensory and Receptor-Based Methods. *J. Agric. Food Chem.* **2016**, *64*, 2619–2626. [CrossRef] [PubMed]
48. Yang, J.; Wu, X.; Yu, H.; Teng, L. Tangeretin Inhibits Neurodegeneration and Attenuates Inflammatory Responses and Behavioural Deficits in 1-Methyl-4-Phenyl-1,2,3,6-Tetrahydropyridine (MPTP)-Induced Parkinson's Disease Dementia in Rats. *Inflammopharmacology* **2017**, *25*, 471–484. [CrossRef] [PubMed]

49. Kansy, M.; Senner, F.; Gubernator, K. Physicochemical High Throughput Screening: Parallel Artificial Membrane Permeation Assay in the Description of Passive Absorption Processes. *J. Med. Chem.* **1998**, *41*, 1007–1010. [CrossRef]
50. Grüneberg, S.; Güssregen, S. Drug Bioavailability. Estimation of Solubility, Permeability, Absorption and Bioavailability. (Series: Methods and Principles in Medicinal Chemistry, Vol. 18; Series Editors: R. Mannhold, H. Kubinyi, and G. Folkers). Edited by Han van de Waterbeemd, Hans. *Angew. Chem. Int. Ed.* **2004**, *43*, 146–147. [CrossRef]
51. Nielsen, S.E.; Breinholt, V.; Justesen, U.; Cornett, C.; Dragsted, L.O. In Vitro Biotransformation of Flavonoids by Rat Liver Microsomes. *Xenobiotica* **1998**, *28*, 389–401. [CrossRef] [PubMed]
52. Breinholt, V.M.; Rasmussen, S.E.; Brøsen, K.; Friedberg, T.H. In Vitro Metabolism of Genistein and Tangeretin by Human and Murine Cytochrome P450s. *Pharmacol. Toxicol.* **2003**, *93*, 14–22. [CrossRef] [PubMed]
53. Nielsen, S.E.; Breinholt, V.; Cornett, C.; Dragsted, L.O. Biotransformation of the Citrus Flavone Tangeretin in Rats. Identification of Metabolites with Intact Flavane Nucleus. *Food Chem. Toxicol.* **2000**, *38*, 739–746. [CrossRef] [PubMed]
54. Kurowska, E.M.; Manthey, J.A. Hypolipidemic Effects and Absorption of Citrus Polymethoxylated Flavones in Hamsters with Diet-Induced Hypercholesterolemia. *J. Agric. Food Chem.* **2004**, *52*, 2879–2886. [CrossRef] [PubMed]
55. Vanhoecke, B.W.; Delporte, F.; Van Braeckel, E.; Heyerick, A.; Depypere, H.T.; Nuytinck, M.; De Keukeleire, D.; Bracke, M.E. A Safety Study of Oral Tangeretin and Xanthohumol Administration to Laboratory Mice. *In Vivo* **2005**, *19*, 103–107. [PubMed]
56. Hung, W.-L.; Chang, W.-S.; Lu, W.-C.; Wei, G.-J.; Wang, Y.; Ho, C.-T.; Hwang, L.S. Pharmacokinetics, Bioavailability, Tissue Distribution and Excretion of Tangeretin in Rat. *J. Food Drug Anal.* **2018**, *26*, 849–857. [CrossRef] [PubMed]
57. Elhennawy, M.; Lin, H.-S. Determination of Tangeretin in Rat Plasma: Assessment of Its Clearance and Absolute Oral Bioavailability. *Pharmaceutics* **2017**, *10*, 3. [CrossRef] [PubMed]
58. Manthey, J.A.; Cesar, T.B.; Jackson, E.; Mertens-Talcott, S. Pharmacokinetic Study of Nobiletin and Tangeretin in Rat Serum by High-Performance Liquid Chromatography−Electrospray Ionization−Mass Spectrometry. *J. Agric. Food Chem.* **2011**, *59*, 145–151. [CrossRef]
59. Ting, Y.; Chiou, Y.-S.; Jiang, Y.; Pan, M.-H.; Lin, Z.; Huang, Q. Safety Evaluation of Tangeretin and the Effect of Using Emulsion-Based Delivery System: Oral Acute and 28-Day Sub-Acute Toxicity Study Using Mice. *Food Res. Int.* **2015**, *74*, 140–150. [CrossRef]
60. Delaney, B.; Phillips, K.; Vasquez, C.; Wilson, A.; Cox, D.; Wang, H.-B.; Manthey, J. Genetic Toxicity of a Standardized Mixture of Citrus Polymethoxylated Flavones. *Food Chem. Toxicol.* **2002**, *40*, 617–624. [CrossRef]
61. Delaney, B.; Phillips, K.; Buswell, D.; Mowry, B.; Nickels, D.; Cox, D.; Wang, H.-B.; Manthey, J. Immunotoxicity of a Standardized Citrus Polymethoxylated Flavone Extract. *Food Chem. Toxicol.* **2001**, *39*, 1087–1094. [CrossRef] [PubMed]
62. Shabab, T.; Khanabdali, R.; Moghadamtousi, S.Z.; Kadir, H.A.; Mohan, G. Neuroinflammation Pathways: A General Review. *Int. J. Neurosci.* **2017**, *127*, 624–633. [CrossRef]
63. DiSabato, D.J.; Quan, N.; Godbout, J.P. Neuroinflammation: The Devil Is in the Details. *J. Neurochem.* **2016**, *139*, 136–153. [CrossRef] [PubMed]
64. Shi, Z.-M.; Han, Y.-W.; Han, X.-H.; Zhang, K.; Chang, Y.-N.; Hu, Z.-M.; Qi, H.-X.; Ting, C.; Zhen, Z.; Hong, W. Upstream Regulators and Downstream Effectors of NF-κB in Alzheimer's Disease. *J. Neurol. Sci.* **2016**, *366*, 127–134. [CrossRef]
65. Singh, S.S.; Rai, S.N.; Birla, H.; Zahra, W.; Rathore, A.S.; Singh, S.P. NF-κB-Mediated Neuroinflammation in Parkinson's Disease and Potential Therapeutic Effect of Polyphenols. *Neurotox. Res.* **2020**, *37*, 491–507. [CrossRef]
66. Ding, Z.-B.; Song, L.-J.; Wang, Q.; Kumar, G.; Yan, Y.-Q.; Ma, C.-G. Astrocytes: A Double-Edged Sword in Neurodegenerative Diseases. *Neural Regen. Res.* **2021**, *16*, 1702. [CrossRef]
67. Lee, H.-J.; Suk, J.-E.; Patrick, C.; Bae, E.-J.; Cho, J.-H.; Rho, S.; Hwang, D.; Masliah, E.; Lee, S.-J. Direct Transfer of α-Synuclein from Neuron to Astroglia Causes Inflammatory Responses in Synucleinopathies. *J. Biol. Chem.* **2010**, *285*, 9262–9272. [CrossRef] [PubMed]
68. Phatnani, H.; Maniatis, T. Astrocytes in Neurodegenerative Disease. *Cold Spring Harb. Perspect. Biol.* **2015**, *7*, a020628. [CrossRef]
69. Nagele, R.G.; D'Andrea, M.R.; Lee, H.; Venkataraman, V.; Wang, H.-Y. Astrocytes Accumulate Aβ42 and Give Rise to Astrocytic Amyloid Plaques in Alzheimer Disease Brains. *Brain Res.* **2003**, *971*, 197–209. [CrossRef]
70. Kim, E.K.; Choi, E.-J. Compromised MAPK Signaling in Human Diseases: An Update. *Arch. Toxicol.* **2015**, *89*, 867–882. [CrossRef]
71. Cianciulli, A.; Porro, C.; Calvello, R.; Trotta, T.; Lofrumento, D.D.; Panaro, M.A. Microglia Mediated Neuroinflammation: Focus on PI3K Modulation. *Biomolecules* **2020**, *10*, 137. [CrossRef] [PubMed]
72. Yang, D.; Elner, S.G.; Bian, Z.-M.; Till, G.O.; Petty, H.R.; Elner, V.M. Pro-Inflammatory Cytokines Increase Reactive Oxygen Species through Mitochondria and NADPH Oxidase in Cultured RPE Cells. *Exp. Eye Res.* **2007**, *85*, 462–472. [CrossRef] [PubMed]
73. Bakunina, N.; Pariante, C.M.; Zunszain, P.A. Immune Mechanisms Linked to Depression via Oxidative Stress and Neuroprogression. *Immunology* **2015**, *144*, 365–373. [CrossRef] [PubMed]
74. Akbar, M.; Essa, M.M.; Daradkeh, G.; Abdelmegeed, M.A.; Choi, Y.; Mahmood, L.; Song, B.-J. Mitochondrial Dysfunction and Cell Death in Neurodegenerative Diseases through Nitroxidative Stress. *Brain Res.* **2016**, *1637*, 34–55. [CrossRef] [PubMed]
75. Consilvio, C.; Vincent, A.M.; Feldman, E.L. Neuroinflammation, COX-2, and ALS—A Dual Role? *Exp. Neurol.* **2004**, *187*, 1–10. [CrossRef] [PubMed]
76. Rocha, N.P.; de Miranda, A.S.; Teixeira, A.L. Insights into Neuroinflammation in Parkinson's Disease: From Biomarkers to Anti-Inflammatory Based Therapies. *Biomed Res. Int.* **2015**, *2015*, 628192. [CrossRef] [PubMed]
77. Wang, Q.; Liu, Y.; Zhou, J. Neuroinflammation in Parkinson's Disease and Its Potential as Therapeutic Target. *Transl. Neurodegener.* **2015**, *4*, 19. [CrossRef]

78. Duque, E.d.A.; Munhoz, C.D. The Pro-Inflammatory Effects of Glucocorticoids in the Brain. *Front. Endocrinol.* **2016**, *7*, 78. [CrossRef]
79. Ali, M.M.; Ghouri, R.G.; Ans, A.H.; Akbar, A.; Toheed, A. Recommendations for Anti-Inflammatory Treatments in Alzheimer's Disease: A Comprehensive Review of the Literature. *Cureus* **2019**, *11*, e4620. [CrossRef]
80. de Ceballos, M.L. Cannabinoids for the Treatment of Neuroinflammation. In *Cannabinoids in Neurologic and Mental Disease*; Elsevier: Amsterdam, The Netherlands, 2015; pp. 3–14.
81. Morris, G.; Walker, A.J.; Berk, M.; Maes, M.; Puri, B.K. Cell Death Pathways: A Novel Therapeutic Approach for Neuroscientists. *Mol. Neurobiol.* **2018**, *55*, 5767–5786. [CrossRef]
82. Jung, Y.J.; Tweedie, D.; Scerba, M.T.; Greig, N.H. Neuroinflammation as a Factor of Neurodegenerative Disease: Thalidomide Analogs as Treatments. *Front. Cell Dev. Biol.* **2019**, *7*, 313. [CrossRef] [PubMed]
83. Mai, N.; Knowlden, S.A.; Miller-Rhodes, K.; Prifti, V.; Sims, M.; Grier, M.; Nelson, M.; Halterman, M.W. Effects of 9-t-Butyl Doxycycline on the Innate Immune Response to CNS Ischemia-Reperfusion Injury. *Exp. Mol. Pathol.* **2021**, *118*, 104601. [CrossRef]
84. Ginwala, R.; Bhavsar, R.; Chigbu, D.I.; Jain, P.; Khan, Z.K. Potential Role of Flavonoids in Treating Chronic Inflammatory Diseases with a Special Focus on the Anti-Inflammatory Activity of Apigenin. *Antioxidants* **2019**, *8*, 35. [CrossRef] [PubMed]
85. Chen, J.; Wang, Y.; Zhu, T.; Yang, S.; Cao, J.; Li, X.; Wang, L.-S.; Sun, C. Beneficial Regulatory Effects of Polymethoxyflavone—Rich Fraction from Ougan (*Citrus reticulata* cv. Suavissima) Fruit on Gut Microbiota and Identification of Its Intestinal Metabolites in Mice. *Antioxidants* **2020**, *9*, 831. [CrossRef] [PubMed]
86. Kou, G.; Hu, Y.; Jiang, Z.; Li, Z.; Li, P.; Song, H.; Chen, Q.; Zhou, Z.; Lyu, Q. *Citrus aurantium* L. Polymethoxyflavones Promote Thermogenesis of Brown and White Adipose Tissue in High-Fat Diet Induced C57BL/6J Mice. *J. Funct. Foods* **2020**, *67*, 103860. [CrossRef]
87. Manthey, J.; Guthrie, N.; Grohmann, K. Biological Properties of Citrus Flavonoids Pertaining to Cancer and Inflammation. *Curr. Med. Chem.* **2001**, *8*, 135–153. [CrossRef] [PubMed]
88. Li, G.; Tan, F.; Zhang, Q.; Tan, A.; Cheng, Y.; Zhou, Q.; Liu, M.; Tan, X.; Huang, L.; Rouseff, R.; et al. Protective Effects of Polymethoxyflavone-Rich Cold-Pressed Orange Peel Oil against Ultraviolet B-Induced Photoaging on Mouse Skin. *J. Funct. Foods* **2020**, *67*, 103834. [CrossRef]
89. Suzuki, T.; Shimizu, M.; Yamauchi, Y.; Sato, R. Polymethoxyflavones in Orange Peel Extract Prevent Skeletal Muscle Damage Induced by Eccentric Exercise in Rats. *Biosci. Biotechnol. Biochem.* **2021**, *85*, 440–446. [CrossRef] [PubMed]
90. Bao, Y.; Fenwick, R. *Phytochemicals in Health and Disease*; CRC Press: New York, NY, USA, 2004; ISBN 9780429215421.
91. Tung, Y.-C.; Chou, Y.-C.; Hung, W.-L.; Cheng, A.-C.; Yu, R.-C.; Ho, C.-T.; Pan, M.-H. Polymethoxyflavones: Chemistry and Molecular Mechanisms for Cancer Prevention and Treatment. *Curr. Pharmacol. Rep.* **2019**, *5*, 98–113. [CrossRef]
92. Zeng, S.-L.; Li, S.-Z.; Xiao, P.-T.; Cai, Y.-Y.; Chu, C.; Chen, B.-Z.; Li, P.; Li, J.; Liu, E.-H. Citrus Polymethoxyflavones Attenuate Metabolic Syndrome by Regulating Gut Microbiome and Amino Acid Metabolism. *Sci. Adv.* **2020**, *6*, eaax6208. [CrossRef]
93. Borowiec, K.; Michalak, A. Flavonoids from Edible Fruits as Therapeutic Agents in Neuroinflammation—A Comprehensive Review and Update. *Crit. Rev. Food Sci. Nutr.* **2022**, *62*, 6742–6760. [CrossRef] [PubMed]
94. Alexander, G.E. Biology of Parkinson's Disease: Pathogenesis and Pathophysiology of a Multisystem Neurodegenerative Disorder. *Dialogues Clin. Neurosci.* **2004**, *6*, 259–280. [CrossRef] [PubMed]
95. Stykel, M.G.; Ryan, S.D. Nitrosative Stress in Parkinson's Disease. *NPJ Park. Dis.* **2022**, *8*, 104. [CrossRef] [PubMed]
96. Meng-zhen, S.; Ju, L.; Lan-chun, Z.; Cai-feng, D.; Shu-da, Y.; Hao-fei, Y.; Wei-yan, H. Potential Therapeutic Use of Plant Flavonoids in AD and PD. *Heliyon* **2022**, *8*, e11440. [CrossRef] [PubMed]
97. Shu, Z.; Yang, B.; Zhao, H.; Xu, B.; Jiao, W.; Wang, Q.; Wang, Z.; Kuang, H. Tangeretin Exerts Anti-Neuroinflammatory Effects via NF-KB Modulation in Lipopolysaccharide-Stimulated Microglial Cells. *Int. Immunopharmacol.* **2014**, *19*, 275–282. [CrossRef] [PubMed]
98. Ghribi, O.; Herman, M.M.; Pramoonjago, P.; Savory, J. MPP + Induces the Endoplasmic Reticulum Stress Response in Rabbit Brain Involving Activation of the ATF-6 and NF-KB Signaling Pathways. *J. Neuropathol. Exp. Neurol.* **2003**, *62*, 1144–1153. [CrossRef] [PubMed]
99. Hashida, K.; Kitao, Y.; Sudo, H.; Awa, Y.; Maeda, S.; Mori, K.; Takahashi, R.; Iinuma, M.; Hori, O. ATF6alpha Promotes Astroglial Activation and Neuronal Survival in a Chronic Mouse Model of Parkinson's Disease. *PLoS ONE* **2012**, *7*, e47950. [CrossRef] [PubMed]
100. Fatima, A.; Khanam, S.; Rahul, R.; Jyoti, S.; Naz, F.; Ali, F.; Siddique, Y.H. Protective Effect of Tangeritin in Transgenic Drosophila Model of Parkinson's Disease. *Front. Biosci. (Elite Ed.)* **2017**, *9*, 44–53. [CrossRef]
101. Gruntz, K.; Bloechliger, M.; Becker, C.; Jick, S.S.; Fuhr, P.; Meier, C.R.; Rüegg, S. Parkinson Disease and the Risk of Epileptic Seizures. *Ann. Neurol.* **2018**, *83*, 363–374. [CrossRef]
102. Guo, X.; Cao, Y.; Hao, F.; Yan, Z.; Wang, M.; Liu, X. Tangeretin Alters Neuronal Apoptosis and Ameliorates the Severity of Seizures in Experimental Epilepsy-Induced Rats by Modulating Apoptotic Protein Expressions, Regulating Matrix Metalloproteinases, and Activating the PI3K/Akt Cell Survival Pathway. *Adv. Med. Sci.* **2017**, *62*, 246–253. [CrossRef]
103. Hung, W.-L.; Chiu, T.-H.; Wei, G.-J.; Pan, M.-H.; Ho, C.-T.; Hwang, L.S.; Hsu, B.-Y. Neuroprotective Effects of Nobiletin and Tangeretin against Amyloid B1-42-Induced Toxicity in Cultured Primary Rat Neurons. *Nutrire* **2023**, *48*, 56. [CrossRef]
104. Matsuzaki, K.; Nakajima, A.; Guo, Y.; Ohizumi, Y. A Narrative Review of the Effects of Citrus Peels and Extracts on Human Brain Health and Metabolism. *Nutrients* **2022**, *14*, 1847. [CrossRef] [PubMed]

105. Testai, L.; Calderone, V. Nutraceutical Value of Citrus Flavanones and Their Implications in Cardiovascular Disease. *Nutrients* **2017**, *9*, 502. [CrossRef] [PubMed]
106. Wu, C.; Zhao, J.; Chen, Y.; Li, T.; Zhu, R.; Zhu, B.; Zhang, Y. Tangeretin Protects Human Brain Microvascular Endothelial Cells against Oxygen-glucose Deprivation-induced Injury. *J. Cell. Biochem.* **2019**, *120*, 4883–4891. [CrossRef] [PubMed]
107. You, G.; Zheng, L.; Zhang, Y.; Zhang, Y.; Wang, Y.; Guo, W.; Liu, H.; Tatiana, P.; Vladimir, K.; Zan, J. Tangeretin Attenuates Cerebral Ischemia–Reperfusion-Induced Neuronal Pyroptosis by Inhibiting AIM2 Inflammasome Activation via Regulating NRF2. *Inflammation* **2024**, *47*, 145–158. [CrossRef] [PubMed]
108. Sedik, A.A.; Elgohary, R. Neuroprotective Effect of Tangeretin against Chromium-Induced Acute Brain Injury in Rats: Targeting Nrf2 Signaling Pathway, Inflammatory Mediators, and Apoptosis. *Inflammopharmacology* **2023**, *31*, 1465–1480. [CrossRef] [PubMed]
109. Chalovich, E.M.; Zhu, J.; Caltagarone, J.; Bowser, R.; Chu, C.T. Functional Repression of CAMP Response Element in 6-Hydroxydopamine-Treated Neuronal Cells. *J. Biol. Chem.* **2006**, *281*, 17870–17881. [CrossRef] [PubMed]
110. Ljungberg, M.C.; Ali, Y.O.; Zhu, J.; Wu, C.-S.; Oka, K.; Zhai, R.G.; Lu, H.-C. CREB-Activity and Nmnat2 Transcription Are down-Regulated Prior to Neurodegeneration, While NMNAT2 over-Expression Is Neuroprotective, in a Mouse Model of Human Tauopathy. *Hum. Mol. Genet.* **2012**, *21*, 251–267. [CrossRef]
111. Nagase, H.; Omae, N.; Omori, A.; Nakagawasai, O.; Tadano, T.; Yokosuka, A.; Sashida, Y.; Mimaki, Y.; Yamakuni, T.; Ohizumi, Y. Nobiletin and Its Related Flavonoids with CRE-Dependent Transcription-Stimulating and Neuritegenic Activities. *Biochem. Biophys. Res. Commun.* **2005**, *337*, 1330–1336. [CrossRef]
112. Kawahata, I.; Yoshida, M.; Sun, W.; Nakajima, A.; Lai, Y.; Osaka, N.; Matsuzaki, K.; Yokosuka, A.; Mimaki, Y.; Naganuma, A.; et al. Potent Activity of Nobiletin-Rich Citrus Reticulata Peel Extract to Facilitate CAMP/PKA/ERK/CREB Signaling Associated with Learning and Memory in Cultured Hippocampal Neurons: Identification of the Substances Responsible for the Pharmacological Action. *J. Neural Transm.* **2013**, *120*, 1397–1409. [CrossRef]
113. Halliwell, B. Oxidative Stress and Neurodegeneration: Where Are We Now? *J. Neurochem.* **2006**, *97*, 1634–1658. [CrossRef] [PubMed]
114. Uttara, B.; Singh, A.; Zamboni, P.; Mahajan, R. Oxidative Stress and Neurodegenerative Diseases: A Review of Upstream and Downstream Antioxidant Therapeutic Options. *Curr. Neuropharmacol.* **2009**, *7*, 65–74. [CrossRef] [PubMed]
115. Melo, A.; Monteiro, L.; Lima, R.M.F.; de Oliveira, D.M.; de Cerqueira, M.D.; El-Bachá, R.S. Oxidative Stress in Neurodegenerative Diseases: Mechanisms and Therapeutic Perspectives. *Oxid. Med. Cell. Longev.* **2011**, *2011*, 467180. [CrossRef] [PubMed]
116. Albarracin, S.L.; Stab, B.; Casas, Z.; Sutachan, J.J.; Samudio, I.; Gonzalez, J.; Gonzalo, L.; Capani, F.; Morales, L.; Barreto, G.E. Effects of Natural Antioxidants in Neurodegenerative Disease. *Nutr. Neurosci.* **2012**, *15*, 1–9. [CrossRef] [PubMed]
117. Adams, J., Jr.; Chang, M.-L.; Klaidman, L. Parkinsons Disease—Redox Mechanisms. *Curr. Med. Chem.* **2001**, *8*, 809–814. [CrossRef] [PubMed]
118. Dias, V.; Junn, E.; Mouradian, M.M. The Role of Oxidative Stress in Parkinson's Disease. *J. Park. Dis.* **2013**, *3*, 461–491. [CrossRef] [PubMed]
119. Zhu, J.; Chu, C.T. Mitochondrial Dysfunction in Parkinson's Disease. *J. Alzheimer's Dis.* **2010**, *20*, S325–S334. [CrossRef] [PubMed]
120. Hastings, T.G. The Role of Dopamine Oxidation in Mitochondrial Dysfunction: Implications for Parkinson's Disease. *J. Bioenerg. Biomembr.* **2009**, *41*, 469–472. [CrossRef]
121. Chinta, S.J.; Kumar, M.J.; Hsu, M.; Rajagopalan, S.; Kaur, D.; Rane, A.; Nicholls, D.G.; Choi, J.; Andersen, J.K. Inducible Alterations of Glutathione Levels in Adult Dopaminergic Midbrain Neurons Result in Nigrostriatal Degeneration. *J. Neurosci.* **2007**, *27*, 13997–14006. [CrossRef]
122. Cuadrado, A.; Moreno-Murciano, P.; Pedraza-Chaverri, J. The Transcription Factor Nrf2 as a New Therapeutic Target in Parkinson's Disease. *Expert. Opin. Ther. Targets* **2009**, *13*, 319–329. [CrossRef]
123. Wu, J.; Cui, Y.; Yang, Y.; Jung, S.; Hyun, J.W.; Maeng, Y.; Park, D.; Lee, S.; Kim, S.; Eun, S. Mild Mitochondrial Depolarization Is Involved in a Neuroprotective Mechanism of Citrus Sunki Peel Extract. *Phytother. Res.* **2013**, *27*, 564–571. [CrossRef] [PubMed]
124. Yoon, J.H.; Lim, T.-G.; Lee, K.M.; Jeon, A.J.; Kim, S.Y.; Lee, K.W. Tangeretin Reduces Ultraviolet B (UVB)-Induced Cyclooxygenase-2 Expression in Mouse Epidermal Cells by Blocking Mitogen-Activated Protein Kinase (MAPK) Activation and Reactive Oxygen Species (ROS) Generation. *J. Agric. Food Chem.* **2011**, *59*, 222–228. [CrossRef] [PubMed]
125. Lakshmi, A.; Subramanian, S. Chemotherapeutic Effect of Tangeretin, a Polymethoxylated Flavone Studied in 7, 12-Dimethylbenz(a)Anthracene Induced Mammary Carcinoma in Experimental Rats. *Biochimie* **2014**, *99*, 96–109. [CrossRef] [PubMed]
126. Lakshmi, A.; Subramanian, S.P. Tangeretin Ameliorates Oxidative Stress in the Renal Tissues of Rats with Experimental Breast Cancer Induced by 7,12-Dimethylbenz[a]Anthracene. *Toxicol. Lett.* **2014**, *229*, 333–348. [CrossRef] [PubMed]
127. Sundaram, R.; Shanthi, P.; Sachdanandam, P. Tangeretin, a Polymethoxylated Flavone, Modulates Lipid Homeostasis and Decreases Oxidative Stress by Inhibiting NF-κB Activation and Proinflammatory Cytokines in Cardiac Tissue of Streptozotocin-Induced Diabetic Rats. *J. Funct. Foods* **2015**, *16*, 315–333. [CrossRef]
128. Yin, J.; Valin, K.L.; Dixon, M.L.; Leavenworth, J.W. The Role of Microglia and Macrophages in CNS Homeostasis, Autoimmunity, and Cancer. *J. Immunol. Res.* **2017**, *2017*, 5150678. [CrossRef] [PubMed]

129. Femminella, G.D.; Dani, M.; Wood, M.; Fan, Z.; Calsolaro, V.; Atkinson, R.; Edginton, T.; Hinz, R.; Brooks, D.J.; Edison, P. Microglial Activation in Early Alzheimer Trajectory Is Associated with Higher Gray Matter Volume. *Neurology* **2019**, *92*, e1331–e1343. [CrossRef]
130. George, S.; Rey, N.L.; Tyson, T.; Esquibel, C.; Meyerdirk, L.; Schulz, E.; Pierce, S.; Burmeister, A.R.; Madaj, Z.; Steiner, J.A.; et al. Microglia Affect α-Synuclein Cell-to-Cell Transfer in a Mouse Model of Parkinson's Disease. *Mol. Neurodegener.* **2019**, *14*, 34. [CrossRef]
131. Angelopoulou, E.; Paudel, Y.N.; Shaikh, M.F.; Piperi, C. Fractalkine (CX3CL1) Signaling and Neuroinflammation in Parkinson's Disease: Potential Clinical and Therapeutic Implications. *Pharmacol. Res.* **2020**, *158*, 104930. [CrossRef]
132. Calvani, R.; Picca, A.; Landi, G.; Marini, F.; Biancolillo, A.; Coelho-Junior, H.J.; Gervasoni, J.; Persichilli, S.; Primiano, A.; Arcidiacono, A.; et al. A Novel Multi-Marker Discovery Approach Identifies New Serum Biomarkers for Parkinson's Disease in Older People: An EXosomes in PArkiNson Disease (EXPAND) Ancillary Study. *Geroscience* **2020**, *42*, 1323–1334. [CrossRef]
133. Agrawal, M.; Agrawal, A.K. Pathophysiological Association Between Diabetes Mellitus and Alzheimer's Disease. *Cureus* **2022**, *14*, e29120. [CrossRef] [PubMed]
134. de Bem, A.F.; Krolow, R.; Farias, H.R.; de Rezende, V.L.; Gelain, D.P.; Moreira, J.C.F.; Duarte, J.M.d.N.; de Oliveira, J. Animal Models of Metabolic Disorders in the Study of Neurodegenerative Diseases: An Overview. *Front. Neurosci.* **2021**, *14*, 604150. [CrossRef] [PubMed]
135. Camargo Maluf, F.; Feder, D.; Alves De Siqueira Carvalho, A. Analysis of the Relationship between Type II Diabetes Mellitus and Parkinson's Disease: A Systematic Review. *Park. Dis.* **2019**, *2019*, 4951379. [CrossRef] [PubMed]
136. Wang, S.Y.; Wu, S.L.; Chen, T.C.; Chuang, C. sen Antidiabetic Agents for Treatment of Parkinson's Disease: A Meta-Analysis. *Int. J. Environ. Res. Public Health* **2020**, *17*, 4805. [CrossRef] [PubMed]
137. Yu, H.; Sun, T.; He, X.; Wang, Z.; Zhao, K.; An, J.; Wen, L.; Li, J.-Y.; Li, W.; Feng, J. Association between Parkinson's Disease and Diabetes Mellitus: From Epidemiology, Pathophysiology and Prevention to Treatment. *Aging Dis.* **2022**, *13*, 1591–1605. [CrossRef]
138. Cardoso, S.; Moreira, P.I. Antidiabetic Drugs for Alzheimer's and Parkinson's Diseases: Repurposing Insulin, Metformin, and Thiazolidinediones. *Int. Rev. Neurobiol.* **2020**, *155*, 37–64. [CrossRef]
139. Sundaram, R.; Shanthi, P.; Sachdanandam, P. Effect of Tangeretin, a Polymethoxylated Flavone on Glucose Metabolism in Streptozotocin-Induced Diabetic Rats. *Phytomedicine* **2014**, *21*, 793–799. [CrossRef]
140. Liu, Y.; Han, J.; Zhou, Z.; Li, D. Tangeretin Inhibits Streptozotocin-Induced Cell Apoptosis via Regulating NF-KB Pathway in INS-1 Cells. *J. Cell Biochem.* **2019**, *120*, 3286–3293. [CrossRef] [PubMed]
141. Chen, F.; Ma, Y.; Sun, Z.; Zhu, X. Tangeretin Inhibits High Glucose-Induced Extracellular Matrix Accumulation in Human Glomerular Mesangial Cells. *Biomed. Pharmacother.* **2018**, *102*, 1077–1083. [CrossRef]
142. Liu, T.W.; Chen, C.M.; Chang, K.H. Biomarker of Neuroinflammation in Parkinson's Disease. *Int. J. Mol. Sci.* **2022**, *23*, 4148. [CrossRef]
143. Swanson, C.R.; Joers, V.; Bondarenko, V.; Brunner, K.; Simmons, H.A.; Ziegler, T.E.; Kemnitz, J.W.; Johnson, J.A.; Emborg, M.E. The PPAR-γ Agonist Pioglitazone Modulates Inflammation and Induces Neuroprotection in Parkinsonian Monkeys. *J. Neuroinflam.* **2011**, *8*, 91. [CrossRef] [PubMed]
144. Breidert, T.; Callebert, J.; Heneka, M.T.; Landreth, G.; Launay, J.M.; Hirsch, E.C. Protective Action of the Peroxisome Proliferator-Activated Receptor-Gamma Agonist Pioglitazone in a Mouse Model of Parkinson's Disease. *J. Neurochem.* **2002**, *82*, 615–624. [CrossRef] [PubMed]
145. Dehmer, T.; Heneka, M.T.; Sastre, M.; Dichgans, J.; Schulz, J.B. Protection by Pioglitazone in the MPTP Model of Parkinson's Disease Correlates with I Kappa B Alpha Induction and Block of NF Kappa B and INOS Activation. *J. Neurochem.* **2004**, *88*, 494–501. [CrossRef] [PubMed]
146. Schintu, N.; Frau, L.; Ibba, M.; Caboni, P.; Garau, A.; Carboni, E.; Carta, A.R. PPAR-Gamma-Mediated Neuroprotection in a Chronic Mouse Model of Parkinson's Disease. *Eur. J. Neurosci.* **2009**, *29*, 954–963. [CrossRef] [PubMed]
147. Yoshizaki, N.; Fujii, T.; Masaki, H.; Okubo, T.; Shimada, K.; Hashizume, R. Orange Peel Extract, Containing High Levels of Polymethoxyflavonoid, Suppressed UVB-Induced COX-2 Expression and PGE2 Production in HaCaT Cells through PPAR-γ Activation. *Exp. Dermatol.* **2014**, *23*, 18–22. [CrossRef] [PubMed]
148. Kim, M.S.; Hur, H.J.; Kwon, D.Y.; Hwang, J.T. Tangeretin Stimulates Glucose Uptake via Regulation of AMPK Signaling Pathways in C2C12 Myotubes and Improves Glucose Tolerance in High-Fat Diet-Induced Obese Mice. *Mol. Cell Endocrinol.* **2012**, *358*, 127–134. [CrossRef]
149. Li, R.W.; Theriault, A.G.; Au, K.; Douglas, T.D.; Casaschi, A.; Kurowska, E.M.; Mukherjee, R. Citrus Polymethoxylated Flavones Improve Lipid and Glucose Homeostasis and Modulate Adipocytokines in Fructose-Induced Insulin Resistant Hamsters. *Life Sci.* **2006**, *79*, 365–373. [CrossRef]
150. Kurowska, E.M.; Manthey, J.A.; Casaschi, A.; Theriault, A.G. Modulation of HepG2 Cell Net Apolipoprotein B Secretion by the Citrus Polymethoxyflavone, Tangeretin. *Lipids* **2004**, *39*, 143–151. [CrossRef]

Disclaimer/Publisher's Note: The statements, opinions and data contained in all publications are solely those of the individual author(s) and contributor(s) and not of MDPI and/or the editor(s). MDPI and/or the editor(s) disclaim responsibility for any injury to people or property resulting from any ideas, methods, instructions or products referred to in the content.

Article

Evodiamine Alleviates 2,4-Dinitro-1-Chloro-Benzene-Induced Atopic Dermatitis-like Symptoms in BALB/c Mice

So-Young Han and Dong-Soon Im *

Department of Fundamental Pharmaceutical Sciences, Graduate School, Kyung Hee University, Seoul 02447, Republic of Korea; wlsruddk62@khu.ac.kr
* Correspondence: imds@khu.ac.kr; Tel.: +82-2-961-9377

Abstract: Evodiamine is an alkaloid found in Evodia fruits, a traditional Chinese medicine. Preclinical studies have demonstrated its anti-inflammatory and neuroprotective properties. The 2,4-dinitro-1-chloro-benzene (DNCB) was used to test the effects of evodiamine on a chemically induced atopic dermatitis-like model in BALB/c mice. Evodiamine significantly lowered serum immunoglobulin E levels, which increased as an immune response to the long-term application of DNCB. Several atopic dermatitis-like skin symptoms induced by DNCB, including skin thickening and mast cell accumulation, were suppressed by evodiamine therapy. DNCB induced higher levels of pro-inflammatory cytokines in type 2 helper T (Th2) cells (IL-4 and IL-13), Th1 cells (IFN-γ and IL-12A), Th17 cells (IL-17A), Th22 cells (IL-22), and chemokines (IL-6 and IL-8). These increases were suppressed in the lymph nodes and skin following evodiamine treatment. The results of our study indicate that evodiamine suppresses atopic dermatitis-like responses in mice and may therefore be useful in treating these conditions.

Keywords: *Evodia rutaecarpa*; atopy; evodiamine; dermatitis; eczema; anti-atopy

Citation: Han, S.-Y.; Im, D.-S. Evodiamine Alleviates 2,4-Dinitro-1-Chloro-Benzene-Induced Atopic Dermatitis-like Symptoms in BALB/c Mice. *Life* 2024, 14, 494. https://doi.org/10.3390/life14040494

Academic Editor: Ahmed H. E. Hassan

Received: 4 March 2024
Revised: 26 March 2024
Accepted: 10 April 2024
Published: 11 April 2024

Copyright: © 2024 by the authors. Licensee MDPI, Basel, Switzerland. This article is an open access article distributed under the terms and conditions of the Creative Commons Attribution (CC BY) license (https:// creativecommons.org/licenses/by/ 4.0/).

1. Introduction

Chronic atopic dermatitis is an inflammatory, pruritic skin condition characterized by persistent skin inflammation that often occurs in families with other atopic diseases [1,2]. Various factors, including defects in skin barrier function due to structural protein formation such as filaggrin, abnormal lipid metabolism, environmental factors, and genetic factors, contribute to dysregulated immune responses [1,3]. Pediatric atopic dermatitis patients comprise the majority (1–3% of adults and up to 20% of children) in most countries [4,5]. A hallmark of atopic dermatitis is the presence of widespread eczematous lesions [6]. Atopic dermatitis is often the first step in the development of other atopic diseases, such as allergic rhinoconjunctivitis and/or bronchial asthma [7]. Although elevation of total or allergen-specific immunoglobulin E (IgE) levels in the serum is not always observed in all individuals, atopic dermatitis is characterized by type 2 immune responses and consequently increased IgE levels [8,9]. In addition to the Th2 response, the Th1 and Th17 responses also play key roles in pathogenesis [10–12].

Current therapeutics include glucocorticoids as a first-line anti-inflammatory treatment, topically applied to the skin for pruritus or new flares [5]. In adults, topical application of tacrolimus and pimecrolimus is permitted for eczema treatment [13]. Since mast cell mediators, such as tryptase and histamine, play important roles in pruritus induction in atopic dermatitis, the application of mast cell stabilizers has also been considered [14]. However, owing to the side effects of long-term glucocorticoid use, alternative therapeutics must be developed [15]. Traditional Chinese medicine can provide a variety of resources for the development of new drugs for atopic dermatitis [16].

The dried, unripe fruit of *Evodia rutaecarpa* Bentham (Rutaceae) has been used to treat epilepsy, emesis, dermatophytosis, dysentery, gastrointestinal disorders, and headaches [17].

As alkaloidal compounds found in Evodia fruits, evodiamine and rutaecarpine have shown a number of pharmacological properties in preclinical models, including anti-inflammatory, anti-obesity, anti-bacterial, anti-cancer, anti-cardiovascular, anti-diabetic, insecticide, and neuroprotective effects [17]. Shin et al. found that evodiamine and rutaecarpine inhibit the production of pro-inflammatory cytokines in rat RBL-2H3 cells induced by immunoglobulin E-antigen complexes, resulting in the antagonism of passive cutaneous anaphylaxis [18]. In mouse RAW 264.7 cells, evodiamine inhibited lipopolysaccharide-induced prostaglandin E_2 synthesis at concentrations of 1–10 mM [19]. Human A549 cells treated with evodiamine produced less monocyte chemoattractant protein-1 (MCP-1) after exposure to H1N1 and recruited fewer macrophages toward chemokine (C-C motif) ligand 5 (CCL5) and MCP-1 [20]. A concentration-dependent inverse relationship has been observed between evodiamine and endothelial cell cyclooxygenase 2 (COX-2) and nitric oxide synthase (iNOS) expression under hypoxic conditions [21]. The release of calcitonin gene-related peptides by evodiamine inhibits guinea pig heart anaphylaxis induced by bovine albumin [22]. According to Pearce et al., evodiamine is a new class of agonist of the rat transient receptor potential subfamily V member 1 (TRPV1) [23]. However, its potency is about 3–19-fold lower than that of capsaicin. In vivo studies have shown that evodiamine inhibits ovalbumin (OVA)-induced allergic asthma [24]. However, the efficacy of evodiamine in atopic dermatitis has not yet been examined. This study aimed to explore whether evodiamine acts as an anti-atopic agent and to investigate the mechanism of action of evodiamine in a dinitrochlorobenzene-induced atopic dermatitis-like mouse model.

2. Results

2.1. Evodiamine Suppresses Atopic Dermatitis Symptoms in the Ears of Mice

According to Evodia fruits, evodiamine is a major alkaloid that exerts anti-inflammatory effects on mast cells, macrophages, and lung epithelial cells; the cytokines tumor necrosis factor-α (TNF-α) and IL-4 were suppressed in RBL-2H3 mast cells [18], prostaglandin E_2 synthesis was inhibited in RAW 264.7 macrophages [19], and MCP-1 production was inhibited in lung epithelial cells A549 [20]. Evodiamine was hypothesized to be a therapeutic candidate for atopic dermatitis based on its anti-inflammatory effects, and we examined its effect on 2,4-dinitro-1-chloro-benzene (DNCB)-induced atopic dermatitis responses in mice. From days 7 to 48 after sensitization by DNCB intraperitoneal injection, DNCB were topically challenged every other day [25]. Evodiamine and DEX were administered by intraperitoneal injection 30 min before the DNCB challenge on day 19. Mice were sacrificed on day 49 (Figure 1A).

Scaling, erythema, and erosions were observed in the ears after exposure to DNCB (Figure 1B). Evodiamine-treated mice showed improvements in atopic dermatitis-like symptoms, whereas DEX-treated mice showed more significant improvements, especially swelling of the ears, a characteristic atopic dermatitis-like symptom. As shown in Figure 2A, cutaneous hyperplasia is clearly visible. Evodiamine and DEX significantly reduced DNCB-induced epidermal hyperplasia in response to DNCB (Figure 2A), which was measured as the thickness of the ear (Figure 2B). The suppressive effects of evodiamine (20 mg/kg) were equivalent to those of DEX (Figure 2B). In addition to acanthosis, an increase in epidermal thickness was observed by hematoxylin and eosin (H&E) staining, as well as exocytosis and increased immune cell accumulation (Figure 2A). By counting mast cells using toluidine blue O staining, evodiamine-treated atopic dermatitis mice displayed a significant suppression of immune cell accumulation (Figures 2A and 3A).

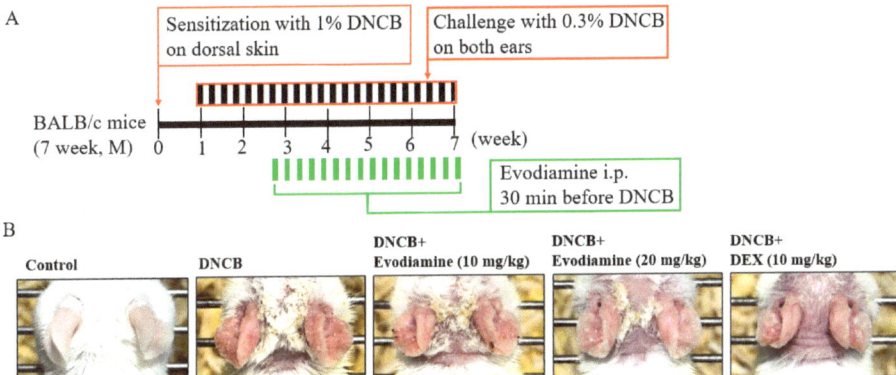

Figure 1. Effect of evodiamine on DNCB-induced atopic dermatitis in the ears. Evaluation of the effect of evodiamine on atopic dermatitis induced by DNCB in ears. (**A**) An outline of the protocol for induction of atopic dermatitis and treatment with evodiamine. On day 0, DNCB was applied to the skin to sensitize it. Following repeated DNCB challenges on days 7–48, atopic dermatitis-like phenotypes were induced. In addition to DNCB, vehicles were applied topically to BALB/c mice, while evodiamine and DEX were injected intraperitoneally 30 min before DNCB exposure (n = 5). (**B**) A view of the ear in its entirety.

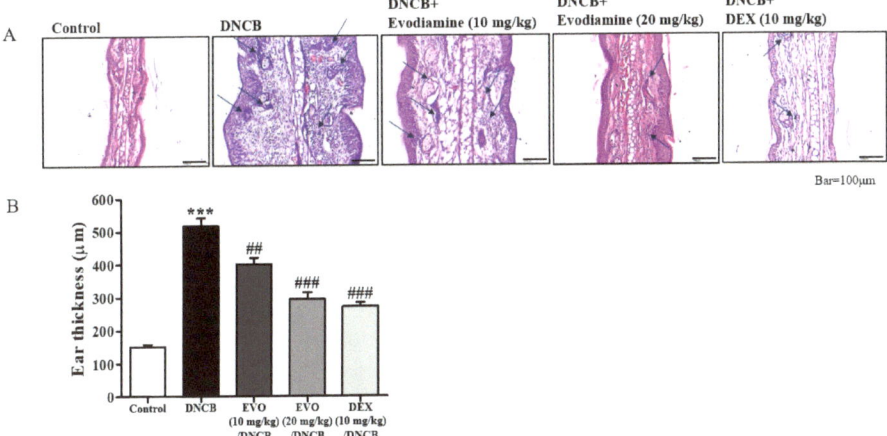

Figure 2. Effect of evodiamine on ear thickness. These are representative histological findings from cutaneous tissue sections taken on day 49. In order to stain the ear tissue, hematoxylin and eosin stain was applied (**A**). Blue arrows indicated immune cells. A comparison of the ear thickness between the groups was made in (**B**). Five mice were used for each group. Statistical significance: *** $p < 0.001$ vs. the vehicle-treated control group; ### $p < 0.001$, ## $p < 0.01$ vs. the DNCB-treated group. A magnification of 200× was used.

In response to allergens, mast cells release pro-inflammatory cytokines and histamine, which contribute to atopic dermatitis symptoms. Toluidine blue O staining revealed mast cell infiltration of the dermis (Figure 3A). DNCB treatment resulted in a dramatic increase in mast cell numbers and hyperplasia (Figure 3A). In evodiamine-treated mice, the number of mast cells significantly decreased in a dose-dependent manner (Figure 3B). The number of mast cells was significantly suppressed in mice treated with DEX, similarly to that observed in mice treated with evodiamine (Figure 3B).

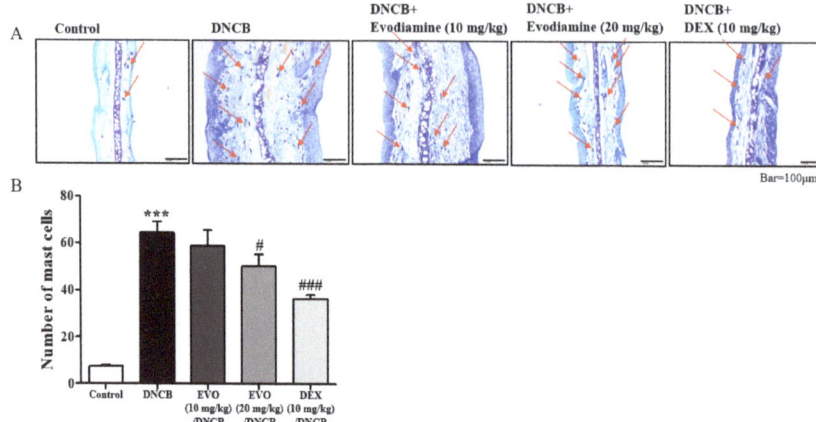

Figure 3. Effect of evodiamine on mast cell count in the ears. In order to identify mast cells, the skin was stained with toluidine blue O. (**A**) Histological findings on day 49 of representative cutaneous tissue sections. Mast cells are indicated with red arrows. (**B**) The number of mast cells in the ear tissues is shown in the histogram ($n = 5$). Statistical significance: *** $p < 0.001$ vs. the vehicle-treated control group; ### $p < 0.001$, # $p < 0.05$ vs. the DNCB-treated group. A magnification of 200× was used.

2.2. Evodiamine Suppresses Pro-Inflammatory Cytokine Expressions in the Ears of Mice

In atopic dermatitis pathogenesis, inflammatory cytokines of the Th2, Th17, Th1, and Th22 subclasses, such as IL-4, IL-13, IL-17A, INF-γ, IL-12A, and IL-22, play important roles [26–28]. In addition, the chemokines IL-6 and IL-8 attract inflammatory immune cells to the site of an atopic dermatitis [26–28]. There was an increase in cytokine and chemokine mRNA expression levels in the ear sections of atopic dermatitis-induced mice (Figure 4). In this study, both evodiamine and DEX significantly reduced the elevated mRNA levels of cytokines, particularly IL-13, IL-17A, IFN-γ, and IL-8 (Figure 4).

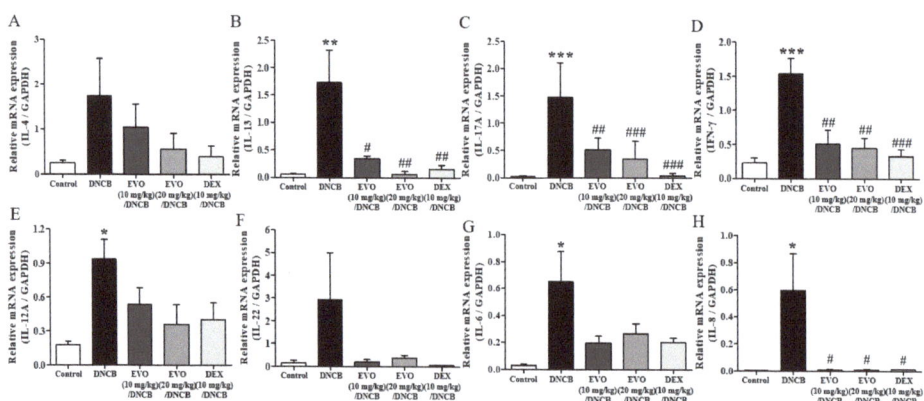

Figure 4. Effect of evodiamine on pro-inflammatory cytokine expression in the ears. Based on mouse ear tissue mRNA isolated from the ears of mice, qRT-PCR analysis of the Th2 cytokines IL-4 (**A**) and IL-13 (**B**) cytokines, Th17 cytokine, IL-17A (**C**), and Th1 cytokine, INF-γ (**D**) and IL-12A (**E**), Th22 cytokine, IL-22 (**F**), and chemokines IL-6 (**G**) and IL-8 (**H**) was conducted ($n = 5$). Statistical significance: *** $p < 0.001$, ** $p < 0.01$, * $p < 0.05$ vs. the vehicle-treated control group; ### $p < 0.001$, ## $p < 0.01$, # $p < 0.05$ vs. the DNCB-treated group. Normalization was also performed by comparing mRNA levels to GAPDH mRNA levels.

2.3. Evodiamine Suppresses DNCB-Induced Immune Responses in Lymph Nodes

Next, the lymph nodes of the cervical region were examined for immune responses (Figure 5A). This response was significantly suppressed (Figure 5B). Significant inhibition was observed after treatment with 20 mg/kg evodiamine, but not as much as after DEX treatment (Figure 5B).

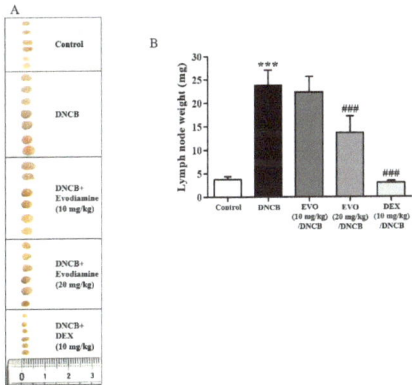

Figure 5. Effect of evodiamine on the size of lymph nodes. (**A**) Photographs were taken of the lymph nodes to measure their morphological changes. (**B**) A measurement of lymph node weight was also conducted ($n = 5$). Statistical significance: *** $p < 0.001$ vs. the vehicle-treated control group; ### $p < 0.001$ vs. the DNCB-treated group.

Additionally, DNCB-induced atopic dermatitis mice exhibited significantly increased Th2 (IL-4 and IL-13), Th1 (IFN-γ), and Th17 (IL-17A) cytokine mRNA levels in the cervical lymph nodes (Figure 6A–D). The treatment efficacy of evodiamine was shown by the substantial suppression of the mRNA levels of cytokines, especially IL-17A (Figure 6C). In summary, the findings of the current study indicate that evodiamine reduces lymph node enlargement and the expression of inflammatory Th17 cytokines.

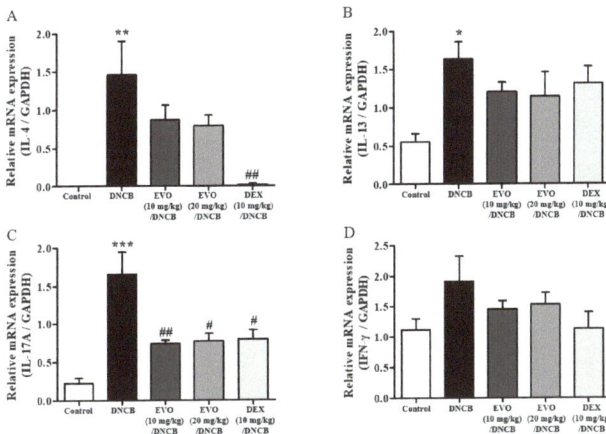

Figure 6. Effect of evodiamine on the expression of pro-inflammatory cytokines in lymph nodes. Based on mouse lymph node tissue mRNA isolated from the lymph nodes of mice, qRT-PCR analysis of the Th2 cytokines IL-4 (**A**) and IL-13 (**B**) cytokines, Th17 cytokine, IL-17A (**C**), and Th1 cytokine, IFN-γ (**D**) was conducted ($n = 5$). A normalization was also performed by comparing mRNA levels to GAPDH mRNA levels. Statistical significance: *** $p < 0.001$, ** $p < 0.01$, * $p < 0.05$ vs. the vehicle-treated control group; ## $p < 0.01$, # $p < 0.05$ vs. the DNCB-treated group.

2.4. Evodiamine Suppresses Pro-Inflammatory Cytokine Expressions in the Ears of Mice

IgE is a key player in the pathogenesis of atopic dermatitis, despite the considerable heterogeneity of the condition. A significant elevation in IgE levels is a key characteristic of atopic dermatitis and is associated with disease severity. Therefore, ELISA was used to test IgE levels in the serum obtained on day 49. The PBS group exhibited low serum IgE levels (Figure 7). However, the DNCB challenge can easily elevate IgE levels. DNCB plus evodiamine treatment significantly reduced IgE levels in a dose-dependent manner, but the efficacy of evodiamine was lower than that in the DEX-treated groups (Figure 7).

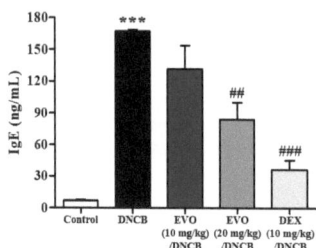

Figure 7. Effect of evodiamine on serum immunoglobulin E levels. Day 49 was the day on which serum was collected from the animals. An enzyme-linked immunosorbent assay was used to measure serum IgE levels ($n = 5$). Statistical significance: *** $p < 0.001$ vs. the vehicle-treated control group; ### $p < 0.001$, ## $p < 0.01$ vs. the DNCB-treated group.

3. Discussion

A chemically induced animal model was used to determine how evodiamine affects atopic dermatitis, which is a chronic inflammatory skin condition. Based on previous studies, the doses of evodiamine used in this study were determined to be between 10 and 20 mg/kg. This is the first study to demonstrate that evodiamine is effective against atopic dermatitis, as indicated by the suppression of ear thickening, mast cell accumulation, lymph node enlargement, IgE levels in the serum, and Th1/Th2/Th17/Th22 cytokines and chemokines. It has previously been demonstrated that evodiamine (25 mM) suppresses passive cutaneous anaphylactic symptoms in mice and exerts inhibitory effects on RBL-2H3 mast cells in vitro [18]. By injecting exogenous IgE directly into the skin, the model delivered antigen-specific IgE [29]. Therefore, the effect of a compound on the suppression of anaphylactic cutaneous reactions can be evaluated without active immune responses from antigen sensitization. Instead, the present atopic dermatitis model was induced by sensitization and challenge with antigens, resulting in more relevant conditions similar to in vivo atopic dermatitis. Similarly, Wang et al. [24] found that evodiamine (40 and 80 mg/kg) protected Sprague-Dawley rats from ovalbumin-induced asthma. In a rat model of allergic asthma, evodiamine inhibited allergic responses, which is in agreement with the results of our current study [24]. It was possible to observe the entire immune response to ovalbumin-induced asthma due to both antigen sensitization and challenge. Evodiamine reduced the thickness of the airway wall of the small bronchioles in the asthma group [24], which was similar to the reduction in the thickness of the ear and lymph nodes in this study. It was found that evodiamine caused downregulation of toll-like receptor-4, MyD88, nuclear factor κ-light-chain-enhancer of activated B cells (NF-κB), and high mobility group box 1 (HMGB1) mRNA in lung tissues [24]. It would be worth investigating whether evodiamine suppresses toll-like receptor 4 (TLR-4), myeloid differentiation primary response gene 88 (Myd88), NF-κB, and HMGB1 mRNA expression in the skin.

A primary mode of action of evodiamine may be its anti-inflammatory effect, as evodiamine was found to have anti-inflammatory properties in mast cells, macrophages, and lung epithelial cells [18–20]. Previous studies have shown that evodiamine inhibits cytokine production in RBL-2H3 mast cells [18], prostaglandin E_2 synthesis in RAW 264.7 macrophages [19], and the production of chemotactic MCP-1 in A549 lung epithelial cells [20]. In the bronchoalveolar lavage fluid and lung tissue, evodiamine inhibits IL-4, IL-13, and IL-17 [24]. As a result of evodiamine treatment, TNF-α and IL-4 levels were suppressed in RBL-2H3 mast cells [18]. Treatment of rutaecarpine, another indoloquinazoline alkaloid from the fruits of *Evodia rutaecarpa*, reduced plasma levels of IL-4 and IgE but enhanced plasma IFN-γ levels, which is contrasting to our results [30]. Evodiamine suppressed expression of pro-inflammatory COX-2 and iNOS as well as prostaglandin E_2 release under hypoxic conditions [21] and suppressed cardiac anaphylaxis in isolated guinea-pig hearts [22].

This study indicated that evodiamine may have anti-inflammatory effects on several immune cells that affect atopic dermatitis, as in allergic asthma models. Because of its high binding affinity to TLR-4, evodiamine can be considered a molecular target of TLR-4, since molecular docking studies of evodiamine with TLR-4 have shown its drug-binding affinity [24]. Although TRPV1 is also considered a potential target of evodiamine for pain perception [31], it is unlikely to be responsible for its anti-atopic dermatitis effects.

In addition to evodiamine, rutaecarpine also has anti-inflammatory properties [19]. A previous study showed that rutaecarpine inhibited the COX-2-dependent generation of prostaglandin D_2 in bone marrow-derived mast cells and reduced carrageenan-induced paw swelling [32]. Rutaecarpine treatment promoted macrophage immune training activators, including fos-related antigen 2 (FOSL2), SWI/SNF-related, matrix-associated, actin-dependent chromatin regulator subfamily a, member 4 (SMARCA4), and signal transducer and activator of transcription 3 (STAT3) [33], resulting in anti-inflammatory effects. Furthermore, rutaecarpine reduced the symptoms of atopic dermatitis in mice carrying the NC/Nga genotype [30].

We demonstrated the anti-atopic dermatitis effects of evodiamine in a murine model, demonstrating that the effectiveness of evodiamine was caused by the inhibition of mast cell accumulation in the skin, along with the inhibition of pro-inflammatory cytokines (IL-17A, IL-4, IL-13, IFN-γ, and IL-12A) and chemokines (IL-8 and IL-6) in the lymph nodes and epidermis. Based on the results of this study, evodiamine appears to have the potential to treat atopic dermatitis.

4. Materials and Methods

4.1. Chemicals

The 2,4-dinitro-1-chloro-benzene (DNCB) and evodiamine (Cat No. E3531, purity: ≥98% in HPLC) were purchased from Sigma-Aldrich (St. Louis, MO, USA).

4.2. Mouse Strain

We purchased male, 7-week-old BALB/c mice from Daehan Biolink (Seoul, Republic of Korea). Animal protocols were reviewed and approved by the Kyung Hee University-Institutional Animal Care Committee (KHSASP-23-012). The mice were housed in two per plastic cage under conditions of temperature at 22–24 °C, humidity at 60 ± 5%, and alternating light/dark cycles. We provided standard laboratory chow and water ad libitum.

4.3. Induction of DNCB-Induced Atopic Dermatitis in BALB/c Mice

Eight-week-old BALB/c mice were randomized into five groups ($n = 5$): a vehicle-treated control group, a DNCB-treated group, a DNCB-treated and evodiamine (10 mg/kg)-administered group, a DNCB-treated and evodiamine (20 mg/kg)-administered group, and a DNCB-treated and dexamethasone (DEX, 10 mg/kg)-administered group. To induce atopic dermatitis-like symptoms, we used DNCB as previously described [34]. Dorsal skin was shaved, and 300 μL of 1% DNCB in acetone/olive (3:1) was spread to the dorsal skin on day 0, which is sensitization. On day 7, 200 μL of 0.3% DNCB was applied to both ears every other day for up to 48 days. From day 19 until completion of the experiment, the evodiamine/DNCB-treated group was administered evodiamine (10 or 20 mg/kg body weight) by intraperitoneal injection 30 min prior to the challenge. The mice were sacrificed on day 49.

4.4. Total Immunoglobulin E (IgE) Levels in Serum

Total serum IgE levels were assessed using an ELISA kit (eBioscience, San Diego, CA, USA). In brief, 96-well plates (NUNC, Penfield, NY, USA) were coated with eBioscience's IgE capture antibody (88-50460-88, San Diego, CA, USA) and incubated overnight at 4 °C. After the washing, the plates were incubated at room temperature for 2 h with blocking buffer. Standard IgE was serially diluted and added to the appropriate wells to generate a calibration curve. Serum samples were added to each well. The plates were incubated at room temperature for 2 h on a shaker, followed by two washes. Each well was incubated with a biotinylated antibody designed to detect IgE (cat. 88–50460–88, eBioscience, San Diego, CA, USA). The plates were shaken for 1 h at room temperature. Each well was treated with avidin-horseradish peroxidase (HRP) after four washes. The incubation was performed in a shaker at room temperature for 30 min. After washing the plates four times, the substrate solution was applied, and the plates were incubated for 15 min at room temperature. The absorbance was measured at 450 nm after the addition of the stop solution.

4.5. Mast Cell Count in the Skin

Ear sections from mice in different experimental groups were analyzed after sacrifice on day 49. A 10% formalin fixative was used along with a 30% sucrose solution to dehydrate the ears, and an O.C.T. compound was used to embed the ears. To visualize mast cells in the skin, eight micrometer sections were stained with toluidine blue O (cat. T3260, Sigma-Aldrich, St. Louis, MO, USA). In each group, the tissues of five mice were examined. After removing the O.C.T. compound from the sections, the toluidine blue O reagent was added to the sections and soaked for 2 min. The sections were then rinsed, dehydrated, and mounted on coverslips. The number of mast cells was counted by analyzing photographs taken after toluidine blue O staining. Toluidine blue O staining was used to detect mast cells. The mast cells were counted twice in 50 optical fields, and the average was used [34].

4.6. Histologic Analysis and Measurement of Ear Thickness

The ear skin of each mouse was excised on day 49. The ears from different groups were fixed with neutral-buffered formalin (10%), dehydrated in a sucrose solution (30%), and embedded in the O.C.T. compound. The sections (8 µm) were stained with hematoxylin and eosin (H&E). For H&E staining, the sections were washed under running tap water for 5 min and counterstained with a hematoxylin solution for 90 s. After washing under running tap water, the sections were stained with eosin solution for 10 s, rinsed, dehydrated, and coverslipped with Permount mounting medium. Ear thickness was determined using H&E-stained photomicrographs. The average value was calculated by measuring five measurements per mouse. ImageJ software (version 1.54g, National Institutes of Health, Bethesda, MD, USA) was used to assess skin thickness by measuring the distance.

4.7. Quantitative Real-Time PCR

The expression levels of inflammatory markers in the ears of the mice were measured using qRT-PCR. Total RNA in the lymph nodes and skin tissue was isolated using Trizol® (Invitrogen, Waltham, MA, USA). The RNA was reverse transcribed into cDNA using MMLV reverse transcriptase (Promega, Madison, WI, USA). Thunderbird Next SYBR qPCR Mix was used for qRT-PCR on a CFX Connect Real-Time System (Bio-Rad, Hercules, CA, USA). The PCR program consisted of a cycle at 95 °C for 4 min, 40 cycles at 95 °C for 30 s, 57 °C for 30 s, and the last at 95 °C for 30 s. With the help of CFX Maestro Software version 2.3 (Bio-Rad Laboratories, Hercules, CA, USA), the obtained data were analyzed using the $2^{-\Delta\Delta Ct}$ method. The primer sequences are listed in Table 1. The results were normalized to GAPDH gene expression levels [35].

Table 1. Quantitative real-time PCR primers.

Mouse Primers		Sequence
Il-6	forward	5′-TTC TTG GGA CTG ATG CTG GT-3′
	reverse	5′-CTG TGA AGT CTC CTC TCC GG-3′
Il-8	forward	5′-AAC TCC TTG GTG ATG CTG GT-3′
	reverse	5′-CCA GGT TCA GCA GGT AGA CA-3′
Il-12A	forward	5′-GAA GCT CTG CAT CCT GCT TC-3′
	reverse	5′-CAG ATA GCC CAT CAC CCT GT-3′
IFN-γ	forward	5′-CAC GGC ACA GTC ATT GAA AG-3′
	reverse	5′-GTC ACC ATC CTT TTG CCA GT-3′
Il-4	forward	5′-TCT CGA ATG TAC CAG GAG CC-3′
	reverse	5′-CCT TCT CCT GTG ACC TCG TT-3′
Il-13	forward	5′-GCA GCA TGG TAT GGA GTG TG-3′
	reverse	5′-AGG CCA TGC AAT ATC CTC TG-3′
Il-22	forward	5′-GTC AAC CGC ACC TTT ATG CT-3′
	reverse	5′-GTT GAG CAC CTG CTT CAT CA-3′
Il-17A	forward	5′-TCC AGC AAG AGA TCC TGG TC-3′
	reverse	5′-AGC ATC TTC TCG ACC CTG AA-3′
Gapdh	forward	5′-AGA ACA TCA TCC CTG CAT CC-3′
	reverse	5′-CAC ATT GGG GGT AGG AAC AC-3′

4.8. Statistics

We expressed the results as the mean ± standard error of the mean (SEM) of five determinations for each animal group. The statistical significance of differences was determined by analysis of variance (ANOVA) and Tukey's multiple comparison test. GraphPad Prism software version 5 (GraphPad Software, Inc., La Jolla, CA, USA) was used for data analysis. Statistical significance was accepted for $p < 0.05$: * $p < 0.05$, ** $p < 0.01$, and *** $p < 0.001$, vs. the vehicle-treated control group; # $p < 0.05$, ## $p < 0.01$, and ### $p < 0.001$ vs. the DNCB-treated group.

Author Contributions: Conceptualization, S.-Y.H. and D.-S.I.; methodology, S.-Y.H.; formal analysis and data curation, S.-Y.H.; writing—original draft preparation, D.-S.I.; writing—review and editing, D.-S.I.; supervision, D.-S.I. All authors have read and agreed to the published version of the manuscript.

Funding: This research was supported by the Basic Science Research Program of the Korean National Research Foundation, funded by the Korean Ministry of Science, ICT, and Future Planning (NRF-2023R1A2C2002380).

Institutional Review Board Statement: This study was conducted according to the guidelines of the Declaration of Helsinki and approved by the Kyung Hee University Institutional Animal Care Committee with respect to the ethics of the procedures and animal care (KHSASP-23-12).

Informed Consent Statement: Not applicable for not involving humans.

Data Availability Statement: Data are contained within the article.

Conflicts of Interest: The authors declare no conflicts of interest.

References

1. Labib, A.; Yosipovitch, G. An evaluation of abrocitinib for moderate-to-severe atopic dermatitis. *Expert Rev. Clin. Immunol.* **2022**, *18*, 1107–1118. [CrossRef] [PubMed]
2. Sroka-Tomaszewska, J.; Trzeciak, M. Molecular Mechanisms of Atopic Dermatitis Pathogenesis. *Int. J. Mol. Sci.* **2021**, *22*, 4130. [CrossRef] [PubMed]
3. Yang, G.; Seok, J.K.; Kang, H.C.; Cho, Y.Y.; Lee, H.S.; Lee, J.Y. Skin Barrier Abnormalities and Immune Dysfunction in Atopic Dermatitis. *Int. J. Mol. Sci.* **2020**, *21*, 2867. [CrossRef] [PubMed]
4. Silverberg, J.I.; Barbarot, S.; Gadkari, A.; Simpson, E.L.; Weidinger, S.; Mina-Osorio, P.; Rossi, A.B.; Brignoli, L.; Saba, G.; Guillemin, I. Atopic dermatitis in the pediatric population: A cross-sectional, international epidemiologic study. *Ann. Allergy Asthma Immunol.* **2021**, *126*, 417–428.e412. [CrossRef] [PubMed]

5. Wollenberg, A.; Barbarot, S.; Bieber, T.; Christen-Zaech, S.; Deleuran, M.; Fink-Wagner, A.; Gieler, U.; Girolomoni, G.; Lau, S.; Muraro, A.; et al. Consensus-based European guidelines for treatment of atopic eczema (atopic dermatitis) in adults and children: Part I. *J. Eur. Acad. Dermatol. Venereol.* **2018**, *32*, 657–682. [CrossRef] [PubMed]
6. Paller, A.S.; Simpson, E.L.; Siegfried, E.C.; Cork, M.J.; Wollenberg, A.; Arkwright, P.D.; Soong, W.; Gonzalez, M.E.; Schneider, L.C.; Sidbury, R. Dupilumab in children aged 6 months to younger than 6 years with uncontrolled atopic dermatitis: A randomised, double-blind, placebo-controlled, phase 3 trial. *Lancet* **2022**, *400*, 908–919. [CrossRef] [PubMed]
7. Lee, J.H.; Son, S.W.; Cho, S.H. A Comprehensive Review of the Treatment of Atopic Eczema. *Allergy Asthma Immunol. Res.* **2016**, *8*, 181–190. [CrossRef] [PubMed]
8. Graff, P. Potential Drivers of the Atopic March-Unraveling the Skin-Lung Crosstalk. Master's Thesis, Free University of Berlin, Berlin, Germany, 2022. [CrossRef]
9. Criado, P.R.; Miot, H.A.; Ianhez, M. Eosinophilia and elevated IgE serum levels: A red flag: When your diagnosis is not a common atopic eczema or common allergy. *Inflamm. Res.* **2023**, *72*, 541–551. [CrossRef]
10. Kim, J.Y.; Jeong, M.S.; Park, M.K.; Lee, M.K.; Seo, S.J. Time-dependent progression from the acute to chronic phases in atopic dermatitis induced by epicutaneous allergen stimulation in NC/Nga mice. *Exp. Dermatol.* **2014**, *23*, 53–57. [CrossRef] [PubMed]
11. Koga, C.; Kabashima, K.; Shiraishi, N.; Kobayashi, M.; Tokura, Y. Possible pathogenic role of Th17 cells for atopic dermatitis. *J. Investig. Dermatol.* **2008**, *128*, 2625–2630. [CrossRef]
12. Muraro, A.; Lemanske, R.F., Jr.; Hellings, P.W.; Akdis, C.A.; Bieber, T.; Casale, T.B.; Jutel, M.; Ong, P.Y.; Poulsen, L.K.; Schmid-Grendelmeier, P.; et al. Precision medicine in patients with allergic diseases: Airway diseases and atopic dermatitis-PRACTALL document of the European Academy of Allergy and Clinical Immunology and the American Academy of Allergy, Asthma & Immunology. *J. Allergy Clin. Immunol.* **2016**, *137*, 1347–1358. [CrossRef]
13. Ashcroft, D.M.; Dimmock, P.; Garside, R.; Stein, K.; Williams, H.C. Efficacy and tolerability of topical pimecrolimus and tacrolimus in the treatment of atopic dermatitis: Meta-analysis of randomised controlled trials. *BMJ* **2005**, *330*, 516. [CrossRef] [PubMed]
14. Brown, J.M.; Wilson, T.M.; Metcalfe, D.D. The mast cell and allergic diseases: Role in pathogenesis and implications for therapy. *Clin. Exp. Allergy* **2008**, *38*, 4–18. [CrossRef] [PubMed]
15. Oray, M.; Abu Samra, K.; Ebrahimiadib, N.; Meese, H.; Foster, C.S. Long-term side effects of glucocorticoids. *Expert Opin. Drug Saf.* **2016**, *15*, 457–465. [CrossRef] [PubMed]
16. Wang, Z.; Wang, Z.Z.; Geliebter, J.; Tiwari, R.; Li, X.M. Traditional Chinese medicine for food allergy and eczema. *Ann. Allergy Asthma Immunol.* **2021**, *126*, 639–654. [CrossRef] [PubMed]
17. Xiao, S.-J.; Xu, X.-K.; Chen, W.; Xin, J.-Y.; Yuan, W.-L.; Zu, X.-P.; Shen, Y.-H. Traditional Chinese medicine Euodiae Fructus: Botany, traditional use, phytochemistry, pharmacology, toxicity and quality control. *Nat. Prod. Bioprospect.* **2023**, *13*, 6. [CrossRef] [PubMed]
18. Shin, Y.-W.; Bae, E.-A.; Cai, X.F.; Lee, J.J.; Kim, D.-H. In vitro and in vivo antiallergic effect of the fructus of Evodia rutaecarpa and its constituents. *Biol. Pharm. Bull.* **2007**, *30*, 197–199. [CrossRef] [PubMed]
19. Choi, Y.H.; Shin, E.M.; Kim, Y.S.; Cai, X.F.; Lee, J.J.; Kim, H.P. Anti-inflammatory principles from the fruits of Evodia rutaecarpa and their cellular action mechanisms. *Arch. Pharm. Res.* **2006**, *29*, 293–297. [CrossRef] [PubMed]
20. Chiou, W.F.; Ko, H.C.; Wei, B.L. Evodia rutaecarpa and Three Major Alkaloids Abrogate Influenza A Virus (H1N1)-Induced Chemokines Production and Cell Migration. *Evid. Based Complement. Altern. Med.* **2011**, *2011*, 750513. [CrossRef] [PubMed]
21. Liu, Y.N.; Pan, S.L.; Liao, C.H.; Huang, D.Y.; Guh, J.H.; Peng, C.Y.; Chang, Y.L.; Teng, C.M. Evodiamine represses hypoxia-induced inflammatory proteins expression and hypoxia-inducible factor 1alpha accumulation in RAW264.7. *Shock* **2009**, *32*, 263–269. [CrossRef] [PubMed]
22. Rang, W.Q.; Du, Y.H.; Hu, C.P.; Ye, F.; Tan, G.S.; Deng, H.W.; Li, Y.J. Protective effects of calcitonin gene-related peptide-mediated evodiamine on guinea-pig cardiac anaphylaxis. *Naunyn Schmiedebergs Arch. Pharmacol.* **2003**, *367*, 306–311. [CrossRef] [PubMed]
23. Pearce, L.V.; Petukhov, P.A.; Szabo, T.; Kedei, N.; Bizik, F.; Kozikowski, A.P.; Blumberg, P.M. Evodiamine functions as an agonist for the vanilloid receptor TRPV1. *Org. Biomol. Chem.* **2004**, *2*, 2281–2286. [CrossRef] [PubMed]
24. Wang, Q.; Cui, Y.; Wu, X.; Wang, J. Evodiamine protects against airway remodelling and inflammation in asthmatic rats by modulating the HMGB1/NF-κB/TLR-4 signalling pathway. *Pharm. Biol.* **2021**, *59*, 192–199. [CrossRef] [PubMed]
25. Riedl, R.; Kühn, A.; Rietz, D.; Hebecker, B.; Glowalla, K.G.; Peltner, L.K.; Jordan, P.M.; Werz, O.; Lorkowski, S.; Wiegand, C.; et al. Establishment and Characterization of Mild Atopic Dermatitis in the DNCB-Induced Mouse Model. *Int. J. Mol. Sci.* **2023**, *24*, 2325. [CrossRef] [PubMed]
26. Lee, J.E.; Choi, Y.W.; Im, D.S. Inhibitory effect of α-cubebenoate on atopic dermatitis-like symptoms by regulating Th2/Th1/Th17 balance in vivo. *J. Ethnopharmacol.* **2022**, *291*, 115162. [CrossRef] [PubMed]
27. Kong, L.; Liu, J.; Wang, J.; Luo, Q.; Zhang, H.; Liu, B.; Xu, F.; Pang, Q.; Liu, Y.; Dong, J. Icariin inhibits TNF-α/IFN-γ induced inflammatory response via inhibition of the substance P and p38-MAPK signaling pathway in human keratinocytes. *Int. Immunopharmacol.* **2015**, *29*, 401–407. [CrossRef]
28. Brunner, P.M.; Pavel, A.B.; Khattri, S.; Leonard, A.; Malik, K.; Rose, S.; Jim On, S.; Vekaria, A.S.; Traidl-Hoffmann, C.; Singer, G.K.; et al. Baseline IL-22 expression in patients with atopic dermatitis stratifies tissue responses to fezakinumab. *J. Allergy Clin. Immunol.* **2019**, *143*, 142–154. [CrossRef]
29. Ovary, Z. Passive cutaneous anaphylaxis in the mouse. *J. Immunol.* **1958**, *81*, 355–357. [CrossRef] [PubMed]

30. Tong, M.; Guo, Y.; Zhang, G. Effect and mechanisms of rutaecarpine on treating atopic dermatitis in mice. *Sichuan Da Xue Xue Bao Yi Xue Ban = J. Sichuan Univ. Med. Sci. Ed.* **2011**, *42*, 234–236, 263.
31. Zhang, W.D.; Chen, X.Y.; Wu, C.; Lian, Y.N.; Wang, Y.J.; Wang, J.H.; Yang, F.; Liu, C.H.; Li, X.Y. Evodiamine reduced peripheral hypersensitivity on the mouse with nerve injury or inflammation. *Mol. Pain* **2020**, *16*, 1744806920902563. [CrossRef] [PubMed]
32. Moon, T.C.; Murakami, M.; Kudo, I.; Son, K.H.; Kim, H.P.; Kang, S.S.; Chang, H.W. A new class of COX-2 inhibitor, rutaecarpine from Evodia rutaecarpa. *Inflamm. Res.* **1999**, *48*, 621–625. [CrossRef] [PubMed]
33. John, S.P.; Singh, A.; Sun, J.; Pierre, M.J.; Alsalih, L.; Lipsey, C.; Traore, Z.; Balcom-Luker, S.; Bradfield, C.J.; Song, J.; et al. Small-molecule screening identifies Syk kinase inhibition and rutaecarpine as modulators of macrophage training and SARS-CoV-2 infection. *Cell Rep.* **2022**, *41*, 111441. [CrossRef] [PubMed]
34. Kang, J.; Lee, J.H.; Im, D.S. Topical Application of S1P(2) Antagonist JTE-013 Attenuates 2,4-Dinitrochlorobenzene-Induced Atopic Dermatitis in Mice. *Biomol. Ther.* **2020**, *28*, 537–541. [CrossRef] [PubMed]
35. Lee, J.E.; Im, D.S. Suppressive Effect of Carnosol on Ovalbumin-Induced Allergic Asthma. *Biomol. Ther.* **2021**, *29*, 58–63. [CrossRef] [PubMed]

Disclaimer/Publisher's Note: The statements, opinions and data contained in all publications are solely those of the individual author(s) and contributor(s) and not of MDPI and/or the editor(s). MDPI and/or the editor(s) disclaim responsibility for any injury to people or property resulting from any ideas, methods, instructions or products referred to in the content.

Article

Magnolol Reduces Atopic Dermatitis-like Symptoms in BALB/c Mice

Ju-Hyun Lee [1] and Dong-Soon Im [1,2,*]

[1] Department of Biomedical and Pharmaceutical Sciences, Graduate School, Kyung Hee University, Seoul 02447, Republic of Korea; ljh0620@khu.ac.kr
[2] Department of Fundamental Pharmaceutical Sciences, Graduate School, Kyung Hee University, Seoul 02447, Republic of Korea
* Correspondence: imds@khu.ac.kr; Tel.: +82-2-961-9377

Abstract: In traditional Korean medicines, *Magnolia officinalis* is commonly included for the remedy of atopic dermatitis, and magnolol is a major constituent of *Magnolia officinalis*. Its pharmacological effects include anti-inflammatory, hepatoprotective, and antioxidant effects. Using BALB/c mice repeatedly exposed to 1-chloro-2,4-dinitrobenzene (DNCB), magnolol was evaluated in atopic dermatitis-like lesions. Administration of magnolol (10 mg/kg, intraperitoneal injection) markedly relieved the skin lesion severity including cracking, edema, erythema, and excoriation, and significantly inhibited the increase in IgE levels in the peripheral blood. A DNCB-induced increase in mast cell accumulation in atopic dermatitis skin lesions was reversed by magnolol administration, as well as a rise in expression levels of pro-inflammatory Th2/Th17/Th1 cytokines' (IL-4, IL-13, IL-17A, IFN-γ, IL-12A, TARC, IL-8, and IL-6) mRNAs in the lymph nodes and skin (n = 5 per group). In lymph nodes, magnolol reversed DNCB's increase in $CD4^+ROR\gamma t^+$ Th17 cell fraction and decrease in $CD4^+FoxP3^+$ regulatory T cell fraction. The results also showed that magnolol suppressed T cell differentiation into Th17 and Th2 cells, but not Th1 cells. Magnolol suppresses atopic dermatitis-like responses in the lymph nodes and skin, suggesting that it may be feasible to use it as a treatment for atopic dermatitis through its suppression of Th2/Th17 differentiation.

Keywords: *Magnolia officinalis*; atopy; magnolol; dermatitis; magnolia; anti-atopy; atopic dermatitis

Citation: Lee, J.-H.; Im, D.-S. Magnolol Reduces Atopic Dermatitis-like Symptoms in BALB/c Mice. *Life* **2024**, *14*, 339. https://doi.org/10.3390/life14030339

Academic Editor: Stanislav Miertus

Received: 5 February 2024
Revised: 2 March 2024
Accepted: 2 March 2024
Published: 5 March 2024

Copyright: © 2024 by the authors. Licensee MDPI, Basel, Switzerland. This article is an open access article distributed under the terms and conditions of the Creative Commons Attribution (CC BY) license (https://creativecommons.org/licenses/by/4.0/).

1. Introduction

Chronic inflammatory immune responses evoke atopic dermatitis, a skin disease [1]. Dysregulated immune responses are the leading cause, resulting from barrier defects, environmental factors, and genetic defects [1]. Atopic dermatitis reportedly affects 13.5~41.9% of the pediatric population, as per a global survey in 18 countries [2]. Widespread eczematous lesions are a cardinal characteristic of atopic dermatitis [3]. As in other atopic diseases, like allergic asthma, allergic rhinitis, and rhinoconjunctivitis, elevated IgE levels and type 2 immune responses are immunologic features of atopic dermatitis [4]. The Th2 response is not the only factor involved in the pathogenesis of atopic dermatitis, as the Th1 and Th17 responses are also deeply involved in the chronic phase of atopic dermatitis [5–7].

Administration of Gwakhyangjeonggi-san, a traditional Korean medicine, suppresses 1-chloro-2,4-dinitrobenzene (DNCB)-induced atopic dermatitis-like symptoms in mice [8]. Another Korean medicine, Pyeongwee-San (KMP6), also has a suppressive effect against atopic dermatitis in mice [9,10]. Gwakhyangjeonggi-san and KMP6 contain the cortex of *Magnolia officinalis*, in percentages of 8.7% and 21.5%, respectively. Magnolol (5, 5'-diallyl-2, 2'-dihydroxybiphenyl) is a main constituent in *Magnolia obovata* and *M. officinalis* [11]. The percentages of magnolol in *M. obovate* and *M. officinalis* were found to be 4.4% and 2.6%, respectively [12]. The magnolol dosage was estimated to be 1.4 mg/kg since Gwakhyangjeonggi-san contained 8.7% of *M. officialis*, and 0.6 g/kg of it was administered in the atopic dermatitis study [8]. Similarly, because KMP-6 contains 21.5% of *M. officialis*,

1 g/kg was orally treated in the atopic dermatitis study [9], the dose of magnolol was estimated as 5.6 mg/kg. Therefore, three scientific references provided evidence of traditional use of two Korean medicines, Gwakhyangjeonggi-san and KMP-6, for atopic dermatitis therapy. Among the used plants, the bark of *Magnolia officinalis* is commonly included, and magnolol is the active constituent of *Magnolia officinalis*.

Previously, we found that honokiol (3,5′-diallyl-4,2′-dihydroxybiphenyl), a positional isomer of magnolol found in the bark of *Magnolia officinalis*, showed inhibitory activity on atopic dermatitis [10]. By administering honokiol, DNCB-induced mast cell accumulation and inflammation in skin tissues were significantly reduced [10]. Also, a significant reduction in inflammation-induced cytokines in skin and lymph nodes was observed after honokiol administration in combination with DNCB treatment [10]. In the present study, we investigated magnolol, a bioactive lignan found in the bark stripped from the roots, stems, and branches of the Houpu magnolia [13], a traditional Chinese medicine [14]. Magnolol has a variety of pharmacological effects, such as anti-anxiety, anti-cancer, anti-depressant, anti-inflammatory, antioxidant, and hepatoprotective properties [11,13,15–20]. Magnolol showed inhibitory effects on passive cutaneous anaphylactic reactions in mice [21] as well as on ovalbumin-induced allergic asthma by regulating T cell cytokines [22,23]. *Magnolia* has been commonly included in two traditional Korean herbal medicines (Gwakhyangjeonggi-san and KMP6) for atopic dermatitis treatment [9,10,21], and magnolol is the major constituent of *Magnolia*. However, it has not yet been explored in vivo whether magnolol is effective against atopic dermatitis. In the present study, thus, we evaluated whether magnolol could reduce atopic dermatitis symptoms using a murine DNCB-induced atopic dermatitis model and investigated its mechanism of action.

2. Results

2.1. Magnolol Suppresses Atopic Dermatitis-like Symptoms in the Ears of Mice

Magnolia officinalis has been included in two Korean herbal medicine concoctions for atopic dermatitis therapy [9,10,21], and magnolol is a major lignan found in the bark of the Houpu magnolia, a traditional Chinese medicine. Here, we evaluated the efficacy of magnolol against overall atopic dermatitis responses in a murine in vivo atopic dermatitis model caused by DNCB. The sensitization and challenges were made with DNCB, and magnolol or dexamethasone (DEX) was delivered 30 min before the DNCB treatment via intraperitoneal injection. We sacrificed the mice on day 49.

As a result of DNCB, atopic dermatitis symptoms such as scaling, erythema, and erosion in the ears were experienced (Figure 1A). As a result of magnolol treatment, atopic dermatitis symptoms were ameliorated in the ears of DNCB-induced mice, as well as cutaneous hyperplasia, which is a key feature of atopic dermatitis-like lesions. For each treatment group, we measured the average thickness of the skin. The ears of magnolol-treated atopic dermatitis mice showed less swelling than their control counterparts (Figure 1A). A high level of epidermal hyperplasia is observed in atopic dermatitis skin lesions, along with an accumulation of immune cells. Hematoxylin and eosin (H&E) staining showed the anti-atopic dermatitis effects of magnolol (Figure 1B). Histologically, the infiltration of inflammatory cells into the dermis (exocytosis) and thickening of the epidermis (acanthosis) were reproduced in atopic dermatitis mice. In magnolol-treated atopic dermatitis mice, acanthosis and exocytosis were significantly suppressed (Figure 1B). Figure 1B shows a significantly lower ear thickness in magnolol-treated atopic dermatitis mice than in control atopic dermatitis mice. The measurement of ear thickness showed marked effects of magnolol, which were similar to the effects of DEX-treated mice (positive control) (Figure 1C). Figure 1D shows the chemical structure of magnolol.

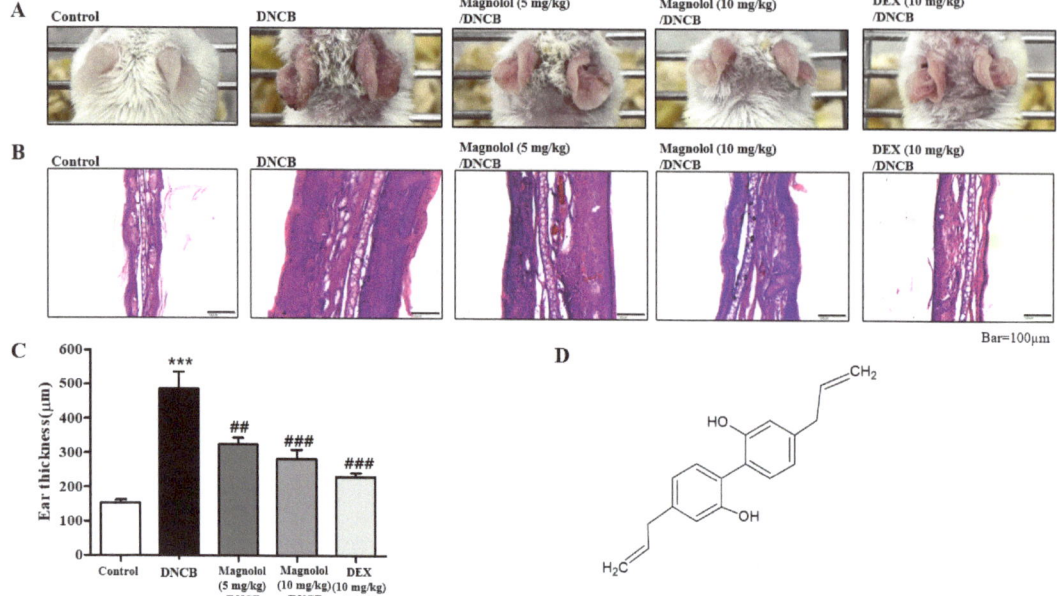

Figure 1. Effect of magnolol on 1-chloro-2, 4-dinitrobenzene [DNCB]-induced atopic dermatitis in ears of mice. (A) Photographs of the ears. (B) H&E staining. (C) Histogram of ear thickness. (D) Chemical structure of magnolol. *** $p < 0.001$ vs. the vehicle-treated group, ### $p < 0.001$, ## $p < 0.01$ vs. the DNCB-treated group. A magnification of ×200 was used.

The Th2 cytokines and IgE have a fundamental role in the pathogenesis of atopic dermatitis regardless of its high heterogeneity. A significant elevation in IgE levels is a cardinal character of atopic dermatitis and is linked with atopic dermatitis severity. Therefore, IgE levels in the serum obtained on day 49 were assessed by ELISA. Figure 2 shows that the basal level of serum IgE in the PBS group was about 40 ng/mL. A DNCB challenge, however, readily led to an increase in IgE to approximately 140 ng/mL. DNCB plus 10 mg/kg magnolol treatment, however, resulted in significant reductions in IgE levels compared to DNCB alone. It should be noted, however, that magnolol was less effective than DEX (Figure 2).

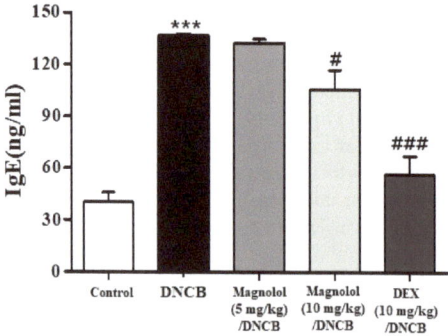

Figure 2. Effect of magnolol on the levels of serum immunoglobulin E. An ELISA was used to measure serum IgE levels. *** $p < 0.001$ vs. the vehicle-treated group, ### $p < 0.001$, # $p < 0.05$ vs. the DNCB-treated group.

Mast cells are primarily involved in detrimental allergic disorders. Mast cell infiltration into the dermis was determined in toluidine blue O staining (Figure 3A). A significant increase in mast cell counts was observed following DNCB administration (Figure 3A). However, a dose-dependent decrease in the numbers was shown in atopic dermatitis mice treated with magnolol (Figure 3B). A similar decrease in mast cell numbers was shown in atopic dermatitis mice treated with DEX (Figure 3B).

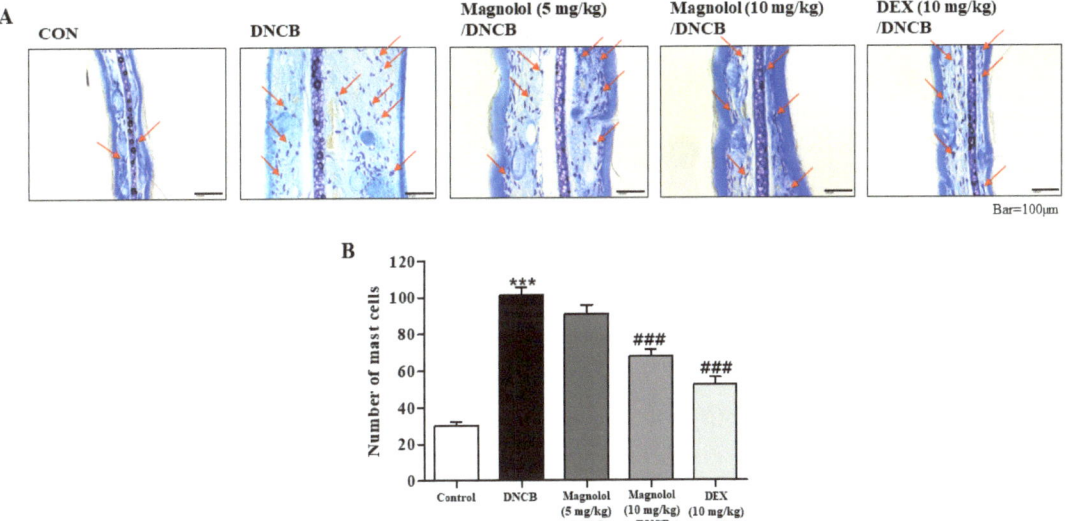

Figure 3. Magnolol reduces mast cell count in the ears. (**A**) Toluidine blue O staining of ear sections. Red arrows indicate mast cells. (**B**) Histogram of counts of mast cells (n = 5). *** $p < 0.001$ vs. the vehicle-treated group, ### $p < 0.001$ vs. the DNCB-treated group. ×200 magnification.

Atopic dermatitis is defined as an immune disorder driven by Th2 cells and characterized by elevated levels of interleukin (IL)-4, IL-13, and IL-5. In addition, the inflammatory pathogenesis of atopic dermatitis is influenced by Th1 and Th17 cytokines, interferons (IFN)-γ, IL-12A, and IL-17A. Thus, we measured pro-inflammatory cytokines from three different types of cells, Th2 (IL-4 and IL-13), Th1 (IFN-γ and IL-12A), and Th17 (IL-17A) [24]. An increase in mRNA expression levels of five cytokines was observed in the ear sections of atopic dermatitis-induced mice (Figure 4A–E). Pro-inflammatory cytokine gene expression was suppressed by magnolol or DEX (Figure 4A–E). As a chemokine, thymus and activation-regulated chemokine (TARC) plays an imperative role in the homing of skin-specific T cells [25], whereas IL-8 and IL-6 are involved in the recruitment of various cells into atopic dermatitis skin [26]. Therefore, the levels of TARC, IL-8, and IL-6 mRNA expression were measured. Each was significantly elevated in the ears of atopic dermatitis mice (Figure 4F–H). It was noted that both magnolol and DEX demonstrated a significant suppression of TARC, IL-8, and IL-6 mRNA expression levels (Figure 4F–H). Based on these results, magnolol appears to reduce the local inflammatory response caused by DNCB. All of these findings suggest that there is a significant attenuation of atopic dermatitis responses, hypertrophy, mast cell infiltration, and pro-inflammatory cytokine levels following magnolol administration in response to DNCB.

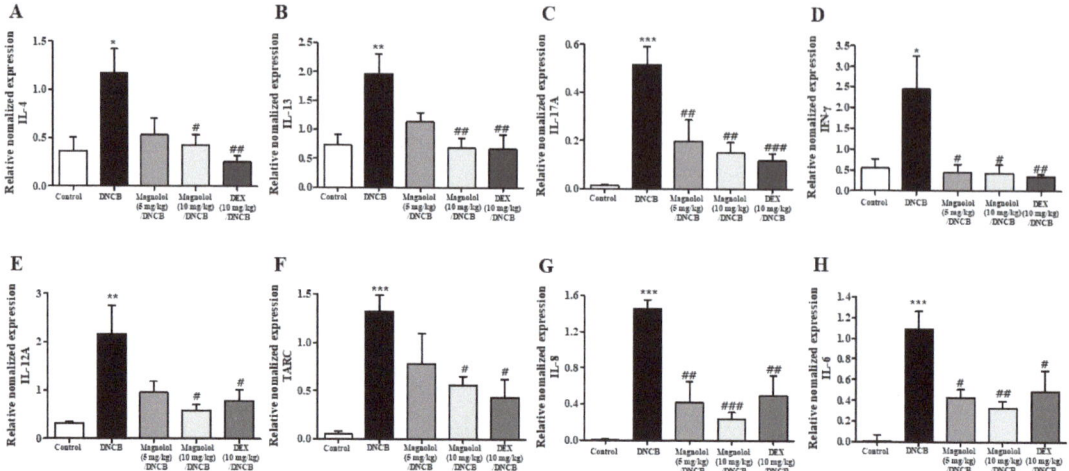

Figure 4. Effect of magnolol on the expression levels of pro-inflammatory cytokines mRNAs in the ears. (**A**) IL-4, (**B**) IL-13, (**C**) IL-17A, (**D**) IFN-γ, (**E**) IL-12A, (**F**) TARC, (**G**) IL-8, and (**H**) IL-6 mRNA expression levels in skin tissues (n = 5). *** $p < 0.001$, ** $p < 0.01$, * $p < 0.05$ vs. the vehicle-treated group, ### $p < 0.001$, ## $p < 0.01$, # $p < 0.05$ vs. the DNCB-treated group.

2.2. Magnolol Suppresses DNCB-Induced Enlargement of Lymph Node and Increases in mRNA Expression of Pro-Inflammatory Cytokines

Next, immune responses were investigated in the cervical lymph nodes. DNCB treatment enlarged the size of cervical lymph nodes (Figure 5), which was significantly suppressed dose-dependently following magnolol administration (Figure 5). A stronger suppression was observed in DEX-treated lymph nodes than in magnolol (Figure 5).

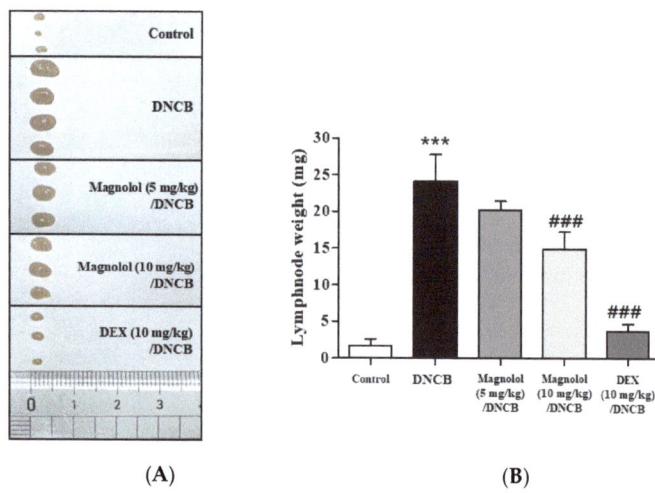

Figure 5. Effect of magnolol on immune responses induced by DNCB in lymph nodes. (**A**) Sizes of cervical lymph nodes. (**B**) Lymph node weight (n = 5). *** $p < 0.001$ vs. the vehicle-treated group, ### $p < 0.001$ vs. the DNCB-treated group.

Flow cytometry determined the proportion of CD4$^+$RORγt$^+$ Th17 cells and CD4$^+$FoxP3$^+$ Treg cells in cervical lymph nodes, as these cells play key roles in the pathogenesis of atopic dermatitis, particularly in the chronic phase. In atopic dermatitis mice, CD4$^+$RORγt$^+$

Th17 cell numbers in lymph nodes increased, which was reversed by magnolol treatments (Figure 6A,C). As contrast, CD4$^+$FoxP3$^+$ Treg cells were decreased in atopic dermatitis mice, and magnolol restoration of this decrease was found (Figure 6B,D). A DNCB-induced increase in CD4$^+$RORγt$^+$ Th17 cell fraction was reversed by DEX administration in the atopic dermatitis model but not a decrease in CD4$^+$FoxP3$^+$ Treg cell fraction (Figure 6C,D).

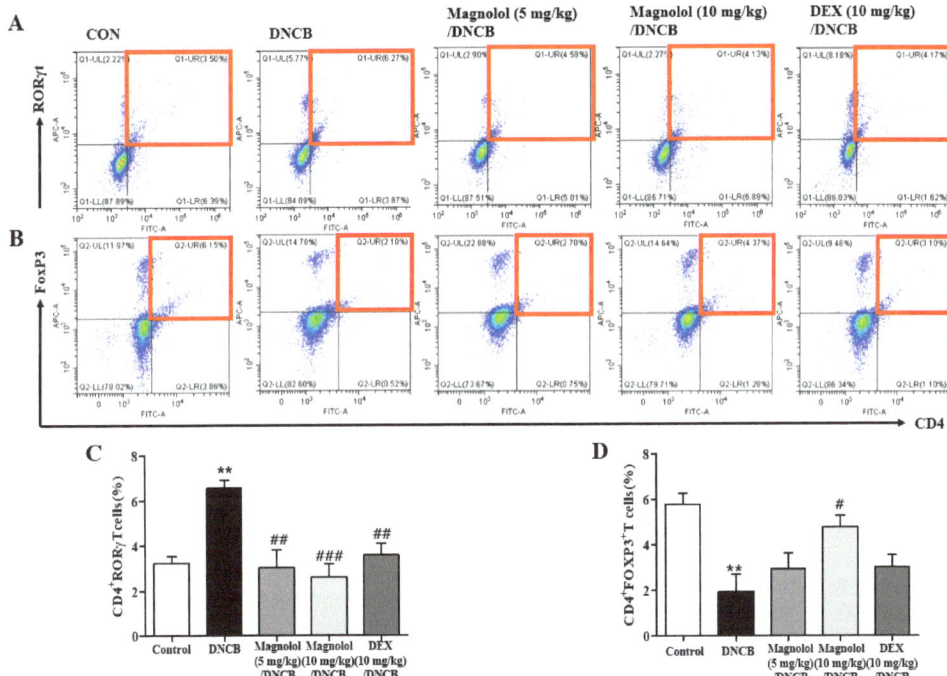

Figure 6. Suppressive effect of magnolol on the proportion of CD4$^+$RORγt$^+$ Th17 cells and CD4$^+$FoxP3$^+$ Treg cells in lymph nodes. (**A**) Dot plots of CD4$^+$RORγt$^+$ Th17 cells. (**B**) Dot plots of CD4$^+$FoxP3$^+$ Treg cells. (**C**) Percentage of CD4$^+$RORγt$^+$ Th17 cells. (**D**) Percentage of CD4$^+$FoxP3$^+$ Treg cells (n = 5). Double-positive cells are shown in the red-lined boxes. ** $p < 0.01$ vs. the vehicle-treated group, ### $p < 0.001$, ## $p < 0.01$, # $p < 0.05$ vs. the DNCB-treated group.

Additionally, we found that the levels of mRNA expression of Th2 (IL-4 and IL-13), Th1 (IFN-γ), and Th17 (IL-17A) cytokines were significantly increased in the cervical lymph nodes of atopic dermatitis mice induced with DNCB, as shown in Figure 7A–D. Figure 7A–D showed that magnolol treatment significantly suppressed the increased mRNA levels of these cytokines, especially IL-4 and IL-17A. Aside from reducing lymph node enlargement and suppressing inflammatory cytokines, magnolol reversed changes in CD4$^+$RORγt$^+$ Th17 and CD4$^+$FoxP3$^+$ Treg cell proportions caused by DNCB.

2.3. Magnolol Suppresses T Cells Differentiation into Th17 and Th2 Cells

A DNCB-induced increase in IL-17A expression was markedly suppressed by magnolol in the lymph nodes and skin, whereas a reduction in CD4$^+$RORγt$^+$ Th17 cell fraction in lymph nodes was reversed by magnolol. Therefore, we investigated whether magnolol could suppress the T cell differentiation into Th17 cells. As shown in Figure 8, magnolol treatment of Th17 differentiation media significantly suppressed the differentiation into IL-17A$^+$ Th17 cells in a concentration-dependent manner in vitro (Figure 8A,B). Furthermore, we assessed the effects of magnolol on the differentiation into IFN-γ$^+$ Th1 or IL-4$^+$ Th2 cells and found that it did not affect IFN-γ$^+$ Th1 cell generation (Figure 8C,D) but inhibited IL-4+ Th2 cell production (Figure 8E,F). In summary, magnolol treatment reduces the differentiation into IL-4+ Th2 cells and IL-17A+ Th17 cells.

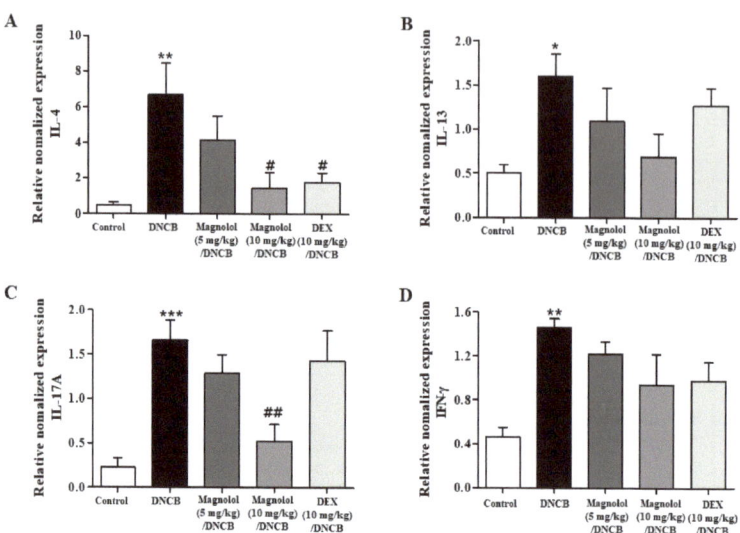

Figure 7. Effect of magnolol on the expression levels of pro-inflammatory cytokines mRNAs in lymph nodes. (A) IL-4, (B) IL-13, (C) IL-17A, and (D) IFN-γ mRNA expression levels in the lymph node tissues (n = 5). *** $p < 0.001$, ** $p < 0.01$, * $p < 0.05$ vs. the vehicle-treated group, ## $p < 0.01$, # $p < 0.05$ vs. the DNCB-treated group.

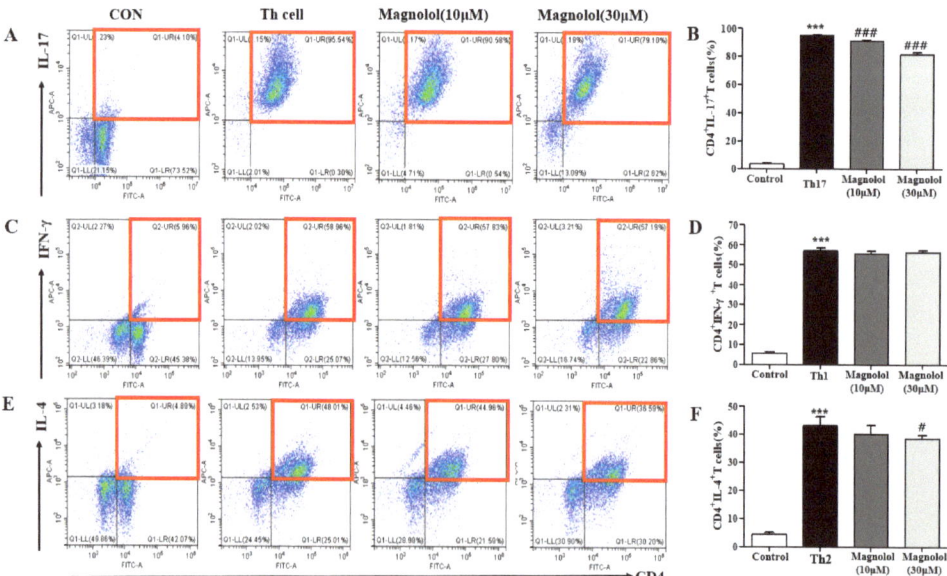

Figure 8. Suppressive effect of magnolol on T cell differentiation into Th17/Th1/Th2 cells. CD4+ T cells isolated from splenocytes were cultured in each differentiation media for Th17, Th1, or Th2 cells for 5 days in plates precoated with antibody to mouse CD3. (A,C,E) Flow cytometry results for (A) Th17 differentiation, (C) Th1 differentiation, and (E) Th2 differentiation. (B,D,F). Histograms of

the cell percentage of (**B**) CD4⁺IL-17A⁺ cell population, (**D**) CD4⁺IFN-γ⁺ cell population, and (**F**) CD4⁺IL-4⁺ cell population (n = 5). Double-positive cells are shown as the red-lined boxes. *** $p < 0.001$ vs. the vehicle-treated group, ### $p < 0.001$, # $p < 0.05$ vs. the differentiation media-treated group.

3. Discussion

A murine atopic dermatitis model was evaluated to determine the effect of magnolol, since *Magnolia officinalis* is included in two traditional Korean herbal formulations for atopic dermatitis therapy (Gwakhyangjeonggi-san and KMP6) [9–11,21]. Previous studies revealed that magnolol contents in *M. obovate* and *M. officialis* were 4.4% and 2.6%, respectively [10]. As *M. officialis* constitutes 8.7% and 21.5% of Gwakhyangjeonggi-san and Pyeongwee-San (KMP-6), respectively, 53.7 mg/kg and 5.6 mg/kg of magnolol were estimated to be used in each atopic dermatitis experiment [9,10,21]. This study administered 5 mg/kg and 10 mg/kg magnolol in response to previous findings that 25–55 μM of magnolol inhibited lymphocytes and RBL-2H3 mast cells in vitro [21]. This study, for the first time, reports the anti-atopic efficacy of magnolol in a murine model. In the ears, mast cell accumulation and ear thickening were suppressed following magnolol treatment. In addition, magnolol treatment suppressed lymph node enlargement in magnolol-treated mice. Aside from suppressing serum IgE, magnolol decreased Th2/Th17/Th1 cytokines and chemokines in the ears and lymph nodes. An increase in CD4⁺FoxP3⁺ Treg cells and a reduction in CD4⁺RORγt⁺ Th17 cells in lymph nodes were observed in vivo following magnolol administration. Magnolol's suppressive effects on T cell differentiation into IL-17A⁺ Th17 cells in vitro may be responsible for the decrease in CD4⁺RORγt⁺ Th17 cells in the lymph nodes in vivo.

Previous studies revealed that the contents of magnolol and honokiol in *M. obovate* were 4.4% and 1.17%, respectively, and in *M. officinalis*, 2.6% and 0.92%, respectively [10]. Considering that the contents of magnolol are higher than those of honokiol, magnolol may contribute mainly to the anti-atopic dermatitis efficacy of magnolia bark [27]. The potencies of magnolol and honokiol are quite similar in multiple parameters, including ear thickness, mast cell accumulation in skin tissues, serum IgE levels, size of lymph nodes, and inflammation-induced cytokines in the skin and lymph nodes [12]. The efficacy of magnolol on IL-17A expression levels is higher than that of honokiol with a statistically significant suppression by magnolol but not significant by honokiol [12]. Magnolol is an antioxidative molecule [28] and possesses antibacterial and anti-inflammatory activities [9,15–20], which could contribute to its anti-atopic dermatitis effects. Magnolol treatment suppressed significantly a lipopolysaccharide-induced increase in pro-inflammatory mRNA expression levels of TNF-α, MCP-1, NF-κB, and iNOS in the small intestine of rats [16]. Also, magnolol ameliorated an injury of mastitis tissues and reduced lipopolysaccharide-induced phosphorylation of p38, ERK, JNK, IκBα, and p65 in mouse mammary epithelial cells [17]. Magnolol significantly attenuated a lipopolysaccharide-induced lung injury via reducing immune cells in the bronchoalveolar lavage fluid and NF-κB signaling pathways in mice [18]. In dextran sulphate sodium-induced colitis mice, magnolol ameliorated disease activities index and suppressed expression levels of IL-1β, IL-12, and TNF-α via the regulation of NF-κB and peroxisome proliferator-activated receptor-γ pathways [19]. In imiquimod-induced psoriasis-like animal models, magnolol significantly inhibited pro-inflammatory cytokine expression [29], and it activated peroxisome proliferator-activated receptor-γ [30,31]. The anti-inflammatory properties of magnolol are attributed to several mechanisms, including redox-sensitive transcription factor NF-κB inhibition and PI3K/Akt signaling inhibition [29,32]. A reduction in eicosanoid mediator levels, including leukotrienes, as well as nitric oxide, TNF-α, and IL-4 appears to be required for magnolol to exert its effect [20,28,33]. Therefore, magnolol exerts its anti-inflammatory effects in multiple animal models, including acute lung injury, septic injury in ileum and mastitis, colitis, and psoriatic dermatitis, through suppressing NF-κB, peroxisome proliferator-activated receptor-γ pathways, and the suppression of pro-inflammatory mediators including leukotrienes, TNF-α, IL-4, IL-1β, IL-12, MCP-1, and iNOS.

Magnolol and magnolia bark extract inhibited the release of histamine from mast cells as well [27]. Mechanistically, the suppressive effects of magnolol on mast cells could be regarded as a possible mode of action underlying its anti-atopic dermatitis effects in vivo [27]. The anti-inflammatory actions of magnolol in murine models of asthma and allergic rhinitis have been demonstrated [22,23]. Furthermore, the suppressive effects of magnolol against IgE-mediated passive skin anaphylaxis [21] firmly support the findings of our study. As a mechanism of action of magnolol, we propose that magnolol possibly exerts its anti-inflammatory effects by suppressing Th17/Th2 differentiation and reversing changes in $CD4^+RORγt^+$ Th17 cell and $CD4^+FoxP3^+$ Treg cell populations in the lymph nodes. In the present study, we did not check whether magnolol affected T cell differentiation into $FoxP3^+$ Treg cells. Given that magnolol reversed the DNCB-induced decrease in $CD4^+FoxP3^+$ Treg cell populations in the lymph nodes, magnolol may inhibit the suppressive effect of DNCB or enhance the naïve T cell differentiation into $FoxP3^+$ Treg cells. Therefore, it is necessary to investigate whether magnolol affects T cell differentiation into $FoxP3^+$ Treg cells. In addition, it would be worth investigating how magnolol differentially regulate T cell differentiation into each cell type. Considering the structural similarity between magnolol and honokiol, a comparative study on the effects of honokiol on T cell differentiation could help to elucidate the mode of action of magnolol.

Considering that magnolol was administered via intraperitoneal injection, it is worth investigating if the topical application of magnolol is as effective as intraperitoneal administration. A previous study on the psoriasis model already applied magnolol topically, and positive effects were observed [29], encouraging the topical application of magnolol. Furthermore, 2-O-acetyl-2′-O-methylmagnolol was found to be a potent and safe candidate for the treatment of cutaneous inflammation [34]. It was concluded that magnolol inhibited mast cell accumulation in a murine atopic dermatitis model and that the efficacy was likely due to the inhibition of pro-inflammatory cytokines in the skin (IL-17A, IL-4, IL-13, IFN-γ, and IL-12A) as well as chemokines (TARC, IL-8, and IL-6) in lymph nodes and epidermis. The anti-inflammatory properties of magnolol in the lymph nodes and epidermis may result from the inhibition of differentiation into Th17/Th2 cells. As a conclusion, this study suggests that magnolol is an active phytochemical with efficacy against atopic dermatitis and is a potential therapeutic constituent of *Magnolia officinalis* for atopic dermatitis therapy.

4. Materials and Methods

4.1. Chemicals

1-Chloro-2, 4-dinitrobenzene (DNCB), DEX (cat no. D1756, purity: ≥95% in HPLC), and magnolol (cat no. M3445, purity: ≥95% in HPLC) were purchased from Sigma-Aldrich (St. Louis, MO, USA).

4.2. BALB/c Mice

All experimental protocols for animal care and standard guidelines were performed in accordance with the rules and regulations of the Animal Ethics Committee of Kyung Hee University (KHSASP-22-407). In this study, Daehan Biolink provided 7-week-old male BALB/c mice housed in a room maintained at 22–24 °C and 60 + 5% relative humidity. The laboratory provided ad libitum chow and water.

4.3. Chemical-Induced Atopic Dermatitis Model and Magnolol Treatment

According to previous descriptions, DNCB causes atopic dermatitis-like symptoms [35]. The sensitization step was performed by spreading 1% DNCB on the dorsal skin on day 0, and the challenge step was performed by applying 0.3% DNCB to the ears every other day from day 7 through day 42. We randomly grouped mice into five groups (n = 5 per group): (1) PBS-treated controls, (2) DNCB-treated mice, (3) DNCB + magnolol (5 mg/kg)-treated mice, (4) DNCB + magnolol (10 mg/kg)-treated mice, and (5) DNCB + dexamethasone (10 mg/kg)-treated mice.

4.4. Serum Immunoglobulin E (IgE) Levels

Blood samples were collected from the mice on day 49. A centrifuge was used to obtain serum samples at 4 °C for 10 min. Samples were stored at 80 °C until use. Serum IgE levels were measured using sandwich ELISA with an IgE Mouse Uncoated ELISA Kit (cat. 88-50460-88; Invitrogen) and a 96-well plate (cat. 442404; Thermo Scientific, Waltham, MA, USA), according to the manufacturer's instructions.

4.5. Mast Cell Count in the Skin

Detection of mast cells was facilitated by toluidine blue O staining. The counting of mast cells twice in 50 optical fields was conducted, and an average was calculated [35].

4.6. Histological Analysis of the Skin

On day 49, the ear skin of mice was evaluated for the severity of dermatitis. The ear tissues of mice were embedded in 10% formalin. After fixation, the ears were dehydrated in a 30% sucrose solution and embedded in O.C.T. compound. Using ImageJ software version 5, we measured skin thickness in the sections (8 μm) stained with H&E.

4.7. Quantitative Real-Time PCR

A total RNA sample was isolated from lymph nodes and skin using TRIzol (Invitrogen, Waltham, MA, USA). Promega's MMLV reverse transcriptase was used to reverse transcribe the RNA into cDNA. For qRT-PCR, a CFX Connect Real-Time system (Bio-rad, Hercules, CA, USA) was used with Thunderbird Next SYBR qPCR Mix. There were 40 cycles of 95 °C for 30 s and 57 °C for 30 s in the PCR program. The last cycle at 95 °C for 30 s followed the cycle at 95 °C for 4 min. With the help of CFX Maestro Software version 2.3, the obtained data were analyzed using the $2^{-\Delta\Delta Ct}$ method. The primer sequences are listed in Table 1. Using GAPDH gene expression as the standard, results were normalized [36].

Table 1. Quantitative real-time PCR primers.

Mouse Primers		Sequence
Il-6	Forward	5′-TTC TTG GGA CTG ATG CTG GT-3′
	Reverse	5′-CTG TGA AGT CTC CTC TCC GG-3′
Il-8	Forward	5′-AAC TCC TTG GTG ATG CTG GT-3′
	Reverse	5′-CCA GGT TCA GCA GGT AGA CA-3′
Il-12A	Forward	5′-GAA GCT CTG CAT CCT GCT TC-3′
	Reverse	5′-CAG ATA GCC CAT CAC CCT GT-3′
IFN-g	Forward	5′-CAC GGC ACA GTC ATT GAA AG-3′
	Reverse	5′-GTC ACC ATC CTT TTG CCA GT-3′
TARC	Forward	5′-AAT GTA GGC CGA GAG TGC TG-3′
	Reverse	5′-CAT CCC TGG AAC ACT CCA CT-3′
Il-17A	Forward	5′-AGC TGG ACC ACC ACA TGA AT-3′
	Reverse	5′-AGC ATC TTC TCG ACC CTG AA-3′
Gapdh	Forward	5′-AAC TTT GGC ATT GTG GAA GG-3′
	Reverse	5′-GGATGCAGGGATGATGTTCT-3′

4.8. FACS Analysis

Cervical lymph nodes were separated, and single-cell suspensions were prepared using collagenase type II (Gibco, Grand Island, NY, USA). FITC-labeled rat antibodies against CD4 (cat. 11-0041-82, eBioscience) were used to stain single lymph node cells for 15 min at 4 °C to detect Th17 cell populations. It was then used to stain the permeabilized cells with anti-RORγt or anti-FoxP3 APC-labeled rat antibodies at 22 °C for one hour after fixation using IC fixation buffer. The cell sorting was performed on a CytoFLEX flow cytometer (Beckman Coulter, Brea, CA, USA).

4.9. Differentiation of Naïve T Cells into Th17/Th2/Th1 Cells

Naïve CD4$^+$ T cells isolated from the spleens were differentiated into Th17, Th2, or Th1 CD4$^+$ cells. Naïve CD4$^+$ T cells in 12-well plates coated with anti-mouse CD3 were cultured in X-VIVO 15 (Lonza, Basel, Switzerland) for Th17 differentiation; in RPMI1640 media with recombinant mouse IL-12 (R&D 419-ML-010), recombinant human IL-2 (Peprotech 200-02), and anti-IL-4 (BioXcell BE0045) for Th1 differentiation; or in RPMI1640 media with recombinant human IL-2 (Peprotech 200-02), recombinant mouse IL-4 (Peprotech 214-14), and anti-mouse IFN-γ (BioXcell BE0055) for Th2 differentiation for 3 days [37]. The medium was replenished on day 3. The cells were collected on day 5 and analyzed for each differentiation using flow cytometry. Magnolol (10 and 30 µM) was added to each differentiation medium.

4.10. Statistics

A mean and standard error of mean (SEM) were calculated for all data (n = 5). An analysis of statistical significance was performed using version 5 of GraphPad Prism (GraphPad Prism). There was a p value of 0.05 set as the threshold for statistical significance. * $p < 0.05$, ** $p < 0.01$, and *** $p < 0.001$ vs. the vehicle-treated group, and # $p < 0.05$, ## $p < 0.01$, and ### $p < 0.001$ vs. the DNCB-treated group or differentiation media-treated group.

Author Contributions: Conceptualization, J.-H.L. and D.-S.I.; methodology, J.-H.L.; formal analysis and data curation, J.-H.L.; writing—original draft preparation, D.-S.I.; writing—review and editing, D.-S.I.; supervision, D.-S.I. All authors have read and agreed to the published version of the manuscript.

Funding: This research was supported by the Basic Science Research Program of the Korean National Research Foundation funded by the Korean Ministry of Science, ICT, and Future Planning (NRF-2023R1A2C2002380).

Institutional Review Board Statement: This study was conducted according to the guidelines of the Declaration of Helsinki and approved by the Kyung Hee University Institutional Animal Care Committee with respect to the ethics of the procedures and animal care (KHSASP-22-407, approval data: 8 September 2022).

Informed Consent Statement: Not applicable as no humans were involved.

Data Availability Statement: Data are contained within the article.

Conflicts of Interest: The authors declare no conflicts of interest.

References

1. Labib, A.; Yosipovitch, G. An evaluation of abrocitinib for moderate-to-severe atopic dermatitis. *Expert Rev. Clin. Immunol.* **2022**, *18*, 1107–1118. [CrossRef]
2. Silverberg, J.I.; Barbarot, S.; Gadkari, A.; Simpson, E.L.; Weidinger, S.; Mina-Osorio, P.; Rossi, A.B.; Brignoli, L.; Saba, G.; Guillemin, I. Atopic dermatitis in the pediatric population: A cross-sectional, international epidemiologic study. *Ann. Allergy Asthma Immunol.* **2021**, *126*, 417–428.e2. [CrossRef]
3. Paller, A.S.; Simpson, E.L.; Siegfried, E.C.; Cork, M.J.; Wollenberg, A.; Arkwright, P.D.; Soong, W.; Gonzalez, M.E.; Schneider, L.C.; Sidbury, R. Dupilumab in children aged 6 months to younger than 6 years with uncontrolled atopic dermatitis: A randomised, double-blind, placebo-controlled, phase 3 trial. *Lancet* **2022**, *400*, 908–919. [CrossRef]
4. Graff, P. Potential Drivers of the Atopic March-Unraveling the Skin-Lung Crosstalk. Ph.D. Thesis, Freie University, Berlin, Germany, 2022.
5. Kim, J.Y.; Jeong, M.S.; Park, M.K.; Lee, M.K.; Seo, S.J. Time-dependent progression from the acute to chronic phases in atopic dermatitis induced by epicutaneous allergen stimulation in NC/Nga mice. *Exp. Dermatol.* **2014**, *23*, 53–57. [CrossRef]
6. Koga, C.; Kabashima, K.; Shiraishi, N.; Kobayashi, M.; Tokura, Y. Possible pathogenic role of Th17 cells for atopic dermatitis. *J. Investig. Derm.* **2008**, *128*, 2625–2630. [CrossRef]
7. Muraro, A.; Lemanske, R.F., Jr.; Hellings, P.W.; Akdis, C.A.; Bieber, T.; Casale, T.B.; Jutel, M.; Ong, P.Y.; Poulsen, L.K.; Schmid-Grendelmeier, P.; et al. Precision medicine in patients with allergic diseases: Airway diseases and atopic dermatitis-PRACTALL document of the European Academy of Allergy and Clinical Immunology and the American Academy of Allergy, Asthma & Immunology. *J. Allergy Clin. Immunol.* **2016**, *137*, 1347–1358. [CrossRef]

8. Son, M.-J.; Lee, S.-M.; Park, S.-H.; Kim, Y.-E.; Jung, J.-Y. The effects of orally administered Gwakhyangjeonggi-san on DNCB-induced atopic dermatitis like mice model. *J. Korean Med. Ophthalmol. Otolaryngol. Dermatol.* **2019**, *32*, 94–106.
9. Jin, S.-E.; Moon, P.-D.; Koh, J.-H.; Lim, H.-S.; Kim, H.-M.; Jeong, H.-J. Inhibition of chemokine, interleukin-8 expression in an atopic milieu by Pyeongwee-San extract (KMP6). *Orient. Pharm. Exp. Med.* **2011**, *11*, 71–81. [CrossRef]
10. Lee, J.-H.; Im, D.-S. Honokiol suppresses 2, 6-dinitrochlorobenzene-induced atopic dermatitis in mice. *J. Ethnopharmacol.* **2022**, *2022*, 115023. [CrossRef] [PubMed]
11. Park, J.; Lee, J.; Jung, E.; Park, Y.; Kim, K.; Park, B.; Jung, K.; Park, E.; Kim, J.; Park, D. In vitro antibacterial and anti-inflammatory effects of honokiol and magnolol against *Propionibacterium* sp. *Eur. J. Pharmacol.* **2004**, *496*, 189–195. [CrossRef]
12. Fujita, M.; Itokawa, H.; Sashida, Y. Studies on the components of Magnolia obovata Thunb. 3. Occurrence of magnolol and hõnokiol in M. obovata and other allied plants. *Yakugaku Zasshi* **1973**, *93*, 429–434. [CrossRef] [PubMed]
13. Lee, Y.J.; Lee, Y.M.; Lee, C.K.; Jung, J.K.; Han, S.B.; Hong, J.T. Therapeutic applications of compounds in the Magnolia family. *Pharmacol. Ther.* **2011**, *130*, 157–176. [CrossRef] [PubMed]
14. Zhang, R.; Zhang, H.; Shao, S.; Shen, Y.; Xiao, F.; Sun, J.; Piao, S.; Zhao, D.; Li, G.; Yan, M. Compound traditional Chinese medicine dermatitis ointment ameliorates inflammatory responses and dysregulation of itch-related molecules in atopic dermatitis. *Chin. Med.* **2022**, *17*, 1–19. [CrossRef]
15. Zuo, G.-Y.; Zhang, X.-J.; Han, J.; Li, Y.-Q.; Wang, G.-C. In vitro synergism of magnolol and honokiol in combination with antibacterial agents against clinical isolates of methicillin-resistant Staphylococcus aureus (MRSA). *BMC Complement. Altern. Med.* **2015**, *15*, 425. [CrossRef] [PubMed]
16. Yang, T.-C.; Zhang, S.-W.; Sun, L.-N.; Wang, H.; Ren, A.-M. Magnolol attenuates sepsis-induced gastrointestinal dysmotility in rats by modulating inflammatory mediators. *World J. Gastroenterol. WJG* **2008**, *14*, 7353. [CrossRef]
17. Wei, W.; Dejie, L.; Xiaojing, S.; Tiancheng, W.; Yongguo, C.; Zhengtao, Y.; Naisheng, Z. Magnolol inhibits the inflammatory response in mouse mammary epithelial cells and a mouse mastitis model. *Inflammation* **2015**, *38*, 16–26. [CrossRef]
18. Yunhe, F.; Bo, L.; Xiaosheng, F.; Fengyang, L.; Dejie, L.; Zhicheng, L.; Depeng, L.; Yongguo, C.; Xichen, Z.; Naisheng, Z. The effect of magnolol on the toll-like receptor 4/nuclear factor kappa B signaling pathway in lipopolysaccharide-induced acute lung injury in mice. *Eur. J. Pharmacol.* **2012**, *689*, 255–261. [CrossRef]
19. Shen, P.; Zhang, Z.; He, Y.; Gu, C.; Zhu, K.; Li, S.; Li, Y.; Lu, X.; Liu, J.; Zhang, N.; et al. Magnolol treatment attenuates dextran sulphate sodium-induced murine experimental colitis by regulating inflammation and mucosal damage. *Life Sci.* **2018**, *196*, 69–76. [CrossRef]
20. Wang, J.-P.; Hsu, M.-F.; Raung, S.-Z.; Chen, C.-C.; Kuo, J.-S.; Teng, C.-M. Anti-inflammatory and analgesic effects of magnolol. *Naunyn-Schmiedeberg's Arch. Pharmacol.* **1992**, *346*, 707–712. [CrossRef] [PubMed]
21. Han, S.J.; Bae, E.-A.; Trinh, H.T.; Yang, J.-H.; Youn, U.-J.; Bae, K.-H.; Kim, D.-H. Magnolol and honokiol: Inhibitors against mouse passive cutaneous anaphylaxis reaction and scratching behaviors. *Biol. Pharm. Bull.* **2007**, *30*, 2201–2203. [CrossRef]
22. Huang, Q.; Han, L.; Lv, R.; Ling, L. Magnolol exerts anti-asthmatic effects by regulating Janus kinase-signal transduction and activation of transcription and Notch signaling pathways and modulating Th1/Th2/Th17 cytokines in ovalbumin-sensitized asthmatic mice. *Korean J. Physiol. Pharmacol.* **2019**, *23*, 251–261. [CrossRef]
23. Phan HT, L.; Nam, Y.R.; Kim, H.J.; Woo, J.H.; NamKung, W.; Nam, J.H.; Kim, W.K. In-vitro and in-vivo anti-allergic effects of magnolol on allergic rhinitis via inhibition of ORAI1 and ANO1 channels. *J. Ethnopharmacol.* **2022**, *289*, 115061. [CrossRef]
24. Lee, J.E.; Choi, Y.W.; Im, D.S. Inhibitory effect of α-cubebenoate on atopic dermatitis-like symptoms by regulating Th2/Th1/Th17 balance in vivo. *J. Ethnopharmacol.* **2022**, *291*, 115162. [CrossRef]
25. Hijnen, D.; de Bruin-Weller, M.; Oosting, B.; Lebre, C.; De Jong, E.; Bruijnzeel-Koomen, C.; Knol, E. Serum thymus and activation-regulated chemokine (TARC) and cutaneous T cell–attracting chemokine (CTACK) levels in allergic diseases: TARC and CTACK are disease-specific markers for atopic dermatitis. *J. Allergy Clin. Immunol.* **2004**, *113*, 334–340. [CrossRef] [PubMed]
26. Kong, L.; Liu, J.; Wang, J.; Luo, Q.; Zhang, H.; Liu, B.; Xu, F.; Pang, Q.; Liu, Y.; Dong, J. Icariin inhibits TNF-α/IFN-γ induced inflammatory response via inhibition of the substance P and p38-MAPK signaling pathway in human keratinocytes. *Int. Immunopharmacol.* **2015**, *29*, 401–407. [CrossRef] [PubMed]
27. Ikarashi, Y.; Yuzurihara, M.; Sakakibara, I.; Nakai, Y.; Hattori, N.; Maruyama, Y. Effects of the extract of the bark of Magnolia obovata and its biphenolic constituents magnolol and honokiol on histamine release from peritoneal mast cells in rats. *Planta Med.* **2001**, *67*, 709–713. [CrossRef] [PubMed]
28. Shen, J.-L.; Man, K.-M.; Huang, P.-H.; Chen, W.-C.; Chen, D.-C.; Cheng, Y.-W.; Liu, P.-L.; Chou, M.-C.; Chen, Y.-H. Honokiol and magnolol as multifunctional antioxidative molecules for dermatologic disorders. *Molecules* **2010**, *15*, 6452–6465. [CrossRef] [PubMed]
29. Guo, J.W.; Cheng, Y.P.; Liu, C.Y.; Thong, H.Y.; Lo, Y.; Wu, C.Y.; Jee, S.H. Magnolol may contribute to barrier function improvement on imiquimod-induced psoriasis-like dermatitis animal model via the downregulation of interleukin-23. *Exp. Med.* **2021**, *21*, 448. [CrossRef]
30. Dreier, D.; Latkolik, S.; Rycek, L.; Schnürch, M.; Dymáková, A.; Atanasov, A.G.; Ladurner, A.; Heiss, E.H.; Stuppner, H.; Schuster, D.; et al. Linked magnolol dimer as a selective PPARγ agonist—Structure-based rational design, synthesis, and bioactivity evaluation. *Sci. Rep.* **2017**, *7*, 13002. [CrossRef]
31. Dreier, D.; Resetar, M.; Temml, V.; Rycek, L.; Kratena, N.; Schnürch, M.; Schuster, D.; Dirsch, V.M.; Mihovilovic, M.D. Magnolol dimer-derived fragments as PPARγ-selective probes. *Org. Biomol. Chem.* **2018**, *16*, 7019–7028. [CrossRef]

32. Tanaka, K.; Hasegawa, J.; Asamitsu, K.; Okamoto, T. Magnolia ovovata extract and its active component magnolol prevent skin photoaging via inhibition of nuclear factor kappaB. *Eur. J. Pharm.* **2007**, *565*, 212–219. [CrossRef] [PubMed]
33. Hsu, M.F.; Lu, M.C.; Tsao, L.T.; Kuan, Y.H.; Chen, C.C.; Wang, J.P. Mechanisms of the influence of magnolol on eicosanoid metabolism in neutrophils. *Biochem. Pharm.* **2004**, *67*, 831–840. [CrossRef] [PubMed]
34. Lin, C.F.; Hung, C.F.; Aljuffali, I.A.; Huang, Y.L.; Liao, W.C.; Fang, J.Y. Methylation and Esterification of Magnolol for Ameliorating Cutaneous Targeting and Therapeutic Index by Topical Application. *Pharm. Res.* **2016**, *33*, 2152–2167. [CrossRef] [PubMed]
35. Kang, J.; Lee, J.H.; Im, D.S. Topical Application of S1P(2) Antagonist JTE-013 Attenuates 2,4-Dinitrochlorobenzene-Induced Atopic Dermatitis in Mice. *Biomol. Ther.* **2020**, *28*, 537–541. [CrossRef]
36. Lee, J.E.; Im, D.S. Suppressive Effect of Carnosol on Ovalbumin-Induced Allergic Asthma. *Biomol. Ther.* **2021**, *29*, 58–63. [CrossRef]
37. Flaherty, S.; Reynolds, J.M. Mouse naive CD4+ T cell isolation and in vitro differentiation into T cell subsets. *JoVE (J. Vis. Exp.)* **2015**, *2015*, e52739.

Disclaimer/Publisher's Note: The statements, opinions and data contained in all publications are solely those of the individual author(s) and contributor(s) and not of MDPI and/or the editor(s). MDPI and/or the editor(s) disclaim responsibility for any injury to people or property resulting from any ideas, methods, instructions or products referred to in the content.

Article

Cytotoxic Effects of Ardisiacrispin A from *Labisia pumila* on A549 Human Lung Cancer Cells

Yeong-Geun Lee [1], Tae Hyun Kim [1], Jeong Eun Kwon [1], Hyunggun Kim [2,*] and Se Chan Kang [1,*]

[1] Department of Oriental Medicine Biotechnology, College of Life Sciences, Kyung Hee University, Yongin 17104, Gyeonggi, Republic of Korea; lyg629@nate.com (Y.-G.L.); silsoo96@naver.com (T.H.K.); jjung@nmr.kr (J.E.K.)

[2] Department of Biomechatronic Engineering, Sungkyunkwan University, Suwon 16419, Gyeonggi, Republic of Korea

* Correspondence: hkim.bme@skku.edu (H.K.); sckang@khu.ac.kr (S.C.K.); Tel.: +82-31-290-7827 (H.K.); +82-31-201-5637 (S.C.K.)

Abstract: Background: Lung cancer is the predominant cause of cancer-related fatalities. This prompted our exploration into the anti-lung cancer efficacy of *Labisia pumila*, a species meticulously selected from the preliminary screening of 600 plants. Methods: Through the strategic implementation of activity-guided fractionation, ardisiacrispin A (1) was isolated utilizing sequential column chromatography. Structural characterization was achieved employing various spectroscopic methods, including nuclear magnetic resonance (NMR), mass spectrometry (MS), and infrared spectroscopy (IR). Results: *L. pumila* 70% EtOH extract showed significant toxicity in A549 lung cancer cells, with an IC_{50} value of 57.04 ± 10.28 µg/mL, as well as decreased expression of oncogenes and induced apoptosis. Compound **1**, ardisiacrispin A, induced a 50% cell death response in A549 cells at a concentration of 11.94 ± 1.14 µg/mL. Conclusions: The present study successfully investigated ardisiacrispin A extracted from *L. pumila* leaves, employing a comprehensive spectroscopic approach encompassing NMR, IR, and MS analyses. The anti-lung cancer efficacy of ardisiacrispin A and *L. pumila* extract was successfully demonstrated for the first time, to the best of our knowledge.

Keywords: *Labisia pumila*; ardisiacrispin A; A549; natural product; triterpene

1. Introduction

Cancer obviously represents one of the biggest challenges to global human health [1]. Lung cancer is the predominant cause of cancer-related fatalities on a worldwide scale [2]. In 2020, lung cancers constituted 11.4% of newly diagnosed cancer cases, placing the lung as the second most prevalent site of incident cancers, and the majority of cases, approximately 85%, were attributed to a group of histological subtypes collectively known as non-small-cell lung cancer [2,3]. Epidermal growth factor receptor (EGFR) tyrosine kinase inhibitors such as gefitinib and erlotinib are typically the first chemotherapy treatments for lung cancer. Unfortunately, the prognosis of advanced and recurrent lung cancer remains suboptimal, and standard treatments utilizing cytotoxic anticancer drugs demonstrate limited therapeutic effectiveness [4]. In the current landscape of cancer management, diverse treatment options such as chemotherapy, surgery, and radiotherapy exist [5,6]. Despite their efficacy, these therapeutic interventions are often associated with severe side effects, posing a substantial risk to patients, or imposing exacting prerequisites for their implementation.

Given the imperative to advance anti-lung cancer drug development, we conducted a screening of potential candidates from a repository of plants within our institute (data not presented). *Labisia pumila* (Myrsinaceae), grown in southeast Asia, emerged as a promising anti-lung cancer candidate in our preliminary investigation [7]. Traditionally employed for the maintenance of female reproductive health and postpartum care, this botanical specimen holds promise in anti-cancer applications. The investigation of this plant has revealed a number of secondary metabolites, exhibiting phytoestrogenic, anti-bacterial, anti-fungal, anti-oxidant, anti-carcinogenic, and anti-aging effects [7–9]. Historical uses of this botanical entity include the enhancement of stamina and treatment for various conditions, such as dysentery, rheumatism, gonorrhea, excessive flatulence, and cancers, particularly those affecting the breast and uterus [10,11]. Despite the well-established anti-cancer properties of L. pumila, limited information is available concerning its efficacy in the context of lung cancer. Therefore, in the present study, the natural product, L. pumila, and its active compound were rigorously investigated to assess their potential as a candidate for lung cancer treatment.

2. Materials and Methods

2.1. Reagents and Instrumentation

The equipment and chemicals for isolation and structural elucidation of the anti-cancer component were referred to in our previous investigations [11,12]. Briefly, SiO_2 (Kieselgel 60, Merck, Darmstadt, Germany) and ODS (Lichroprep RP-18, 40–60 µm, Merck) were used as resins for column chromatography (c.c.). The separated compound was detected using a UV lamp (Spectroline Model ENF-240 C/F, Spectronics Corporation, Westbury, NY, USA) following application on Kieselgel 60 F254 (Merck) and Kieselgel RP-18 F254S (Merck) plates, subsequent to spraying with a 10% aqueous H_2SO_4 solution. Nuclear magnetic resonance (NMR) spectra were recorded employing a Bruker Avance 600 (Billerica, MA, USA), and melting points were precisely determined using a Fisher-John Melting Point Apparatus (Fisher Scientific, Miami, FL, USA). Deuterium solvents for measurement of NMR and standard organic solvents for extraction were purchased from Sigma Aldrich Co., Ltd. (St. Louis, MO, USA) and Daejung Chemical Ltd. (Seoul, Republic of Korea), respectively.

Tryptic Soy Broth (TSB) was purchased from KisanBio (Seoul, Republic of Korea). Doxorubicin, dimethyl sulfoxide (DMSO), and MTT [3-(4,53-(4,5-dimethylthiazol-2-yl)-2,5-diphenyltetrazolium bromide)] were purchased from Sigma-Aldrich (St. Louis, MO, USA). Dulbecco's modification of Eagle's media (DMEM) with 4.5 g/L glucose, L-glutamine, and sodium pyruvate, and RPMI1640 with L-glutamine were purchased from Mediatech (Manassas, VA, USA). Fetal bovine serum (FBS), bovine calf serum (BCS), trypsin-EDTA, and penicillin–streptomycin were purchased from Thermo Fisher Scientific (Waltham, MA, USA).

2.2. Plants

In this study, a comprehensive set of 600 plant extract samples was sourced from the International Biological Material Research Center (IBMRC, Cheongju, Republic of Korea). Voucher specimens, uniquely identified by the codes KHU-BMRI-2017-001 through KHU-BMRI-2017-600, were deposited at the Bio-Medical Research Institute, Yongin, Kyung Hee University. The process for isolating potential anti-cancer candidates involved the extraction of plant materials using 100% methanol (MeOH), and the extracts were subsequently solubilized to a concentration of 10 mg/mL in dimethylsulfoxide (DMSO).

2.3. Isolation of the Anti-Cancer Component from L. pumila

Dried leaves of L. pumila (180 g), purchased from Malaysia, were chopped into small pieces and subjected to extraction using 70% aqueous EtOH (3 L × 3) for 24 h at room temperature. The EtOH extract (11 g), obtained through filtration and subsequent concentration in vacuo, was reconstituted in 200 mL of H_2O. Subsequent sequential extractions were performed three times using n-hexane (150 mL), dichloromethane (DCM, 150 mL), ethyl acetate (EtOAc, 150 mL), and n-BuOH (150 mL), resulting in distinct fractions: n-hexane (LPH, 2.44 g), DCM (LPD, 1.07 g), EtOAc (LPE, 390 mg), n-BuOH (LPB, 930 mg), and aqueous (LPA, 6.17 g) fractions. Among the fractions obtained, LPD (1.07 g), identified through activity-guided fractionation, underwent further fractionation using SiO_2 cc

(⌀ 4 × 15 cm, CHCl$_3$:MeOH:H$_2$O = 50:3:1→30:3:1→20:3:1→10:3:1→7:3:1→5:3:1, 1 L of each), resulting in the isolation of 21 fractions (LPD-1 to LPD-21). Fraction LPD-13 (82.0 mg, elution volume/total volume (V$_e$/V$_t$) 0.644–0.751) underwent additional fractionation employing ODS cc (⌀ 1 × 5 cm, acetone: H$_2$O = 1:3 → 3:1, 150 mL of each), yielding 6 fractions (LPD-13-1 to LPD-13-6). Ardisiacrispin A (1) was successfully isolated from LPD-13-3 (40.0 mg) using TLC (SiO$_2$) with elution in CHCl$_3$: MeOH: H$_2$O (5:3:1) within the range of 0.160–0.220 and TLC (ODS) with acetone:H$_2$O (1:1) within the range of 0.270–0.360.

Ardisiacrispin A (1): white amorphous powder; negative FAB/MS m/z 1059 [M−H]$^-$; negative HR-FAB/MS m/z 1059.5375 [M−H]$^-$ (calculated for C$_{52}$H$_{83}$O$_{22}$, 1059.5376); melting point: 229–230 °C; IR (KBr, ν_{max}, cm^{-1}): 3415, 3455, 3570 (OH), 1710 (CHO); ^{1}H-NMR (600 MHz, DMSO-d_6, δ_H, J in Hz) 9.61 (1H, s, H-30), 5.36 (1H, d, J = 7.2 Hz, H-glc′-1), 4.97 (1H, d, J = 7.8 Hz, H-glc″-1), 4.96 (1H, d, J = 7.2 Hz, H-xyl-1), 4.76 (1H, d, J = 5.2 Hz, H-ara-1), 4.58 (1H, br.d, J = 9.6 Hz, H-ara-5a), 4.48 (1H, br.dd, J = 11.4, 8.4 Hz, H-glc′-6a), 4.44 (1H, ddd, J = 10.4, 7.8 Hz, H-xyl-5a), 4.44 (1H, dd, J = 9.6, 5.2 Hz, H-ara-2), 4.39 (1H, br.dd, J = 11.4, 8.4 Hz, H-glc″-6a), 4.31 (1H, ddd, J = 11.4, 8.4, 4.8 Hz, H-glc′-6b), 4.26 (1H, ddd, J = 11.4, 8.4, 4.8 Hz, H-glc″-6b), 4.23 (1H, m, H-ara-4), 4.22 (1H, dd, J = 7.8, 7.8 Hz, H-xyl-3), 4.21 (1H, m, H-16), 4.20 (1H, dd, J = 7.8, 7.8 Hz, H-glc″-3), 4.17 (1H, m, H-xyl-4), 4.13 (1H, dd, J = 8.4, 7.2 Hz, H-glc′-4), 4.11 (1H, dd, J = 9.6, 9.6 Hz, H-ara-3), 4.11 (1H, dd, J = 8.4, 7.8 Hz, H-glc″-4), 4.02 (1H, dd, J = 7.8, 7.2 Hz, H-glc′-2), 3.96 (1H, dd, J = 7.2, 7.2 Hz, H-glc′-3), 3.96 (1H, br.d, J = 8.4 Hz, H-glc′-5), 3.93 (1H, dd, J = 7.2, 7.8 Hz, H-xyl-2), 3.83 (1H, br.d, J = 8.4 Hz, H-glc″-5), 3.78 (1H, dd, J = 7.8, 7.8 Hz, H-glc″-2), 3.77 (1H, br.d, J = 9.6 Hz, H-ara-5b), 3.52 (1H, ddd, J = 10.4, 4.2 Hz, H-xyl-5b), 3.51 (1H, d, J = 7.8 Hz, H-28a), 3.16 (1H, dd, J = 12.0, 4.2 Hz, H-3), 3.16 (1H, d, J = 7.8 Hz, H-28b), 2.78 (1H, dd, J = 14.4, 4.2 Hz, H-19a), 2.50 (2H, dd, J = 13.2, 4.2 Hz, H-21), 2.14 (1H, br.d, J = 10.8 Hz, H-15a), 2.10 (1H, t, J = 7.2 Hz, H-12a), 2.06 (1H, br.d, J = 14.4 Hz, H-19b), 1.99 (1H, overlapped, H-22a), 1.96 (1H, br.d, J = 12.0 Hz, H-2a), 1.79 (2H, overlapped, H-11), 1.79 (1H, dd, J = 12.0, 4.2 Hz, H-2b), 1.60 (1H, br.d, J = 12.6 Hz, H-1a), 1.53 (1H, overlapped, H-22b), 1.50 (3H, s, H-27), 1.50 (1H, overlapped, H-15b), 1.40 (1H, overlapped, H-12b), 1.37 (1H, overlapped, H-6a), 1.36 (1H, dd, J = 14.4, 4.2 Hz, H-18), 1.26 (3H, s, H-26), 1.22 (1H, overlapped, H-6b), 1.20 (1H, overlapped, H-9), 1.18 (2H, overlapped, H-7), 1.17 (3H, s, H-23), 1.03 (3H, s, H-24), 1.00 (3H, s, H-29), 0.81 (3H, s, H-25), 0.81 (1H, br.d, J = 12.6 Hz, H-1b), 0.65 (1H, d, J = 10.8 Hz, H-5); ^{13}C-NMR (150 MHz, DMSO-d_6, δ_C) 207.4 (C-30), 105.7 (C-xyl-1), 104.1 (C-glc′-1), 104.0 (C-ara-1), 102.9 (C-glc″-1), 88.3 (C-3), 85.6 (C-13), 83.7 (C-glc″-2), 79.2 (C-ara-2), 78.1 (C-ara-4), 77.1 (C-glc″-5), 76.9 (C-glc′-5), 76.8 (C-glc′-3), 76.2 (C-glc″-3), 76.1 (C-16), 76.0 (C-28), 75.9 (C-xyl-3), 75.1 (C-glc′-2), 74.3 (C-xyl-2), 71.7 (C-xyl-4), 70.8 (C-ara-3), 70.7 (C-glc′-4), 70.2 (C-glc″-4), 64.2 (C-ara-5), 62.8 (C-xyl-5), 61.3 (C-glc″-6), 61.1 (C-glc′-6), 54.9 (C-5), 52.6 (C-18), 49.6 (C-9), 47.6 (C-20), 43.8 (C-14), 43.1 (C-17), 41.7 (C-8), 38.9 (C-4), 38.5 (C-1), 36.1 (C-10), 35.7 (C-15), 33.8 (C-7), 32.6 (C-19), 31.7 (C-22), 31.4 (C-12), 29.7 (C-21), 27.4 (C-23), 25.9 (C-2), 23.4 (C-29), 19.1 (C-27), 18.7 (C-11), 18.1 (C-26), 17.3 (C-6), 15.7 (C-24), 15.3 (C-25).

2.4. Cell Viability and Cytotoxicity Assay

A549 human cell lines, obtained from the Korean Cell Line Bank (KCLB), were cultured in RPMI1640 supplemented with 10% fetal bovine serum (FBS) and 1% penicillin/streptomycin [13]. The cells were cultivated in a humidified incubator at 37 °C with a CO$_2$ concentration of 5%. Cell viability was evaluated utilizing the MTT assay, with seeding densities of 5 × 10^3 cells/well in a 96-well plate. Following a 24 h incubation period, the culture medium was replaced, and the samples were subjected to experimental treatments. After 24 h, the cells were stained using MTT solution in PBS, resulting in a final concentration of 0.5 mg/mL. The cells were subjected to a 4 h incubation at 37 °C. Upon completion of this incubation period, the supernatant was removed and 100 μL of DMSO was added. Utilizing a microplate reader (Tecan, Switzerland), absorbance was measured at 540 nm. The cell cytotoxicity rates were calculated based on the optical density readings, expressed as percentages relative to the vehicle control, and this procedure was repeated for accuracy.

2.5. Tali^{TM} Cell Cycle Assay

Human A549 cells were seeded in 6-well plates at a density of 1×10^5 cells/well. After 24 h incubation, the culture medium was replaced, and the cells were subjected to specific treatments. Following fixation, the cells were treated with the optimized Tali^{TM} cell cycle reagent (Thermo, Middlesex, MA, USA) and incubated in darkness for 30 min. The Tali® image cytometer (Thermo, Middlesex, MA, USA) was employed for cell cycle analysis. The acquired cell cycle data from the Tali^{TM} image-based cytometer were analyzed both on the instrument and through dedicated cell cycle modeling software.

2.6. Statistical Analysis

All data are presented as mean ± standard error of the mean (SEM). The significance of differences between groups was assessed using one-way analysis of variance (ANOVA). Statistical significance was defined as $p < 0.05$.

3. Results and Discussion

3.1. Determination of the Anti-Cancer Agent and Its Optimal Extraction Condition

In pursuit of identifying a natural anti-cancer candidate agent, a pilot study was conducted through cell viability assays to determine IC_{50} values in A549 cells. From an extensive pool of MeOH extracts obtained from a diverse collection of over 600 plants, *L. pumila* emerged as a potent anti-lung cancer candidate. Following this selection, a systematic evaluation was carried out to establish the optimal concentration of the *L. pumila* extract. The evaluation revealed the significant toxicity demonstrated by the 70% EtOH extract in A549 lung cancer cells, with an IC_{50} value of 57.04 ± 10.28 μg/mL (Table 1).

Table 1. Cytotoxic effect (IC_{50}, μg/mL) of *L. pumila* extract against A549 human lung cancer cells at varying EtOH concentrations.

EtOH Concentration (%)	IC_{50} (μg/mL)
10	<100
20	<100
30	<100
40	<100
50	<100
60	<100
70	57.04 ± 10.28
80	76.94 ± 5.56
90	80.91 ± 4.31
100	84.09 ± 8.65

3.2. Structural Evaluation of the Anti-Lung Cancer Component from L. pumila

The dried leaves of *L. pumila* were subjected to extraction using 70% aqueous EtOH, and the resulting concentrate was fractioned into *n*-hexane (LPH), dichloromethane (LPD), ethyl acetate (LPE), *n*-BuOH (LPB), and H_2O (LPW) fractions. A series of activity-guided fractionation steps, employing SiO_2, ODS, and Sephadex LH-20 column chromatography (c.c.) for the LPD fraction, led to the isolation of a singular triterpenoid saponin (1). Elucidation of its chemical structure was achieved through a comprehensive analysis of spectroscopic data, including mass spectrometry (MS), infrared spectroscopy (IR), and NMR (both 1D and 2D).

Compound **1**, a white amorphous powder (MeOH), showed IR absorbance bands of hydroxyl (3415, 3455, and 3570 cm^{-1}) and formyl groups (1710 cm^{-1}). The molecular formula of Compound **1** was determined to be $C_{52}H_{83}O_{22}$ through negative fast atom bombardment mass spectrometry (FAB/MS) *m/z* 1059 [M − H]$^-$ and negative high-resolution

FAB/MS m/z 1059.5375 [M − H]$^-$ (calcd for $C_{52}H_{83}O_{22}$, 1059.5376). The ^{1}H-NMR spectrum (600 MHz, DMSO-d_6) showed proton signals due to six singlet methyls [δ_H 1.50 (3H, s, H-27), δ_H 1.26 (3H, s, H-26), δ_H 1.17 (3H, s, H-23), δ_H 1.03 (3H, s, H-24), δ_H 1.00 (3H, s, H-29), and δ_H 0.81 (3H, s, H-25)]; one formyl [δ_H 9.61 (1H, s, H-30)]; two oxygenated methines [δ_H 4.21 (1H, m, H-16) and δ_H 3.16 (1H, dd, J = 12.0, 4.2 Hz, H-3)]; two oxygenated methylenes [δ_H 3.51 (1H, d, J = 7.8 Hz, H-28a) and δ_H 3.16 (1H, d, J = 7.8 Hz, H-28b)]; three methines [δ_H 1.36 (1H, dd, J = 14.4, 4.2 Hz, H-18), δ_H 1.20 (1H, overlapped, H-9), and δ_H 0.65 (1H, d, J = 10.8 Hz, H-5)]; and ten methylenes [δ_H 2.78 (1H, dd, J = 14.4, 4.2 Hz, H-19a), δ_H 2.50 (2H, dd, J = 13.2, 4.2 Hz, H-21), δ_H 2.14 (1H, br.d, J = 10.8 Hz, H-15a), δ_H 2.10 (1H, t, J = 7.2 Hz, H-12a), δ_H 2.06 (1H, br.d, J = 14.4 Hz, H-19b), δ_H 1.99 (1H, overlapped, H-22a), δ_H 1.96 (1H, br.d, J = 12.0 Hz, H-2a), δ_H 1.79 (2H, overlapped, H-11), δ_H 1.79 (1H, dd, J = 12.0, 4.2 Hz, H-2b), δ_H 1.60 (1H, br.d, J = 12.6 Hz, H-1a), δ_H 1.53 (1H, overlapped, H-22b), δ_H 1.50 (1H, overlapped, H-15b), δ_H 1.40 (1H, overlapped, H-12b), δ_H 1.37 (1H, overlapped, H-6a), δ_H 1.22 (1H, overlapped, H-6b), δ_H 1.18 (2H, overlapped, H-7), and δ_H 0.81 (1H, br.d, J = 12.6 Hz, H-1b)]. The proton signals indicated the aglycone of Compound **1** to be an oleanane-type triterpenoid possessing a formyl group. Also, four hemiacetals [δ_H 5.36 (1H, d, J = 7.2 Hz, H-glc'-1), δ_H 4.97 (1H, d, J = 7.8 Hz, H-glc''-1), δ_H 4.96 (1H, d, J = 7.2 Hz, H-xyl-1), δ_H 4.76 (1H, d, J = 5.2 Hz, H-ara-1)]; fourteen oxygenated methines [δ_H 4.44 (1H, dd, J = 9.6, 5.2 Hz, H-ara-2), δ_H 4.23 (1H, m, H-ara-4), δ_H 4.22 (1H, dd, J = 7.8, 7.8 Hz, H-xyl-3), δ_H 4.20 (1H, dd, J = 7.8, 7.8 Hz, H-glc'-3), δ_H 4.17 (1H, m, H-xyl-4), δ_H 4.13 (1H, dd, J = 8.4, 7.2 Hz, H-glc'-4), δ_H 4.11 (1H, dd, J = 9.6, 9.6 Hz, H-ara-3), δ_H 4.11 (1H, dd, J = 8.4, 7.8 Hz, H-glc''-4), δ_H 4.02 (1H, dd, J = 7.8, 7.2 Hz, H-glc'-2), δ_H 3.96 (1H, dd, J = 7.2, 7.2 Hz, H-glc'-3), δ_H 3.96 (1H, br.d, J = 8.4 Hz, H-glc'-5), δ_H 3.93 (1H, dd, J = 7.2, 7.8 Hz, H-xyl-2), δ_H 3.83 (1H, br.d, J = 8.4 Hz, H-glc''-5), and δ_H 3.78 (1H, dd, J = 7.8, 7.8 Hz, H-glc''-2)]; and four germinal oxygenated methylene proton [δ_H 4.58 (1H, br.d, J = 9.6 Hz, H-ara-5a), δ_H 4.48 (1H, br.dd, J = 11.4, 8.4 Hz, H-glc'-6a), δ_H 4.44 (1H, ddd, J = 10.4, 7.8 Hz, H-xyl-5a), δ_H 4.39 (1H, br.dd, J = 11.4, 8.4 Hz, H-glc'-6a), δ_H 4.31 (1H, ddd, J = 11.4, 8.4, 4.8 Hz, H-glc'-6b), δ_H 4.26 (1H, ddd, J = 11.4, 8.4, 4.8 Hz, H-glc''-6b), δ_H 3.77 (1H, br.d, J = 9.6 Hz, H-ara-5b), δ_H 3.52 (1H, ddd, J = 10.4, 4.2 Hz, H-xyl-5b)] signals were observed as the proton signals of four hexoses. The coupling constants of the anomer proton signals of three sugars (glc'-1, glc-''-1, xyl-1; J = 7.8 or 7.2 Hz) and one sugar (ara-1; J = 5.2 Hz) confirmed the axial–axial and axial–equatorial configurations of the anomer hydroxyl groups, respectively.

The ^{13}C-NMR data exhibited a total of 30 carbon signals corresponding to the aglycone along with 22 carbons derived from four hexoses, indicating Compound **1** to be a triterpenoid with four hexoses. The ^{13}C-NMR (150 MHz, DMSO-d_6) spectrum showed one formyl carbon signal, δ_C 207.4 (C-30); one oxygenated quaternary, δ_C 85.6 (C-13); one oxygenated methylene, δ_C 76.0 (C-28); two oxygenated methines [δ_C 88.3 (C-3) and δ_C 76.1 (C-16)]; six quaternaries [δ_C 47.6 (C-20), δ_C 43.8 (C-14), δ_C 43.1 (C-17), δ_C 41.7 (C-8), δ_C 38.9 (C-4), and δ_C 36.1 (C-10)]; three methines [δ_C 54.9 (C-5), δ_C 52.6 (C-18), and δ_C 49.6 (C-9)]; ten methylenes [δ_C 38.5 (C-1), δ_C 35.7 (C-15), δ_C 33.8 (C-7), δ_C 32.6 (C-19), δ_C 31.7 (C-22), δ_C 31.4 (C-12), δ_C 29.7 (C-21), δ_C 25.9 (C-2), δ_C 18.7 (C-11), and δ_C 17.3 (C-6)]; and six methyls [δ_C 27.4 (C-23), δ_C 23.4 (C-29), δ_C 19.1 (C-27), δ_C 18.1 (C-26), δ_C 15.7 (C-24), and δ_C 15.3 (C-25)]. Based on the chemical shifts of the sugar carbon signals, we observed four hemiacetals [δ_C 105.7 (C-xyl-1), δ_C 104.1 (C-glc'-1), δ_C 104.0 (C-ara-1), and δ_C 102.9 (C-glc''-1)]; fourteen oxygenated methines [δ_C 83.7 (C-glc''-2), δ_C 79.2 (C-ara-2), δ_C 78.1 (C-ara-4), δ_C 77.1 (C-glc''-5), δ_C 76.9 (C-glc'-5), δ_C 76.8 (C-glc'-3), δ_C 76.2 (C-glc''-3), δ_C 75.9 (C-xyl-3), δ_C 75.1 (C-glc'-2), δ_C 74.3 (C-xyl-2), δ_C 71.7 (C-xyl-4), δ_C 70.8 (C-ara-3), δ_C 70.7 (C-glc'-4), and δ_C 70.2 (C-glc''-4)]; and four oxygenated methylenes [δ_C 64.2 (C-ara-5), δ_C 62.8 (C-xyl-5), δ_C 61.3 (C-glc''-6), and δ_C 61.1 (C-glc'-6)]; these sugars were determined to be two β-glucopyranoses, one α-xylopyranose, and one α-arabinopyranose, respectively. The oxygenated methine resonances of aglycone (C-3), glucose (C-glc''-2), and arabinose (C-ara-2 and C-ara-4) were detected at lower magnetic fields (δ_C 88.3, δ_C 83.7, δ_C 79.2, and δ_C 78.1) than the commonly detected chemical shift (δ_C 78, δ_C 75, δ_C 71, and δ_C 71).

This discrepancy, attributed to glycosidation-induced shifts, provided conclusive evidence confirming the precise positions of the glycosidic linkage.

In the gHMBC spectrum, one formyl [δ_H 9.61 (1H, s, H-30)] proton signal showed cross-peaks with the quaternary carbon signal δ_C 47.6 (C-20), and oxygenated methylenes [δ_H 3.51 (1H, d, J = 7.8 Hz, H-28a) and δ_H 3.16 (1H, d, J = 7.8 Hz, H-28b)] showed cross-peaks with the oxygenated quaternary carbon δ_C 85.6 (C-13) and oxygenated methine carbon δ_C 76.1 (C-16) signals. Furthermore, the four anomer proton signals of two glucoses [δ_H 5.36 (1H, d, J = 7.2 Hz, H-glc'-1) and δ_H 4.97 (1H, d, J = 7.8 Hz, H-glc''-1)], one xylose [δ_H 4.96 (1H, d, J = 7.2 Hz, H-xyl-1)], and one arabinose [δ_H 4.76 (1H, d, J = 5.2 Hz, H-ara-1)] showed cross-peaks with four oxygenated methine carbon signals [δ_C 88.3 (C-3), δ_C 83.7 (C-glc''-2), δ_C 79.2 (C-ara-2), and δ_C 78.1 (C-ara-4)]. This spectral evidence strongly suggests the positioning of arabinose at the C-3 of the aglycone, xylose at the C-glc''-2 of the glucopyranose moiety, and the two glucoses at the C-ara-2 and C-ara-4 of the arabinose moiety, respectively. Taken together, ardisiacrispin A (3β-O-[α-L-xylopyranosyl-(1→2)-O-β-D-glucopyranosyl-(1→4)-[O-β-D-glucopyranosyl-(1→2)]-α-L-arabinopyranosyl]-16α-hydroxy-13β,28-epoxyolean-30-al; **1**) was identified as the chemical structure of Compound **1** (Figure 1). Another study also investigated the spectroscopic parameters of Compound **1** [14]. Compound **1** was isolated from the *L. pumila* leaves for the first time in this study. Originally documented in 1987 from *Ardisia crispa*, Compound **1** has been recognized for its cytotoxic efficacy against diverse cancer cell lines, including NCI-H46; SF-268; MCF-7; melanoma WM793, HTB140, and A375 (skin panel); prostate cancer Du145 and PC3 and normal prostate epithelial PNT2 (prostate panel); colon cancer Caco2 and HT29; and HepG2 (gastrointestinal panel) liver cells [15–18].

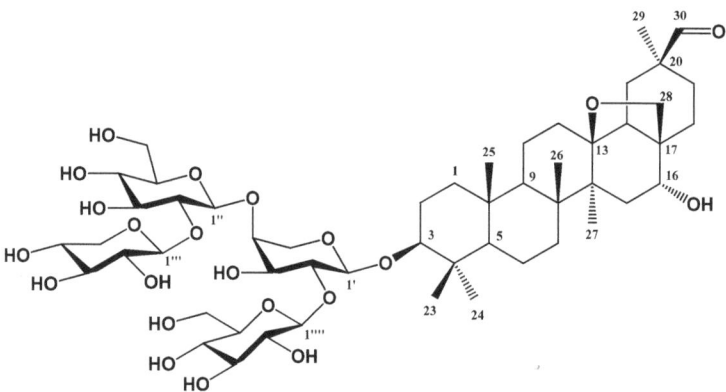

Figure 1. Chemical structure of ardisiacrispin A (**1**) extracted from *L. pumila* leaves.

3.3. Regulation of Cell Cycle by L. pumila

Apoptosis, an intricate mechanism of programmed cell death inherent to multicellular organisms, serves as a pivotal process for the elimination of undesirable and defective cells [19]. This orchestrated cellular death not only facilitates the removal of superfluous entities but also mitigates the risk of inciting undesirable inflammatory responses. Apoptosis is a ubiquitous phenomenon, actively participating during normal development and cellular turnover, as well as extending to various pathological conditions.

The cell cycle, a highly conserved mechanism, orchestrates the replication of eukaryotic cells. The regulation of cell death is intricately linked to genes governing cell cycle progression. Cumulative evidence has underscored the impact of cell cycle manipulation on modulating apoptosis reactions, contingent upon the specific cellular context [20,21].

Concentration-dependent reductions in the G0/G1, G2/M, and S phases were discerned in response to the *L. pumila* extract. Conversely, an elevation in the Sub G1 phase was observed upon treatment with the *L. pumila* extract. It is well documented that an augmentation in the Sub G1 phase corresponds to the onset of apoptosis. Thus, the observed elevation in the Sub G1 phase following treatment with the *L. pumila* extract can be attributed to the induction of apoptosis (Figure 2).

μg/mL	-	12.5	25	50	75	100
Sub G1	6.67 ± 0.88	6.00 ± 1.15	9.67 ± 0.33	14.00 ± 1.73**	37.33 ± 1.45**	47.33 ± 0.88**
G0/G1	53.00 ± 1.73	58.00 ± 1.00	55.00 ± 0.58	58.67 ± 0.67	49.00 ± 0.58	37.33 ± 1.20**
S	10.67 ± 0.33	8.33 ± 0.33	9.00 ± 1.15	7.67 ± 0.33*	4.00 ± 0.58**	5.33 ± 0.33**
G2/M	29.67 ± 0.88	28.00 ± 1.53	26.33 ± 0.88	20.00 ± 1.15**	10.00 ± 0.58**	10.67 ± 0.33**

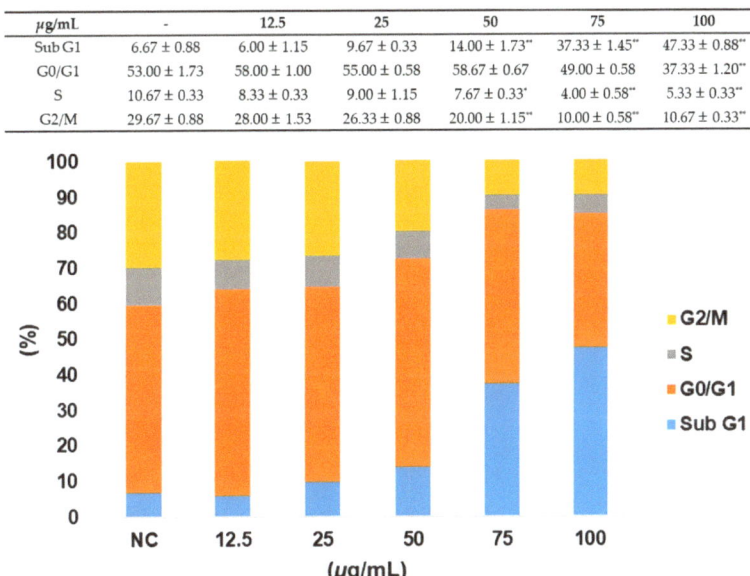

Figure 2. Regulation of cell cycle by *L. pumila* extract. A549 cells were treated with varying concentrations (0–100 μg/mL) of *L. pumila* extract for 24 h, and the effects were assessed using a Tali™ image-based cytometer. * $p < 0.05$, ** $p < 0.01$.

3.4. Evaluation of the Anti-Lung Cancer Efficacy of L. pumila and Ardisiacrispin A

Compound **1** induced a 50% cell death response in A549 cells at a concentration of 11.94 ± 1.14 μg/mL (Figure 3). Natural products are inexpensive and relatively safe compared with synthetics as extensive research studies have been conducted on the development of anti-cancer candidates for decades [22,23]. In particular, compared with dehydrocostuslactone (15 μg/mL) from *Aucklandia lappa*; resveratrol (2.04 ± 0.3 μg/mL) from grape; isochaihulactone (7.37 ± 2.03 μg/mL) from *Bupleurum scorzonerifolium*, ixerinoside (29 μg/mL), and ixerin Z (25 μg/mL); and 3-hydroxydehydroleucodin (15 μg/mL) from *Ixeris sonchifolia*, the compound we investigated in this study, ardisiacrispin A, revealed a significant inhibition efficacy (11.94 ± 1.14 μg/mL) against A549 cells [23–26]. Although this compound has a lower inhibition effect than the synthetics vincristine (1.81 ± 0.25 ng/mL) and paclitaxel (2.22 ± 0.43 ng/mL), *L. pumila* and ardisiacrispin A are anti-lung cancer candidates thanks to their advantage of having no or minor adverse effects [26].

Extracellular signal-regulated kinase (ERK) plays a crucial role in tumorigenesis [27]. ERK activity is associated with the promotion of apoptotic pathways, including the induction of mitochondrial cytochrome C release, caspase-8 activation, permanent cell cycle arrest, and autophagic vacuolization. The active state of ERK is characterized by its phosphorylated form [28–30]. Figure 4 demonstrates that the p-ERK/ERK ratio in ardisiacrispin A-treated cells was lower than the normal control. In addition, *L. pumila* revealed a dose-dependent decrease in the p-ERK/ERK ratio.

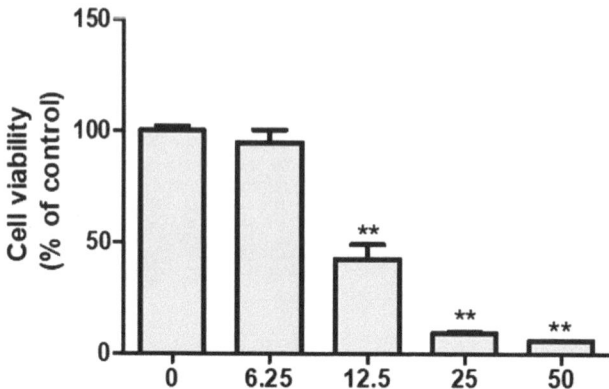

Figure 3. Cytotoxicity against A549 cells of ardisiacrispin A (**1**) from *L. pumila* leaves. Data are presented as the mean ± standard deviation. $n = 3$, ** $p < 0.01$.

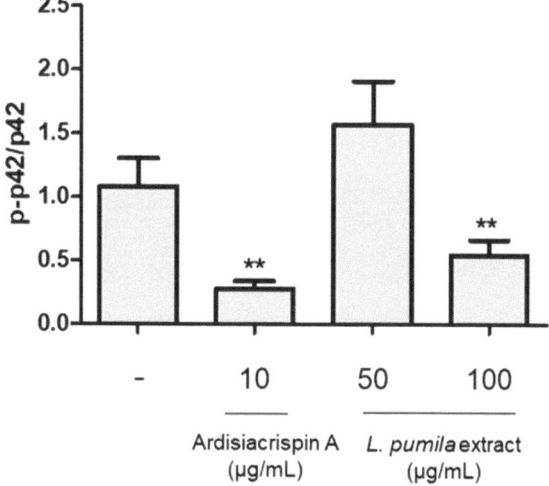

Figure 4. Changes in protein expression levels (p-ERK and ERK) associated with cell cycle regulation. Data are presented as the mean ± standard deviation. $n = 3$, ** $p < 0.01$.

4. Conclusions

The present study successfully extracted ardisiacrispin A (**1**) from *L. pumila* leaves, employing a comprehensive spectroscopic approach encompassing NMR, IR, and MS analyses. Compound **1** was isolated from *L. pumila* leaves for the first time in this study. *L. pumila* and its active compound, ardisiacrispin A (**1**), demonstrated potential in suppressing the proliferation and metastasis of lung cancer cells. This inhibitory effect is attributed to the modulation of oncogenic signaling pathways related to EGFR and FGER in lung cancer. Although further apoptosis studies are needed to investigate the involved apoptotic pathways, this is the first report to demonstrate the anti-lung cancer efficacy of ardisiacrispin A (**1**) and *L. pumila* extract, to the best of our knowledge. Consequently, these findings underscore the feasibility of utilizing ardisiacrispin A (**1**) and *L. pumila* extract as anti-lung cancer agents. To validate their efficacy as anti-lung cancer candidates, further investigations encompassing the elucidation of the mode of action and preclinical trials are imperative.

Author Contributions: Conceptualization, Y.-G.L., H.K. and S.C.K.; methodology, Y.-G.L. and T.H.K.; software, Y.-G.L.; validation, T.H.K. and J.E.K.; formal analysis, Y.-G.L.; investigation, Y.-G.L., T.H.K. and J.E.K.; resources, J.E.K.; data curation, J.E.K.; writing—original draft preparation, Y.-G.L.; writing—review and editing, H.K. and S.C.K.; visualization, Y.-G.L., J.E.K. and H.K.; supervision, H.K. and S.C.K.; project administration, S.C.K.; funding acquisition, S.C.K. All authors have read and agreed to the published version of the manuscript.

Funding: This research received no external funding.

Institutional Review Board Statement: Not applicable.

Informed Consent Statement: Not applicable.

Data Availability Statement: The data presented in this study are available on reasonable request from the corresponding authors. All data generated or analyzed during this study are included in the manuscript.

Conflicts of Interest: The authors declare no conflicts of interest.

References

1. Bray, F.; Ferlay, J.; Soerjomataram, I.; Siegel, R.L.; Torre, L.A.; Jemal, A. Global cancer statistics 2018: Globocan estimates of incidence and mortality worldwide for 36 cancers in 185 countries. *CA Cancer J. Clin.* **2018**, *68*, 394–424. [CrossRef]
2. Sung, H.; Ferlay, J.; Siegel, R.L.; Laversanne, M.; Soerjomataram, I.; Jemal, A.; Bray, F. Global cancer statistics 2020: Globocan estimates of incidence and mortality worldwide for 36 cancers in 185 countries. *CA Cancer J. Clin.* **2021**, *71*, 209–249. [CrossRef]
3. Herbst, R.S.; Morgenszstern, D.; Boshoff, C. The biology and management of non-small cell lung cancer. *Nature* **2018**, *553*, 446–454. [CrossRef]
4. Onoi, K.; Chihara, Y.; Uchino, J.; Shimamoto, T.; Morimoto, Y.; Iwasaku, M.; Kaneko, Y.; Yamada, T.; Takayama, K. Immune checkpoint inhibitors for lung cancer treatment: A review. *J. Clin. Med.* **2020**, *9*, 1362. [CrossRef]
5. Charmsaz, Z.; Collins, D.M.; Perry, A.S.; Prencipe, M. Novel strategies for cancer treatment: Highlights from the 55th IACR annual conference. *Cancer* **2019**, *11*, 1125. [CrossRef]
6. Debela, D.T.; Muzazu, S.G.; Heraro, K.D.; Ndalama, M.T.; Mesele, B.W.; Haile, D.C.; Kitui, S.K.; Manyazewal, T. New approaches and procedures for cancer treatment: Current perspectives. *SAGE Open Med.* **2021**, *9*, 20503121211034366. [CrossRef] [PubMed]
7. Chua, L.S.; Lee, S.Y.; Abdullah, N.; Sarmidi, M.R. Review on *Labisia pumila* (Kacip Fatimah): Bioactive phytochemicals and skin collagen synthesis promoting herb. *Fitoterapia* **2012**, *83*, 1322–1335. [CrossRef] [PubMed]
8. Bdullah, N.; Chermahini, S.H.; Suan, C.L.; Sarmidi, M.R. *Labisia pumila*: A review on its traditional, phytochemical and biological uses. *World Appl. Sci. J.* **2013**, *27*, 1297–1306.
9. Choi, H.; Kim, D.H.; Kim, J.W.; Ngadiran, S.; Sarmidi, M.R.; Park, C.S. *Labisia pumila* extract protects skin cells from photoaging caused by UVB irradiation. *J. Biosci. Bioeng.* **2010**, *109*, 291–296. [CrossRef] [PubMed]
10. Shuid, A.N.; Ping, L.L.; Muhammad, N.; Mohamed, N.; Soelaiman, I.N. The effects of *Labisia pumila* var. Alata on bone markers and bone calcium in a rat model of post-menopausal osteoporosis. *J. Ethnopharmacol.* **2011**, *133*, 538–542. [CrossRef] [PubMed]
11. Lee, Y.G.; Kang, K.W.; Hong, W.; Kim, Y.H.; Oh, J.T.; Park, D.W.; Ko, M.; Bai, Y.F.; Seo, Y.J.; Lee, S.M.; et al. Potent antiviral activity of *Agrimonia pilosa, Galla rhois*, and their components against SARS-CoV-2. *Bioorg. Med. Chem.* **2021**, *45*, 116329. [CrossRef] [PubMed]
12. Lee, Y.G.; Lee, H.; Ryuk, J.A.; Hwang, J.T.; Kim, H.G.; Lee, D.S.; Kim, Y.J.; Yang, D.C.; Ko, B.S.; Baek, N.I. 6-methoxyflavonols from the aerial parts of *Tetragonia tetragonoides* (pall.) kuntze and their anti-inflammatory activity. *Bioorg. Chem.* **2019**, *88*, 102922. [CrossRef] [PubMed]
13. Park, B.K.; Moon, H.R.; Yu, J.R.; Kook, J.; Chai, J.Y.; Lee, S.H. Comparative susceptibility of different cell lines for culture of toxoplasma gondii in vitro. *Korean J. Parasitol.* **1993**, *31*, 215–222. [CrossRef] [PubMed]
14. Ali, Z.; Khan, I.A. Alkyl phenols and saponins from the roots of *Labisia pumila* (kacip fatimah). *Phytochemistry* **2011**, *72*, 2075–2080. [CrossRef]
15. Jansakul, C.; Baumann, H.; Kenne, L.; Samuelsson, G. Ardisiacrispin a and b, two utero-contracting saponins from *Ardisia crispa*. *Planta Med.* **1987**, *53*, 405–409. [CrossRef] [PubMed]
16. Podolak, I.; Zuromska-Witek, B.; Grabowska, K.; Zebrowska, S.; Galanty, A.; Hubicka, U. Comparative quantitative study of ardisiacrispin a in extracts from *Ardisia crenata* sims varieties and their cytotoxic activities. *Chem. Biodivers.* **2021**, *18*, e2100335. [CrossRef]
17. Tian, Y.; Tang, H.F.; Qiu, F.; Wang, X.J.; Chen, X.L.; Wen, A.D. Triterpenoid saponins from *Ardisia pusilla* and their cytotoxic activity. *Planta Med.* **2009**, *75*, 70–75. [CrossRef]
18. Wen, P.; Zhang, X.M.; Yang, Z.; Wang, N.L.; Yao, X.S. Four new triterpenoid saponins from *Ardisia gigantifolia* stapf. And their cytotoxic activity. *J. Asian Nat. Prod. Res.* **2008**, *10*, 873–880. [CrossRef]
19. Portt, L.; Norman, G.; Clapp, C.; Greenwood, M.; Greenwood, M.T. Anti-apoptosis and cell survival: A review. *BBA-Mol. Cell Res.* **2011**, *1813*, 238–259. [CrossRef]

20. Evan, G.I.; Brown, L.; Whyte, M.; Harrington, E. Apoptosis and the cell cycle. *Curr. Opin Cell Biol.* **1995**, *7*, 825–834. [CrossRef]
21. Pucci, B.; Kasten, M.; Giordano, A. Cell cycle and apoptosis. *Neoplasia* **2000**, *2*, 291–299. [CrossRef] [PubMed]
22. Huang, C.Y.; Ju, D.T.; Chang, C.F.; Reddy, P.M.; Velmurugan, B.K. A review on the effects of current chemotherapy drugs and natural agents in treating non–small cell lung cancer. *Biomedicine* **2017**, *7*, 23. [CrossRef]
23. Kim, C.; Kim, B. Anti-cancer natural products and their bioactive compounds inducing ER stress-mediated apoptosis: A review. *Nutrients* **2018**, *10*, 1021. [CrossRef] [PubMed]
24. Park, J.-W.; Woo, K.J.; Lee, J.-T.; Lim, J.H.; Lee, T.-J.; Kim, S.H.; Choi, Y.H.; Kwon, T.K. Resveratrol induces pro-apoptotic endoplasmic reticulum stress in human colon cancer cells. *Oncol. Rep.* **2007**, *18*, 1269–1273. [CrossRef]
25. Chen, Y.L.; Lin, S.Z.; Chang, J.Y.; Cheng, Y.L.; Tsai, N.M.; Chen, S.P.; Chang, W.L.; Harn, H.J. In vitro and in vivo studies of a novel potential anticancer agent of isochaihulactone on human lung cancer A549 cells. *Biochem. Pharmacol.* **2006**, *72*, 308–319. [CrossRef]
26. Zhang, Y.C.; Zhou, L.; Ng, K.Y. Sesquiterpene lactones from *Ixeris sonchifolia* Hance and their cytotoxicities on A549 human non-small cell lung cancer cells. *J. Asian Nat. Prod. Res.* **2009**, *11*, 294–298. [CrossRef] [PubMed]
27. Guo, Y.J.; Pan, W.W.; Liu, S.B.; Shen, Z.F.; Xu, Y.; Hu, L.L. Erk/mapk signalling pathway and tumorigenesis. *Exp. Ther. Med.* **2020**, *19*, 1997–2007. [CrossRef] [PubMed]
28. Cagnol, S.; Chambard, J. ERK and cell death: Mechanisms of ERK-induced cell death—Apoptosis, autophagy and senescence. *FEBS J.* **2010**, *277*, 2–21. [CrossRef]
29. Park, M.T.; Choi, J.A.; Kim, M.J.; Um, H.D.; Bae, S.W.; Kang, C.M.; Cho, C.W.; Kang, S.; Chung, H.Y.; Lee, Y.S.; et al. Suppression of extracellular signal-related kinase and activation of p38 MAPK are two critical events leading to caspase-8 and mitochondria-mediated cell death in phytosphingosine-treated human cancer cells. *J. Biol. Chem.* **2003**, *278*, 50624–50634. [CrossRef]
30. Yue, J.; LÓpez, J.M. Understanding MAPK signaling pathways in apoptosis. *Int. J. Mol. Sci.* **2020**, *21*, 2346. [CrossRef]

Disclaimer/Publisher's Note: The statements, opinions and data contained in all publications are solely those of the individual author(s) and contributor(s) and not of MDPI and/or the editor(s). MDPI and/or the editor(s) disclaim responsibility for any injury to people or property resulting from any ideas, methods, instructions or products referred to in the content.

Article

Antioxidant Activity and Anticarcinogenic Effect of Extracts from *Bouvardia ternifolia* (Cav.) Schltdl.

Carmen Valadez-Vega [1,*], Olivia Lugo-Magaña [2,*], Lorenzo Mendoza-Guzmán [1], José Roberto Villagómez-Ibarra [3], Raul Velasco-Azorsa [4], Mirandeli Bautista [5], Gabriel Betanzos-Cabrera [6], José A. Morales-González [7] and Eduardo Osiris Madrigal-Santillán [7]

1 Área Académica de Medicina, Instituto de Ciencias de la Salud, Universidad Autónoma del Estado de Hidalgo, Ex-Hacienda de la Concepción, Tilcuautla, San Agustín Tlaxiaca 42080, Mexico; lencho_1095@hotmail.com
2 Preparatoria Número 1, Universidad Autónoma del Estado de Hidalgo, Av. Benito Juárez S/N, Constitución, Pachuca de Soto 42060, Mexico
3 Área Académica de Química, Instituto de Ciencias Básicas e Ingeniería, Universidad Autónoma del Estado del Hidalgo, Ciudad del Conocimiento, Mineral de la Reforma 42184, Mexico; jrvi@uaeh.edu.mx
4 Área Académica de Biología, Instituto de Ciencias Básicas e Ingeniería, Universidad Autónoma del Estado del Hidalgo, Ciudad del Conocimiento, Mineral de la Reforma 42184, Mexico; raul_velasco@uaeh.edu.mx
5 Área Académica de Farmacia, Instituto de Ciencias de la Salud, Universidad Autónoma del Estado de Hidalgo, Ex-Hacienda de la Concepción, Tilcuautla, San Agustín Tlaxiaca 42080, Mexico; mibautista@uaeh.edu.mx
6 Área Académica de Nutrición, Instituto de Ciencias de la Salud, Universidad Autónoma del Estado de Hidalgo, Ex-Hacienda de la Concepción, Tilcuautla, San Agustín Tlaxiaca 42080, Mexico; gbetanzo@uaeh.edu.mx
7 Laboratorio de Medicina de Conservación, Escuela Superior de Medicina, Instituto Politécnico Nacional, México, Plan de San Luis y Díaz Mirón, Col. Casco de Santo Tomás, Del. Miguel Hidalgo, Ciudad de México 11340, Mexico; jmorales101@yahoo.com.mx (J.A.M.-G.); eomsmx@yahoo.com.mx (E.O.M.-S.)
* Correspondence: marynavaladez@hotmail.com (C.V.-V.); olivia_lugo@uaeh.edu.mx (O.L.-M.)

Citation: Valadez-Vega, C.; Lugo-Magaña, O.; Mendoza-Guzmán, L.; Villagómez-Ibarra, J.R.; Velasco-Azorsa, R.; Bautista, M.; Betanzos-Cabrera, G.; Morales-González, J.A.; Madrigal-Santillán, E.O. Antioxidant Activity and Anticarcinogenic Effect of Extracts from *Bouvardia ternifolia* (Cav.) Schltdl. *Life* **2023**, *13*, 2319. https://doi.org/10.3390/life13122319

Academic Editors: Seung Ho Lee and Stefania Lamponi

Received: 31 October 2023
Revised: 30 November 2023
Accepted: 8 December 2023
Published: 10 December 2023

Copyright: © 2023 by the authors. Licensee MDPI, Basel, Switzerland. This article is an open access article distributed under the terms and conditions of the Creative Commons Attribution (CC BY) license (https://creativecommons.org/licenses/by/4.0/).

Abstract: According to the available ethnobotanical data, the *Bouvardia ternifolia* plant has long been used in Mexican traditional medicine to relieve the symptoms of inflammation. In the present study, the cytotoxic effect of extracts obtained from the flowers, leaves and stems of *B. ternifolia* using hexane, ethyl acetate (AcOEt) and methanol (MeOH) was evaluated by applying them to the SiHa and MDA-MB-231 cancer cell lines. An MTT reduction assay was carried out along with = biological activity assessments, and the content of total phenols, tannins, anthocyanins, betalains and saponins was quantified. According to the obtained results, nine extracts exhibited a cytotoxic effect against both the SiHa and MDA lines. The highest cytotoxicity was measured for leaves treated with the AcOEt (ID$_{50}$ of 75 μg/mL was obtained for MDA and 58.75 μg/mL for SiHa) as well as inhibition on ABTS•$^+$ against DPPH• radical, while MeOH treatment of stems and AcOEt of flowers yielded the most significant antioxidant capacity (90.29% and 90.11% ABTS•$^+$ radical trapping). Moreover, the highest phenolic compound content was measured in the stems (134.971 ± 0.294 mg EAG/g), while tannins were more abundant in the leaves (257.646 mg eq cat/g) and saponins were most prevalent in the flowers (20 ± 0 HU/mg). Screening tests indicated the presence of flavonoids, steroids, terpenes and coumarins, as well as ursolic acid, in all the studied extracts. These results demonstrate the biological potential of *B. ternifolia*.

Keywords: antioxidant; *Bouvardia ternifolia*; anticarcinogenic; bioactive

1. Introduction

In Mexico, traditional medicine is still widely used, and such ancestral knowledge of medicinal plants is passed from one generation to another. Although different plant parts are used, depending on the species and the target ailment, leaves and flowers predominate in the traditional recipes, some of which also require the stem or root. Throughout history, herbal remedies have been employed to alleviate symptoms and enhance human health.

Even today, in numerous regions worldwide, herbal therapy remains the primary, and at times the sole, option [1].

Medicinal plants are consumed directly or can be prepared as infusions or homeopathic remedies either as a complement to Western medicine or as a stand-alone treatment. While medicinal plants have long been used by indigenous peoples across the globe, their value is increasingly being recognized by the medical profession. Accordingly, evidence of their effectiveness in treating skin conditions, hair loss, herpes, scabies, toothaches and headaches, as well as diseases that weaken the circulatory, digestive, endocrine, nervous, reproductive, respiratory and urinary systems, as well as diseases of cultural affiliation, has grown considerably in recent decades [2,3].

In this context, the *Rubiaceae* family is particularly relevant, as it comprises approximately 650 genera and more than 13,500 species distributed throughout the world, many of which are used in traditional medicine to alleviate headaches and pain during childbirth, as well as lessen the symptoms of autoimmune diseases [4]. In Mexican traditional medicine, the *Bouvardia* genus features most prominently, owing to its anti-inflammatory properties. It can be either ingested or applied as an infusion or compress, to treat intoxication, colic, diarrhea, erysipelas and insect bites, among other conditions [2,5,6]. The firecracker bush or scarlet bouvardia, trumpet bush or clove bush is a 0.5 to 1.5 m. shrub with lustrous, oval, dark green leaves and tubular, bugle-shaped, red flowers, 5 cm long, with the edge flared into four segments. The flowers are arranged in clusters at the ends of numerous erect branches [7].

Consequently, this genus has been subjected to a significant number of phytochemical studies, the findings of which confirm the presence of peptide compounds, such as bouvardin, as well as deoxy-bouvardin and its methylated derivatives, in different plant components [8]. Some secondary metabolites, such as ursolic acid (triterpene) (UA), have also been reported, and flavonoids (such as rutin, quercetin and kaempferol) have been isolated from the aerial part [9,10] of the root, where triterpenes, oleanolic acid (OA) and ursolic acid are also found in high concentrations [11].

The available data also indicate that *B. ternifolia* extracts and different compounds isolated from the plant exhibit cytotoxic activity against some cell lines, causing cell cycle arrest and the inhibition of protein synthesis. In traditional *B. ternifolia* medicinal practices, the plant's upper parts, encompassing the leaves, stems and flowers, are employed to address conditions such as genital ulcers, dysentery, rabies, cold sweat pains, tumors, fever, and joint pain, and are utilized as a fortifying agent. In addition, they have sedative, analgesic and antispasmodic properties, and are applied for the treatment of snake, bee, scorpion and spider bites [12].

Their cytotoxic effect on some malignant cell lines has also been demonstrated, along with the enzymatic inhibition of acetylcholinesterase and anti-inflammatory, analgesic, sedative and hepatoprotective activities. The evidence yielded by in vivo studies further indicates that the *B. ternifolia* extracts can reduce inflammation of the pinna, as well as impart a nootropic effect, thus protecting the nervous system [8,9,13,14]. The aim of the current study was to analyze their antioxidant activity and anticarcinogenic effects on breast cancer cell lines (MDA-MB-231) and cervical cancer cell lines (SiHa). Extracts of the stems, leaves, and flowers of the plant *Bouvardia ternifolia* (Cav.) Schltdl. were evaluated using three organic solvents.

2. Materials and Methods

2.1. Plant Material

The plant material required for the current study was collected in the town Epazoyucan in Hidalgo State (north latitude $20°$, $01'$ and $05''$, west longitude $98°$, $08'$ and $03''$) and all specimens were placed on absorbent paper for 20 days to dry at room temperature (20–26 °C). The plant was submitted for identification to the herbarium of the Autonomous University of the State of Hidalgo for classification under the accession number 010. Next, the plant parts (flowers, leaves and stems) were separated and were ground in an electric grinder (analytical mill, 4301-00, Cole Palmer, Vernon Hills, IL, USA) until the powder was sufficiently fine to pass through a 40-mesh. The samples were stored in bags to be preserved until their utilization.

2.2. Extract Preparation

The required extracts were obtained from 148 g of flowers, 1450 g of leaves and 1.28 g of ground stems via maceration for 15 days at room temperature in darkness with hexane, ethyl acetate (AcOEt) and methanol (MeOH), followed by filtering on Whatman number 2 paper. The resulting samples were rotaevaporated to dryness (BÜCHI Walter Bath B-480) at 40 °C (330, 240 and 337 mBar), after which the extracts were stored in opaque vials at room temperature until use.

2.3. Antioxidant Capacity

ABTS•+ assay: The ABTS•+ assay was performed according to the methodology described in the extant literature [15], using 2,2′ azinobis-(3-ethylbenzothiazoline)-6-sulfonic acid (ABTS•+, Sigma Chemical Co., St Louis, MO, USA) and 6-hydroxy-2, 5, 7, 8-tetramethylcroo-2-carboxylic acid (Trolox, Sigma Chemical Co., St Louis, MO, USA) as the standards. The extracts from *B. ternifolia* were prepared at 100 mg/mL in ethanol. To determine the Trolox concentration (TEAC), 900 µL of the ABTS•+ solution was added to 100 µL of the extract and was reacted for 5 min in darkness, after which the absorbance was measured in a BioTek Epoch spectrophotometer at $\lambda = 734$ nm. The results were reported as the percent entrapment and Trolox equivalents in mg/g of the sample (TEAC mg/g) using Trolox as the standard.

DPPH assay: For the DPPH assay, the method developed by Schenk and Brown [16] based on the reduction of the 2,2-diphenyl-1-picrylhydrazyl radical (DPPH• Sigma Chemical Co., St Louis, MO, USA) was used. The extracts were prepared in ethanol at 100 mg/mL, whereby 900 µL of DPPH• solution was added to 100 µL of each extract and the sample was left to react in darkness for 60 min before reading the absorbance at $\lambda = 734$ nm (BioTek Epoch instrument, Santa Clara, CA, USA). The results were reported as the Trolox equivalent antioxidant capacity in mg/g of the sample (TEAC mg/g) using a calibration curve with Trolox as the standard.

2.4. Phytochemical Analysis

Total phenol content: For total phenol determination, the spectrophotometric method described by Singleton [17] was employed. Extracts from *B. ternifolia* were used. To determine the total phenolic content, 100 µL of a 1 mg/mL sample from each extract, 500 µL of Folin's solution and 400 µL of Na_2CO_3 were added. The reaction mixture was allowed to react in darkness for 30 min and then measured at a wavelength of 765 nm using a microplate reader (BioTek Epoch instrument). The results were reported in mg gallic acid equivalents per gram of sample (mg EGA/g sample), using gallic acid as the standard (Sigma Chemical Co., St Louis, MO, USA).

Tannin content: The extracts were prepared in methanol at a concentration of 4 mg/mL, by shaking for 1 h. Then, 0.5 mL of the extract was mixed with 2.5 mL of a 0.5% vanillin solution (Sigma Chemical Co., St. Louis, MO, USA). It was reacted for 45 min at room temperature and the absorbance was measured at $\lambda = 500$ nm (BioTek Epoch instrument). A calibration curve was performed using (+) catechin (Sigma Chemical Co., St. Louis, MO, USA) as a standard and the results were expressed as mg CATE/g (milligram catechin equivalents per gram of sample) [18].

Anthocyanin content: For determining the anthocyanin content, organic extracts (0.3 mg/mL) were left to react in acidified ethanol (0.2%) overnight in darkness, after which the samples were filtered and diluted in ethanol-HCl at 4 °C. The total content of monomeric anthocyanins was quantified using the differential pH method (109), for which extract solutions were prepared at a pH of 1 (KCl 0.1 M) and a pH of 4.5 (CH_3COOH/CH_3COO-). Finally, the absorbance was measured in the $\lambda = 515$–700 nm wavelength range (BioTek Epoch instrument) [19].

Betalain content: The betalain content was determined using the spectrophotometric method described by Elbe [20]. For this purpose, betalains were obtained from each *B. ternifolia* extract in phosphate buffer at a pH if 6.5 (17 mg/mL), after which the samples were vortexed and centrifuged at 5000 rpm for 15 min at 4 °C. Next, the supernatant was filtered in Phenomenex 0.45 μm, and the absorbance was measured at λ = 538 and 483 nm (Epoch-BioTek Instrument). The betacyanin and betaxanthin content was quantified using the method described in the extant literature [21]. The betacyanin content was expressed as mg betanin/100 g sample (mg BE/100 g), and the betaxanthin content was expressed as mg vulgaxanthin-I/100 g sample (mg VE/100 g) [21].

Saponin content: To determine the saponin content, the extracts were subjected to the methodology described by Valadez, extracted for 1 h from 10 mg of the extract sample using 85:15 (%) methanol–water solution. The solvents were removed via evaporation, and the extracted saponins were diluted in a NaCl solution. Employing a serial dilution method with human erythrocytes type O in a U-shaped microtiter 96-well plate, the saponin-containing solution underwent a 2-fold serial dilution. The volume of each sample in the wells was adjusted to 50 μL with NaCl (0.9%), and the resulting diluted samples were combined with 50 μL of a 4% erythrocyte suspension. The reaction mixture underwent a 1 h incubation at room temperature, and the maximum dilution demonstrating hemolysis was subsequently observed. The analyses were conducted in triplicate, and the results were quantified and reported as hemolytic units [22].

2.5. Cytotoxicity Assay

The assay was performed as described by Valadez-Vega [23]. The MDA-MB-231 (human breast adenocarcinoma) and SiHa (cervix squamous cancer cells) cell lines used for this purpose were obtained from the American Type Culture Collection (ATCC, Rockville, MD, USA). The cells were propagated in Dubelco's Modified Eagle Medium (DMEM) supplemented with 10% fetal bovine serum (FBS) and 0.1% antibiotic (a combination of streptomycin and penicillin). Incubation was carried out at 37 °C in a humidified atmosphere with 5% CO_2. Passages between 10 to 25 were routinely employed for both cell lines [24].

The analysis was performed using the colorimetric method described by Mosmann, focusing on the mitochondrial functionality of the treated cells [25]. MTT assay is a colorimetric test that utilizes a tetrazolium dye known as 3-(4,5-dimethylthiazol-2-yl)-2-5-diphenyltetrazolium bromide (MTT) to assess the viability of cell lines. This assay relies on the capacity of active mitochondria to facilitate the conversion of MTT into a solid formazan, the amount of which can be quantified via spectrophotometric means. Prior to the measurements, both cell lines were cultured in 96-well microplates (1×10^4 cell/well) in a culture medium containing fetal bovine serum and antibiotics. After 24 h incubation, the culture medium was removed and was replaced by the extracts in 0–2000 μg/mL concentration. Following further 24 h incubation, the solution containing the cells was replaced with MTT (5 mg/mL) and was incubated at 37 °C for 3 h. Then, the culture medium was removed and 200 μL of dimethyl sulfoxide was added to each well to dissolve the formazan compound produced by the cells. The absorbance of each well was measured at λ = 540 nm (BioTek Epoch instrument) and the cell viability percentage was calculated considering the blank as 100% viability.

2.6. Statistical Analyses

An ANOVA and Tukey's test ($p < 0.05$) were conducted to determine the differences between extracts in terms of the antioxidant capacity, scavenging capacity, phenol content, tannin content and saponin content. All statistical analyses were performed using StatGraphics Centurion version 19.1 (StatGraphics, The Plains, VA, USA).

3. Results

3.1. Antioxidant Capacity

The antioxidant capacity of the studied *B. ternifolia* extracts from flowers, leaves and stems was evaluated by applying the ABTS•+ and DPPH• assays. The obtained findings are depicted in Figure 1 and reported in Table 1, revealing significant differences between these two approaches with respect to all three morphological parts ($p < 0.005$) with the ABTS•+ assay consistently yielding greater values.

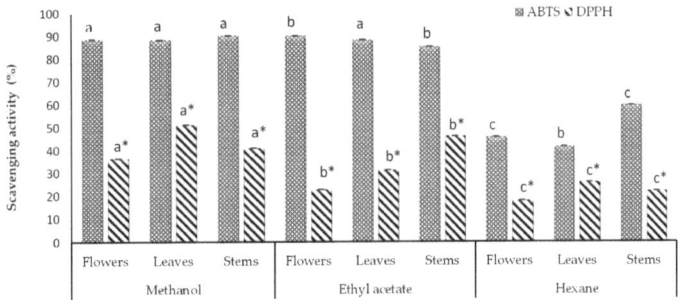

Figure 1. Scavenging capacity of *B. ternifolia* extracts based on the findings yielded by ABTS•+ and DPPH• assays based on 6.66 mg/mL extract concentration. The statistically significant differences ($p < 0.05$) in values obtained from different plant parts based on the ABTS•+ and DPPH• assay are denoted with a, b and c, and a*, b* and c*, respectively. The average of three independent assays ± standard deviation is shown.

Table 1. Antioxidant capacity of *B. ternifolia* extracts.

Solvent	Parts of Plants	Antioxidant Capacity	
		DPPH•	ABTS•+
		mg TEAC/g	
Methanol	Flowers	16.909 ± 0.040 [a]	22.166 ± 0.043 [a]
	Leaves	23.142 ± 0.054 [a]	22.123 ± 0.043 [a]
	Stems	18.875 ± 0.054 [a]	22.551 ± 0.043 [a]
Ethyl acetate	Flowers	11.168 ± 0.060 [b]	22.509 ± 0.043 [a]
	Leaves	14.737 ± 0.040 [b]	22.08 ± 0.042 [a]
	Stems	21.013 ± 0.045 [b]	22.737 ± 0.025 [a]
Hexane	Flowers	9.134 ± 0.040 [c]	12.165 ± 0.049 [b]
	Leaves	12.522 ± 0.015 [c]	11.194 ± 0.042 [b]
	Stems	10.866 ± 0.040 [c]	15.394 ± 0.042 [b]

ABTS•+ and DPPH• results are expressed as Trolox equivalent antioxidant activity (TEAC). The average of three independent replicates ± standard deviation is reported and the a, b and c superscripts indicate statistically significant differences ($p < 0.05$) across different plants depending on the solvent used.

3.2. Total Phenol, Tannin and Saponin Content

The total phenol, tannin and saponin content in the extracts obtained from the *B. ternifolia* flowers, leaves and stems was determined using methanol, ethyl acetate and hexane as solvents, and the findings are reported in Table 2.

Table 2. The total phenol, tannin and saponin content in the extracts obtained from the *B. ternifolia* flowers, leaves and stems.

Parts of Plant	Solvent	Phenols (mg GAE/g)	Tannins (mg CATE/g)	Saponins (HU/mg)
Flowers	Methanol	78.794 ± 0.294 [a]	43.721 ± 5.747 [b]	20
	Ethyl acetate	37.618 ± 0.304 [b]	34.143 ± 4.619 [a]	ND
	Hexane	51.147 ± 0.284 [c]	37.974 ± 1.437 [ab]	ND
Leaves	Methanol	133.402 ± 0.34 [a]	134.717 ± 6.789 [a]	ND
	Ethyl acetate	71.441 ± 0.294 [b]	257.646 ± 27.201 [b]	ND
	Hexane	43.696 ± 0.170 [c]	32.706 ± 2.991 [c]	ND
Stems	Methanol	117.324 ± 0.29 [a]	58.089 ± 2.488 [a]	ND
	Ethyl acetate	134.971 ± 0.89 [b]	61.920 ± 3.616 [a]	ND
	Hexane	8.794 ± 0.09 [c]	32.706 ± 5.982 [b]	ND

Total phenols are expressed in gallic acid equivalents (GAE), tannins are expressed in catechin equivalents (CATE) and saponins are presented in hemolytic units per milligram (HU/mg). ND: Not detected. The average of three independent replicates ± standard deviation is reported. The a, b and c superscripts indicate statistically significant differences ($p < 0.05$) across different plants depending on the solvent used.

3.3. Cytotoxicity of the Extracts Obtained from the B. ternifolia Flowers, Leaves and Stems against MDA and SiHa Cells

As shown in Figures 2 and 3, the extracts have a cytotoxic effect on the MDA and SiHa cell lines, which is more pronounced in the SiHa case. As can be seen from the tabulated results, while all extracts had a dose-dependent effect, the extracts from the leaves and stems obtained using MeOH and AcOEt exhibited the highest cytotoxicity on the SiHa cell line, while the lowest cytotoxicity was measured in the extracts from flowers and stems obtained using hexane.

It can also be concluded from Figure 2A that the flowers extract in ethyl acetate had the greatest cytotoxic effect on the SiHa cell line, while the leaf extract obtained using AcOEt exhibited the highest inhibition of cell viability (Figure 2B) and the methanolic stem extract showed the greatest effect on this cell line (Figure 2C).

The corresponding findings related to the MDA cell line are shown in Figure 3, revealing that the flower extract obtained with ethyl acetate has the greatest cytotoxic effect on the SiHa cell line (Figure 3A), while the AcOEt- and MeOH-based extracts from the leaves were the most effective in reducing cell viability (Figure 3B). On the other hand, the hexane extract caused the greatest inhibition of cell viability at high concentrations (Figure 3C).

When the same tests were performed with the MDA cell line, the ethyl acetate extract presented the greatest inhibition of cell viability irrespective of the concentration used.

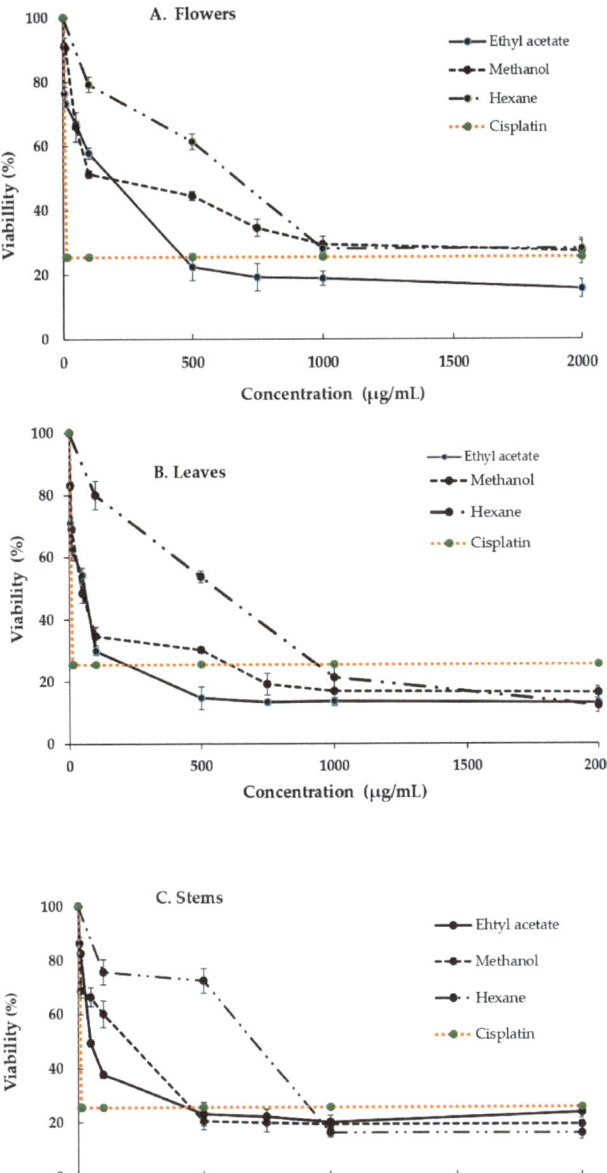

Figure 2. Cytotoxic effect of *B. ternifolia* extracts on the SiHa cell line (**A**) Flowers, (**B**) Leaves and (**C**) stems in three different solvents. Cells were exposed for 24 h to different concentrations of extracts, and, after incubation, cell viability was measured using the MTT technique. All values are expressed as the mean ± standard deviation of three independent experiments.

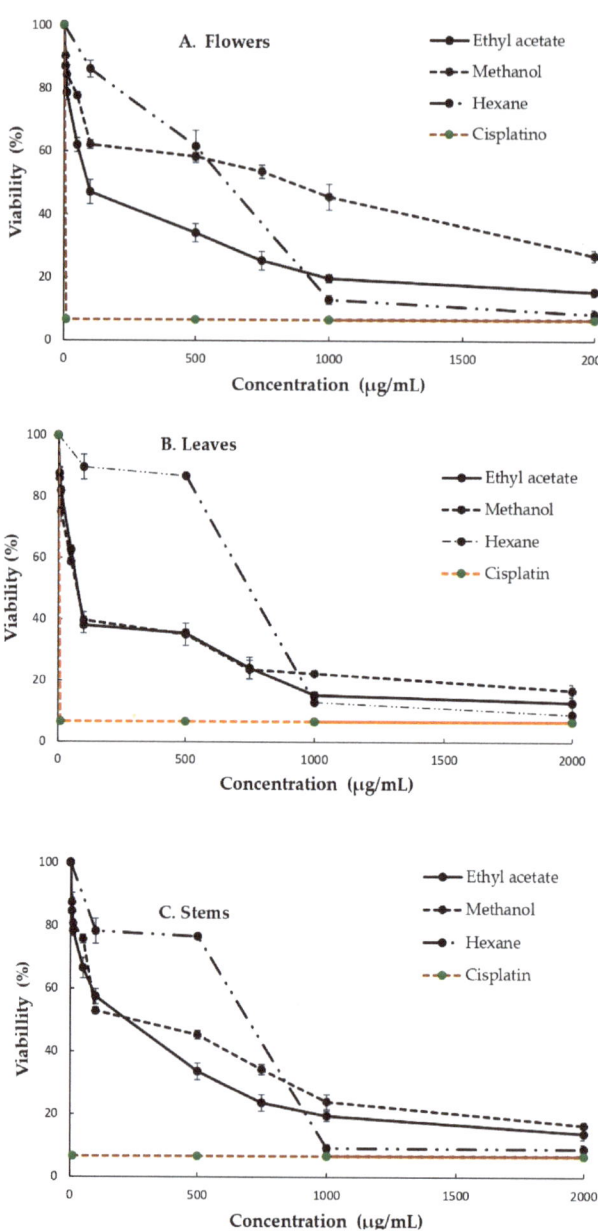

Figure 3. Cytotoxic effect of *B. ternifolia* extracts on the MDA cell line (**A**) Flowers, (**B**) Leaves and (**C**) stems in three different solvents. Cells were exposed for 24 h to different concentrations of extracts, and, after incubation, cell viability was measured using the MTT technique. All values are expressed as the mean ± standard deviation of three independent experiments.

4. Discussion

The results obtained in the present study indicate that the ABTS•+ method shows a greater scavenging capacity than DPPH• for the MeOH, AcOEt and hexane extracts from different parts of the *B. ternifolia* plant. The ABTS•+ assay produced greater values when applied to the extracts: the ABTS•⁺ method was considered to show higher sensitivity when applied to plant extracts containing hydrophilic, lipophilic and highly pigmented antioxidant compounds compared to the DPPH method [26].

Mosquera, focusing on species belonging to the families *Euphorbiaceae* and *Asteraceae*, found significantly higher scavenging percentages when MeOH extracts were tested in DPPH• above 50%; similar results were found in this investigation (22.13–51.32%), where a value of more than 25% is considered indicative of active antioxidants [27]. In general, highly pigmented and hydrophilic antioxidants, such as those at the focus of the current investigation, tend to have a better response in ABTS•⁺ assays than in DPPH• [28], suggesting the presence of phenolic acids, flavonoids and other polyphenols that are closely related to the antioxidant capacity [29,30]. Still, it is worth noting that the DPPH• technique is more selective, as ABTS•⁺ does not react with flavonoids that lack hydroxyl groups in the B-ring, or with aromatic acids that contain a single hydroxyl group [31]. Irrespective of these differences, in the present study, the antioxidant capacity and the content of phenolic compounds obtained using both methods correlated with the content of total phenolic compounds [32].

When the ABTS•⁺ assay was adopted, the greatest amount of total phenols in the AcOEt extracts from the *B. ternifolia* stems and MeOH from the leaves was measured, along with the highest antioxidant activity, supporting the previous reports related to the extracts of the same polarity of *Palicourea guianensis* [33]. As expected, AcOEt extracts from *B. ternifolia* flowers and hexane from the stems presented a lower phenol content and antioxidant capacity, as the total phenol content is closely related to the antioxidant activity [34]. The chemical complexity of plant extracts makes it difficult to explain and interpret antioxidant activity.

Due to the complexity of the oxidation–antioxidation processes, it is obvious that no single test method is able to provide a complete picture of the antioxidant profile of a sample under study [35].

The ABTS•⁺ establishing the relationship between the total phenol content and antioxidant activity was compared to DPPH•. As noted by Javanmardi et al. [36], the antioxidant capacity of phenolic compounds mainly stems from their redox properties, which play an important role in neutralizing free radicals at the cellular level. Accordingly, in several members of the *Rubiaceae* family—such as *Palicourea petiolaris wemh*, *Palicourea andaluciana* and *Palicourea thyrsiglora*—antioxidant activity has already been confirmed [27]. Given the intricate nature of plant extracts, elucidating antioxidant activity remains a complex task. It is evident that no single test method can comprehensively depict the complete antioxidant profile of a given sample. The ABTS•+ assay proves to be particularly valuable in establishing the relationship between total phenol content and antioxidant activity, surpassing the capabilities of the DPPH• method. This underscores the significance of the redox properties of phenolic compounds in neutralizing free radicals at the cellular level, as evidenced by previous studies in certain members of the *Rubiaceae* family. It is worth noting that, as of now, there are no existing reports on the total phenols in species of the genus *Bouvardia*.

In the present study, the MeOH and AcOEt leaf extracts contained a higher amount of tannins, 134.71 and 257.64 mg CATE/g, respectively. These values are lower than those reported for *Simira mexicana* and *Randia echinocarpa* leaf extracts, namely 7.58 and 1.06 mg CATE/g, respectively. However, phenol content is not indicative of tannin production, as phenols can give rise to other compounds such as phenolic acids, stilbenes, lignans, phenolic alcohols and flavonoids [37,38].

Tannins are found in all organs or parts of the plant (stems, wood, leaves, seeds and domes) and can contribute to 2–7% of the fresh weight of the plant, depending on the species, climate, soil, temperature and other factors [38].

In the present study, when the saponin assay was employed, only the MeOH extract obtained from flowers showed hemolytic activity at 20 HU/mg, which was expected as saponins are amphipathic and glycosidic in nature. Accordingly, they can be found in solvents of higher polarity [39,40]. Previously, Giraldo and Ramírez reported that extracts with higher polarity have a higher saponin content for *Palicourea guianensis* (Rubiaceae) leaves, supporting the findings reported in this work [33]. On the other hand, in the 250–370 nm and 500–545 nm wavelength ranges, the anthocyanin quantification method failed to provide any evidence of these compounds [41].

Several factors can influence anthocyanin degradation, including temperature, light, solvents, storage time and pH [21]. Therefore, the absence or low amount of betalains in the studied extracts is expected given that these compounds have only been observed in families of the order Caryophyllales, and *Bouvardia* belongs to the order Gentianales [42].

Compounds derived from plants, along with their semisynthetic and synthetic counterparts, constitute a prominent reservoir of pharmaceuticals for treating human diseases. Within the realm of cancer therapy, a key focus lies in identifying plant proteins that exhibit robust cytotoxic activity while maintaining low toxicity, and that exert diverse mechanisms of action on tumors. The cell viability results revealed the greater cytotoxic effect of *B. ternifolia* leaf and stem extracts compared to flower extracts irrespective of the solvent used or the cell line. Still, the MeOH leaf extract had the greatest cytotoxic effect on the MDA-MB-231 cell line, whereas the SiHa cell line was more susceptible to the cytotoxicity of the MeOH and AcOEt leaf extracts. On the other hand, the AcOEt extracts from the flowers and leaves and MeOH extract from the stems were the most inhibitory to the viability of the SiHa cell line.

Currently, no cytotoxicity studies on *B. ternofolia* have been reported; however, Rupachandra and Sarada studied= the seeds of *Borreria hispida*, a member of the Rubiaceae family, which revealed that the seeds have cytotoxic activity against lung (A549) and cervical (HeLa) cancer cell lines [43].

Within the Rubaceae family, Thai noni/Yor (*Morinda citrifolia* Linn.) were extracted using several methods and evaluated against human cancer cell lines: KB (human epidermoid carcinoma), HeLa (human cervical carcinoma), MCF-7 (human breast carcinoma) and HepG2 (human hepatocellular carcinoma) cell lines, as well as a Vero (African green monkey kidney) cell line, employing the MTT colorimetric method, and the dichloromethane extract of the leaves had a higher inhibitory effect on the HeLa cells [44].

However, as the cytotoxic activity of *B. ternifolia* on the cervical carcinoma cell line SiHa has never been reported, direct comparisons of the results obtained in this work are not possible.

In the MDA-MB-231 cells, the greatest inhibition of viability was achieved by the hexane extracts from flowers and stems and AcOEt extracts obtained from leaves. In other studies, the cytotoxic activity of *Chiococca alba*, *Hamelia patens* extracts showed no cytotoxic activity in Hep-G2 and MDA-MB-231 cells, while *Posokeria latifolia* extracts yielded 10.56% inhibition [45]. The ID_{50} of AcOEt and MeOH leaf extracts (with the values of 75.66 and 72.87 µg/mL, respectively) reflected higher cytotoxic activity. Similar results were observed for compounds isolated from the branches and leaves of *Heinsa crinita*, indicating cytotoxic activity in HL-60, SMMC-7721, A-549, MCF-7 and SW-480 cells, with ID_{50} in the 3.11–20.12 µg/mL range [46].

MeOH extracts from the leaves and branches of *Pavetta indica* L. were shown by Nguyen et al. to exhibit a cytotoxic effect on MDA-MB-231 cells, whereby $ID_{50} = 25.2$ µg/mL was measured at 24 h. Likewise, extracts from the bark of *Hymenodictyon excelsum*, belonging to the Rubiaceae family, had significant cytotoxic effects on both MCF-7 and MDA-MB-231 cells, with an ID_{50} of 80 and 440 µg/mL, respectively [47].

In previous studies, *Morinda citrifolia* Linn extract in ethyl acetate also showed a cytotoxic effect on MCF-7 and MDA-MB-231 cells with an ID_{50} of 25 and 35 µg/mL, as well as increased apoptosis in these cell lines, while it arrested the cell cycle in the G1/S phase in MCF-7 and G0/G1 in MDA-MB-231 -MB-231 [43]. According to the prior analyses of bouvardin (BVB), deoxybouvardin and its methylated derivatives, which have already been identified and isolated from *B. ternifolia*, are capable of inducing the proliferation of Chinese hamster ovary (CHO) cells, in which BVB reduced the ability to pass through the cell cycle [13]. Furthermore, BVB is effective in inhibiting protein synthesis in leukemia (P388) and melanoma (B16) cells [48].

For future investigations, it is proposed to explore the specific bioactive compounds within the extracts demonstrating cytotoxic effects, employing techniques such as chromatography and spectroscopy to isolate and identify these compounds. Additionally, mechanistic studies on cytotoxicity should be conducted to elucidate the pathways through which the identified compounds exert cytotoxic effects on SiHa and MDA-MB-231 cancer cell lines. This may involve the application of molecular biology techniques to investigate phenomena like cell cycle arrest, apoptosis, or autophagy induction. Finally, the transition from in vitro studies to in vivo experiments is needed to validate the observed cytotoxic effects in cell lines. Animal models must be utilized to assess the safety, bioavailability and efficacy of *B. ternifolia* extracts within a more complex biological context.

5. Conclusions

The observed cytotoxic effects of the studied extracts exhibited distinct characteristics in comparison to the findings reported by other researchers, particularly evident in their significant impact on both cell lines. The variability in outcomes could potentially be attributed to differences in the concentration and composition of secondary metabolites present in the *Bouvardia ternifolia* plant.

The phytochemical analyses conducted revealed the presence of saponins, tannins and phenolic compounds. Notably, extracts from the leaves exhibited the highest concentrations of tannins and phenols, suggesting a potential influential role in cellular viability.

It is noteworthy that this study represents the inaugural exploration utilizing SiHa and MDA-MB-231 cell lines to assess the cytotoxic effects of *Bouvardia ternifolia*, thus contributing novel insights to Mexican traditional medicine. Furthermore, this research lays the groundwork for future investigations, both biological and phytochemical, aimed at comprehensively understanding the therapeutic potential and chemical constituents of this plant species.

Author Contributions: All authors contributed significantly to the research conceptualization and original draft preparation, C.V.-V. and O.L.-M. were responsible for the writing—review and editing. L.M.-G. carried out the experimental part, while M.B., R.V.-A., J.A.M.-G., E.O.M.-S., G.B.-C. and J.R.V.-I. collaborated in the review and conceptualization. All authors have read and agreed to the published version of the manuscript.

Funding: This research received no external funding.

Institutional Review Board Statement: All procedures were approved with by the Research and Ethics Committee of Autonomous University of Hidalgo State by number CEEI- 7/2021, approval date 10 October 2021.

Informed Consent Statement: Informed consent was obtained from all subjects involved in the study.

Data Availability Statement: The data presented in this study are available in this article.

Acknowledgments: Thanks to Autonomous University of Hidalgo State for the technical services during this study. Lorenzo Mendoza-Guzman thanks the National Council of Humanities, Sciences and Technology for a graduate scholarship.

Conflicts of Interest: The authors declare no conflict of interest.

References

1. Pan, L.; Lezama-Davila, C.M.; Isaac-Marquez, A.P.; Calomeni, E.P.; Fuchs, J.R.; Satoskar, A.R.; Kinghorn, A.D. Sterols with Antileishmanial Activity Isolated from the Roots of Pentalinon Andrieuxii. *Phytochemistry* **2012**, *82*, 128–135. [CrossRef]
2. Guzmán Maldonado, S.H.; Diaz Huacuz, R.S.; González Chavira, M.M. Plantas Medicinales La Realidad de Una Tradicion Ancestral. 2017. Available online: https://vun.inifap.gob.mx/VUN_MEDIA/BibliotecaWeb/_media/_folletoinformativo/1044_4729_Plantas_medicinales_la_realidad_de_una_tradici%C3%B3n_ancestral.pdf (accessed on 20 November 2023).

3. Chamorro-Cevallos, G.; Mojica-Villegas, M.A.; García-Martínez, Y.; Pérez-Gutiérrez, S.; Madrigal-Santillán, E.; Vargas-Mendoza, N.; Morales-González, J.A.; Cristóbal-Luna, J.M. A Complete Review of Mexican Plants with Teratogenic Effects. *Plants* **2022**, *11*, 1675. [CrossRef]
4. Fernandez, M. Familia Rubiaceae. Importancia En La Flora de Cuba. *Acta Bot. Cuba* **2017**, *216*, 62–64.
5. Blackwell, W.H. Revision of Bouvardia (Rubiaceae). *Ann. Mo. Bot. Gard.* **1968**, *55*, 1–30. [CrossRef]
6. García-Morales, G.; Huerta-Reyes, M.; González-Cortazar, M.; Zamilpa, A.; Jiménez-Ferrer, E.; Silva-García, R.; Román-Ramos, R.; Aguilar-Rojas, A. Anti-Inflammatory, Antioxidant and Anti-Acetylcholinesterase Activities of *Bouvardia Ternifolia*: Potential Implications in Alzheimer's Disease. *Arch. Pharm. Res.* **2015**, *38*, 1369–1379. [CrossRef]
7. Texas, U. of. Lady Bird Johnson Wildflower Center. Available online: https://www.wildflower.org/plants/result.php?id_plant=BOTE2 (accessed on 20 November 2023).
8. Shivanand, D.J.; Hoffmann, J.J.; Torrance, S.J.; Wiedhopf, R.M.; Cole, J.R.; Arora, S.K.; Bates, R.B.; Gargiulo, R.L.; Kriek, G.R. Bouvardin and Deoxybouvardin, Antitumor Cyclic Hexapeptides from *Bouvardia Ternifolia* (Rubiaceae). *J. Am. Chem. Soc.* **1977**, *99*, 8040–8044. [CrossRef]
9. Herrera-Ruiz, M.; García-Morales, G.; Zamilpa, A.; González-Cortazar, M.; Tortoriello, J.; Ventura-Zapata, E.; Jiménez-Ferrer, E. Inhibition of Acetylcholinesterase Activity by Hidroalcoholic Extract and Their Fractions of *Bouvardia Ternifolia* (Cav.) Shcltdl (Rubiaceae). *Bol. Latinoam. Caribe Plantas Med. Aromat.* **2012**, *11*, 526–541.
10. Cornejo Garrido, J.; Chamorro Cevallos, G.A.; Garduño Siciliano, L.; Hernández Pando, R.; Jimenez Arellanes, M.A. Acute and Subacute Toxicity (28 Days) of a Mixture of Ursolic Acid and Oleanolic Acid Obtained from *Bouvardia Ternifolia* in Mice. *Bol. Latinoam. Caribe Plantas Med. Aromat.* **2012**, *11*, 91–102.
11. Perez, G.R.M.; Perez, G.C.; Perez, G.S.; Zavala, S.M.A. Effect of Triterpenoids of *Bouvardia Terniflora* on Blood Sugar Levels of Normal and Alloxan Diabetic Mice. *Phytomedicine* **1998**, *5*, 475–478. [CrossRef]
12. Zapata Lopera, Y.M.; Jiménez-Ferrer, E.; Herrera-Ruiz, M.; Zamilpa, A.; González-Cortazar, M.; Rosas-Salgado, G.; Santillán-Urquiza, M.A.; Trejo-Tapia, G.; Jiménez-Aparicio, A.R. New Chromones from *Bouvardia Ternifolia* (Cav.) Schltdl with Anti-Inflammatory and Immunomodulatory Activity. *Plants* **2023**, *12*, 1. [CrossRef] [PubMed]
13. Tobey, R.A.; Orlicky, D.J.; Deaven, L.L.; Rail, L.B.; Kissane, R.J. Effects of Bouvardin (NSC 259968), a Cyclic Hexapeptide from *Bouvardia Ternifolia*, on the Progression Capacity of Cultured Chinese Hamster Cells. *Cancer Res.* **1978**, *38*, 4415–4421.
14. Jiménez-Ferrer, E.; Reynosa-Zapata, I.; Pérez-Torres, Y.; Tortoriello, J. The Secretagogue Effect of the Poison from Centruroides Limpidus Limpidus on the Pancreas of Mice and the Antagonistic Action of the *Bouvardia Ternifolia* Extract. *Phytomedicine* **2005**, *12*, 65–71. [CrossRef]
15. Re, R.; Pellegrini, N.; Proteggente, A.; Pannala, A.; Yang, M.; Rice-Evans, C. Antioxidant Activity Applying an Improved ABTS Radical Cation Decolorization Assay. *Free Radic. Biol. Med.* **1999**, *26*, 1231–1237. [CrossRef]
16. Schenk, G.H.; Brown, D.J. Free Radical Oxidation of Dihydric Phenols with Diphenylpicrylhydrazyl. *Talanta* **1967**, *14*, 257–261. [CrossRef]
17. Singleton, V.L.; Rossi, J.A. Colorimetry of Total Phenolics with Phosphomolybdic-Phosphotungstic Acid Reagents. *Am. J. Enol. Vitic.* **1965**, *16*, 144–158. [CrossRef]
18. Price, M.L.; Van Scoyoc, S.; Butler, L.G. A Critical Evaluation of the Vanillin Reaction as an Assay for Tannin in Sorghum Grain. *J. Agric. Food Chem.* **1978**, *26*, 1214–1218. [CrossRef]
19. Wrolstad, R.E.; Durst, R.W.; Lee, J. Tracking Color and Pigment Changes in Anthocyanin Products. *Trends Food Sci. Technol.* **2005**, *16*, 423–428. [CrossRef]
20. Elbe, J.H.; Schwartz, S.J.; Hildenbrand, B.E. Loss and Regeneration of Betacyanin Pigments During Processing of Red Beets. *J. Food Sci.* **1981**, *46*, 1713–1715. [CrossRef]
21. Hurtado, N.H.; Pérez, M. Identificación, Estabilidad y Actividad Antioxidante de Las Antocianinas Aisladas de La Cáscara Del Fruto de Capulí (*Prunus serotina* Spp Capuli (Cav) Mc. Vaug Cav). *Inf. Tecnol.* **2014**, *25*, 131–140. [CrossRef]
22. Giron-Martinez, M.C. Determinacion Semicuantitativa de Saponinas En Muestras Vegetales Aprovechando Su Capacidad Hemolitica, Universidad Nacional Autónoma de México. 1992. Available online: https://ru.dgb.unam.mx/handle/20.500.14330/TES01000183303 (accessed on 20 November 2023).
23. Valadez-Vega, C.; Alvarez-Manilla, G.; Riverón-Negrete, L.; García-Carrancá, A.; Morales-González, J.A.; Zuñiga-Pérez, C.; Madrigal-Santillán, E.; Esquivel-Soto, J.; Esquivel-Chirino, C.; Villagómez-Ibarra, R.; et al. Detection of Cytotoxic Activity of Lectin on Human Colon Adenocarcinoma (Sw480) and Epithelial Cervical Carcinoma (C33-A). *Molecules* **2011**, *16*, 2107–2118. [CrossRef] [PubMed]
24. Valadez-Vega, C.; Morales-González, J.A.; Sumaya-Martínez, M.T.; Delgado-Olivares, L.; Cruz-Castañeda, A.; Bautista, M.; Sánchez-Gutiérrez, M.; Zuñiga-Pérez, C. Cytotoxic and Antiproliferative Effect of *Tepary Bean* Lectins on C33-A, MCF-7, SKNSH, and SW480 Cell Lines. *Molecules* **2014**, *19*, 9610–9627. [CrossRef]
25. Mosmann, T. Rapid Colorimetric Assay for Cellular Growth and Survival: Application to Proliferation and Cytotoxicity Assays. *J. Immunol. Methods* **1983**, *65*, 55–63. [CrossRef] [PubMed]
26. Gaviria, A.; Correa, C.E.; Mosquera, O.M.; Niño, J.; Correa, Y.M. Evaluación de Las Actividades Antioxidante y Antitopoisomerasa de Extractos de Plantas de La Ecorregión Cafetera Colombiana. *Rev. Fac. Cienc. Básicas* **2015**, *11*, 86–101. [CrossRef]
27. Mosquera, O.M.; Correra, Y.M.; Niño, J. Antioxidant Activity of Plant Extracts from Colombian Flora. *Rev. Bras. Farmacogn.* **2009**, *19*, 382–387. [CrossRef]

28. Floegel, A.; Kim, D.O.; Chung, S.J.; Koo, S.I.; Chun, O.K. Comparison of ABTS/DPPH Assays to Measure Antioxidant Capacity in Popular Antioxidant-Rich US Foods. *J. Food Compos. Anal.* **2011**, *24*, 1043–1048. [CrossRef]
29. Kumazawa, S.; Hamasaka, T.; Nakayama, T. Antioxidant Activity of Propolis of Various Geographic Origins. *Food Chem.* **2004**, *84*, 329–339. [CrossRef]
30. Ishige, K.; Schubert, D.; Sagara, Y. Flavonoids Protect Neuronal Cells from Oxidative Stress by Three Distinct Mechanisms. *Free Radic. Biol. Med.* **2001**, *30*, 433–446. [CrossRef]
31. Roginsky, V.; Lissi, E.A. Review of Methods to Determine Chain-Breaking Antioxidant Activity in Food. *Food Chem.* **2005**, *92*, 235–254. [CrossRef]
32. Muñoz-Jáuregui, A.; Ramos-Escudero, D.F.; Alvarado-Ortiz Ureta, C.; Castañeda-Castañeda, B. Evaluación de la capacidad antioxidante y contenido de compuestos fenólicos en recursos vegetales promisorios. *Revista de La Sociedad Química Del Perú*. Sociedad Química del Perú. 2007. Volume 73. Available online: http://www.scielo.org.pe/scielo.php?pid=S1810-634X20070003 00003&script=sci_abstract (accessed on 20 November 2023).
33. Giraldo Vásquez, L.M.; Ramírez Aristizabal, L.S. Evaluación de La Actividad Antioxidante de Extractos de *Palicourea Guianensis* (Rubiaceae). *Rev. Cuba. Farm.* **2013**, *47*, 483–491.
34. Dudonné, S.; Vitrac, X.; Coutiére, P.; Woillez, M.; Mérillon, J.M. Comparative Study of Antioxidant Properties and Total Phenolic Content of 30 Plant Extracts of Industrial Interest Using DPPH, ABTS, FRAP, SOD, and ORAC Assays. *J. Agric. Food Chem.* **2009**, *57*, 1768–1774. [CrossRef]
35. Koleva, I.I.; Van Beek, T.A.; Linssen, J.P.H.; De Groot, A.; Evstatieva, L.N. Screening of Plant Extracts for Antioxidant Activity: A Comparative Study on Three Testing Methods. *Phytochem. Anal.* **2002**, *13*, 8–13. [CrossRef]
36. Javanmardi, J.; Stushnoff, C.; Locke, E.; Vivanco, J.M. Antioxidant Activity and Total Phenolic Content of Iranian *Ocimum* Access. *Food Chem.* **2003**, *83*, 547–550. [CrossRef]
37. González-Gómez, J.C.; Ayala-Burgos, A.; Gutiérrez-Vázquez, E. Determinación de Fenoles Totales y Taninos Condensados En Especies Arbóreas Con Potencial Forrajero de La Región de Tierra Caliente Michoacán, México. *Livest. Res. Rural Dev.* **2006**, *18*.
38. Haslam, E. Vegetable Tannins—Lessons of a Phytochemical Lifetime. *Phytochemistry* **2007**, *68*, 2713–2721. [CrossRef]
39. Oda, K.; Matsuda, H.; Murakami, T.; Katayama, S.; Ohgitani, T.; Yoshikawa, M. Relationship between Adjuvant Activity and Amphipathic Structure of Soyasaponins. *Vaccine* **2003**, *21*, 2145–2151. [CrossRef] [PubMed]
40. Lorent, J.H.; Quetin-Leclercq, J.; Mingeot-Leclercq, M.P. The Amphiphilic Nature of Saponins and Their Effects on Artificial and Biological Membranes and Potential Consequences for Red Blood and Cancer Cells. *Org. Biomol. Chem.* **2014**, *12*, 8803–8822. [CrossRef] [PubMed]
41. Ortega, G.M.; Guerra, M. Separación, Caracterización Estructural y Cuantificación de Antocianinas Mediante Métodos Químico-Físicos. *ICIDCA. Sobre Deriv. Caña Azúcar* **2006**, *XL*, 3–11.
42. Rahimi, P.; Abedimanesh, S.; Mesbah-Namin, S.A.; Ostadrahimi, A. Betalains, the Nature-Inspired Pigments, in Health and Diseases. *Crit. Rev. Food Sci. Nutr.* **2019**, *58*, 2949–2978. [CrossRef]
43. Rupachandra, S.; Sarada, D.V.L. Induction of Apoptotic Effects of Antiproliferative Protein from the Seeds of *Borreria Hispida* on Lung Cancer (A549) and Cervical Cancer (HeLa) Cell Lines. *Biomed Res. Int.* **2014**, *2014*, 179836. [CrossRef]
44. Thani, W.; Vallisuta, O.; Siripong, P.; Ruangwises, N. Anti-Proliferative and Antioxidant Activities of Thai Noni/Yor (*Morinda Citrifolia* Linn.) Leaf Extract. *Southeast Asian J. Trop. Med. Public Health* **2010**, *41*, 482–489.
45. Setzer, M.C.; Moriarity, D.M.; Lawton, R.O.; Setzer, W.N.; Gentry, G.A.; Haber, W.A. Phytomedicinal Potential of Tropical Cloudforest Plants from Monteverde, Costa Rica. *Rev. Biol. Trop.* **2003**, *51*, 647–673.
46. Wu, X.D.; He, J.; Li, X.Y.; Dong, L.B.; Gong, X.; Gao, X.; Song, L.D.; Li, Y.; Peng, L.Y.; Zhao, Q.S. Triterpenoids and Steroids with Cytotoxic Activity from *Emmenopterys henryi*. *Planta Med.* **2013**, *79*, 1356–1361. [CrossRef] [PubMed]
47. Akter, R.; Uddin, S.J.; Grice, I.D.; Tiralongo, E. Cytotoxic Activity Screening of Bangladeshi Medicinal Plant Extracts. *J. Nat. Med.* **2014**, *68*, 246–252. [CrossRef] [PubMed]
48. Chitnis, M.P.; Alate, A.D.; Menon, R.S. Effect of Bouvardin (NSC 259968) on the Growth Characteristics and Nucleic Acids and Protein Syntheses Profiles of P388 Leukemia Cells. *Chemotherapy* **1981**, *27*, 126–130. [CrossRef] [PubMed]

Disclaimer/Publisher's Note: The statements, opinions and data contained in all publications are solely those of the individual author(s) and contributor(s) and not of MDPI and/or the editor(s). MDPI and/or the editor(s) disclaim responsibility for any injury to people or property resulting from any ideas, methods, instructions or products referred to in the content.

Article

Sibjotang Protects against Cardiac Hypertrophy In Vitro and In Vivo

Chan-Ok Son [1], Mi-Hyeon Hong [2], Hye-Yoom Kim [2], Byung-Hyuk Han [2], Chang-Seob Seo [3], Ho-Sub Lee [2], Jung-Joo Yoon [2,*] and Dae-Gill Kang [2,4,*]

[1] Department of Ophthalmology, Konkuk University School of Medicine, Gwangjin-gu, Seoul 05030, Republic of Korea; study0815@naver.com
[2] Hanbang Cardio-Renal Syndrome Research Center, Wonkwang University, 460, Iksan-daero, Iksan 54538, Republic of Korea; mihyeon123@naver.com (M.-H.H.); hyeyoomc@naver.com (H.-Y.K.); arum0924@naver.com (B.-H.H.); host@wku.ac.kr (H.-S.L.)
[3] KM Science Research Division, Korea Institute of Oriental Medicine, Daejeon 34054, Republic of Korea; csseo0914@kiom.re.kr
[4] College of Oriental Medicine, Wonkwang University, 460, Iksan-daero, Iksan 54538, Republic of Korea
* Correspondence: mora16@naver.com (J.-J.Y.); dgkang@wku.ac.kr (D.-G.K.)

Citation: Son, C.-O.; Hong, M.-H.; Kim, H.-Y.; Han, B.-H.; Seo, C.-S.; Lee, H.-S.; Yoon, J.-J.; Kang, D.-G. Sibjotang Protects against Cardiac Hypertrophy In Vitro and In Vivo. *Life* 2023, *13*, 2307. https://doi.org/10.3390/life13122307

Academic Editor: Stefania Lamponi

Received: 4 September 2023
Revised: 19 October 2023
Accepted: 26 October 2023
Published: 7 December 2023

Copyright: © 2023 by the authors. Licensee MDPI, Basel, Switzerland. This article is an open access article distributed under the terms and conditions of the Creative Commons Attribution (CC BY) license (https:// creativecommons.org/licenses/by/ 4.0/).

Abstract: Cardiac hypertrophy is developed by various diseases such as myocardial infarction, valve diseases, hypertension, and aortic stenosis. Sibjotang (十棗湯, Shizaotang, SJT), a classic formula in Korean traditional medicine, has been shown to modulate the equilibrium of body fluids and blood pressure. This research study sought to explore the impact and underlying process of Sibjotang on cardiotoxicity induced by DOX in H9c2 cells. In vitro, H9c2 cells were induced by DOX (1 μM) in the presence or absence of SJT (1–5 μg/mL) and incubated for 24 h. In vivo, SJT was administrated to isoproterenol (ISO)-induced cardiac hypertrophy mice (n = 8) at 100 mg/kg/day concentrations. Immunofluorescence staining revealed that SJT mitigated the enlargement of H9c2 cells caused by DOX in a dose-dependent way. Using SJT as a pretreatment notably suppressed the rise in cardiac hypertrophic marker levels induced by DOX. SJT inhibited the DOX-induced ERK1/2 and p38 MAPK signaling pathways. In addition, SJT significantly decreased the expression of the hypertrophy-associated transcription factor GATA binding factor 4 (GATA 4) induced by DOX. SJT also decreased hypertrophy-associated calcineurin and NFAT protein levels. Pretreatment with SJT significantly attenuated DOX-induced apoptosis-associated proteins such as Bax, caspase-3, and caspase-9 without affecting cell viability. In addition, the results of the in vivo study indicated that SJT significantly reduced the left ventricle/body weight ratio level. Administration of SJT reduced the expression of hypertrophy markers, such as ANP and BNP. These results suggest that SJT attenuates cardiac hypertrophy and heart failure induced by DOX or ISO through the inhibition of the calcineurin/NFAT/GATA4 pathway. Therefore, SJT may be a potential treatment for the prevention and treatment of cardiac hypertrophy that leads to heart failure.

Keywords: heart failure; heart diseases; apoptosis; H9c2 cell; isoproterenol

1. Introduction

Heart failure is a disease caused by insufficient blood supply to the body due to cardiac dysfunction and is well known as a leading cause of morbidity and mortality worldwide [1]. Cardiac remodeling has characteristics of heart failure, including myocardial hypertrophy, fibrosis, and progressive ventricular expansion [2]. Cardiac hypertrophy is developed by various diseases such as myocardial infarction, valve diseases, hypertension, and aortic stenosis [3]. Cardiac hypertrophy is well known for its characteristics including the thickening of heart muscles and is associated with chronic heart failure [4]. It is established that hypertrophy and apoptosis in cardiomyocytes occur in the progression of heart failure and arrhythmias [3]. Therefore, myocardial hypertrophy is one of the most frequent causes of heart failure and is a critical sign of a negative prognosis.

Doxorubicin (DOX), the most potent and widely used anthracycline, has been used to treat malignancies like leukemias, lymphomas, soft tissue sarcomas, and other solid tumors [5]. Anthracyclines have been known to induce cancer cell death by a variety of mechanisms including inhibition of the topoisomerase II enzyme, damage to cell membranes, and generation of free radicals [6–8]. In addition, anthracyclines have been shown to modulate various signaling pathways including the apoptosis pathway [9]. However, the anti-neoplastic application of DOX is limited by its side effects, which result in cardiac hypertrophy, fibrosis, and cell death [10–12]. DOX-induced cardiotoxicity has been attributed to apoptosis, oxidative stress, mitochondrial impairment, DNA strand breaks, sarcomere structural alterations, and calcium overloading [13,14].

In the mammalian heart, cardiomyocytes and cardiac fibroblasts account for 90% of the cells in the myocardium. Cardiomyocytes are key cells not only in the maintenance of cardiac structure but also in the contraction function of heart. The contraction function of heart is associated with cytoskeletal structures, which act in myosin-based contractile systems [15,16]. Stress fibers, a key component of cytoskeletal structures, are seen in almost all prominent cardiovascular disorders such as hypertension, cardiac fibrosis, cardiomyopathy, valvular heart diseases, and heart failure [17]. Stress fibers are contracted under stress and are composed of a filamentous actin (F-actin). In addition, it is well established that stress fiber formation is discovered in cardiac hypertrophy [18].

Atrial natriuretic peptide (ANP) and brain natriuretic peptide (BNP) are major hormones produced by the heart, and these hormones play a role in regulating the heart and blood pressure [19,20]. Previous studies have demonstrated the clinical utility of ANP and BNP in the assessment of the severity of heart failure, particularly left ventricular systolic dysfunction [21,22]. Functionally distinct, the two types of cardiac MHC isoforms, alpha-myosin heavy chain (α-MHC) and beta-myosin heavy chain (β-MHC), are expressed influenced by developmental processes and hormonal factors [23]. The expression ratio of these two isoforms can vary in conditions like heart failure or cardiac hypertrophy [24]. The increased expression of β-MHC can serve as an early sensitive indicator of cardiac hypertrophic response [25]. Myosin light chain-2v (MLC-2v), a ventricular myosin light chain, is essential for the formation of thick filaments in heart muscle cells and is linked to inherited hypertrophic cardiomyopathy [26,27].

MAPK (Mitogen-Activated Protein Kinase) is an important kinase family responsible for mediating various signals within cells. Among this family, ERK1/2 (Extracellular Signal-Regulated Kinase 1/2) primarily plays a role in transmitting signals related to cell growth, division, and survival. It was established that the ERK1/2 MAPK pathways participate in hypertrophic signaling in cardiac myocytes [28,29]. ERK1/2 MAPK is stimulated by diverse neurohumoral triggers and by the stretching of cardiac myocytes, both in cultured cells and in vivo [30]. Previous studies have demonstrated that ERK1/2 also plays a role in the regulation of GATA4 phosphorylation, a transcription factor [31]. In the heart, p38 MAPK is a part of a crucial signaling pathway involved in cellular stress responses and inflammation reactions. p38 MAPK is activated by various external signals, such as cellular stress, cytokines, and cardiac damage. Activation of p38 MAPKs is necessary for the binding of GATA4 to the BNP gene induced by hypertrophic agonists and is sufficient for GATA-dependent BNP gene expression [31,32]. GATA4, a transcription factor, serves as a key regulator of cardiac genes in normal cardiac development and pathological hypertrophy [33]. In cardiac myocytes, GATA4 is primarily phosphorylated in response to agonist stimulation through the MEK1-ERK1/2 pathway, while the JNK1/2 or p38 MAPK pathways have weaker effects [34]. Furthermore, GATA4 controls the expression of various genes, encompassing α-MHC, β-MHC, cardiac troponin-C, ANP, and BNP [35]. Excessive calcium accumulation increases the activation of calcineurin which also dephosphorylates the nuclear factor of activated T cells (NFAT) and triggers its translocation from the cytosol to the nucleus causing hypertrophy [36,37]. The nuclear factor of activated T cells (NFAT) is a transcription factor family that mediates various intracellular signaling pathways. The NFAT family comprises several members (NFAT1, NFAT2, NFAT3, NFAT4, and NFAT5), each having specific functions in different cell types and tissues. NFAT is primarily activated through pathways associated with calcium signaling and plays crucial roles in various cellular functions, notably in cell activation, differentiation, and survival [36]. Among them, NFAT3 (NFATc4) is expressed especially in the heart. The Calcineurin–NFAT (Calcineurin–Nuclear Factor of Activated T-cells) pathway is one of the primary regulatory mechanisms for cardiac hypertrophy, modulating the expression of specific genes that influence the size and function of cardiac cells. In particular, the activation of this pathway promotes the increased expression of cardiac hypertrophy-related genes such as SKA (skeletal α-actin), β-MHC, and BNP [38,39].

In response to apoptotic stimuli, pro-apoptotic members of the Bax and the Bcl-2 family are activated on the mitochondria to induce the release of cytochrome c. Released cytochrome c results in the formation of the apoptosome and the apoptosome activates caspase-9, which leads to the activation of caspase-3. This process leads to the same type of apoptotic response as observed for the extrinsic pathway [40–42].

In traditional Korean medicine, various herbal prescriptions have been used for the treatment of heart diseases. Sibjotang (SJT, Shizaotang in Chinese, Jyusoto in Japanese) was recorded in a traditional Chinese medical book named *"Shanghan Lun"* and a traditional Korean medical book named *"Donguibogam"*. SJT has been recorded to be used extensively for symptoms accompanied by edema. SJT is composed of four component herbal medicines: *Euphorbia kansui*, *Euphorbia pekinensis*, *Daphne genkwa*, and *Ziziphus jujube*. Recently, SJT has been known to have anti-inflammatory, anti-tumor, and anti-allergy effects [43]. In a previous study, compounds of SJT were identified as salvianolic acid B, rosmarinic acid, apigenin 7-O-β-glucuronide, apigenin, and yankanin by analysis of 1D and 2D NMR. Additionally, it is reported that SJT increased the positive inotropic effect in rabbit atria [44]. However, research on whether SJT improves cardiac dysfunction caused by cardiac hypertrophy has not yet been conducted. This study investigated the protective effect of SJT on cardiac hypertrophy through the regulation of the calcineurin/NFAT/GATA4 pathway.

2. Materials and Methods

2.1. Chemicals and Materials

Dulbecco's Modified Eagle Medium (DMEM), fetal bovine serum (FBS), 0.05% trypsin-EDTA, antibiotic-antimycotic, and Alexa Fluor™ 488 Phalloidin (A12379) were purchased from Thermo Fisher Scientific (Waltham, MA, USA). Doxorubicin (sc-280681), primary antibody p38 (1:1000; sc-7972), ERK1/2 (1:1000; sc-135900), p-ERK1/2 (1:1000; sc-7383), p-GATA4 (1:500; sc-377543), Bcl-2 (1:500; sc-7382), Lamin B1 (1:2000; sc-374015), and β-actin (1:2000; sc-47778) were purchased from Santa Cruz Biotechnology (Dallas, TX, USA); p-p38 (1:1000; #9211), JNK (1:1000; #9252), p-JNK (1:1000; #9251), NFAT3 (1:1000; #2188), Caspase-9 (1:1000; #9502), Caspase-3 (1:1000; #9665), Bax (1:1000; #2772), and α-tubulin (1:1000; #2144) were purchased from Cell signaling technology (Danvers, MA, USA); Calcineurin (1:1000; 610259) was purchased from BD Bioscience (San Jose, CA, USA).

2.2. Preparation of Sibjotang

SJT consists of four herbs, *Euphorbia kansui*, *Euphorbia pekinensis*, *Daphne genkwa*, and *Ziziphus jujube*. The botanical ingredients *Euphorbia kansui*, *Euphorbia pekinensis*, *Daphne genkwa*, and *Ziziphus jujube* were sourced from the Herbal Medicine Cooperative Association in Iksan, Jeonbuk, Korea. Reference samples of SJT, labeled as HBG132, were cataloged and stored at the Herbarium of the Professional Graduate School of Oriental Medicine at Wonkwang University, located in Iksan, Korea. The manufacturing method follows the regimen based on a Sanghanlun scale. *Euphorbia kansui*, *Euphorbia pekinensis*, and *Daphne genkwa* were scored the same weight of 37.5 g and *Ziziphus jujube* was scored as 85 g. The dried pre-SJT herb (500 g) was subjected to extraction using 2000 mL of distilled water at a temperature of 100 °C for a duration of 4 h. Post-extraction, the mixture was passed through

Whatman No. 5 filter paper and later centrifuged at a speed of 3000 rpm for a period of 10 min at a temperature of 4 °C. The supernatant was then condensed using a rotary vacuum evaporator (Model N-11, manufactured by Tokyo Rikakikai, Tokyo, Japan). Finally, the concentrated extract was freeze-dried using a lyophilizer (PVTFD10RS, IlsinBioBase, Yangju, Republic of Korea) and retained at −70 °C until required. The powdered form of Sibjotang was dissolved in PBS at a 1000-fold higher concentration before being used in in vitro experiments and prepared as a stock solution. For instance, to achieve a final concentration of 5 µg/mL for STJ treatment, we prepared a stock solution of SJT with a concentration of 5 mg/mL.

2.3. HPLC Fingerprinting Analysis of SJT

HPLC fingerprinting analysis of the SJT and its authentic standards (apigenin, apigenin 7-O-β-glucuronide, rosmarinic acid, and salvianolic acid B) was conducted using a Shimadzu Prominence LC-20A series system (Shimadzu Co., Kyoto, Japan) coupled with a photodiode array (PDA) detector and an analytical column (SunFire C18, 250 × 4.6 mm, 5 µm, Waters, Milford, MA, USA). The SJT and its standard solution were analyzed using a sequential gradient mobile phase system of 0.1% (v/v) aqueous formic acid (mobile phase A) and 0.1% (v/v) formic acid in acetonitrile (mobile phase B), namely, 10% B (initial), 50% B (25 min; hold for 5 min), and 10% B (40 min; hold for 10 min). The flow rate was 1.0 mL/min and the injection volume was 10.0 µL. The column thermostat and auto-sampler were maintained at 40 °C and room temperature, respectively. The chromatographic data were processed by the LabSolutions software (version 5.117, Shimadzu Co.). Each standard stock solution for quantitative determination was prepared at 1.0 mg/mL using methanol. The SJT solution for quantitative analysis was dissolved in 70% (v/v) methanol at a concentration of 100 mg/10 mL and then subjected to ultrasonic extraction for 1 h. To quantify two components (apigenin 7-O-β-glucuronide and salvianolic acid B), the prepared SJT solution was diluted 10-fold and then analyzed.

2.4. Cell Culture

The rat-derived H9c2 cardiomyocytes (CRL-1446) were purchased from the American Type Culture Collection (Manassas, VI, USA). Cells were cultured in DMEM supplemented with 1% penicillin-streptomycin antibiotic-antimycotic mixture, 2 mM glutamine, 1.5 g/L sodium bicarbonate, 3.5 g/L glucose, and 10–15% fetal bovine serum (FBS). Culturing was maintained in a humidified incubator at 37 °C with 95% air and 5% CO. When cell confluence reached approximately 80–90%, cells were detached using Trypsin-EDTA solution, and the culture medium was replaced every three days.

2.5. Cell Viability Assay

Cytotoxicity was assessed using the 3-(4,5-dimethylthiazol-2-yl)-2,5-diphenyltetrazolium bromide (MTT) assay. H9c2 cells were seeded in a 96-well culture plate at a density of 1×10^4 cells per well. They were then cultured in serum-free DMEM with various concentrations of SJT for 24 h. Subsequently, 10 µL of MTT solution (0.5 mg/mL) was added to each well, and the plates were incubated for an additional 4 h at 37 °C. After that, the MTT solution was removed and 100 µL of dimethyl sulfoxide (DMSO, Amresco Inc., Dallas, TX, USA) was added to each well. The absorbance of the solubilized formazan was measured at 595 nm using a spectrofluorometer (F-2500, Hitachi, Tokyo, Japan). The absorbance served as an indicator of cell viability, with it being standardized to cells cultured in the control medium, which were deemed to be 100% alive.

2.6. Animals and Treatment

Male ICR mice (23–25 g) were purchased from Chengdu Da Shuo Experimental Animal Co., Ltd. (Chengdu, China). The mice were bred in standard plastic cages with a controlled temperature (22 ± 3 °C) and a 12:12 light–dark cycle, and they were provided with free access to food and water in the animal facility. The experiment was carried out by caring for three animals in a cage dedicated to animals. SJT pretreatment (100 mg/kg) was administered for 7 days, then ISO (30 mg/kg body weight) dissolved in saline was injected subcutaneously for 7 consecutive days. The doses administrated were comparable to those used in mice studies when normalized by body surface area. The control and treatment groups were randomly divided into groups, and the order of randomization was determined by randomization by the researcher with all animals placed in large cages. Control group: fed normal diet and subcutaneous injection of PBS ($n = 6$). ISO group: normal diet and subcutaneous injection of ISO (30 mg/kg·day, $n = 6$). SJT 100 group: normal diet, SJT treatment 100 mg/kg·day, and subcutaneous injection of ISO ($n = 6$). PRO group: normal diet, propranolol treatment 10 mg/kg·day, and subcutaneous injection of ISO ($n = 6$). The number of animals used in the experiment was 24. The ratio of heart weight to body weight was used as an index of cardiac hypertrophy. Mice were anesthetized with 4% isoflurane using an N_2O & O_2 flowmeter system (Harvard Apparatus, Small Animal Ventilator, Harvard, MA, USA) mounted on an Anesthesia Tabletop Bracket. The abdominal artery was then incised to sacrifice the animals. All management and use of experimental animals were conducted in accordance with the guidelines of the National Institute of Health and were approved by the Animal Experimental and Utilization Committee of Wonkwang University School of Medicine (approval number: WKU 20-115). Animal cages were repositioned twice a week to minimize potential confounding factors such as treatment and measurement order or animal/cage location. One researcher conducted one analysis, and the experiment was conducted at the same time with the minimum required time. Information on group assignments, etc., at different stages of the experiment (during assignment, conduct of experiment, evaluation of results, and data analysis) was known only to the research director and was not disclosed at the time of analysis. All animals used in the experiment were subjected to the study after a one-week adaptation period. All research procedures were randomized and designed to generate groups of equal size using blinded analysis methods.

2.7. Cell Size and Stress Fiber Formation

H9c2 cells were plated in a 6-well plate at a density of 4.5×10^5 cells per well and grown at 37 °C. To assess the full effect of SJT, it was incubated in serum-free media for 14 h before processing. H9c2 cells were treated with SJT (1, 5 g/mL), and 30 min later, they were treated with DOX. After Doxorubicin treatment, they were incubated for 24 h or an appropriate time for each experiment. H9c2 cells were fixed with a 3.7% formaldehyde solution in PBS buffer for 10 min at room temperature. The cells were then permeabilized with 0.1% Triton X-100 for 5 min at room temperature and incubated in PBS containing 1% BSA for 30 min at 37 °C. To visualize F-actin, the cells were stained with phalloidin-Alexa488 (A12379, Thermo Fisher, Waltham, MA, USA) for 20 min at room temperature. Fluorescent images of the fixed samples were acquired using an inverted microscope, specifically the EVOS M5000 Cell Imaging System (Thermo Fisher Scientific, Waltham, MA, USA).

2.8. Immunocytochemical Stain

H9c2 cells were cultured in a 6-well plate at a density of 4.5×10^5 cells per well at 37 °C. The cells were incubated in serum-free media for 14 h before processing SJT. The cells were treated with SJT (5 g/mL), and 30 min later, they were treated with Doxorubicin. After Doxorubicin treatment, they were incubated for 24 h. H9c2 cells were fixed with 3.7% formaldehyde solution in PBS buffer for 10 min at room temperature. Cells were permeabilized in 0.1% Triton X-100 for 5 min at RT and incubated in PBS containing 1% BSA for 30 min at 37 °C. Cells were stained with primary antibody overnight at 4 °C to visualize phospho-GATA4 (1:500; sc-377543). After washing, the corresponding secondary antibody was labeled with Alexa Fluor 488 (1:500; Molecular Probes, Eugene, OR) for 1 h at RT. Fluorescent images of fixed samples were acquired on an inverted microscope, using an Eclipse Ti-U inverted microscope (Nikon, Minato, Tokyo, Japan).

2.9. RNA Isolation and Real-Time PCR

Total cellular RNA was extracted using TRIzol reagent (Ambion, Carlsbad, CA, USA). cDNA synthesis was performed using the extracted mRNA through a reverse transcription reaction using the SimpliAmp Thermal Cycler (Life Technologies, Carlsbad, CA, USA). The sequences of primers and probes were as follows: ANP (forward: 5′-GCT CGA GCA GAT CGC AAA AG-3′, reverse: 5′-GAG TGG GAG AGG TAA GGC CT-3′), BNP (forward: 5′-AGC CAG TCT CCA GAA CAA TCC A-3′, reverse: 5′-TGT GCC ATC TTG GAA TTT CGA-3′), β-MHC (forward: 5′-CAG AAC ACC AGC CTC ATC AA-3′, reverse: 5′-CCT CTG CGT TCC TAC ACT CC-3′, α-tubulin (forward: 5′-GAC CAA GCG TAC CAT CCA GT-3′, reverse: 5′-CCA CGT ACC AGT GCA CAA AG-3′). Real-Time PCR was conducted using the Step-One Real-Time PCR System, initiating with a denaturation step at 95 °C for 10 min, followed by 40 cycles of 15 s at 95 °C and 60 s at 60 °C (Product No. 4376600, Applied Biosystems, Foster City, CA, USA). Each RNA sample was assessed three times. The resulting mRNA abundance data were normalized against α-tubulin mRNA abundance. Melting curve analysis was performed on PCR products, and the relative mRNA expression was determined using the $2^{-\Delta\Delta Ct}$ method. GAPDH was employed as the internal reference gene, and the mean Ct values were normalized to GAPDH.

2.10. Western Blot Analysis

Briefly, an average of 0.13 g of heart tissue was chopped and ground using a glass homogenizer on ice, followed by mixing it with 500 μL of RIPA buffer for 30 min on ice. The solution was then centrifuged at $13,000\times g$ for 30 min at 4 °C, and the resulting supernatant was collected. H9C2 cells were cultured in a 100 mm dish at a density of 2.5×10^6 cells per well at 37 °C. They were treated with SJT (5 μg/mL), and 30 min later, they were treated with Doxorubicin. Doxorubicin treatment was followed by a 24 h incubation of the cells or until the necessary expression occurred. Protein lysis was performed using WSE-7420 EzRIPA Lysis buffer (EzWestLumi plus, ATTO Technology, Amherst, New York, NY, USA), supplemented with Protease Inhibitor and Phosphatase Inhibitor. The cell lysate (containing 30 μg of protein) was separated using 10% SDS-polyacrylamide gel electrophoresis (PAGE) and transferred onto a nitrocellulose membrane. The membrane was blocked with 5% BSA powder in TBS-T buffer (10 mM Tris-HCl, pH 7.6, 150 mM NaCl, 0.05% Tween-20) for 1 h, following the supplier's recommendation. It was then incubated with an appropriate primary antibody in a recommended dilution solution. After washing, the primary antibody was detected using a horseradish peroxidase-conjugated secondary antibody against rabbit IgG (A120-101P) and mouse IgG (A90-116P), and the bands were visualized using an enhanced chemiluminescence system (EzWestLumi plus, ATTO Technology, Amherst, New York, NY, USA). The protein expression levels were determined by analyzing the captured signals on the nitrocellulose membrane using a Chemi-doc image analyzer (iBright FL100, Thermo Fisher Scientific, Waltham, MA, USA).

2.11. Preparation of Cytoplasmic and Nuclear Extracts

Nuclear extraction followed the manuals for NE-PER Nuclear and Cytoplasmic Extraction Reagents (Thermo Scientific™, Waltham, MA, USA). For cell culture preparation, adherent cells were harvested using trypsin-EDTA and centrifuged at $500\times g$ for 5 min, while suspension cells were collected by centrifugation at $500\times g$ for 5 min. After washing with PBS, 1–10×10^6 cells were transferred to a 1.5 mL microcentrifuge tube, pelleted at $500\times g$ for 2–3 min, and the supernatant was carefully removed. Then, ice-cold CER I was added to the cell pellet, and the procedure for Cytoplasmic and Nuclear Protein Extraction was followed using the reagent volumes specified in the reagent manual. In this extraction process, the cell pellet was vigorously vortexed, incubated on ice, and mixed with CER II before centrifugation to obtain the cytoplasmic extract. The insoluble pellet fraction containing nuclei was suspended in ice-cold NER, vortexed intermittently, and centrifuged to yield the nuclear extract. Both extracts were stored at −80 °C until further use.

2.12. Histopathological Analysis

For histopathological analysis, the hearts were randomly selected from each group and euthanized using isoflurane. Samples were collected from the hearts of other groups and fixed in 10% saline solution for 24 h. The samples were then washed with tap water, dehydrated using a series of alcohol dilutions, cleared in xylene, and embedded in paraffin at 56 °C for 24 h in a hot water oven. Paraffin wax tissue blocks were prepared for sectioning at a thickness of 4 µm using a microtome. The generated tissue sections were collected on glass slides, deparaffinized, stained with hematoxylin and eosin, and examined using an optical microscope (EVOS™ M5000, Thermo Fisher Scientific, Bothell, WA, USA).

2.13. Statistical Analysis

All experiments were repeated at least three times and statistical analyses were performed with the t-test. Results were expressed as mean ± standard error (S.E.), and data were analyzed using one-way analysis of variance followed by Student's t-test to determine any significant differences. $p < 0.05$ was considered as a statistically significant difference.

3. Results

3.1. HPLC Fingerprinting Analysis of SJT

The typical three-dimensional HPLC chromatogram of SJT is shown in Figure 1. The authentic standards (apigenin, apigenin 7-O-β-glucuronide, rosmarinic acid, and salvianolic acid B) were already known to be isolated from each single extract of SJT. Authentication of these four constituents in SJT was performed by comparison of retention time and the specific absorption spectra pattern in between peaks of authentic standards and SJT. The four peaks observed in the SJT chromatogram and its standards apigenin, apigenin 7-O-β-glucuronide, rosmarinic acid, and salvianolic acid B perfectly matched in the parallel chromatogram. Moreover, the specific UV absorption spectra pattern of SJT and its four standards showed the same patterns in the 190–400 nm wavelength range with maximum and minimum absorption at those spectral scanning ranges. Therefore, the four major peaks observed in the SJT chromatogram were identified as rosmarinic acid, apigenin 7-O-β-glucuronide, salvianolic acid B, and apigenin. The calibration curve was measured in the ranges of 0.63–10.00 (apigenin), 1.25–20.00 µg/mL (apigenin 7-O-β-glucuronide), and 2.50–40.00 µg/mL (rosmarinic acid and salvianolic acid B). The calibration curves for rosmarinic acid, apigenin 7-O-β-glucuronide, salvianolic acid B, and apigenin were y = 29622.35x − 5060.67, y = 37934.35x − 1671.49, y = 12314.68x − 3764.01, and y = 77610.17x − 2204.89, respectively, and the coefficients of determination were all > 0.9999. Quantification of these four constituents in SJT sample was monitored at 290 nm (salvianolic acid B), 330 nm (rosmarinic acid), and 335 nm (apigenin 7-O-β-glucuronide and apigenin). Four constituents (apigenin, apigenin 7-O-β-glucuronide, rosmarinic acid, and salvianolic acid B) were detected at concentrations of 1.74, 6.26, 10.71, and 0.38 mg/lyophilized g, respectively.

3.2. Effect of SJT on DOX-Induced H9c2 Cell Death

MTT assays were performed to evaluate the cytotoxic effect of SJT on H9c2 cells. As shown in Figure 2, SJT did not alter cell viability at the range of 1–5 µg/mL (>90% cell viability). However, incubation with 10–50 µg/mL SJT significantly decreased cell viability. Thus, SJT was experimented with at a non-cytotoxic concentration (less than 10 µg/mL) in H9c2 cells (Figure 2A). Treatment with DOX for 24 h resulted in decreased cell viability in H9c2 cells. Pretreatment with SJT was observed to alleviate cell damage and cell death (Figure 2). Cell viability was assessed using the MTT assay and normalized to the untreated group, presented as a percentage (the control group was considered 100% cell viability). Elevated cell viability was achieved with a high-dose SJT treatment ($p < 0.001$) compared to the Model (DOX only) group (Figure 2B).

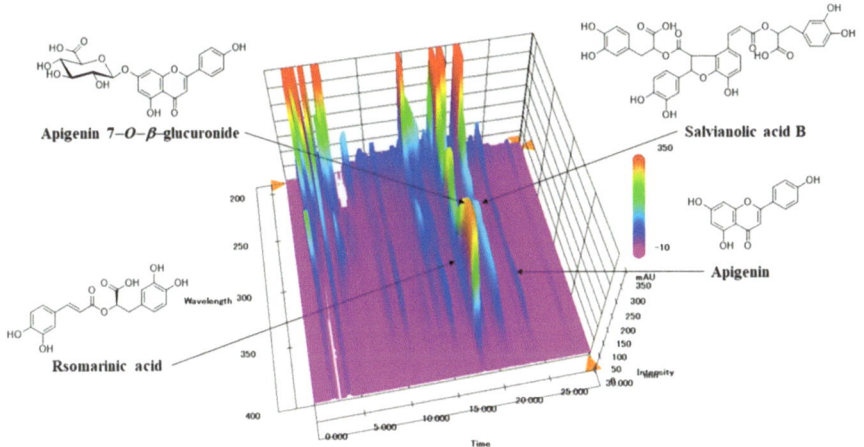

Figure 1. Three-dimensional HPLC chromatogram of SJT.

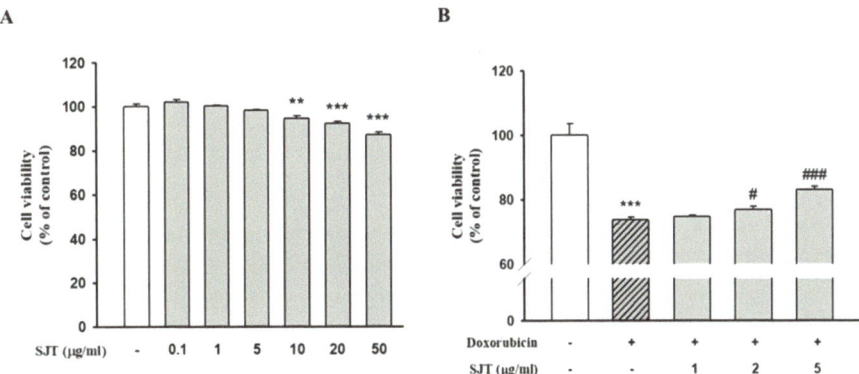

Figure 2. Effect of SJT on DOX–induced H9c2 cell death. (**A**) Cells were treated with concentrations (0–50 μg/mL) of SJT for 24 h. (**B**) H9c2 cells were exposed to doxorubicin (1 μM) while treated with or without SJT at concentrations of 1, 2, and 5 μg/mL for 24 h. Cell viability was measured by Cell Cytotoxicity Assay. The data are expressed as a percentage of basal value and are the means ± S.E of five independent experiments. *** $p < 0.001$, ** $p < 0.01$ vs. control, ### $p < 0.001$, # $p < 0.05$ vs. DOX.

3.3. Effect of SJT on DOX-Induced Cardiac Hypertrophy in H9c2 Cells

Immunofluorescence assays were performed to determine the effect of SJT on cell stress fiber formation using the antibody against F-actin. The results determined that the DOX-increased cell size was completely abolished by pretreatment with SJT (Figure 3A).

To evaluate the effects of SJT on hypertrophy induced by DOX, H9c2 cells were pretreated for 30 min with 1–5 μg/mL SJT prior to 1 μM DOX exposure. As shown in Figure 3B, MLC-2v and β-MHC protein expressions induced by DOX were inhibited by SJT. In addition, SJT significantly inhibited ANP, BNP, and β-MHC mRNA expressions in DOX-induced H9c2 cells (Figure 3C).

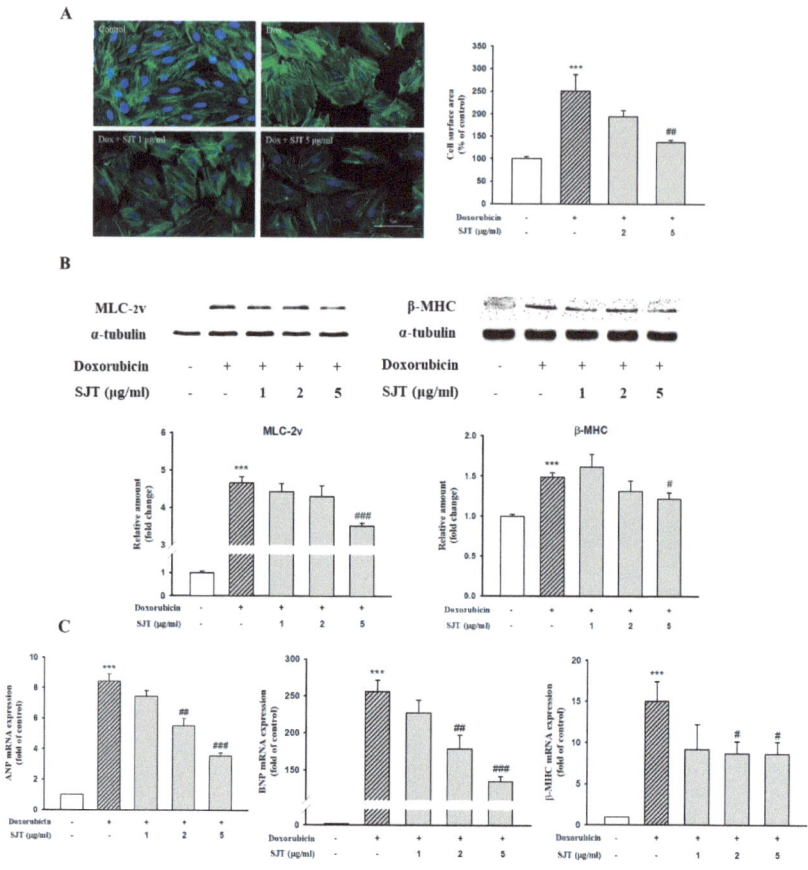

Figure 3. Effect of SJT on DOX–increased cardiac hypertrophy in H9c2 cells. (**A**) H9c2 cells were pretreated with SJT (2, 5 μg/mL) for 30 min and then stimulated with doxorubicin (1 μM) for 24 h. Cells were stained with Alexa Fluor™ 488 Phalloidin F-actin. Cell size was quantified by measuring the surface area of the cells. (**B**) Effect of SJT on DOX–induced hypertrophic protein expression. The protein levels of MLC–2v and β–MHC were determined by Western blot analysis. (**C**) ANP, BNP, and β–MHC mRNA expressions were analyzed using Real–Time PCR. The data are expressed as a percentage of basal value and are the means ± S.E of three independent experiments. *** $p < 0.001$ vs. control, ### $p < 0.001$, ## $p < 0.01$, # $p < 0.05$ vs. doxorubicin.

3.4. Effect of SJT on DOX-Induced Phosphorylation of p38 and ERK1/2

The activation of MAPK signaling pathways is known to be important in cardiac hypertrophy. In this study, we investigated the effect of SJT pretreatment on the activation of MAPK signaling during cardiac hypertrophy, focusing on JNK, ERK1/2, and p38. Among these signaling factors, DOX stimulation led to the activation of phospho-p38. However, pretreatment with SJT completely attenuated the DOX-induced increase in p38 phosphorylation. Similarly, DOX induced the phosphorylation of ERK1/2, but this increase was attenuated by pretreatment (Figure 4A). Stimulation with DOX significantly enhanced the phosphorylation of ERK1/2 in H9c2 cells, and this increase was significantly reduced by pretreatment with SJT (Figure 4B). However, no activation of JNK MAPK was observed following DOX stimulation (Figure 4C).

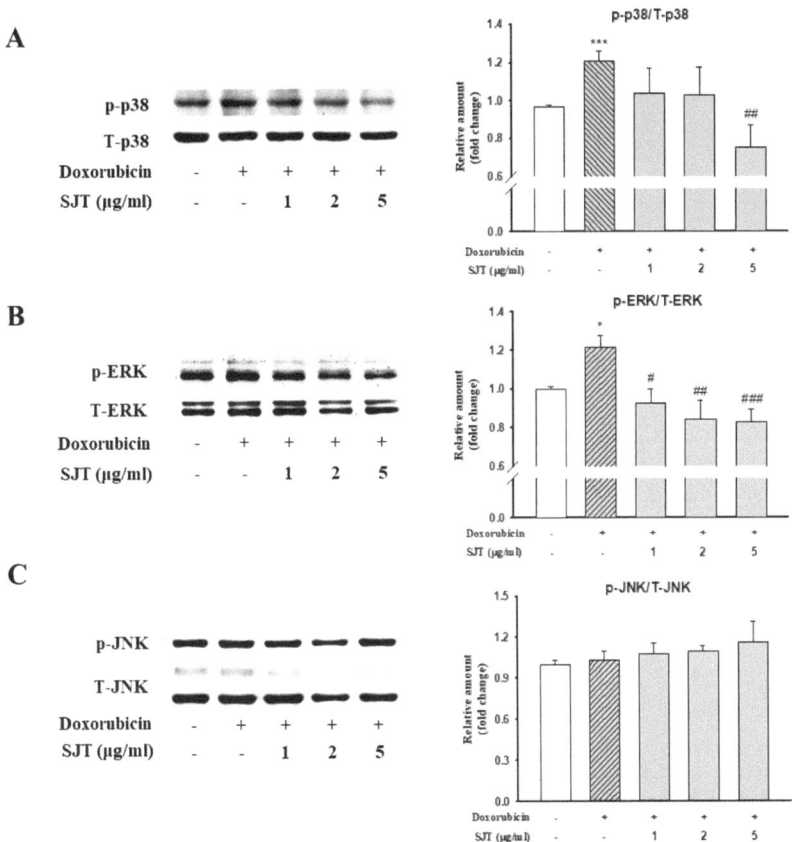

Figure 4. Effect of SJT on DOX–induced phosphorylation of p38 and ERK. H9c2 cells were pretreated with SJT (1–5 μg/mL) for 30 min and then treated with doxorubicin (1 μM) for 1 h. (**A**) p38, (**B**) ERK, and (**C**) JNK MAPK protein expressions were analyzed using Western blot analysis. p38, ERK, and JNK protein levels were used as loading controls. *** $p < 0.001$, * $p < 0.05$ vs. control, ### $p < 0.001$, ## $p < 0.01$, # $p < 0.05$ vs. DOX.

3.5. Effect of SJT on DOX-Induced Calcineurin/NFAT/GATA4 Pathway

GATA4, an important transcription factor related to cardiac hypertrophy, promotes cardiac hypertrophic marker proteins including ANP, BNP, and β-MHC. This result determined whether DOX-induced hypertrophy was associated with p-GATA4 using Western blot analysis and immunostaining. SJT inhibited the DOX-induced phosphorylation protein expression of GATA4 in H9c2 cells (Figure 5A). To confirm the consistency with the results of the Western blot analysis, immunofluorescence staining was performed using the phosphor-GATA4 antibody. Nuclei were stained with DAPI (blue) and p-GATA was stained with Alexa Fluor 488. As a result, DOX increased p-GATA4 phosphorylation in the nucleus, but phosphorylation was significantly decreased in the case of treatment with SJT (Figure 5B). In addition, SJT inhibited DOX-induced calcineurin expression in a dose-dependent manner (Figure 5C).

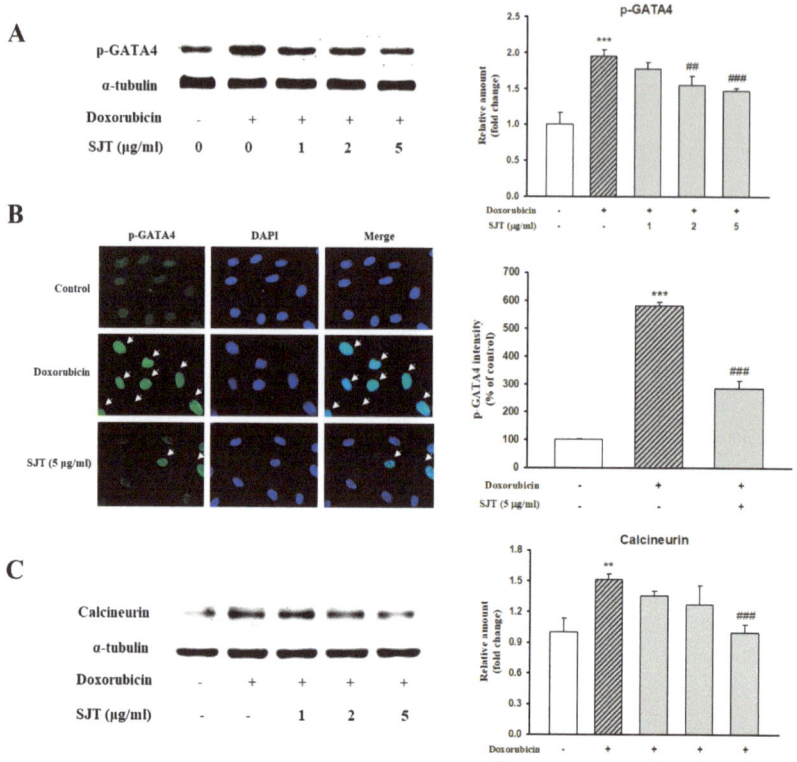

Figure 5. Effect of SJT on DOX–induced GATA–4 and Calcineurin. H9c2 cells were pretreated with SJT (1–5 μg/mL) for 30 min and then treated with DOX (1 μM) for 90 min and 24 h. (**A**) The protein levels of phosphorylated GATA–4 were determined by Western blot analysis. (**B**) Immunofluorescent images of p–GATA–4 nuclear translocation under the laser scanning confocal microscopy are shown (magnification. 400×). Nuclei were stained with DAPI (blue) and p–GATA–4 was stained with Alexa Fluor 488 (green) (immunofluorescence, 200×). (**C**) Calcineurin protein expression was analyzed using Western blot analysis. The results are expressed as the mean ± SE values of three experiments. *** $p < 0.001$, ** $p < 0.01$ vs. control, ### $p < 0.001$, ## $p < 0.01$ vs. DOX.

As shown in Figure 6A, DOX enhanced calcineurin protein expression, whereas it was inhibited by SJT. In addition, NFAT3 levels were also enhanced in DOX-induced H9c2 cells. However, this elevation was significantly attenuated by SJT. Therefore, these results suggest that SJT improves cardiac hypertrophy by regulating the calcineurin/NFAT/GATA4 pathway.

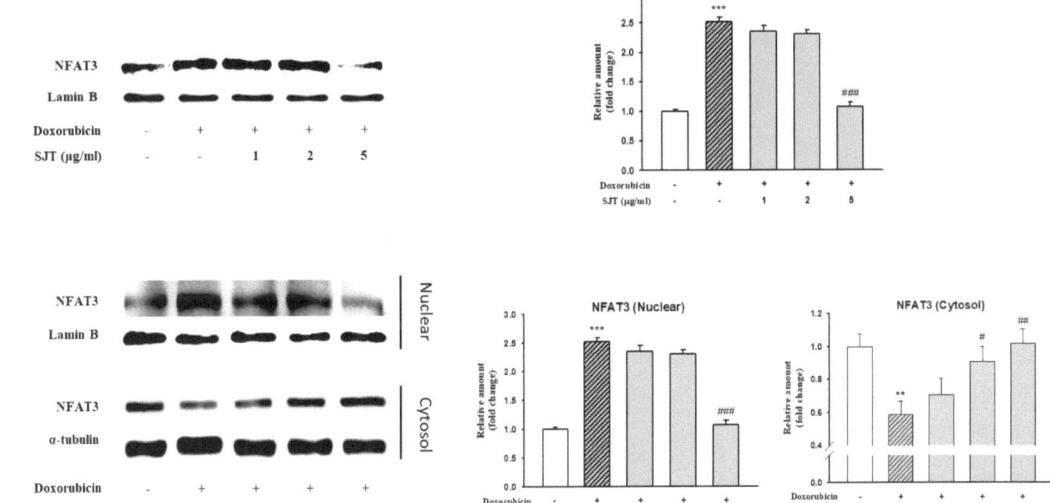

Figure 6. Effect of SJT on DOX–induced NFAT expression. H9c2 cells were pretreated with SJT (1–5 μg/mL) for 30 min and then treated with DOX (1 μM) for 90 min and 24 h. (**A**) NFAT–3 protein expressions were analyzed using Western blot analysis. (**B**) Effect of SJT on nuclear translocation of NFAT–3 was confirmed in H9c2 cells exposed to DOX. The results are expressed as the mean ± SE values of three experiments. *** $p < 0.001$, ** $p < 0.01$ vs. control, ### $p < 0.001$, ## $p < 0.01$, # $p < 0.05$ vs. DOX.

3.6. Effect of SJT on DOX-Induced Cardiac Apoptosis

To clarify the effect of SJT on apoptosis in H9c2 cells exposed to DOX, we performed Western blot analysis in H9c2 cells. DOX induced cleaved caspase-3, cleaved caspase-9, and Bax protein expression, whereas SJT inhibited DOX-induced cleaved caspase-3, cleaved caspase-9, and Bax protein expression (Figure 7A–C). In addition, DOX decreased Bcl-2 protein expression, whereas pretreatment of SJT increased DOX-inhibited Bcl-2 protein expression (Figure 7D). Based on these results, it can be inferred that SJT may improve cardiac hypertrophy-induced heart failure by enhancing the attenuation of DOX-induced cardiac apoptosis through the modulation of apoptosis-related factors.

3.7. The Effect of SJT on Cardiac Hypertrophy in ISO-Treated Mice

To identify whether SJT has a protective effect on cardiac hypertrophy, we pretreated mice with SJT (100 mg/kg/day) and then co-administered ISO (30 mg/kg/day). As shown in Figure 8A,C, the SJT group had reduced heart size and left ventricle/body weight ratio compared to the control group ($p < 0.05$). In addition, heart morphology was also ameliorated by administration of SJT (Figure 8B). We next examined the effects of TGW on the expression of ANP and BNP, which are markers of cardiac hypertrophy. As shown in Figure 8D, in the ISO group, there was a significant increase in the protein expression of cardiac hypertrophic markers in the left ventricle tissue. The SJT group had significantly reduced ISO-increased ANP and BNP protein expression. These results showed that administration of SJT (100 mg/kg/day) inhibited cardiac hypertrophy in ISO-treated mice.

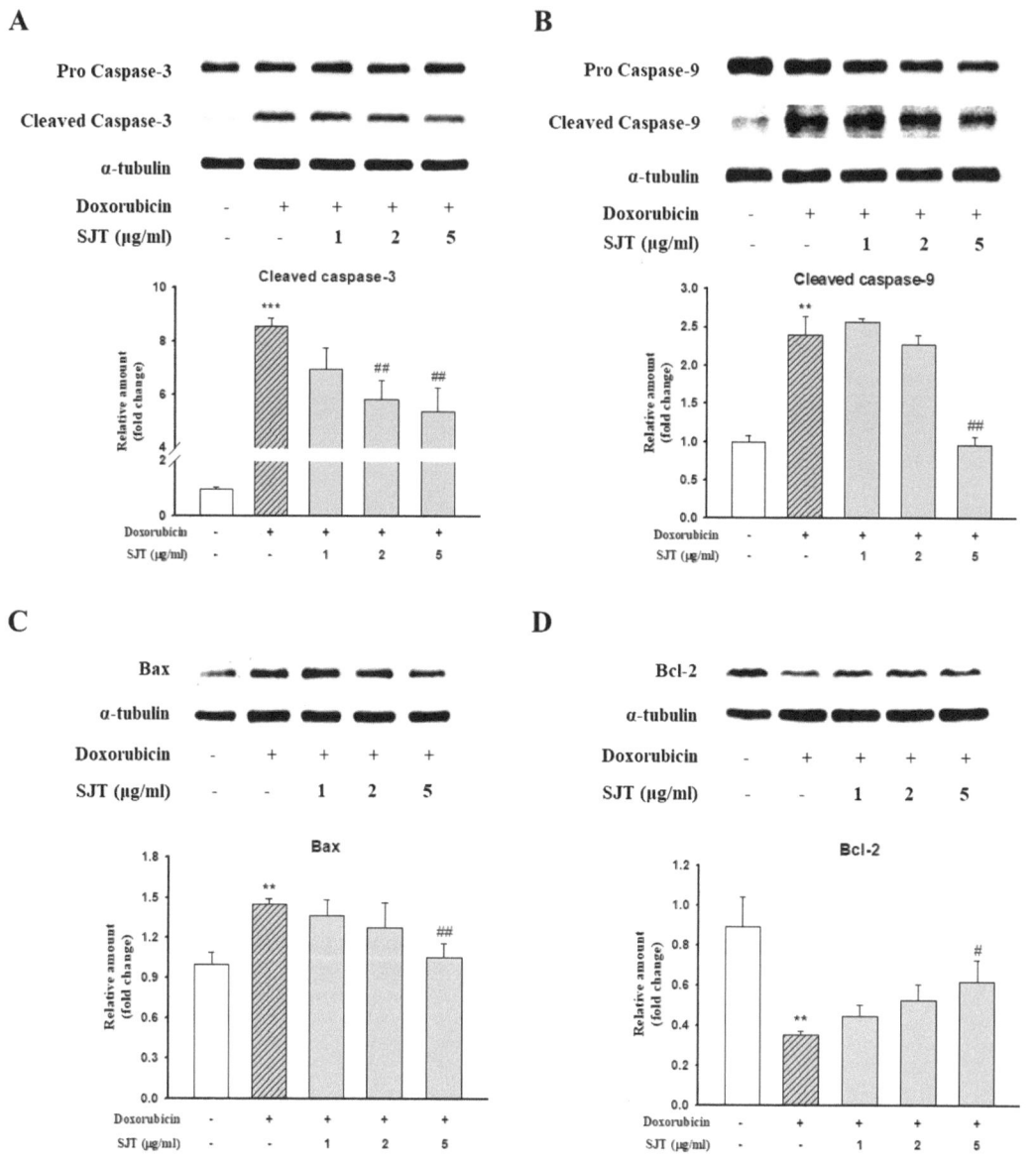

Figure 7. Effect of SJT on DOX–induced cardiac apoptosis. H9c2 cells were pretreated with SJT (1–5 µg/mL) for 30 min and then treated with DOX (1 µM) for 18 h and 12 h. (**A**) Caspase–3, (**B**) Caspase–9, (**C**) Bax, and (**D**) Bcl–2 protein expressions were analyzed using Western blot analysis. The results are expressed as the mean ± SE values of three experiments. *** $p < 0.001$, ** $p < 0.01$ vs. control, ## $p < 0.01$, # $p < 0.05$ vs. doxorubicin.

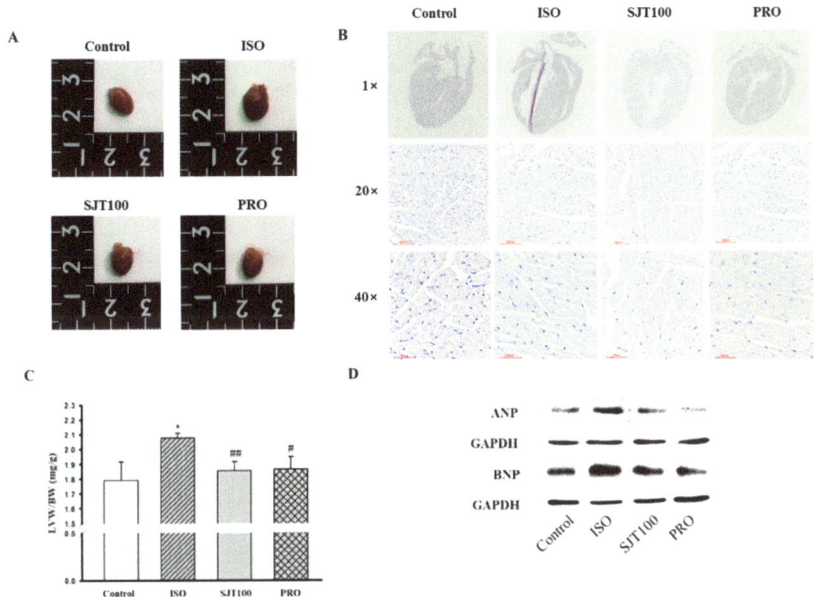

Figure 8. Effect of SJT on ISO–induced cardiac hypertrophy. Effect of SJT on heart size (**A**) and left ventricular/body weight (**B**) in ISO–induced ICR mice. (**C**) Images of the whole hearts of animal models. (**D**) Effect of SJT on cardiomyocyte hypertrophy markers in left ventricular tissues. Mice were infused with PBS (control), ISO, propranolol 30 mg/kg·day with ISO (PRO), and SJT 100 mg/kg·day with ISO (SJT–100). Data are expressed as mean ± SE. There are 6 experimental cases. * $p < 0.05$ vs. Control, ## $p < 0.01$, # $p < 0.05$ vs. ISO. BW, body weight; LVW, left ventricular weight; Iso, isoprenaline; PRO, propranolol.

4. Discussion

Cardiac hypertrophy develops as an adaptive response to various diseases such as myocardial infarction and hypertension and leads to deterioration of cardiac function and chronic heart failure [3,4]. SJT has anti-tumor, anti-inflammatory, and anti-allergy effects [43]. Among the compounds of SJT, salvianolic acid B, rosmarinic acid, and apigenin attenuate cardiac hypertrophy, ischemia, myocardial infarction, and arrhythmias [30,45,46]. However, the effect of SJT in cardiac hypertrophy has been unclear. Therefore, this study has demonstrated for the first time that SJT blocks the cardiac hypertrophy induced by DOX, through the inhibition of the Calcineurin/NFAT/GATA4 pathway.

Stress fibers, commonly known as F-actin and α-actinin, are primarily composed of bundled actin filaments and serve as the contractile system based on actomyosin in cells [15]. Dysregulation of stress fiber formation is commonly observed in nearly all major cardiovascular disorders, including hypertension, myocardial syndrome, heart failure, and cardiac remodeling after myocardial infarction [17]. In the present study, immunofluorescence staining revealed that pretreatment with SJT reduced the DOX-induced cell surface area by decreasing stress fiber formation (F-actin). A previous study has shown that cardiac hypertrophy was attenuated by reducing stress fiber formation [18]. Therefore, these results suggest that SJT decreases cell surface area through attenuating stress fiber formation.

The release of ANP and BNP, cardiac hypertrophic biomarkers, occurs during the development of cardiac hypertrophy [47]. In addition, one of the typical characteristics of cardiac hypertrophy is an increase in the expression of β-MHC and MLC-2v [48,49]. This result was examined to find out whether SJT has an inhibitory effect on the mRNA expression of ANP, BNP, and β-MHC as well as the protein expression of β-MHC and MLC-2v. These results showed that pretreatment with SJT reduced the expression of ANP and BNP in DOX-treated H9c2 cells. In addition, the DOX-induced expression of β-MHC and MLC-2v was inhibited by pretreatment with SJT. The results of this study demonstrated that SJT has an inhibitory effect on the DOX-induced expression of ANP, BNP, β-MHC, and MLC-2v in H9c2 cells. These results suggest that SJT may have a potential role in cardiac hypertrophy.

Evidence has suggested that the MAPK family including ERK1/2, p38, and JNK plays pivotal roles in the development of cardiac hypertrophy [50,51]. This result showed that SJT significantly inhibited the DOX-induced phosphorylation of p38 and ERK1/2. A previous study has shown that the ERK1/2-GATA4 and p38-GATA4 pathways induced by Angiotensin II are associated with cardiac hypertrophy in neonatal rat cardiomyocytes [52]. Thus, to investigate the anti-hypertrophic effect of SJT on DOX-treated H9c2 cells, this study examined the protein expression of p-GATA4. The results of this study revealed that SJT reduced the DOX-induced phosphorylation of GATA4, which has been identified to have an important role in the transcription of the hypertrophic gene. These results suggest that SJT alleviates cardiac hypertrophy by inhibiting the DOX-induced MAPK (p38 and ERK1/2)-GAPA4 pathway.

The Calcineurin/NFAT pathway is one of the key signaling pathways associated with cardiac hypertrophy. Calcineurin, serine-threonine phosphatase, promotes dephosphorylation and translocation of the nuclear factor of activated T cells (NFAT) to the nucleus, leading to cardiac hypertrophy [36]. This study sought to find out whether SJT modulates the expression of Calcineurin and NFAT3 induced by DOX. These data showed that SJT inhibited the DOX-induced expression of Calcineurin and NFAT3 in H9c2 cells. These results suggest that SJT may have a significant protective effect against cardiac hypertrophy by regulation of the calcineurin/NFAT3/GATA4 signaling pathway.

As hypertrophic stimuli persist, cardiomyocytes undergo a loss of cell viability and the disruption of cardiac structure, ultimately leading to heart failure [53]. Numerous experimental studies have revealed that the activation of caspases is a molecular characteristic of apoptosis and they play a crucial regulatory role in the apoptotic cascade [54]. Caspase-9 acts as the initiating enzyme in the mitochondrial apoptotic pathway, while caspase-3 serves as the key effector caspase, which is activated through cleavage by caspase-9 [55]. Thus, this result was performed to determine the expression of the cleaved form of caspase-9 and caspase-3, which is the active form in H9c2 cells. The protein expression of cleaved caspase-9 and cleaved caspase-3 induced by DOX was significantly reversed by pretreatment with SJT. These results indicate that SJT may have a significant protective effect against cardiac hypertrophy by regulating the apoptosis signaling pathway.

This study indicated that SJT attenuated DOX-induced stress fiber formation and regulated the DOX-enhanced MAPK-GATA4, calcineurin-NFAT3, and apoptosis pathway. In addition, SJT alleviated the expression of cardiac hypertrophic biomarkers, the ANP, BNP, β-MHC, and MLC-2v levels induced by the hypertrophy model. In conclusion, these data provide the first piece of evidence that SJT may play an important role in the prevention of cardiac hypertrophy and apoptosis. The findings imply that SJT could be a promising intervention for cardiac hypertrophy and its associated heart failure. The protective effects of SJT, as a traditional herbal medicine, offer valuable insights for the development of novel therapeutic strategies in managing heart failure.

Author Contributions: Conceptualization, J.-J.Y. and D.-G.K.; Formal analysis, C.-O.S. and M.-H.H.; Investigation, C.-O.S. and M.-H.H.; Methodology, H.-Y.K. and B.-H.H.; Project administration, J.-J.Y. and H.-S.L.; Resources, C.-S.S. and H.-S.L.; Supervision, D.-G.K.; Validation, J.-J.Y. and H.-Y.K.; Writing—original draft, C.-O.S.; Writing—review and editing, C.-O.S. and J.-J.Y. All authors have read and agreed to the published version of the manuscript.

Funding: This work was supported by the National Research Foundation of Korea (NRF) grant funded by the Korean government (MSIP) (2017R1A5A2015805) (2021R1C1C2009542).

Institutional Review Board Statement: All experimental procedures were carried out in accordance with the national institute of health guide for the care and use of laboratory animals and were approved by the Institutional Animal Care and Utilization Committee for Medical Science of Wonkwang University (approval number: WKU 20-115, approval date 22 December 2020).

Informed Consent Statement: Not applicable.

Data Availability Statement: Data are available on reasonable request.

Conflicts of Interest: The authors declare no conflict of interest.

References

1. Li, B.; Chi, R.F.; Qin, F.Z.; Guo, X.F. Distinct changes of myocyte autophagy during myocardial hypertrophy and heart failure: Association with oxidative stress. *Exp. Physiol.* **2016**, *101*, 1050–1063. [CrossRef] [PubMed]
2. Wu, M.P.; Zhang, Y.S.; Xu, X.; Zhou, Q.; Li, J.D.; Yan, C. Vinpocetine Attenuates Pathological Cardiac Remodeling by Inhibiting Cardiac Hypertrophy and Fibrosis. *Cardiovasc. Drugs Ther.* **2017**, *31*, 157–166. [CrossRef] [PubMed]
3. Heineke, J.; Molkentin, J.D. Regulation of cardiac hypertrophy by intracellular signalling pathways. *Nat. Rev. Mol. Cell Biol.* **2006**, *7*, 589–600. [CrossRef] [PubMed]
4. Shimizu, I.; Minamino, T. Myocardial hypertrophy is characterized by the thickening of heart muscles without an obvious cause and found to be involved in several pathological conditions, including hypertension, vascular disease and chronic heart failure. Physiological and pathological cardiac hypertrophy. *J. Mol. Cell. Cardiol.* **2016**, *97*, 245–262. [PubMed]
5. Bai, J.; Ma, M.; Cai, M.; Xu, F.; Chen, J.; Wang, G.; Shuai, X.; Tao, K. Inhibition enhancer of zeste homologue 2 promotes senescence and apoptosis induced by doxorubicin in p53 mutant gastric cancer cells. *Cell Prolif.* **2014**, *47*, 211–218. [CrossRef]
6. Zhang, S.; Liu, X.; Bawa-Khalfe, T.; Lu, L.S.; Lye, Y.L.; Liu, L.F.; Yeh, E.T.H. Identification of the molecular basis of doxorubicin-induced cardiotoxicity. *Nat. Med.* **2012**, *18*, 1639–1642. [CrossRef]
7. Vejpongsa, P.; Yeh, E.T. Prevention of anthracycline-induced cardiotoxicity: Challenges and opportunities. *J. Am. Coll. Cardiol.* **2014**, *64*, 938–945. [CrossRef] [PubMed]
8. Efferth, T.; Oesch, F. Oxidative stress response of tumor cells: Microarray-based comparison between artemisinins and anthracyclines. *Biochem. Pharmacol.* **2004**, *68*, 3–10. [CrossRef] [PubMed]
9. Green, P.S.; Leeuwenburgh, C. Mitochondrial dysfunction is an early indicator of doxorubicin-induced apoptosis. *Biochim. Biophys. Acta* **2002**, *1588*, 94–101. [CrossRef]
10. Karagiannis, T.C.; Lin, A.J.; Ververis, K.; Chang, L.; Tang, M.M.; Okabe, J.; El-Osta, A. Trichostatin A accentuates doxorubicin-induced hypertrophy in cardiac myocytes. *Aging* **2010**, *2*, 659–668. [CrossRef]
11. Levick, S.P.; Soto-Pantoja, D.R.; Bi, J.; Hundley, W.G.; Widiapradja, A.; Manteufel, E.J.; Bradshaw, T.W.; Meléndez, G.C. Doxorubicin-Induced Myocardial Fibrosis Involves the Neurokinin-1 Receptor and Direct Effects on Cardiac Fibroblasts. *Heart Lung Circ.* **2018**, *28*, 1598–1605. [CrossRef]
12. Yuan, Y.P.; Ma, Z.G.; Zhang, X.; Xu, S.C.; Zeng, X.F.; Yang, Z.; Deng, W.; Tang, Q.Z. CTRP3 protected against doxorubicin-induced cardiac dysfunction, inflammation and cell death via activation of Sirt1. *J. Mol. Cell. Cardiol.* **2018**, *114*, 38–47. [CrossRef] [PubMed]
13. Ludke, A.R.; Al-Shudiefat, A.A.; Dhingra, S.; Jassal, D.S.; Singal, P.K. A concise description of cardioprotective strategies in doxorubicin-induced cardiotoxicity. *Can. J. Physiol. Pharmacol.* **2009**, *87*, 756–763. [PubMed]
14. Alkreathy, H.; Damanhouri, Z.A.; Ahmed, N.; Slevin, M.; Ali, S.S.; Osman, A.M.M. Aged garlic extract protects against doxorubicin-induced cardiotoxicity in rats. *Food Chem. Toxicol.* **2010**, *48*, 951–956. [CrossRef] [PubMed]
15. Humeres, C.; Frangogiannis, N.G. Fibroblasts in the Infarcted, Remodeling, and Failing Heart. *JACC Basic Transl. Sci.* **2019**, *4*, 449–467. [CrossRef] [PubMed]
16. Swiatlowska, P.; Iskratsch, T. Tools for studying and modulating (cardiac muscle) cell mechanics and mechanosensing across the scales. *Biophys. Rev.* **2021**, *13*, 611–623. [CrossRef]
17. Yu, C.M.; Tipoe, G.L.; Lai, K.W.H.; Lau, C.P. Effects of combination of angiotensin-converting enzyme inhibitor and angiotensin receptor antagonist on inflammatory cellular infiltration and myocardial interstitial fibrosis after acute myocardial infarction. *J. Am. Coll. Cardiol.* **2001**, *38*, 1207–1215. [CrossRef]
18. Oh, K.; Lee, J.; Oh, B.; Mun, J.; Park, B.; Lee, B. Polymyxin B Alleviates Angiotensin II-Induced Stress Fiber Formation and Cellular Hypertrophy. *Pharmacol. Pharm.* **2014**, *5*, 903–910. [CrossRef]
19. Ramos, H.; de Bold, A.J. Gene expression, processing and secretion of natriuretic peptides: Physiologic and diagnostic implications. *Heart Fail. Clin.* **2006**, *2*, 255–268. [CrossRef]
20. Ogawa, T.; de Bold, A.J. The heart as an endocrine organ. *Endocr. Connect.* **2014**, *3*, R31–R44. [CrossRef]
21. Dickstein, K.; Larsen, A.I.; Bonarjee, V.; Thoresen, M.; Aarsland, T.; Hall, C. Plasma proatrial natriuretic factor is predictive of clinical status in patients with congestive heart failure. *Am. J. Cardiol.* **1995**, *76*, 679–683. [CrossRef] [PubMed]
22. Jortani, S.A.; Prabhu, S.D.; Valdes, R., Jr. Strategies for developing biomarkers of heart failure. *Clin. Chem.* **2004**, *50*, 265–278. [CrossRef]

23. Allen, D.L.; Leinwand, L.A. Postnatal myosin heavy chain isoform expression in normal mice and mice null for IIb or IId myosin heavy chains. *Dev. Biol.* **2001**, *229*, 383–395. [CrossRef] [PubMed]
24. Gupta, M.P. Factors controlling cardiac myosin-isoform shift during hypertrophy and heart failure. *J. Mol. Cell. Cardiol.* **2007**, *43*, 388–403. [CrossRef] [PubMed]
25. Cheng, T.H.; Shih, N.L.; Chen, C.H.; Lin, H.; Liu, J.C.; Chao, H.H.; Liou, J.Y.; Chen, Y.L.; Tsai, H.W.; Chen, Y.S.; et al. Role of mitogen-activated protein kinase pathway in reactive oxygen species-mediated endothelin-1-induced beta-myosin heavy chain gene expression and cardiomyocyte hypertrophy. *J. Biomed. Sci.* **2005**, *12*, 123–133. [CrossRef]
26. Rottbauer, W.; Wessels, G.; Dahme, T.; Just, S.; Trano, N.; Hassel, D.; Burns, C.G.; Katus, H.A.; Fishman, M.C. Cardiac myosin light chain-2: A novel essential component of thick-myofilament assembly and contractility of the heart. *Circ. Res.* **2006**, *99*, 323–331. [CrossRef] [PubMed]
27. Aoki, H.; Sadoshima, J.; Izumo, S. Myosin light chain kinase mediates sarcomere organization during cardiac hypertrophy in vitro. *Nat. Med.* **2000**, *6*, 183–188. [CrossRef] [PubMed]
28. Bueno, O.F.; De Windt, L.J.; Tymitz, K.M.; Witt, S.A.; Kimball, T.R.; Klevitsky, R.; Hewett, T.E.; Jones, S.P.; Lefer, D.J.; Peng, C.F.; et al. The MEK1-ERK1/2 signaling pathway promotes compensated cardiac hypertrophy in transgenic mice. *EMBO J.* **2000**, *19*, 6341–6350. [CrossRef] [PubMed]
29. Sugden, P.H. Signaling Pathways in Cardiac Myocyte Hypertrophy. *Ann. Med.* **2001**, *33*, 611–622.
30. Rose, B.A.; Force, T.; Wang, Y. Mitogen-activated protein kinase signaling in the heart: Angels versus demons in a heart-breaking tale. *Physiol. Rev.* **2010**, *90*, 1507–1546.
31. Tenhunen, O.; Sármán, B.; Kerkelä, R.; István Szokodi, I.; Papp, L.; Tóth, M.; Ruskoaho, H. Mitogen-activated protein kinases p38 and ERK1/2 mediate the wall stress-induced activation of GATA-4 binding in adult heart. *J. Biol. Chem.* **2004**, *279*, 24852–24860. [PubMed]
32. Kerkela, R.; Pikkarainen, S.; Majalahti-Palviainen, T.; Tokola, H.; Ruskoaho, H. Distinct roles of mitogen-activated protein kinase pathways in GATA-4 transcription factor-mediated regulation of B-type natriuretic peptide gene. *J. Biol. Chem.* **2002**, *277*, 13752–13760. [PubMed]
33. Liang, Q.; De Windt, L.J.; Witt, S.A.; Kimball, T.R.; Markham, B.E.; Molkentin, J.D. The transcription factors GATA4 and GATA6 regulate cardiomyocyte hypertrophy in vitro and in vivo. *J. Biol. Chem.* **2001**, *276*, 30245–30253.
34. Liang, Q.; Wiese, R.J.; Bueno, O.F.; Dai, Y.S.; Markham, B.C.; Molkentin, J.D. The Transcription Factor GATA4 Is Activated by Extracellular Signal-Regulated Kinase 1- and 2-Mediated Phosphorylation of Serine 105 in Cardiomyocytes. *Mol. Cell. Biol.* **2001**, *21*, 7460–7469. [CrossRef]
35. Van Berlo, J.H.; Elrod, J.W.; Aronow, B.J.; Pu, W.T.; Molkentin, J.D. Serine 105 phosphorylation of transcription factor GATA4 is necessary for stress-induced cardiac hypertrophy in vivo. *Proc. Natl. Acad. Sci. USA* **2011**, *108*, 12331–12336. [CrossRef]
36. Hogan, P.G.; Chen, L.; Nardone, J.; Rao, A. Transcriptional regulation by calcium, calcineurin, and NFAT. *Genes Dev.* **2003**, *17*, 2205–2232.
37. Wilkins, B.J.; Dai, Y.S.; Bueno, O.F.; Parsons, S.A.; Xu, J.; Plank, D.M.; Jones, F.; Kimball, T.R.; Molkentin, J.D. Calcineurin/NFAT coupling participates in pathological, but not physiological. *Card. Hypertrophy* **2004**, *94*, 110–118.
38. Tokudome, T.; Horio, T.; Kishimoto, I.; Soeki, T.; Mori, K.; Kawano, Y.; Kohno, M.; Garbers, D.L.; Nakao, K.; Kangawa, K. Calcineurin-nuclear factor of activated T cells pathway-dependent cardiac remodeling in mice deficient in guanylyl cyclase A, a receptor for atrial and brain natriuretic peptides. *Circulation* **2005**, *111*, 3095–3104. [CrossRef]
39. Wilkins, B.J.; Molkentin, J.D. Calcium-calcineurin signaling in the regulation of cardiac hypertrophy. *Biochem. Biophys. Res. Commun.* **2004**, *322*, 1178–1191.
40. Moe, G.W.; Naik, G.; Konig, A.; Lu, X.; Feng, Q. Early and persistent activation of myocardial apoptosis, bax and caspases: Insights into mechanisms of progression of heart failure. *Pathophysiology* **2002**, *8*, 183–192.
41. Louis, X.L.; Murphy, R.; Thandapilly, S.J.; Yu, L.; Netticadan, T. Garlic extracts prevent oxidative stress, hypertrophy and apoptosis in cardiomyocytes: A role for nitric oxide and hydrogen sulfide. *BMC Complement. Altern. Med.* **2012**, *12*, 140. [CrossRef]
42. Prathapan, A.; Raghu, K.G. Apoptosis in angiotensin II-stimulated hypertrophic cardiac cells-modulation by phenolics rich extract of *Boerhavia diffusa* L. *Biomed. Pharmacother.* **2018**, *108*, 1097–1104.
43. Kai, H.; Koine, T.; Baba, M.; Okuyama, T. Pharmacological effects of Daphne genkwa and Chinese medical prescription, "Jyu-So-To". *Yakugaku Zasshi* **2004**, *124*, 349–354. [CrossRef] [PubMed]
44. Kwon, O.J.; Oh, H.C.; Lee, Y.J.; Kim, H.Y.; Tan, R.; Kang, D.G.; Lee, H.S. Sibjotang Increases Atrial Natriuretic Peptide Secretion in Beating Rabbit Atria. *Evid. Based Complement. Altern. Med.* **2015**, *2015*, 268643. [CrossRef] [PubMed]
45. Javidanpour, S.; Dianat, M.; Badavi, M.; Mard, S.A. The inhibitory effect of rosmarinic acid on overexpression of NCX1 and stretch-induced arrhythmias after acute myocardial infarction in rats. *Biomed. Pharmacother.* **2018**, *102*, 884–893. [CrossRef]
46. Zhu, Z.Y.; Gao, T.; Huang, Y.; Xue, J.; Xie, M.L. Apigenin ameliorates hypertension-induced cardiac hypertrophy and down-regulates cardiac hypoxia inducible factor-lα in rat. *Food Funct.* **2016**, *7*, 1992–1998. [PubMed]
47. Li, X.; Lan, Y.; Wang, Y.; Nie, M.; Lu, Y.; Zhao, E. Telmisartan suppresses cardiac hypertrophy by inhibiting cardiomyocytes apoptosis via the NFAT/ANP/BNP signaling pathway. *Mol. Med. Rep.* **2017**, *15*, 2574–2582. [PubMed]
48. Guo, Z.; Lu, J.; Li, J.; Wang, P.; Li, Z.; Zhong, Y.; Guo, K.; Wang, J.; Ye, J.; Liu, P. JMJD3 inhibition protects against isoproterenol-induced cardiac hypertrophy by suppressing β-MHC expression. *Mol. Cell. Endocrinol.* **2018**, *5*, 1–14. [CrossRef]

49. Ding, P.; Huang, J.; Battiprolu, P.K.; Hill, J.A.; Kamm, K.E.; Stull, J.T. Cardiac Myosin Light Chain Kinase Is Necessary for Myosin Regulatory Light Chain Phosphorylation and Cardiac Performance in Vivo. *J. Biol. Chem.* **2010**, *285*, 40819–40829. [CrossRef]
50. Chen, Y.; Yao, F.; Chen, S.; Huang, H.; Wu, L.; He, J.; Dong, Y. Endogenous BNP attenuates cardiomyocyte hypertrophy induced by Ang II via p38 MAPK/Smad signaling. *Pharmazie* **2014**, *69*, 833–837. [PubMed]
51. Li, C.; Chen, Z.X.; Yang, H.; Luo, F.; Chen, L.; Cai, H.; Li, Y.; You, G.; Long, D.; Li, S.; et al. Selumetinib, an Oral Anti-Neoplastic Drug, May Attenuate Cardiac Hypertrophy via Targeting the ERK Pathway. *PLoS ONE* **2016**, *11*, e0159370. [CrossRef]
52. Tang, W.; Wei, Y.; Le, K.; Li, Z.; Bao, Y.; Gao, J.; Zhang, F.; Cheng, S.; Liu, P. Mitogen-activated protein kinases ERK 1/2- and p38-GATA4 pathways mediate the Ang II-induced activation of FGF2 gene in neonatal rat cardiomyocytes. *Biochem. Pharmacol.* **2001**, *81*, 518–525. [CrossRef]
53. Ho, C.Y. Hypertrophic Cardiomyopathy: For Heart Failure Clinics: Genetics of Cardiomyopathy and Heart Failure. *Heart Fail. Clin.* **2010**, *6*, 141–159. [CrossRef] [PubMed]
54. Yan, X.; Feng, C.; Chen, Q.; Li, W.; Wang, H.; Lv, L.; Smith, G.W.; Wang, J. Effects of sodium fluoride treatment in vitro on cell proliferation, apoptosis and caspase-3 and caspase-9 mRNA expression by neonatal rat osteoblasts. *Arch. Toxicol.* **2009**, *83*, 451–458. [CrossRef] [PubMed]
55. Yi, X.; Wang, F.; Feng, Y.; Zhu, J.; Wu, Y. Danhong injection attenuates doxorubicin-induced cardiotoxicity in rats via suppression of apoptosis: Network pharmacology analysis and experimental validation. *Front. Pharmacol.* **2022**, *13*, 2022. [CrossRef] [PubMed]

Disclaimer/Publisher's Note: The statements, opinions and data contained in all publications are solely those of the individual author(s) and contributor(s) and not of MDPI and/or the editor(s). MDPI and/or the editor(s) disclaim responsibility for any injury to people or property resulting from any ideas, methods, instructions or products referred to in the content.

Article

Antinociceptive Effect of *Dendrobii caulis* in Paclitaxel-Induced Neuropathic Pain in Mice

Keun Tae Park [1,2], Yong Jae Jeon [1], Hyo In Kim [3] and Woojin Kim [1,2,*]

[1] Department of Physiology, College of Korean Medicine, Kyung Hee University, Seoul 02453, Republic of Korea; cerex@naver.com (K.T.P.); ad2353@naver.com (Y.J.J.)
[2] Korean Medicine-Based Drug Repositioning Cancer Research Center, College of Korean Medicine, Kyung Hee University, Seoul 02447, Republic of Korea
[3] Department of Surgery, Beth Israel Deaconess Medical Center, Harvard Medical School, Boston, MA 02115, USA; hkim21@bidmc.harvard.edu
* Correspondence: wjkim@khu.ac.kr

Abstract: Paclitaxel-induced neuropathic pain (PINP) is a serious adverse effect of chemotherapy. *Dendrobii caulis* (*D. caulis*) is a new food source used as herbal medicine in east Asia. We examined the antinociceptive effects of *D. caulis* extract on PINP and clarified the mechanism of action of transient receptor potential vanilloid 1 receptor (TRPV1) in the spinal cord. PINP was induced in male mice using multiple intraperitoneal injections of paclitaxel (total dose, 8 mg/kg). PINP was maintained from D10 to D21 when assessed for cold and mechanical allodynia. Oral administration of 300 and 500 mg/kg *D. caulis* relieved cold and mechanical allodynia. In addition, TRPV1 in the paclitaxel group showed increased gene and protein expression, whereas the *D. caulis* 300 and 500 mg/kg groups showed a significant decrease. Among various substances in *D. caulis*, vicenin-2 was quantified by high-performance liquid chromatography, and its administration (10 mg/kg, i.p.) showed antinociceptive effects similar to those of *D. caulis* 500 mg/kg. Administration of the TRPV1 antagonist capsazepine also showed antinociceptive effects similar to those of *D. caulis*, and *D. caulis* is thought to exhibit antinociceptive effects on PINP by modulating the spinal TRPV1.

Keywords: *Dendrobii caulis*; *Dendrobium officinale*; paclitaxel; transient receptor potential vanilloid 1

Citation: Park, K.T.; Jeon, Y.J.; Kim, H.I.; Kim, W. Antinociceptive Effect of *Dendrobii caulis* in Paclitaxel-Induced Neuropathic Pain in Mice. *Life* **2023**, *13*, 2289. https://doi.org/10.3390/life13122289

Academic Editor: Seung Ho Lee

Received: 26 September 2023
Revised: 25 November 2023
Accepted: 27 November 2023
Published: 30 November 2023

Copyright: © 2023 by the authors. Licensee MDPI, Basel, Switzerland. This article is an open access article distributed under the terms and conditions of the Creative Commons Attribution (CC BY) license (https://creativecommons.org/licenses/by/4.0/).

1. Introduction

Chemotherapy can cause peripheral neuropathic pain, leading to discontinuation or dose reduction [1]. Highly effective anticancer drugs, such as oxaliplatin, vincristine, cisplatin, and paclitaxel, can cause peripheral neuropathic pain [2,3]. Paclitaxel, discovered and isolated from *Taxux brevifolia* [4,5], is an anticancer drug used for breast, lung, and ovarian cancer [6]. Although various side effects such as bone marrow toxicity, myalgia, and arthralgia have been reported, chronic and acute peripheral neuropathy is one of the most serious side effects [7]. Although the prevalence of paclitaxel-induced neuropathic pain (PINP) is 60% [8], PINP treatment remains unsatisfactory despite numerous studies. Furthermore, no specific drugs are available for treatment or prevention [9].

D. caulis is a widely recognized traditional tonic in China, used not only as a supplement for patients but also in food, and is native to China and southeast Asian countries [10,11]. *D. caulis* is the dried stem of *Dendrobium officinale* (*D. officinale*). Fengdou (the twisted dried stem part) and its powder are sold in their representative forms in the market and used in various ways, such as making soups, medicinal liquor, and teas [12]. Although there are no reports on *D. caulis*, several studies have reported curative effects of *D. officinale*. Administration of 200 mg/kg *D. officinale* improved intestinal barrier function and strengthened the antitumor immune response in a colorectal cancer mouse model [13]. Additionally, in an acute colitis model, administration of 200 mg/kg of polysaccharide from *D. officinale* alleviated liver damage by increasing antioxidant enzyme activity and downregulating the tumor necrosis factor-α (TNF-α) signaling pathway [14]. In a rat model of type 2 diabetes, administration of 160 mg/kg improved oxidative stress, inflammation, and liver lipid accumulation [15]. Studies have shown that the glucomannan fraction of *D. officinale* attenuates intestinal damage in colitis mice and regulates intestinal mucosal immunity [16]; in cell studies, it is regulated through the toll-like receptor 4/nuclear factor kappa-light-chain-enhancer of activated B cell (TLR4/NF-kB) signaling pathways in gastric epithelial cells (GES-1) [17].

Chemical analysis of *D. caulis* has shown that flavonoids, phenols, and polysaccharides are the major compounds isolated from the stem [18]. Flavonoids are important compounds in *D. caulis* and have potential anticancer, anti-inflammatory, and antioxidant properties. In particular, *D. caulis* is rich in flavonoid C-glycosides [18], and as a result of high-performance gel permeation chromatography analysis, a total of 26 compounds were identified, of which 14 flavonoids including rutin and vicenin-2 were reported as major substances [19].

Members of the transient receptor potential (TRP) family exist in ion channels at the terminals of nociceptors which act as molecular transducers, where many stimuli depolarize neurons [20,21]. TRP channels constitute a unique superfamily of ion channels and are related to the voltage-gated Ca^{2+}, K^+, and Na^+ superfamily. The transient receptor potential vanilloid receptor (TRPV) subfamily consists of channels involved in thermosensitivity and nociception (TRPV 1-4) and channels involved in Ca^{2+} uptake/reuptake (TRPV 5–6) [22]. Among TRP channels, the most studied is TRPV1. TRPV1 is expressed in sensory ganglia such as the dorsal root ganglia (DRG), vagal and trigeminal ganglia, and in small sensory Aδ- and C fibers which contain various neuropeptides such as calcitonin-related peptide and substance P [23–28]. TRPV1 is also found in central nervous system and non-nervous tissues such as the kidney, mast cells, smooth muscle, bladder, and lung [29–32].

Members of the transient receptor family of TRP channels reportedly contribute to PINP induction [33,34]. TRPV4 and transient receptor potential cation channel subfamily melastatin member 8 (TRPM8) showed sensitivity in the DRG neurons involved in PINP, whereas TRPV1 was considered a mediator of allodynia [35–37]. Additionally, oral administration of HC-030031, a selective TRPA1 antagonist, has been reported to be responsible for the mechanical hypersensitivity observed in neuropathic and inflammatory pain models [38]. On the other hand, in humans, when TRPV1 is activated in muscles, patients treated with paclitaxel experience cramping or deep pain [39]. TRPV1 activation is closely related to what chemotherapy-induced neuropathy patients experience [40,41]. In patients, TRPV1 activation causes a burning sensation [42] in skin nociceptors and pain in deep tissue nociceptors [43]. Paclitaxel sensitizes and activates TRPV1 function, and TRPV1 antagonists exert analgesic effects in chemotherapy-induced neuropathy [34]. Similarly, cisplatin or bortezobim treatment increased the expression of TRPV1 in the spinal cord and dorsal root ganglion of mice [44,45]. Paclitaxel interacts directly with TRPV1 channels to acutely stimulate signaling and also produce long-term channel desensitization changes [46]. Parallel studies have shown that oxaliplatin also sensitizes TRPV1, which is mediated by the G-protein-coupled receptor G2A [47].

In the present study, we investigated the effects of *D. caulis* extract on PINP. Subsequently, changes in the gene and protein expression of TRPV1 in the spinal cord caused by paclitaxel were confirmed. In addition, the effect of *D. caulis* and vicenin-2, its active ingredient, on TRPV1 was demonstrated, and their antinociceptive effect on PINP was evaluated.

2. Materials and Methods

2.1. Animals

Six-week-old male C57BL/6J mice were purchased from Daehan Biolink (Daejeon, South Korea). The mice were kept under controlled conditions for one week to adapt. The lighting cycle was adjusted to 12 h light/12 h dark, and constant temperature (23 ± 2 °C) and humidity (65 ± 5%) were maintained. After adaptation, six animals were randomly assigned to each group. During the experimental period, the mice were provided with a standard diet and freely available water. All experimental protocols were approved by the Kyung Hee University Animal Care and Use Committee (approval NO. KHUASP-23-223, 12 April 2023). In total, 80 mice were used and all experiments were conducted in accordance with the guidelines of the International Association for the Study of Pain [48].

2.2. Dendrobii caulis Extract Preparation

Dried *D. caulis* was purchased from Xiuzheng Pharmaceutical Group Co., Ltd. (Hangzhou, China). *D. caulis* was extracted by 25% ethanol and reflux extraction for 6 h at 80 °C. The extract was concentrated under reduced pressure at 60 °C after filtering the residue through filter paper. The extract was dried using an evaporator and then frozen using a lyophilizer to obtain the extract powder. *D. caulis* was diluted in phosphate-buffered saline (PBS) and administered orally or intrathecally to mice (Specimen N. KWJ-0003).

2.3. Paclitaxel, D. caulis, Vicenin-2, and Capsazepine Administration

Paclitaxel (Sigma-Aldrich, St. Louis, MO, USA) was dissolved in a 1:1 mixture of Cremophor EL (Sigma-Aldrich) and ethanol at a concentration of 6 mg/mL. Paclitaxel was diluted to 0.2 mg/mL in PBS, and 200 µL was intraperitoneally administered to mice. In the vehicle group, the same volume of a 1:1 solution of Cremophor EL and ethanol was administered via the same route. Paclitaxel and vehicle solutions were injected intraperitoneally four times every other day (D0, 2, 4, and 6) at a dose of 2 mg/kg (Figure 1). The *D. caulis* extract was administered orally using a sonde (Jungdo-BNP, Seoul, Republic of Korea) at different doses (100, 300, and 500 mg/kg in distilled PBS). Vicenin-2 (Sigma-Aldrich, St. Louis, MO, USA) was dissolved in 10% dimethyl sulfoxide (DMSO) and 100 µL was administered intraperitoneally at doses of 1 and 10 mg/kg. In the intrathecal injection administration experiment, 10 µg/mouse of capsazepine, a TRPV1 antagonist, was dissolved in 10% DMSO, and *D. caulis* was dissolved in doses of 0.1 and 1 mg/mouse in PBS, and the administered volume was 5 µL. *D. caulis*, vicenin-2, and capsazepine were administered between D10 and D21, the period when pain was maintained, and a behavioral assessment was conducted 1 h after administration.

(A)

(B)
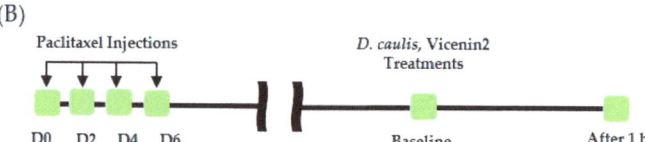

Figure 1. The effect of multiple intraperitoneal paclitaxel injections in mice. Schedule of behavior tests conducted to assess the cold and mechanical allodynia after paclitaxel injection (**A**). Antinociceptive effect evaluation time progress (**B**).

2.4. Behavioral Tests

The acetone drop and von Frey filament methods were used to measure paclitaxel-induced cold and mechanical allodynia. Before the assessment, the mice were placed in a chamber made of steel wire for 30 min for acclimatization. The acetone drop method evaluated behavior by spraying 10 µL of acetone on the hind paw. The numbers of withdrawals, flinches, and licking reactions were recorded for 30 s [49]. The results show the average value of the 'of Response' measured [50,51]. To summarize the mechanical allodynia method, the mechanical allodynia test was started by selecting a strength of 3.22

among several von Frey filaments. The filament was pressed against the mid-plantar part of the hind paw for 2–3 s until it bent. If there was a pain response to the stimulation, a filament of strength one level lower was applied. If there was no pain response, pressure was applied using a filament with strength one level higher [52]. The pain response was marked with an X, otherwise, it was marked with an O. The '50% threshold value' in the result refers to Chaplan's calculation and Dixon's up-down method [53,54]. After the behavioral evaluation, the mice were anesthetized with isoflurane. The mouse heart was perfused with PBS to wash the entire tissue, and lumbar spinal cord sections 4–5 were extracted. PCR and Western blotting were performed on the extracted tissues.

2.5. Quantitative Real-Time Polymerase Chain Reaction (qRT-PCR)

Total ribonucleic acid (RNA) was extracted from spinal tissues using an RNA extraction kit (AccuPrep, Bioneer, Daejeon, Republic of Korea). RNA samples were quantified using Nanodrop (Thermo Scientific, Waltham, MA, USA), and RT-PCR was performed on samples with an OD260/OD280 ratio higher than 2.0. cDNA was synthesized by incubating 0.5 µg of total RNA with the cDNA synthesized kit (Maxime RT Premix, Intron Biotechnology, Seongnam-Si, Republic of Korea), according to the protocol. qPCR amplification of cDNA was performed using a real-time SYBR kit (SensiFAST SYBR, Bio-Rad, Hercules, CA, USA) and CFX Real-Time PCR (Bio-Rad, Hercules, CA, USA). The real-time PCR condition was as follows: one cycle of 95 °C for 5 min, followed by 40 cycles of 95 °C for 20 s, 57 °C for 20 s, and 72 °C for 20 s. The primer sequences used were as follows: Glyceraldehyde 3-phosphate dehydrogenase (*Gapdh*) forward 5'-GTC GTG GAG TCT ACT GGT GTC TTC-3' and reverse 5'-GTC ATC ATA CTT GGC AGG TTT CTC-3', and transient receptor potential vanilloid 1 (*TRPV1*) forward 5'-GGC TGT CTT CAT CAT CCT GCT GCT-3' and reverse 5'-GTT CTT GCT CTC CTG TGC GAT CTT GT-3'. GAPDH was used to normalize the amount of RNA in each sample, and a specific detection threshold (Ct) was selected to calculate the fluorescence produced by amplification. Relative gene expression compared to the control was quantified using the $2^{-\Delta\Delta Ct}$ method [55] with the average Ct value of each gene, and the expression value of the reference group was expressed as 1.

2.6. Western Blot

Western blotting was performed to analyze the protein expression of the TRPV1 receptor. Radioimmunoprecipitation assay buffer was added to liquefy the tissue, which was then centrifuged at 13,000 rpm for 10 min, and the supernatant was used for Western blotting. The protein concentration in the supernatant was quantified using a protein assay kit (Bradford protein assay kit, Bio-rad, Hercules, CA, USA). Then, 20 µg of protein was loaded to sodium dodecyl sulfate–polyacrylamide gel electrophoresis, following which the protein bands were transferred to nitrocellulose membrane. After blocking with 5% non-fat skim milk in tris-buffered saline with tween 20 buffer, the blocked membrane was incubated with primary antibodies overnight at 4 °C. Actin and TRPV1 proteins were detected using rabbit polyclonal anti-actin antibodies (1:1000, cat. PA1-183, Invitrogen, Waltham, MA, USA) and rabbit polyclonal anti-TRPV1 (1:1000, cat. NB100-1617, Novus Biologicals, Littleton, CO, USA), respectively. The membrane was washed with PBS-T and then further incubated with secondary antibody (horseradish peroxidase-conjugated anti-rabbit antibody,1:5000, cat. 31460, Thermo Scientific, Waltham, MA, USA) for 2 h at room temperature. Protein bands were detected using a chemiluminescence detection kit (D-Plus ECL Femto System, Hwaseong, Republic of Korea) and Davinch-Chemi software (Young-Hwa Science, Daejeon, Republic of Korea).

2.7. HPLC Analysis of Rutin and Vicenin-2 in D. caulis Extract

D. caulis was analyzed using high-performance liquid chromatography (HPLC) with a UV detector (Agilent 1260, Santa Clara, CA, USA). The conditions for the rutin, vicenin-2, and *D. caulis* extract analyses are listed in Table 1. Stock solutions of rutin and vicenin-2 were prepared in methanol. The rutin and vicenin-2 standard were purchased from Sigma-

Aldrich (St. Louis, MO, USA) and Chemface (Wuhan, China), respectively. For both rutin and vicenin-2, five dilutions were prepared and analyzed. A total of 100 mg of *D. caulis* was ultrasonically extracted (4 °C, 60 min) using 1 mL of ethanol. After centrifuging (4 °C, 13,000 rpm, 30 min) the diluted solution, the supernatant was filtered through a 0.4 μm filter and used for analysis.

Table 1. Analytical conditions of HPLC for the rutin and vicenin-2 analysis.

Treatment			Conditions Rutin			Vicenin-2
Column			Ymc-Triart C18			Ymc-Triart C18
Flow rate			1.0 mL/mL			1.0 mL/mL
Injection volume			10 μL			10 μL
UV detection			275 nm			335 nm
Run time			30 min			48 min
Rutin			Vicenin-2			Flow
Time (min)	Acetonitrile	0.1% Phosphoric acid	Time (min)	Acetonitrile	0.1% Phosphoric acid	mL/min
0	20	80	0	15	85	1.0
15	20	80	5	15	85	1.0
18	100	0	35	25	75	1.0
23	100	0	37	10	90	1.0
25	20	80	42	10	90	1.0
30	20	80	48	15	85	1.0

2.8. The Effect of Vicenin-2 on Caco-2 Cell and RAW 264.7 Cell Lines Measured by 3-(4,5-Dimethyl Thazolk-2-yl)-2,5-Diphenyl Tetrazolium Bromide (MTT) Viability Assay

Cell viability was evaluated in RAW 264.7 and Caco-2 cells obtained from the Korea Cell Line Bank (ATCC). Raw 274.7 cell lines were grown in Dulbecco's modified Eagle's medium supplemented with 10% fetal bovine serum (FBS), and Caco-2 cell lines were grown in Eagle's minimum essential medium (MEM) supplemented with 10% FBS. Cultures was kept in a humidified atmosphere with 5% CO_2 at 37 °C. The viability of cells were measured using the MTT assay. In a 96-well plate, 1×10^4 cells, suspended to 200 μL of growth medium, were seeded for 24 h. After incubation, each concentration of vicenin-2 previously dissolved in DMSO was treated for 24 h. To evaluate the viability, the medium was discolored for an MTT working solution and incubated for 1 h at 37 °C. The formazan formed in the MTT assay was dissolved in DMSO and the optical density was measured at 540 nm using a microplate reader. The results were expressed as a percentage based on the control (untreated cells) group.

2.9. Statistical Analysis

The statistical analysis and graph work were performed using Prism GraphPad (version 9.0., Graphpad Software Inc., Boston, MA, USA). All data are presented as mean ± SD. Statistical analyses were performed using an unpaired *t*-test for the data shown in paclitaxel-induced pain experiment, whereas the data in *D. caulis*, vicenin-2 and antagonist administration experiments were analyzed using a paired *t*-test. In addition, statistical analyses were performed using one-way ANOVA followed by Tukey's test for gene and protein expressions and cell viability analysis. A level of $p < 0.05$ was considered significant.

3. Results

3.1. Multiple Paclitaxel Injections Induce Cold and Mechanical Allodynia

Several studies have reported that multiple paclitaxel injections induce cold and mechanical allodynia in mice [56–58]. Paclitaxel was injected four times (D0, 2, 4, and 6, total 8 mg/kg), and the vehicle (1:1 ratio of Cremophor EL to ethanol) was administered to the control group. Significant pain was induced in both cold allodynia results obtained using the acetone drop method and mechanical allodynia results obtained using the von Frey filament method (Figure 2A,B). The results showed that paclitaxel injection increased the response to cold pain and decreased the threshold for mechanical stimulation from D10 to D21. All subsequent behavioral assessments were evaluated between D10 and D21.

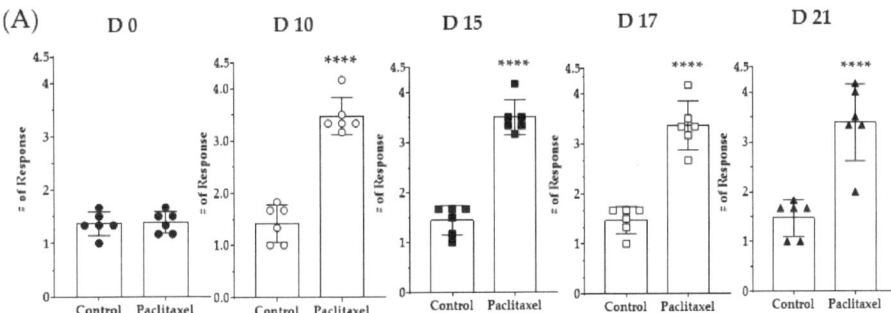

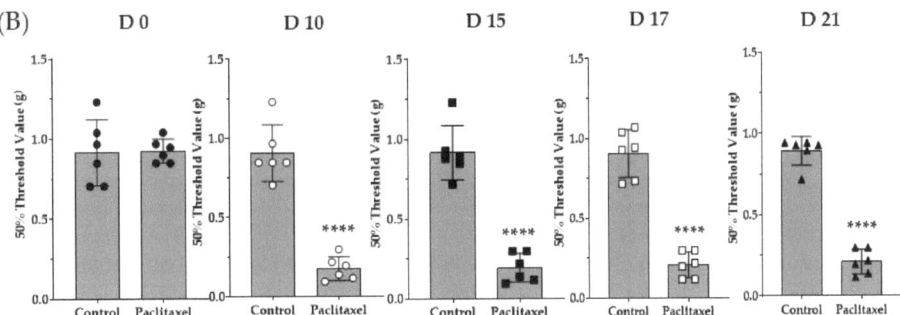

Figure 2. Pain assessment via multiple administrations of paclitaxel in mice. Paclitaxel was administered intraperitoneally at 2 mg/kg every 2 days (D0, D2, D4, and D6), and cold and mechanical allodynia were evaluated using the acetone drop method and the von Frey filament method. Behavioral assessment was conducted on D0, D10, D15, D17, and D21. Behavioral changes observed in the mice were recorded, calculated, and expressed as cold (**A**) and mechanical allodynia (**B**). The black circle, white circle, black square, white square, and black triangle represent D0, D10, D15, D17, and D21, respectively. N = 6 for each group; **** $p < 0.0001$ vs. control group with unpaired t-test.

3.2. Single Oral Administration of D. caulis in Cold and Mechanical Allodynia

The 25% ethanol extract *D. caulis* (100, 300, and 500 mg/kg) was orally administered to mice at different doses to confirm its antinociceptive effects on PINP. Behavioral changes were recorded 1 h after *D. caulis* administration. Cold and mechanical allodynia induced by paclitaxel were significantly alleviated in the *D. caulis* 300 and 500 mg/kg group (Figure 3A,B). When the results obtained one hour after the injection of PBS or 100 mg/kg, 300 mg/kg, or 500 mg/kg of *D. caulis* were analyzed by using one-way ANOVA, the F and p values in ANOVA interaction were F = 44.81, $p < 0.0001$ and F = 56.92, $p < 0.0001$ for cold and mechanical allodynia, respectively.

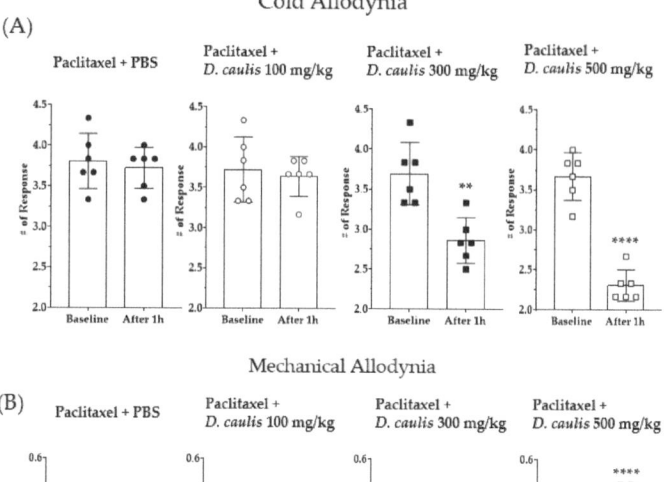

Figure 3. Effect of oral administration of *D. caulis* on cold and mechanical allodynia (**A**,**B**). Behavioral assessments were performed before *D. caulis* administration and 1 h after administration via gavage at concentrations of 100, 300, and 500 mg/kg. PBS was used as a vehicle for *D. caulis* and *D. caulis* or PBS was orally administered. The black circle, white circle, black square, and white square represent PBS control group, *D. caulis* 100 mg/kg group, 300 mg/kg, 500 mg/kg group, respectively. N = 6 for each group; ** $p < 0.01$, *** $p < 0.001$, and **** $p < 0.0001$ vs. control group with paired *t*-test.

3.3. Gene Expression of TRPV1 Channel Using qRT-PCR

TRPV1 gene expression was confirmed in spinal cord lumbar sections 4–5 from paclitaxel-treated mice. In the paclitaxel-administered group, the expression of TRPV1 increased by approximately 123.73% compared with that in the control group. In contrast, the TRPV1 expression which increased with paclitaxel was significantly decreased by 44.68% and 43.77% in the *D. caulis* 300 and 500 mg/kg groups, respectively (Figure 4A).

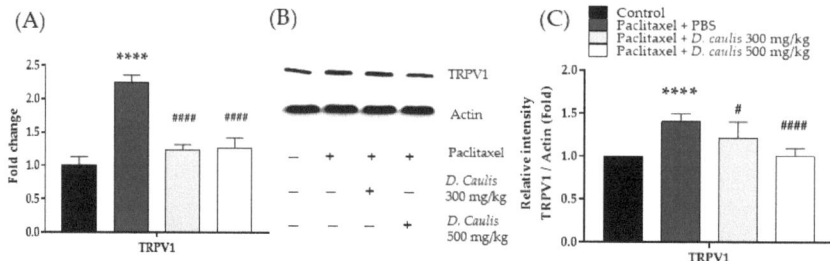

Figure 4. Effect of 300 and 500 mg/kg of *D. caulis* on TRPV1 expression of paclitaxel-induced neuropathic pain in the spinal cord. TRPV1 levels of gene expression (**A**), a representative protein analysis image (**B**), and analyzed relative intensity of TRPV1 protein (**C**). N = 6 per group. **** $p < 0.0001$ vs. control, and # $p < 0.05$ and #### $p < 0.0001$ vs. paclitaxel + PBS with one-way ANOVA followed by Tukey's multiple comparison test.

Western blotting was performed to elucidate the effect of *D. caulis* on TRPV1 protein expression in paclitaxel-induced pain, and the results were consistent with those of the PCR. Similar to the PCR results, TRPV1 protein expression increased by 40.6% after paclitaxel administration and significantly decreased by 13.4% and 28.6% at 300 and 500 mg/kg of *D. caulis*, respectively (Figure 4B,C).

3.4. Identification and Quantification of Rutin and Vicenin-2 in D. caulis

HPLC analysis was conducted to identify rutin and vicenin-2 as the major components of *D. caulis*. The retention times of rutin and vicenin-2 were approximately 12 and 10 min, respectively (Figure 5A–D). Rutin was not detected, and the UV spectrum and retention of vicenin-2 and vicenin-2 standard solution were consistent. The calibration curve of rutin and vicenin-2 showed linearity in the detector over a range of six concentrations (6.25, 12.5, 25, 50, and 100 μg/mL). The rutin regression equation was $y = 17.07092x + 14.48396$ and RSQ = 0.99989. The vicenin-2 regression equation was $y = 9.23568x + 5.89275$ and RSQ = 0.99994. The content of vicenin-2 in *D. caulis* was approximately 0.245%.

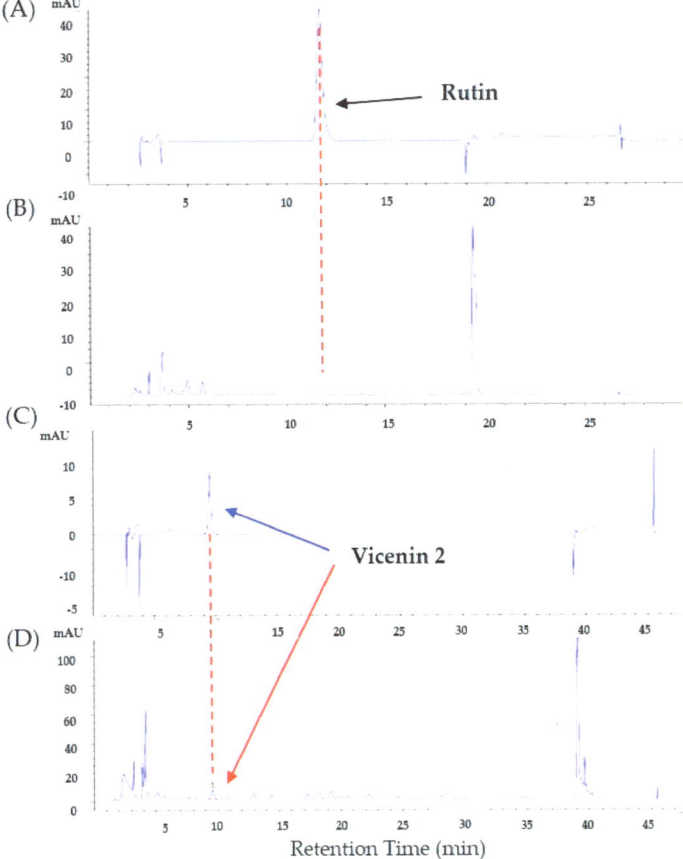

Figure 5. Quantification and identification of vicenin-2 in *D. caulis* by high-performance liquid chromatography (HPLC). HPLC chromatograms of rutin standard (**A**) and *D. caulis* extract for rutin analysis (**B**), and vicenin-2 standard (**C**) and *D. caulis* extract for vicenin-2 analysis (**D**). Black, blue, and red arrows on peaks indicate representative rutin standard, vicenin-2 standard, and vicenin-2 in *D. caulis*, respectively. Retention time and absorbance unit are shown on the X-axis and Y-axis, respectively.

3.5. Administration of Vicenin-2 Alleviates Paclitaxel-Induced Pain

Vicenin-2, a major component of *D. caulis*, has been confirmed to exert an antinociceptive effect on PINP. To evaluate the dose-dependent antinociceptive effect of vicenin-2 on PINP, vicenin-2 was intraperitoneally administered to mice at two concentrations (1 and 10 mg/kg). Behavioral responses were recorded before and 1 h after administration of vicenin-2 in the pain-induced mice. The behavioral assessment showed that cold and mechanical allodynia were greatly alleviated in the 10 mg/kg vicenin-2 administration group (Figure 6). When the results obtained one hour after the injection of PBS or 1 mg/kg or 10 mg/kg of vicenin-2 were analyzed by using one-way ANOVA, the F and p values in ANOVA interaction were F = 20.76 and $p < 0.0001$ and F = 53.50 and $p < 0.0001$ for cold and mechanical allodynia, respectively.

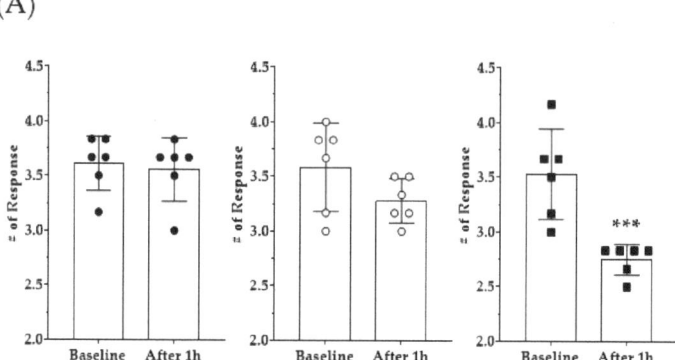

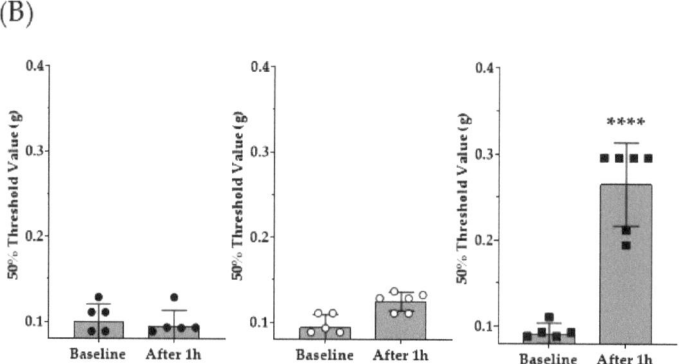

Figure 6. Effects of vicenin-2 on paclitaxel-induced neuropathic pain in mice (**A**,**B**). All groups received multiple injections of paclitaxel (total 8 mg/kg). Doses of 1 and 10 mg/kg of vicenin-2 were injected intraperitoneally in mice. PBS was used as a vehicle for vicenin-2. The black circle, white circle, and black square represent PBS control group, Vicenin-2 1 mg/kg group, and 10 mg/kg group, respectively. N = 6 for each group; *** $p < 0.001$ and **** $p < 0.0001$ vs. control group with paired *t*-test.

3.6. D. caulis Mimics the Role of TRPV1 Antagonist

The TRPV1 antagonist capsazepine and *D. caulis* were administered intrathecally, and their antinociceptive effects were compared. In the groups administered with paclitaxel and PBS, there was no change in the pain response even after 1 h. When 1 μg/mouse of capsazepine, a TRPV1 receptor antagonist, was injected intrathecally, cold and mechanical allodynia were significantly alleviated (Figure 7). In addition, *D. caulis* showed excellent antinociceptive effects when administered intrathecally at 1 mg/kg/mouse, with no significant difference from the antinociceptive effect of capsazepine (Figure 7). When the results obtained one hour after the injection of PBS, 1 µg of capsazepine, or 0.1 mg or 1 mg of *D. caulis* were analyzed by using one-way ANOVA, the F and *p* values in ANOVA interaction were F = 17.09 and $p < 0.0001$ and F = 26.50 and $p < 0.0001$ for cold and mechanical allodynia, respectively.

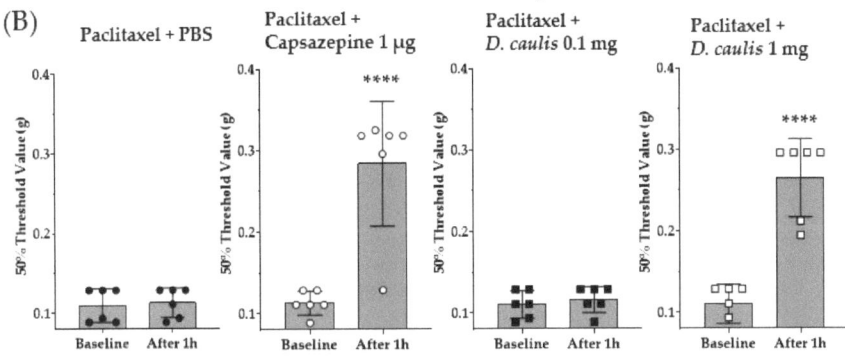

Figure 7. Effects of intrathecal injection of capsazepine and *D. caulis* on paclitaxel-induced neuropathic pain in mice (**A**,**B**). All groups received multiple injections of paclitaxel (total 8 mg/kg). Doses of 1 µg of capsazepine and 0.1 and 1 mg of *D. caulis* were injected intrathecally in mice. PBS was used as a vehicle for *D. caulis*. The black circle, white circle, black square, and white square represent PBS control group, Capsazepine 1 µg, *D. caulis* 0.1 mg, and 1 mg group, respectively. N = 6 for each group; **** $p < 0.0001$ vs. control group with paired *t*-test.

3.7. Evaluation of Cell Viability of Vicenin-2 in Caco-2 and RAW 264.7 Cells

To evaluate the cytotoxic effect of vicenin-2, in vitro studies were conducted using human colonic carcinoma Caco-2 cells and non-cancerous RAW 264.7 cells. The cytotoxic effects of vicenin-2 on the viability of the two cell types are presented as the percentage viabilities of three dependent experiments (Figure 8). Vicenin-2 showed a significant cytotoxic effect on Caco-2 cells in concentrations of 75 µM and 100 µM (decreased to 91.10% and 91.05%, respectively). However, vicenin-2 was not cytotoxic in non-cancerous RAW 264.7 cells up to a concentration of 100 µM.

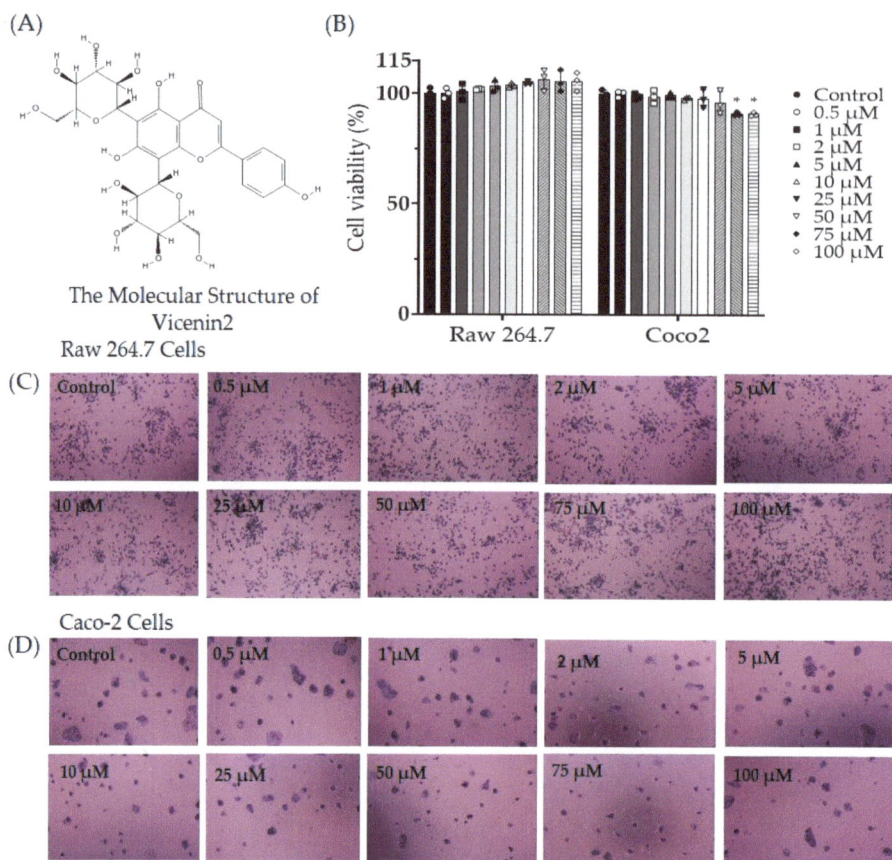

Figure 8. Evaluation of the cytotoxic effect of vicenin-2. The chemical structure of vicenin-2 (**A**). The effect of 10 different doses of vicenin-2 on the viability of RAW 264.7 and Caco-2 cells (**B**) and their morphologies (**C**,**D**). Data were expressed as the mean value ± SD of three independent experiments. N = 3 for each group; * $p < 0.05$ vs. control group with one-way ANOVA followed by Tukey's multiple comparison test.

4. Discussion

Although paclitaxel has excellent efficacy, only a few injections can cause serious neuropathy, which reduces the patient's quality of life and may lead to treatment discontinuation. Several analgesic agents are used, but they also induce side effects which limit their wide use [59–61]. In this study, 100, 300, and 500 mg/kg *D. caulis* was orally administered to paclitaxel-treated mice. In mice treated with paclitaxel, the pain levels of cold and mechanical allodynia increased significantly and decreased in the *D. caulis* 300 and 500 mg/kg groups. Vicenin-2, the active ingredient of *D. caulis*, was subjected to qualitative and quantitative analyses using HPLC and showed a significant antinociceptive effect at 10 mg/kg. Vicenin-2 did not cause significant cell death in non-cancerous Caco-2 cells up to 50 μM. To determine whether *D. caulis* exhibits an antagonistic effect on TRPV1, capsazepine and *D. caulis* were injected intrathecally, and the same antinociceptive effect was confirmed. In this study, all experiments were conducted in male mice only as some papers have reported that paclitaxel-induced neuropathic pain is not sex specific [57,62]. In addition, no significant difference in body weight change was observed between the sexes after paclitaxel injection [63].

Increased TRPV1 expression in the dorsal horn of the spinal cord has been studied in various types of pain, such as paclitaxel-induced pain (4 mg/kg), sciatic nerve injury, and diabetes-induced neuropathy [64–66]. In this study, TRPV1 expression increased in lumbar 4–5 of the spinal cord after paclitaxel administration (total 8 mg/kg), and this upregulation was reduced by *D. caulis*. Furthermore, intrathecal injections of the TRPV1 antagonists capsazepine showed similar analgesic effects to *D. caulis*. Although in this study, the direct effect of *D. caulis* or vicenin-2 on TRPV1 have not been assessed, Kang et al. [67] have reported that 1 h after the intrathecal administration of capsazepine, a significant decrease in TRPV1 protein expression was shown in the spinal cord, suggesting that *D. caulis* may have exerted TRPV1 antagonist-like effects.

Furthermore, early clinical studies have reported that TRPV1 antagonists (AMG-517, 3 mg/kg) increase core body temperature and reduce heat pain recognition in healthy subjects [68]. Preclinical studies have shown that TRPV1 antagonists (XEN-D0501, 5 mg/kg and ABT-102, 4 mg/kg) can block heat hypersensitivity to many proinflammatory agents (e.g., complete Freund's adjuvant and carrageenan) [69,70]. Preclinical studies using surgical nerve injury models have also reported that TRPV1 expression is altered under chronic pain conditions.

In rodent and human studies, the presence of the TRPV1 protein and mRNA has been confirmed in the spinal cord, hippocampus, hypothalamus, locus coeruleus, striatum, and cerebellum [29]. Focusing on the spinal cord, TRPV1 channels were positive in the presynaptic (from the central terminal of sensory neurons) and postsynaptic regions (from dendrites of spinal cord dorsal horn neurons) and were particularly prominent in areas I and II of the superficial laminae, the first relay regions of the pain sensory pathway [71]. In a study on rhizotomy, a spinal nerve root cutting surgery, histology showed that the postsynaptic TRPV1 expression level was highly dependent on peripheral input, indicating that spinal TRPV1 expression and function can be dynamically regulated in a sensory state. In addition, TRPV1 expression was confirmed in the spinal cord and central nervous system even after rhizotomy, suggesting that TRPV1 receptor proteins show a broad and individual distribution pattern in the central nervous system and that TRPV1 receptors exist in brain regions [72,73]. The antinociceptive effects of TRPV1 antagonists have been studied in several pain models. As reported in these studies, capsazepine, SD-705498, and JNJ-17203212 modestly alleviated osteoarthritic and bone cancer pain related to neuropathic pain [74–76]. These results suggest that TRPV1 plays a central role in pain transmission in the spinal cord.

The composition of *D. caulis* was studied using HPLC with electrospray ionization multistage mass spectrometry (HPLC-ESI-MS). The structures of 1 lignan, 6 phenolic acids, and 12 flavonoids have been identified in *D. caulis* [19]. Rutin and vicenin-2 are the major components of *D. caulis*. Rutin has attracted attention as a new antioxidant [77], and vicenin-2 has exhibited anticancer activity by inhibiting the growth of prostate cancer (DU-145, PC-3, and LNCaP cells) cells in vivo and in vitro [78,79] and by inducing apoptosis in HT-29 human colon cancer cells [80]. In addition, vicenin-2 has shown an anti-inflammatory effect by suppressing the protein signaling pathway triggered by TGFβ and inducing H23 cell death through PI3K/Akt/mTOR signaling [81,82]. Vicenin-2 is a major flavonoid component of *D. caulis* [78], and academic consensus suggests that flavonoids can cross the blood–brain barrier (BBB) and reach the central nervous system [83]. Therefore, it has been suggested that vicenin-2 may have a direct effect on the brain and may be used preventively or therapeutically for diseases such as neuropathy.

However, no previous studies have investigated the antinociceptive effects on neuropathic pain induced by anticancer drugs. In this study, rutin was not detected, and vicenin-2 was detected at 0.245%. It has been reported that *D. caulis* varies greatly depending on the harvesting region [84]. Although *D. caulis* and vicenin-2 showed analgesic effects in this study, it cannot be certain that the analgesic effect is entirely dependent on vicenin-2. This is because natural extracts contain many phytochemicals. To confirm the independent effect of vicenin-2, studies should be performed using the Ca^{2+} imaging technique in the TRPV1 overexpression system (e.g., transfected HEK 293 cells) or TRPV1 knock-out mouse. By carefully reviewing the literature relevant to our study, we have found that downregulation of TRPV1 also affects the Ca^{2+} current [85]. This suggests that the gene and protein expression changes demonstrated in our study may also be related to the Ca^{2+} current evoked by spinal TRPV1.

In evaluating the cytotoxicity of vicenin-2 in this study, no cytotoxicity was observed up to 100 μM in non-cancerous cell lines, and toxicity was observed at 75–100 μM in cancer cell lines. In addition, it was reported that vicenin-2 has no cytotoxicity in human dermal fibroblast cells and human umbilical vein endothelial cells up to 400 μM and 100 μM, respectively [80,86]. On the other hand, in human colorectal adenocarcinoma cells (HT-29), which are cancer cells, significant cytotoxicity was observed starting at 25 μM [80]. To date, no adverse effects have been reported for *D. caulis* or vicenin-2. A limitation of this study was the lack of safety data regarding the oral use of *D. caulis*. There is still a lack of safety data owing to limited research, and additional in vivo and cellular studies are needed to confirm its safety. In addition, calcium imaging and electrophysiological studies should be conducted to better understand the correlation between *D. caulis* and TRPV1 channels.

5. Conclusions

D. caulis, the stem of *D. officinale*, is an important herbal medicine and a natural health food. Collectively, these results suggest that *D. caulis* exerts its antinociceptive effect on PINP by modulating the activity of TRPV1, and the effective substance is vicenin-2. This study provided new insights into the mode of action based on the pharmacological effects of *D. caulis*. Although several biological effects of *D. caulis* have been studied, to the best of our knowledge, this is the first study to suggest an antinociceptive effect of *D. caulis* on anticancer drug-induced neuropathic pain. This suggests that it may be a potentially beneficial ingredient in medicines and food.

6. Patent

The content of this article is related to a patent application in Korea (10-2023-0146282).

Author Contributions: Conceptualization, W.K.; methodology, W.K. and K.T.P.; validation, K.T.P. and Y.J.J.; data curation, K.T.P. and Y.J.J.; writing—original draft preparation, K.T.P. and H.I.K.; writing—review and editing, W.K.; supervision, W.K.; project administration, W.K. All authors have read and agreed to the published version of the manuscript.

Funding: This work was supported by the National Research Foundation of Korea (NRF) and the grants were funded by the Korean government (MSIT): No. 2020R1A5A2019413.

Institutional Review Board Statement: All experimental protocols were approved by the Kyung Hee University Animal Care and Use Committee (approval NO. KHUASP-23-223, 12 April 2023).

Informed Consent Statement: Not applicable.

Data Availability Statement: The data presented in this study are available on request from the corresponding author.

Acknowledgments: We thank Jaechul Lee for providing *Dendrobii caulis*.

Conflicts of Interest: The authors declare no conflict of interest.

References

1. Addington, J.; Freimer, M. Chemotherapy-induced peripheral neuropathy: An update on the current understanding. *F1000Res* **2016**, *5*, 1466. [CrossRef] [PubMed]
2. Franconi, G.; Manni, L.; Schröder, S.; Marchetti, P.; Robinson, N. A systematic review of experimental and clinical acupuncture in chemotherapy-induced peripheral neuropathy. *Evid. Based Complement. Alternat. Med.* **2013**, *2013*, 516916. [CrossRef] [PubMed]
3. Gang, J.; Park, K.T.; Kim, S.; Kim, W. Involvement of the Spinal Serotonergic System in the Analgesic Effect of [6]-Shogaol on Oxaliplatin-Induced Neuropathic Pain in Mice. *Pharmaceuticals* **2023**, *16*, 1465. [CrossRef] [PubMed]
4. Bernabeu, E.; Cagel, M.; Lagomarsino, E.; Moretton, M.; Chiappetta, D.A. Paclitaxel: What has been done and the challenges remain ahead. *Int. J. Pharm.* **2017**, *526*, 474–495. [CrossRef] [PubMed]

5. Yang, Y.H.; Mao, J.W.; Tan, X.L. Research progress on the source, production, and anti-cancer mechanisms of paclitaxel. *Chin. J. Nat. Med.* **2020**, *18*, 890–897. [CrossRef] [PubMed]
6. Foland, T.B.; Dentler, W.L.; Suprenant, K.A.; Gupta, M.L., Jr.; Himes, R.H. Paclitaxel-induced microtubule stabilization causes mitotic block and apoptotic-like cell death in a paclitaxel-sensitive strain of Saccharomyces cerevisiae. *Yeast* **2005**, *22*, 971–978. [CrossRef] [PubMed]
7. Markman, M. Managing taxane toxicities. *Support Care Cancer* **2003**, *11*, 144–147. [CrossRef]
8. Seretny, M.; Currie, G.L.; Sena, E.S.; Ramnarine, S.; Grant, R.; MacLeod, M.R.; Colvin, L.A.; Fallon, M. Incidence, prevalence, and predictors of chemotherapy-induced peripheral neuropathy: A systematic review and meta-analysis. *Pain* **2014**, *155*, 2461–2470. [CrossRef]
9. Hershman, D.L.; Lacchetti, C.; Dworkin, R.H.; Lavoie Smith, E.M.; Bleeker, J.; Cavaletti, G.; Chauhan, C.; Gavin, P.; Lavino, A.; Lustberg, M.B.; et al. Prevention and management of chemotherapy-induced peripheral neuropathy in survivors of adult cancers: American Society of Clinical Oncology clinical practice guideline. *J. Clin. Oncol.* **2014**, *32*, 1941–1967. [CrossRef]
10. Zotz, G. The systematic distribution of vascular epiphytes–A critical update. *Bot. J. Linn. Soc.* **2013**, *171*, 453–481. [CrossRef]
11. Ngo, L.T.; Okogun, J.I.; Folk, W.R. 21st century natural product research and drug development and traditional medicines. *Nat. Prod. Rep.* **2013**, *30*, 584–592. [CrossRef] [PubMed]
12. Li, G.; Lu, J.; Chen, X. Some worries about Dendrobium officinale industry. *China J. Chin. Mater. Medica* **2013**, *38*, 469–471.
13. Liang, J.; Li, H.; Chen, J.; He, L.; Du, X.; Zhou, L.; Xiong, Q.; Lai, X.; Yang, Y.; Huang, S.; et al. Dendrobium officinale polysaccharides alleviate colon tumorigenesis via restoring intestinal barrier function and enhancing anti-tumor immune response. *Pharmacol. Res.* **2019**, *148*, 104417. [CrossRef] [PubMed]
14. Liang, J.; Chen, S.; Hu, Y.; Yang, Y.; Yuan, J.; Wu, Y.; Li, S.; Lin, J.; He, L.; Hou, S.; et al. Protective roles and mechanisms of Dendrobium officinal polysaccharides on secondary liver injury in acute colitis. *Int. J. Biol. Macromol.* **2018**, *107*, 2201–2210. [CrossRef] [PubMed]
15. Yang, J.; Chen, H.; Nie, Q.; Huang, X.; Nie, S. Dendrobium officinale polysaccharide ameliorates the liver metabolism disorders of type II diabetic rats. *Int. J. Biol. Macromol.* **2020**, *164*, 1939–1948. [CrossRef] [PubMed]
16. Zhang, L.J.; Huang, X.J.; Shi, X.D.; Chen, H.H.; Cui, S.W.; Nie, S.P. Protective effect of three glucomannans from different plants against DSS induced colitis in female BALB/c mice. *Food Funct.* **2019**, *10*, 1928–1939. [CrossRef] [PubMed]
17. Yang, K.; Lu, T.; Zhan, L.; Zhou, C.; Zhang, N.; Lei, S.; Wang, Y.; Yang, J.; Yan, M.; Lv, G.; et al. Physicochemical characterization of polysaccharide from the leaf of Dendrobium officinale and effect on LPS induced damage in GES-1 cell. *Int. J. Biol. Macromol.* **2020**, *149*, 320–330. [CrossRef] [PubMed]
18. Tang, H.; Zhao, T.; Sheng, Y.; Zheng, T.; Fu, L.; Zhang, Y. Dendrobium officinale Kimura et Migo: A Review on Its Ethnopharmacology, Phytochemistry, Pharmacology, and Industrialization. *Evid. Based Complement. Alternat. Med.* **2017**, *2017*, 7436259. [CrossRef]
19. Ye, Z.; Dai, J.R.; Zhang, C.G.; Lu, Y.; Wu, L.L.; Gong, A.G.W.; Xu, H.; Tsim, K.W.K.; Wang, Z.T. Chemical Differentiation of Dendrobium officinale and Dendrobium devonianum by Using HPLC Fingerprints, HPLC-ESI-MS, and HPTLC Analyses. *Evid. Based Complement. Alternat. Med.* **2017**, *2017*, 8647212. [CrossRef]
20. Price, D.D. Psychological and neural mechanisms of the affective dimension of pain. *Science* **2000**, *288*, 1769–1772. [CrossRef]
21. Costigan, M.; Woolf, C.J. Pain: Molecular mechanisms. *J. Pain* **2000**, *1*, 35–44. [CrossRef] [PubMed]
22. Venkatachalam, K.; Montell, C. TRP channels. *Annu. Rev. Biochem.* **2007**, *76*, 387–417. [CrossRef] [PubMed]
23. Caterina, M.J.; Schumacher, M.A.; Tominaga, M.; Rosen, T.A.; Levine, J.D.; Julius, D. The capsaicin receptor: A heat-activated ion channel in the pain pathway. *Nature* **1997**, *389*, 816–824. [CrossRef] [PubMed]
24. Szallasi, A.; Blumberg, P.M. Vanilloid (Capsaicin) receptors and mechanisms. *Pharmacol. Rev.* **1999**, *51*, 159–212. [PubMed]
25. Szallasi, A.; Nilsson, S.; Farkas-Szallasi, T.; Blumberg, P.M.; Hökfelt, T.; Lundberg, J.M. Vanilloid (capsaicin) receptors in the rat: Distribution in the brain, regional differences in the spinal cord, axonal transport to the periphery, and depletion by systemic vanilloid treatment. *Brain Res.* **1995**, *703*, 175–183. [CrossRef]
26. Szallasi, A.; Blumberg, P.M. Characterization of vanilloid receptors in the dorsal horn of pig spinal cord. *Brain Res.* **1991**, *547*, 335–338. [CrossRef]
27. Fischer, M.J.; Reeh, P.W.; Sauer, S.K. Proton-induced calcitonin gene-related peptide release from rat sciatic nerve axons, in vitro, involving TRPV1. *Eur. J. Neurosci.* **2003**, *18*, 803–810. [CrossRef]
28. Bernardini, N.; Neuhuber, W.; Reeh, P.W.; Sauer, S.K. Morphological evidence for functional capsaicin receptor expression and calcitonin gene-related peptide exocytosis in isolated peripheral nerve axons of the mouse. *Neuroscience* **2004**, *126*, 585–590. [CrossRef]
29. Mezey, E.; Tóth, Z.E.; Cortright, D.N.; Arzubi, M.K.; Krause, J.E.; Elde, R.; Guo, A.; Blumberg, P.M.; Szallasi, A. Distribution of mRNA for vanilloid receptor subtype 1 (VR1), and VR1-like immunoreactivity, in the central nervous system of the rat and human. *Proc. Natl. Acad. Sci. USA* **2000**, *97*, 3655–3660. [CrossRef]
30. Szallasi, A.; Conte, B.; Goso, C.; Blumberg, P.M.; Manzini, S. Characterization of a peripheral vanilloid (capsaicin) receptor in the urinary bladder of the rat. *Life Sci.* **1993**, *52*, PL221–PL226. [CrossRef]
31. Russell, J.A.; Lai-Fook, S.J. Reflex bronchoconstriction induced by capsaicin in the dog. *J. Appl. Physiol.* **1979**, *47*, 961–967. [CrossRef] [PubMed]

32. Lundberg, J.M.; Martling, C.R.; Saria, A. Substance P and capsaicin-induced contraction of human bronchi. *Acta Physiol. Scand.* **1983**, *119*, 49–53. [CrossRef] [PubMed]
33. Chen, Y.; Yang, C.; Wang, Z. Proteinase-activated receptor 2 sensitizes transient receptor potential vanilloid 1, transient receptor potential vanilloid 4, and transient receptor potential ankyrin 1 in paclitaxel-induced neuropathic pain. *Neuroscience* **2011**, *193*, 440–451. [CrossRef] [PubMed]
34. Hara, T.; Chiba, T.; Abe, K.; Makabe, A.; Ikeno, S.; Kawakami, K.; Utsunomiya, I.; Hama, T.; Taguchi, K. Effect of paclitaxel on transient receptor potential vanilloid 1 in rat dorsal root ganglion. *Pain* **2013**, *154*, 882–889. [CrossRef] [PubMed]
35. Anand, U.; Otto, W.R.; Anand, P. Sensitization of capsaicin and icilin responses in oxaliplatin treated adult rat DRG neurons. *Mol. Pain* **2010**, *6*. [CrossRef]
36. Nassini, R.; Gees, M.; Harrison, S.; De Siena, G.; Materazzi, S.; Moretto, N.; Failli, P.; Preti, D.; Marchetti, N.; Cavazzini, A. Oxaliplatin elicits mechanical and cold allodynia in rodents via TRPA1 receptor stimulation. *Pain* **2011**, *152*, 1621–1631. [CrossRef]
37. Chen, K.; Zhang, Z.-F.; Liao, M.-F.; Yao, W.-L.; Wang, J.; Wang, X.-R. Blocking PAR2 attenuates oxaliplatin-induced neuropathic pain via TRPV1 and releases of substance P and CGRP in superficial dorsal horn of spinal cord. *J. Neurol. Sci.* **2015**, *352*, 62–67. [CrossRef]
38. Eid, S.R.; Crown, E.D.; Moore, E.L.; Liang, H.A.; Choong, K.C.; Dima, S.; Henze, D.A.; Kane, S.A.; Urban, M.O. HC-030031, a TRPA1 selective antagonist, attenuates inflammatory- and neuropathy-induced mechanical hypersensitivity. *Mol. Pain* **2008**, *4*, 48. [CrossRef]
39. Witting, N.; Svensson, P.; Gottrup, H.; Arendt-Nielsen, L.; Jensen, T.S. Intramuscular and intradermal injection of capsaicin: A comparison of local and referred pain. *Pain* **2000**, *84*, 407–412. [CrossRef]
40. Boyette-Davis, J.A.; Cata, J.P.; Zhang, H.; Driver, L.C.; Wendelschafer-Crabb, G.; Kennedy, W.R.; Dougherty, P.M. Follow-up psychophysical studies in bortezomib-related chemoneuropathy patients. *J. Pain* **2011**, *12*, 1017–1024. [CrossRef]
41. Boyette-Davis, J.A.; Cata, J.P.; Driver, L.C.; Novy, D.M.; Bruel, B.M.; Mooring, D.L.; Wendelschafer-Crabb, G.; Kennedy, W.R.; Dougherty, P.M. Persistent chemoneuropathy in patients receiving the plant alkaloids paclitaxel and vincristine. *Cancer Chemother. Pharmacol.* **2013**, *71*, 619–626. [CrossRef] [PubMed]
42. Simone, D.A.; Baumann, T.K.; LaMotte, R.H. Dose-dependent pain and mechanical hyperalgesia in humans after intradermal injection of capsaicin. *Pain* **1989**, *38*, 99–107. [CrossRef] [PubMed]
43. Marchettini, P.; Simone, D.A.; Caputi, G.; Ochoa, J. Pain from excitation of identified muscle nociceptors in humans. *Brain Res.* **1996**, *740*, 109–116. [CrossRef] [PubMed]
44. Quartu, M.; Carozzi, V.A.; Dorsey, S.; Serra, M.P.; Poddighe, L.; Picci, C.; Boi, M.; Melis, T.; Del Fiacco, M.; Meregalli, C. Bortezomib treatment produces nocifensive behavior and changes in the expression of TRPV1, CGRP, and substance P in the rat DRG, spinal cord, and sciatic nerve. *BioMed Res. Int.* **2014**, *2014*, 180428. [CrossRef] [PubMed]
45. Ta, L.E.; Bieber, A.J.; Carlton, S.M.; Loprinzi, C.L.; Low, P.A.; Windebank, A.J. Transient Receptor Potential Vanilloid 1 is essential for cisplatin-induced heat hyperalgesia in mice. *Mol. Pain* **2010**, *6*. [CrossRef] [PubMed]
46. Li, Y.; Adamek, P.; Zhang, H.; Tatsui, C.E.; Rhines, L.D.; Mrozkova, P.; Li, Q.; Kosturakis, A.K.; Cassidy, R.M.; Harrison, D.S. The cancer chemotherapeutic paclitaxel increases human and rodent sensory neuron responses to TRPV1 by activation of TLR4. *J. Neurosci.* **2015**, *35*, 13487–13500. [CrossRef]
47. Hohmann, S.W.; Angioni, C.; Tunaru, S.; Lee, S.; Woolf, C.J.; Offermanns, S.; Geisslinger, G.; Scholich, K.; Sisignano, M. The G2A receptor (GPR132) contributes to oxaliplatin-induced mechanical pain hypersensitivity. *Sci. Rep.* **2017**, *7*, 446. [CrossRef]
48. National Research Council (US) Committee for the Update of the Guide for the Care and Use of Laboratory Animals. *Guide for the Care and Use of Laboratory Animals*; The National Academies Press: Washington, DC, USA, 2011.
49. Yoon, C.; Wook, Y.Y.; Sik, N.H.; Ho, K.S.; Mo, C.J. Behavioral signs of ongoing pain and cold allodynia in a rat model of neuropathic pain. *Pain* **1994**, *59*, 369–376. [CrossRef]
50. Lee, J.H.; Ji, H.; Ko, S.G.; Kim, W. JI017 Attenuates Oxaliplatin-Induced Cold Allodynia via Spinal TRPV1 and Astrocytes Inhibition in Mice. *Int. J. Mol. Sci.* **2021**, *22*, 8811. [CrossRef]
51. Lee, J.H.; Min, D.; Lee, D.; Kim, W. Zingiber officinale Roscoe Rhizomes Attenuate Oxaliplatin-Induced Neuropathic Pain in Mice. *Molecules* **2021**, *26*, 548. [CrossRef]
52. Chung, K. Allodynia Test, Mechanical and Cold Allodynia. In *Encyclopedia of Pain*; Schmidt, R.F., Willis, W.D., Eds.; Springer: Berlin/Heidelberg, Germany, 2007; pp. 55–57.
53. Dixon, W.J. Efficient analysis of experimental observations. *Annu. Rev. Pharmacol. Toxicol.* **1980**, *20*, 441–462. [CrossRef] [PubMed]
54. Chaplan, S.R.; Bach, F.W.; Pogrel, J.W.; Chung, J.M.; Yaksh, T.L. Quantitative assessment of tactile allodynia in the rat paw. *J. Neurosci. Methods* **1994**, *53*, 55–63. [CrossRef] [PubMed]
55. Rao, X.; Huang, X.; Zhou, Z.; Lin, X. An improvement of the 2ˆ(-delta delta CT) method for quantitative real-time polymerase chain reaction data analysis. *Biostat. Bioinforma. Biomath.* **2013**, *3*, 71–85. [PubMed]
56. Choi, J.; Jeon, C.; Lee, J.H.; Jang, J.U.; Quan, F.S.; Lee, K.; Kim, W.; Kim, S.K. Suppressive effects of bee venom acupuncture on paclitaxel-induced neuropathic pain in rats: Mediation by spinal α2-adrenergic receptor. *Toxins* **2017**, *9*, 351. [CrossRef] [PubMed]
57. Park, K.-T.; Kim, S.; Choi, I.; Han, I.-H.; Bae, H.; Kim, W. The involvement of the noradrenergic system in the antinociceptive effect of cucurbitacin D on mice with paclitaxel-induced neuropathic pain. *Front. Pharmacol.* **2023**, *13*, 1055264. [CrossRef]
58. Lee, J.H.; Kim, B.; Ko, S.-G.; Kim, W. Analgesic effect of SH003 and Trichosanthes kirilowii Maximowicz in paclitaxel-induced neuropathic pain in mice. *Curr. Issues Mol. Biol.* **2022**, *44*, 718–730. [CrossRef]

59. Riediger, C.; Schuster, T.; Barlinn, K.; Maier, S.; Weitz, J.; Siepmann, T. Adverse Effects of Antidepressants for Chronic Pain: A Systematic Review and Meta-analysis. *Front Neurol.* **2017**, *8*, 307. [CrossRef]
60. Quilici, S.; Chancellor, J.; Löthgren, M.; Simon, D.; Said, G.; Le, T.K.; Garcia-Cebrian, A.; Monz, B. Meta-analysis of duloxetine vs. pregabalin and gabapentin in the treatment of diabetic peripheral neuropathic pain. *BMC Neurol.* **2009**, *9*, 6. [CrossRef]
61. Feighner, J.P. The role of venlafaxine in rational antidepressant therapy. *J. Clin. Psychiatry.* **1994**, *55* (Suppl. A), 62–68.
62. Liu, X.; Tonello, R.; Ling, Y.; Gao, Y.-J.; Berta, T. Paclitaxel-activated astrocytes produce mechanical allodynia in mice by releasing tumor necrosis factor-α and stromal-derived cell factor 1. *J. Neuroinflammation* **2019**, *16*, 209. [CrossRef]
63. Hwang, B.-Y.; Kim, E.-S.; Kim, C.-H.; Kwon, J.-Y.; Kim, H.-K. Gender differences in paclitaxel-induced neuropathic pain behavior and analgesic response in rats. *Korean J. Anesthesiol.* **2012**, *62*, 66–72. [CrossRef] [PubMed]
64. Bishnoi, M.; Bosgraaf, C.A.; Abooj, U.; Zhong, L.; Premkumar, L.S. Streptozotocin-induced early thermal hyperalgesia is independent of glycemic state of rats: Role of transient receptor potential vanilloid 1 (TRPV1) and inflammatory mediators. *Mol. Pain* **2011**, *7*, 1744–8069. [CrossRef] [PubMed]
65. Kamata, Y.; Kambe, T.; Chiba, T.; Yamamoto, K.; Kawakami, K.; Abe, K.; Taguchi, K. Paclitaxel induces upregulation of transient receptor potential vanilloid 1 expression in the rat spinal cord. *Int. J. Mol. Sci.* **2020**, *21*, 4341. [CrossRef] [PubMed]
66. Baba, K.; Kawasaki, M.; Nishimura, H.; Suzuki, H.; Matsuura, T.; Fujitani, T.; Tsukamoto, M.; Tokuda, K.; Yamanaka, Y.; Ohnishi, H. Heat hypersensitivity is attenuated with altered expression level of spinal astrocytes after sciatic nerve injury in TRPV1 knockout mice. *Neurosci. Res.* **2021**, *170*, 273–283. [CrossRef] [PubMed]
67. Kang, S.Y.; Seo, S.Y.; Bang, S.K.; Cho, S.J.; Choi, K.H.; Ryu, Y. Inhibition of Spinal TRPV1 Reduces NMDA Receptor 2B Phosphorylation and Produces Anti-Nociceptive Effects in Mice with Inflammatory Pain. *Int. J. Mol. Sci.* **2021**, *22*, 11177. [CrossRef] [PubMed]
68. Gavva, N.R. Body-temperature maintenance as the predominant function of the vanilloid receptor TRPV1. *Trends Pharmacol. Sci.* **2008**, *29*, 550–557. [CrossRef] [PubMed]
69. Round, P.; Priestley, A.; Robinson, J. An investigation of the safety and pharmacokinetics of the novel TRPV1 antagonist XEN-D0501 in healthy subjects. *Br. J. Clin. Pharmacol.* **2011**, *72*, 921–931. [CrossRef]
70. Rowbotham, M.C.; Nothaft, W.; Duan, W.R.; Wang, Y.; Faltynek, C.; McGaraughty, S.; Chu, K.L.; Svensson, P. Oral and cutaneous thermosensory profile of selective TRPV1 inhibition by ABT-102 in a randomized healthy volunteer trial. *Pain* **2011**, *152*, 1192–1200. [CrossRef]
71. Valtschanoff, J.G.; Rustioni, A.; Guo, A.; Hwang, S.J. Vanilloid receptor VR1 is both presynaptic and postsynaptic in the superficial laminae of the rat dorsal horn. *J. Comp. Neurol.* **2001**, *436*, 225–235. [CrossRef]
72. Roberts, J.C.; Davis, J.B.; Benham, C.D. [3H]Resiniferatoxin autoradiography in the CNS of wild-type and TRPV1 null mice defines TRPV1 (VR-1) protein distribution. *Brain Res.* **2004**, *995*, 176–183. [CrossRef]
73. Micale, V.; Cristino, L.; Tamburella, A.; Petrosino, S.; Leggio, G.M.; Drago, F.; Di Marzo, V. Anxiolytic effects in mice of a dual blocker of fatty acid amide hydrolase and transient receptor potential vanilloid type-1 channels. *Neuropsychopharmacology* **2009**, *34*, 593–606. [CrossRef] [PubMed]
74. Szallasi, A.; Cruz, F.; Geppetti, P. TRPV1: A therapeutic target for novel analgesic drugs? *Trends Mol. Med.* **2006**, *12*, 545–554. [CrossRef] [PubMed]
75. Chizh, B.A.; O'Donnell, M.B.; Napolitano, A.; Wang, J.; Brooke, A.C.; Aylott, M.C.; Bullman, J.N.; Gray, E.J.; Lai, R.Y.; Williams, P.M.; et al. The effects of the TRPV1 antagonist SB-705498 on TRPV1 receptor-mediated activity and inflammatory hyperalgesia in humans. *Pain* **2007**, *132*, 132–141. [CrossRef] [PubMed]
76. Menéndez, L.; Juárez, L.; García, E.; García-Suárez, O.; Hidalgo, A.; Baamonde, A. Analgesic effects of capsazepine and resiniferatoxin on bone cancer pain in mice. *Neurosci. Lett* **2006**, *393*, 70–73. [CrossRef] [PubMed]
77. Zhang, Y.; Zhang, L.; Liu, J.; Liang, J.; Si, J.; Wu, S. Dendrobium officinale leaves as a new antioxidant source. *J. Funct. Foods* **2017**, *37*, 400–415. [CrossRef]
78. Nagaprashantha, L.D.; Vatsyayan, R.; Singhal, J.; Fast, S.; Roby, R.; Awasthi, S.; Singhal, S.S. Anti-cancer effects of novel flavonoid vicenin-2 as a single agent and in synergistic combination with docetaxel in prostate cancer. *Biochem. Pharmacol.* **2011**, *82*, 1100–1109. [CrossRef]
79. Singhal, S.S.; Jain, D.; Singhal, P.; Awasthi, S.; Singhal, J.; Horne, D. Targeting the mercapturic acid pathway and vicenin-2 for prevention of prostate cancer. *Biochim. Biophys. Acta Rev. Cancer* **2017**, *1868*, 167–175. [CrossRef]
80. Yang, D.; Zhang, X.; Zhang, W.; Rengarajan, T. Vicenin-2 inhibits Wnt/β-catenin signaling and induces apoptosis in HT-29 human colon cancer cell line. *Drug Des. Devel. Ther.* **2018**, *12*, 1303–1310. [CrossRef]
81. Lee, W.; Ku, S.K.; Bae, J.S. Ameliorative Effect of Vicenin-2 and Scolymoside on TGFBIp-Induced Septic Responses. *Inflammation* **2015**, *38*, 2166–2177. [CrossRef]
82. Baruah, T.J.; Kma, L. Vicenin-2 acts as a radiosensitizer of the non-small cell lung cancer by lowering Akt expression. *Biofactors* **2019**, *45*, 200–210. [CrossRef]
83. Faria, A.; Meireles, M.; Fernandes, I.; Santos-Buelga, C.; Gonzalez-Manzano, S.; Dueñas, M.; de Freitas, V.; Mateus, N.; Calhau, C. Flavonoid metabolites transport across a human BBB model. *Food Chem.* **2014**, *149*, 190–196. [CrossRef] [PubMed]
84. Lan, Q.; Liu, C.; Wu, Z.; Ni, C.; Li, J.; Huang, C.; Wang, H.; Wei, G. Does the Metabolome of Wild-like Dendrobium officinale of Different Origins Have Regional Differences? *Molecules* **2022**, *27*, 7024. [CrossRef] [PubMed]

85. Krishnan, V.; Baskaran, P.; Thyagarajan, B. Troglitazone activates TRPV1 and causes deacetylation of PPARγ in 3T3-L1 cells. *Biochim. Biophys. Acta BBA-Mol. Basis Dis.* **2019**, *1865*, 445–453. [CrossRef] [PubMed]
86. Tan, W.S.; Arulselvan, P.; Ng, S.F.; Taib, C.N.M.; Sarian, M.N.; Fakurazi, S. Healing Effect of Vicenin-2 (VCN-2) on Human Dermal Fibroblast (HDF) and Development VCN-2 Hydrocolloid Film Based on Alginate as Potential Wound Dressing. *Biomed. Res. Int.* **2020**, *2020*, 4730858. [CrossRef]

Disclaimer/Publisher's Note: The statements, opinions and data contained in all publications are solely those of the individual author(s) and contributor(s) and not of MDPI and/or the editor(s). MDPI and/or the editor(s) disclaim responsibility for any injury to people or property resulting from any ideas, methods, instructions or products referred to in the content.

Systematic Review

Mangifera indica L., By-Products, and Mangiferin on Cardio-Metabolic and Other Health Conditions: A Systematic Review

Giulia Minniti [1], Lucas Fornari Laurindo [1,2,*], Nathalia Mendes Machado [1], Lidiane Gonsalves Duarte [3], Elen Landgraf Guiguer [1,3,4], Adriano Cressoni Araujo [1,3], Jefferson Aparecido Dias [3], Caroline Barbalho Lamas [5], Yandra Crevelin Nunes [2], Marcelo Dib Bechara [1,3], Edgar Baldi Júnior [3], Fabrício Bertoli Gimenes [3] and Sandra Maria Barbalho [1,3,4,*]

1. Department of Biochemistry and Pharmacology, School of Medicine, Universidade de Marília (UNIMAR), Marília 17525-902, SP, Brazil; giulia.minniti@hotmail.com (G.M.); nathaliamendesmachado@gmail.com (N.M.M.); elguiguer@gmail.com (E.L.G.); adrianocressoniaraujo@yahoo.com.br (A.C.A.); dib.marcelo1@gmail.com (M.D.B.)
2. Department of Biochemistry and Pharmacology, School of Medicine, Faculdade de Medicina de Marília (FAMEMA), Marília 17519-030, SP, Brazil; yandracrevelin@hotmail.com
3. Department of Biochemistry and Nutrition, School of Food and Technology of Marília (FATEC), Marília 17500-000, SP, Brazil; lidianegonsalves.duarte@fate.sp.gov.br (L.G.D.); jeffersondias@unimar.br (J.A.D.); reumatoedgar@hotmail.com (E.B.J.); fabricio.bg@gmail.com (F.B.G.)
4. Postgraduate Program in Structural and Functional Interactions in Rehabilitation, School of Medicine, Universidade de Marília (UNIMAR), Marília 17525-902, SP, Brazil
5. Department of Gerontology, School of Gerontology, Universidade Federal de São Carlos (UFSCar), São Carlos 13565-905, SP, Brazil; carolblamas@gmail.com
* Correspondence: lucasffffor@gmail.com (L.F.L.); smbarbalho@gmail.com (S.M.B.)

Citation: Minniti, G.; Laurindo, L.F.; Machado, N.M.; Duarte, L.G.; Guiguer, E.L.; Araujo, A.C.; Dias, J.A.; Lamas, C.B.; Nunes, Y.C.; Bechara, M.D.; et al. *Mangifera indica* L., By-Products, and Mangiferin on Cardio-Metabolic and Other Health Conditions: A Systematic Review. *Life* **2023**, *13*, 2270. https://doi.org/10.3390/life13122270

Academic Editor: Stefania Lamponi

Received: 26 October 2023
Revised: 22 November 2023
Accepted: 25 November 2023
Published: 28 November 2023

Copyright: © 2023 by the authors. Licensee MDPI, Basel, Switzerland. This article is an open access article distributed under the terms and conditions of the Creative Commons Attribution (CC BY) license (https:// creativecommons.org/licenses/by/ 4.0/).

Abstract: Mango and its by-products have traditional medicinal uses. They contain diverse bioactive compounds offering numerous health benefits, including cardioprotective and metabolic properties. This study aimed to explore the impact of mango fruit and its by-products on human health, emphasizing its metabolic syndrome components. PUBMED, EMBASE, COCHRANE, and GOOGLE SCHOLAR were searched following PRISMA guidelines, and the COCHRANE handbook was utilized to assess bias risks. In vivo and in vitro studies have shown several benefits of mango and its by-products. For this systematic review, 13 studies met the inclusion criteria. The collective findings indicated that the utilization of mango in various forms—ranging from fresh mango slices and mango puree to mango by-products, mango leaf extract, fruit powder, and mangiferin—yielded many favorable effects. These encompassed enhancements in glycemic control and improvements in plasma lipid profiles. Additionally, mango reduces food intake, elevates mood scores, augments physical performance during exercise, improves endothelial function, and decreases the incidence of respiratory tract infections. Utilizing mango by-products supports the demand for healthier products. This approach also aids in environmental conservation. Furthermore, the development of mango-derived nanomedicines aligns with sustainable goals and offers innovative solutions for healthcare challenges whilst being environmentally conscious.

Keywords: *Mangifera indica* L.; by-products; antioxidant; anti-inflammatory; diabetes; obesity; cardiovascular disease

1. Introduction

Mangifera indica L., popularly known as mango, is one of the most common tropical fruits of the genus Mangifera, which comprises around 30 species of fruit trees in the *Anacardiaceae* family. It originates from Malaysia and India, has been domesticated and cultivated for more than 4000 years, and is produced in more than 100 countries, including

Pakistan, China, Philippines, Thailand, Nigeria, Israel, Italy, Spain, Mexico, and Brazil, with India as the world's largest producer. Due to its enormous popularity, pleasant flavor, and excellent nutritional value, it ranks fifth in production among perennial fruit trees worldwide and second among the most commercialized tropical fruits, with a production of more than 40 million tons in 2021 [1–3].

Various parts of the mango, such as its fruits, flowers, leaves, roots, and peels, have been commonly used to treat multiple diseases. Its fruits are also rich in vitamin C and amino acids. The fruits, leaves, peels, and seeds are rich in phytochemicals, including polyphenols, terpenoids, carotenoids, and phytosterols. These bioactive compounds provide several health benefits, including anti-inflammatory, immunomodulatory, antibacterial, antiviral, antifungal, and anticancer effects [2–7]. Figure 1 shows the main parts of the tree.

Figure 1. Mango tree. (**A**) tree, (**B**) leaves, (**C**) bark, (**D**) flowers, (**E**) fruit, and (**F**) seed.

Mangoes contain ascorbic acid, gallic acid, protocatechuic acid, chlorogenic acid, and vanillic acid as the major phenolics. Mango pulp contains sugars; vitamins (such as ascorbic acid and carotenoids); polyphenols such as xanthonoids (mangiferin); flavonoids such as catechins, kaempferol, rhamnetin, quercetin, and anthocyanins; tannins such as gallotannins; and phenolic acids and derivatives thereof, such as ellagic acid, gallic acid, protocatechuic acid, methyl gallate, and propyl gallate [8].

Among the most significant phenolic compounds contained in mango peels, mangiferin (1,3,6,7-tetrahydroxyxanthone-C2-β-D-glucoside) is found in various amounts and forms. Although the primary source of this phenolic compound is *Mangifera indica*, it is also found in other species (*Anacardiaceae*, *Gentianaceae*, and *Iridaceae* families). Mangiferin has been investigated for its potential as a health promotion agent. It can exhibit antioxidant, anti-inflammatory, immunomodulatory, antibacterial, and anti-obesity effects. Due to these effects, it is a promising adjuvant therapeutic to chronic disorders such as cardiovascular, renal, and pulmonary diseases; neurodegenerative disorders; obesity; diabetes; and metabolic syndrome [2,9–11]. Table 1 shows the main bioactive compounds of mango and its by-products.

Besides its consumption *in natura*, mango is used to prepare many food products such as jellies, liqueurs, juice, nectar, and vinegar. Furthermore, it is used in the pharmaceutical and cosmetic industries to produce herbal medicines and cosmetics [12,13]. This vast range of uses leads to a substantial economic and environmental impact regarding the generation of by-products, including leaves, peel, seeds, bark, and extracts (from leaves, peel, and bark) [14–16].

These by-products exhibit several bioactive compounds beneficial to human health, including vitamins (A, B, C, and E) and other antioxidants such as mangiferin, benzophenones (iriflophenone 3-C-glucoside), anthocyanins, phenolic acids, gallic acid, coumarin, quercetin, and flavonoids [12,17–24].

Since mango and its by-products have a lot of versatility in terms of the presence of bioactive compounds and can therefore contribute to the prevention and treatment of health conditions, the aim of this study is to investigate the effects of mango, mangiferin, and its by-products on cardiometabolic and other health conditions. Only two other systematic reviews have investigated the impact of mango on human health. Zarasvand et al. [25] investigated the effects of the mango plant on type 2 diabetes mellitus; however, they did not investigate the impact on metabolic syndrome, a condition closely related to hyperglycemia. Lum et al. [26] investigated the effects of mangiferin only on memory impairment. For this reason, and to the best of our knowledge, this is the first systematic review addressing the effects of mango, mangiferin, and by-products on human health, with an emphasis on metabolic syndrome, which is a risk condition for the development of cardiovascular diseases, which are among the leading causes of death in the world.

Table 1. Main bioactive compounds of *M. indica* and their effects on human health.

Phytochemical	Structure	Actions	Ref.
Mangiferin		Antidiabetic, hypolipidemic, antioxidant, and anti-inflammatory	[17,18]
Catechin		Antidiabetic, antioxidant, antimicrobial, and anti-inflammatory	[18,27,28]
Quercetin		Antioxidant and anti-inflammatory	[28]
Ferulic acid		Antioxidant, anti-inflammatory, and photoprotective	[18,28]
Vanillic acid		Antioxidant	[18,29]

Table 1. Cont.

Phytochemical	Structure	Actions	Ref.
4-hydroxybenzoic acid		Antioxidant and anti-inflammatory	[28]
Gallic acid		Antioxidant, anti-inflammatory, antimicrobial, and antiproliferative	[18,28]
Coumarin		Antibiotic, bronchodilator, fungicide, anticoagulant, vasodilator, spasmolytic, and antithrombotic	[18,28]
Iriflophenone 3-C-glucoside		Antioxidant and antiproliferative	[30,31]
Antocyanin		Antioxidant, anti-inflammatory, antidiabetic, and antiproliferative	[32,33]

2. Materials and Methods

This review involved two sections: the first was based on the application of *M. indica* by-products in the food industry. The second part involved a systematic review of the health effects of the mango plant and its derivatives. The strategies for searching and including clinical trials were based on the following aspects:

2.1. Focal Question

The focused question was "Can *M. indica* L. have beneficial effects on health?".

2.2. Language

Only studies in English were selected.

2.3. Databases

This review included studies in MEDLINE–PubMed, COCHRANE, EMBASE, Google Scholar, and Science Direct databases. The mesh terms used were *M. indica*, mango, mangiferin, human health, antioxidant, anti-inflammatory, obesity, diabetes, metabolic syndrome, and cardiovascular disease.

The use of these descriptors helped identify studies related to mango and its health effects. We followed PRISMA (Preferred Reporting Items for a Systematic Review and Meta-Analysis) guidelines [34,35] to perform the search for clinical trials. Moreover, we also consulted in vivo and in vitro studies to help in the discussion section.

2.4. Study Selection

Conferences, abstracts, letters to editors, and other sources were evaluated but not included. The inclusion criteria were only human interventional studies, and the exclusion criteria were reviews, studies not in English, editorials, case reports, and poster presentations.

2.5. Search and Selection of Relevant Articles

The PICO (Patients, Intervention, Comparison, and Outcomes) format was used to perform the systematic review, and the flow diagram shows the selection of the randomized clinical trials (Figure 2) and the inclusion and exclusion processes.

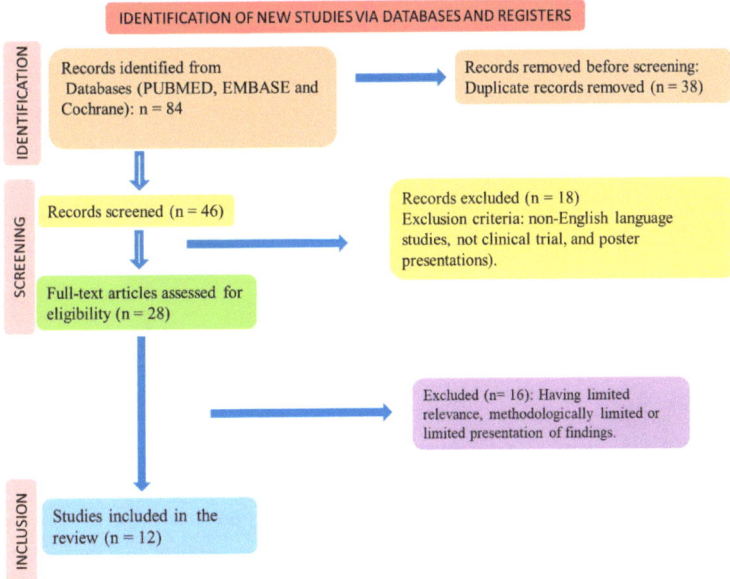

Figure 2. Flow diagram showing the study selection.

2.6. Quality Assessment

We examined the Cochrane Handbook for the systematic review of interventions to evaluate the bias risks related to the selection of randomized clinical trials.

3. Results

Figure 2 shows the study selection according to PRISMA (Preferred Reporting Items for a Systematic Review and Meta-Analysis) guidelines [34,35].

The selection of the included clinical trials is shown in Figure 2, and the results of the selected articles are found in Table 2. Twelve clinical trials were found. The risk of bias for these studies is shown in Table 3. Thirteen studies were included in this review. The findings of these studies revealed that the utilization of mango in different forms, including fresh mango slices, mango puree, mango by-products, mango leaf extract, fruit powder, and mangiferin yielded several positive effects. These effects encompassed improvements in glycemic control and plasma lipid levels, such as reductions in triglycerides, LDL cholesterol, and total cholesterol, as well as an elevation in HDL cholesterol. Additionally, mango consumption was associated with decreased appetite and food intake, enhanced mood scores, improved physical performance during exercise, enhanced endothelial function, and a reduced incidence of respiratory tract infections.

Table 2. Clinical trials showing the effects of *M. indica* L. or its derivatives on human health.

Study Type and Patients	Interventions	Results	Side Effects	Ref.
• Interventional study; • 23 overweight and obese adults, including 15 men and 8 women, 18–55 y; • San Diego, United States.	Participants received 100 kcal of mango snacks	After 45 min of ingestion, significant ↑ in cholecystokinin ($p \leq 0.05$) and adiponectin ($p \leq 0.05$) levels. Additionally, participants reported ↓ hunger and an increase in feelings of satiety. Interestingly, no significant differences were found in the levels of satiety hormones, including leptin, ghrelin, and peptide YY, among the interventions. The consumption improved postprandial glycemic responses.	Not mentioned by the authors.	[8]
• Randomized crossover clinical trial; • 27 overweight/obese patients (16 men and 11 women) without gluten allergies, non-smokers, non-pregnant, and non-lactating; • United States.	The individuals were divided and analyzed in two phases: for 12 weeks, they were offered 100 Kcal of fresh mango/day; for another 12 weeks, they were given 24 g of LFC/day.	Fresh mango: ↓ glycemia ($p = 0.004$) in weeks 4–12. There were no significant modifications in insulin levels. ↓ LDL-c, and additionally a slight increase in adiponectin. ↓ AST and PCR and ↑ total antioxidant capacity. LFC: ↑ TG levels and basal insulin, ↓ adiponectin at the end of the 12 weeks. ↑ PCR but no significant change in total antioxidant capacity. For both interventions, there were no significant differences in HbA1C values, anthropometry, or blood pressure.	Not mentioned by the authors.	[36]
• Interventional trial; • 31 overweight women, 25–45 y (no history of diabetes, hypertension, cardiovascular, or liver diseases); • Pakistan.	Over a period of 84 days, 21 subjects were administered 1 g of mango peel powder 2x/d, 30 min before each meal. 10 subjects received no treatment.	The mango group showed a decrease in weight gain and an increase in antioxidant levels. ↓ Urea, creatinin, TG, TC, and LDL-c. ↑ HDL-c levels when compared to the control group. The mango group also experienced a decrease in appetite compared to the control group.	Not mentioned by the authors.	[37]
• 39 participants: group 1/healthy weight was divided into two subgroups: one subgroup consumed a combination of raspberry and mango, and the other subgroup consumed a combination of passion fruit and mango. Group 2/obesity completed both subgroups; • United Kingdom.	Group A received 162 g of raspberry and 114.2 g of mango in each meal, while group B received 150 g of passion fruit and 114.2 g of mango in each meal.	The average glycemic index for raspberry/mango was higher in the healthy weight group compared to the obesity group. The blood glucose levels of participants with a healthy weight were significantly ↑ after consuming raspberry and mango compared to the non-energy meal. The blood glucose levels in the obese group were significantly ↑ throughout the raspberry/mango arm at 15 and 120 min postprandial compared to the NE meal. ↓ postprandial blood glucose levels when compared to consuming the whole fruit.	Not mentioned by the authors.	[38]
• Randomized, double-blind, parallel-group trial with controlled intervention; • 65 children, 6–8 y (upper respiratory and gastrointestinal tract infections); • Mexico.	The treated group received 2 g of mango by-product (peel and pulp) mixed in 50 mL of water, while the control group received flavored water daily for a period of 2 months.	↓ incidence of diarrhea, nasal congestion, dry cough, gastrointestinal tract infections, and abdominal inflammation. The JBP group showed ↑ counts of red blood cells, hemoglobin, hematocrit, platelets, and neutrophils compared to the control group. The control group experienced ↑ lymphocyte count compared to the treated group.	Not mentioned by the authors.	[39]
• Double-blind trial; • 48 healthy participants, 18 men and 30 women, 18–45 y; • Spain.	1 h before running and every 8 h/24 h (after running), the participants ingested either a placebo or 140 mg of Zynamite® (mangiferin-rich mango leaf extract) with 140 mg of quercetin.	A single dose of 140 mg of Zynamite combined with a similar amount of quercetin attenuated pain and muscle damage caused by running and accelerated the recovery of muscle performance. This analgesic effect may have been mediated by the free-radical scavenging properties of Zynamite and quercetin, since free radicals have been implicated in nociception.	Not mentioned by the authors.	[40]
• Two double-blind, placebo-controlled, randomized crossover clinical trial; • 38 participants, including 16 men and 16 women, 18–40 y; • South Africa.	The participants received 500 mg of MLE at either 60 min (study 1) or 90 min (study 2).	In the first study, MLE improved all mood state profile scores, with a significant improvement in the "fatigue" score. In the second study, there was a trend toward faster reaction time in the MLE group compared to the placebo. The Profile of Mood States (POMS) score for "depression" improved in the caffeine group. In both studies, MLE caused significant changes in cortical brain electrical activity during cognitive challenges, distinct from the attenuated spectral changes induced by caffeine.	Not mentioned by the authors.	[41]
• Double-blind, crossover, counterbalanced design; • 12 healthy and physically active men, 21.3 ± 2.1 y; • Spain.	The placebo group received capsules containing 500 mg of maltodextrin, while the treatment group received capsules containing 50 mg and 100 mg of luteolin, along with 100 mg and 300 mg of mangiferin, for 15 days.	The supplementation improved exercise performance, facilitated muscular oxygen extraction, and improved cerebral oxygenation without increasing VO2 compared to the placebo.	Not mentioned by the authors.	[42]

Table 2. Cont.

Study Type and Patients	Interventions	Results	Side Effects	Ref.
• Monocentric, randomized, placebo-controlled, 3-armed, double-blind trial with a parallel design; • 75 healthy participants, 42 men and 33 women, 40–70 y, non-smokers; • Switzerland.	The participants were assigned to three groups: the first group received 100 mg of a 100% mango fruit powder, the second group received 300 mg of the fruit powder, and the third group received a placebo/4 weeks.	The reactive hyperemic microcirculatory flow increased, particularly in the 100 mg group. The group that consumed 300 mg showed ↓ postprandial blood glucose compared to the placebo. The ingestion of 300 mg significantly improved postprandial endothelial function in individuals with impaired endothelial function following high-dose glucose ingestion.	Not mentioned by the authors.	[43]
• Monocentric, randomized, double-blind, crossover trial; • 10 healthy women, 40–70 y; • United States.	The participants consumed 100 mg followed by 300 mg or 300 mg followed by 100 mg of Careless™ (mangiferin) in a single dose.	After 6 h: ↑ cutaneous blood flow over time with both 100 mg (54% above baseline, $p = 0.0157$) and 300 mg (35% above baseline, $p = 0.209$) of Careless™. The reactive hyperemia response was slightly improved 3 h after ingestion compared to the pre-test with 30 mg of Careless™.	Not mentioned by the authors.	[44]
• Randomized, double-blind, placebo-controlled trial; • 97 overweight patients with hyperlipidemia, 52.1 ± 7.2 y; • China.	The subjects consumed a daily dose of 150 mg of mangiferin or a placebo/12 weeks.	↓ TG and FFA in the mangiferin supplementation group. ↓ HOMA-IR and total serum fatty acids, SFA, MUFA, PUFA, total serum n-3 fatty acids, and total serum n-6 fatty acids compared to the control group. ↑ HDL-c, L-carnitine, β-hydroxybutyrate, and acetoacetate levels.	No side effects or changes in liver enzymes or renal variables were observed.	[45]
• Clinical trial; • 30 patients (gender not specified; 30–69 y); • India.	Group A followed diet and medication; group B: diet, medication, and mango-leaf powder (1–2 teaspoons in 150 mL of water, taken 2x/d on an empty stomach); and Group C: diet, medication, and a placebo.	After approximately 1 month, both group A and group B showed significant reductions in both fasting and postprandial blood sugar levels. However, group C showed only a slight reduction in blood sugar levels.	Not mentioned by the authors.	[46]

Abbreviations: ↑: increase; ↓: decrease; AST: aspartate aminotransferase; FFA: free fatty acids; HbA1C: glycated hemoglobina; HDL-c: high-density lipoprotein cholesterol; HOMA-IR: homeostatic model assessment of insulin resistance; JBP: mango juice by-product; LDL-c: low-density lipoprotein cholesterol; LFC: low-fat cookies; MLE: mango leaf extract; MUFA: monounsaturated fatty acids; NE: nutrient extracted fruit; PCR: c-reactive protein; PUFA: polyunsaturated fatty acids; SFA: saturated fatty acids; TC: total cholesterol; TG: triglycerides; VO2: oxygen consumption.

Table 3. Descriptive table of the biases in the included randomized clinical trials.

Study	Question Focus	Appropriate Randomization	Allocation Blinding	Double Blind	Losses (<20%)	Prognostics or Demographic Characteristics	Outcomes	Intention to Treat Analysis	Sample Calculation	Adequate Follow-Up
[8]	Yes	No	Yes	No	Yes	No	Yes	Yes	No	Yes
[36]	Yes	Yes	Yes	No	Yes	No	Yes	Yes	No	Yes
[37]	Yes	Yes	Yes	No	Yes	No	Yes	Yes	No	Yes
[38]	Yes	Yes	Yes	No	Yes	No	Yes	Yes	No	Yes
[39]	Yes	No	Yes	Yes	Yes	No	Yes	Yes	No	Yes
[40]	Yes	No	Yes	Yes	Yes	No	Yes	Yes	No	Yes
[41]	Yes	Yes	Yes	No	Yes	No	Yes	Yes	No	Yes
[42]	Yes	No	Yes	Yes	Yes	No	Yes	Yes	No	Yes
[43]	Yes	Yes	Yes	Yes	Yes	No	Yes	Yes	No	Yes
[44]	Yes	Yes	Yes	Yes	Yes	No	Yes	Yes	No	Yes
[45]	Yes	Yes	Yes	Yes	Yes	No	Yes	Yes	No	Yes
[46]	Yes	No	Yes	No	Yes	No	Yes	Yes	No	Yes

4. Discussion

4.1. M. indica By-Products

Mango by-products include bark, leaf, peel, and seed/kernel. For instance, about 15–25 million tons of mango peels are produced yearly, leading to harmful environmental effects. The phytocompounds found in these by-products can represent a fantastic source of bioactive compounds.

Leaves and bark present almost 20 flavonoids and four new benzophenones. As examples, it is possible to cite mangiferin, gallic acid, ascorbic acid, quercetin-3-O-β-d glucoside, and protocatechuic acid, which play powerful antioxidant and anti-inflammatory roles [47]. Other bioactive compounds are functional lipids, unsaturated fatty acids, sterols, and soluble (starches and rhamno-galacturonans) and insoluble (hemicellulose and lignin) dietary fibers. The extract produced from the bark is traditionally used to treat hyperglycemia, anemia, diarrhea, cancer, and many other conditions [24,48–50]. The leaves are used in traditional medicine to treat hyperglycemia, obesity, dyslipidemia, cancer, and infectious and inflammatory conditions [12,51]. The most phenolic compounds in the leaves are mangiferin, quercetin, ellagic acid, and syringic acid. Peels are also a good mangiferin source; this compound can be considered heat-stable and pharmacologically active. The actions of mangiferin can include antiinflammation, antioxidant, anti-obesity, anti-type 2 diabetes, antitumor, and immunomodulatory effects [52,53].

Peels are a relevant source of dietary fiber (36–78 g/100 g of dry weight), and mango peel powder can be applied in the production of bakery and pasta products, beverages, snacks, ice cream, and meat products [54]. Peels also contain ascorbic acid, tocopherol, carotenoids, phenolic compounds, gallic acid, and derivatives [55–58]. Mango peel extracts play a significant role in exhibiting 2,2-diphenyl-1-picrylhydrazyl (DPPH) free-radical scavenging capacity and, consequently, can prevent or combat oxidative stress both in food products and in the human body. Polyphenols extracted from peels can be a natural substitute for artificial antioxidants on functional food or dietary supplements [59].

Kernels contain cellulose, starch, and pectin and can be used as an important dietary fiber source. The kernel phytocompounds include catechin, gallotannins, flavonoids, and benzophenones. Other phytocompounds are functional phytosterols, sesquiterpenoids, fatty acids, and tocopherols [48,55,56,60,61].

4.2. M. indica and Health Effects: Results from Clinical Trials

Table 2 summarizes the main findings of the included studies. Pinneo et al. [8] focused their research on analyzing the effect of mango consumption on biochemical parameters and satiety responses. However, the study has a limitation because it measures satiety hormones at only 45 min after consumption, as each hormone may have a different time curve. More frequent blood collections with a longer time window could provide a more accurate picture of the responses.

Rosas et al. [36] found that consuming fresh mango significantly reduces blood glucose levels and LDL-c levels. This is a very interesting study; however, it has several limitations: the small sample size, the authors only used one dose of fresh mango, and the optimal dose cannot be determined.

According to Arshad et al. [37], the use of mango peel powder can improve lipid profiles in obese women. They also found important antioxidant activity, suggesting that mango peels have a considerable potential to control oxidative stress and dyslipidemia in obese individuals. Although they present promising results, there are important limitations: the small sample size and the inclusion of only women, which reduce the possibility of spoiling results.

Alkutbe et al. [38] studied the combination of nutrient extraction from two fruits, raspberry and passion fruit, combined with mango. They demonstrated that the combinations of mango/raspberry and mango/passion fruit can reduce the glycemic index in individuals with a healthy weight and in obese individuals. However, some limitations can be cited: the authors did not measure polyphenol content, including piceatannol (other authors have shown they can vary according to their country of origin) and factors such as insulin responses and changes in other hormones such as gastric inhibitory polypeptide (GIP) or glucagon-like peptide-1 (GLP-1). Only blood glucose responses after the intake of test meals were analyzed. Furthermore, healthy-weight individuals participated in only one of the two arms of the study.

Anaya-Loyola et al. [39] found that by-products of mango juice (peel and pulp) did not prevent or eliminate upper respiratory and gastrointestinal infections. At the end of the study, the children still experienced some infection, but there was a low occurrence of gastrointestinal diseases and upper respiratory tract infection symptoms. Although the results are interesting, this study has some limitations: mango bioavailability was not evaluated, so the amount of bioavailable mango polyphenols is unclear; only children were included, so the findings may not be extended to other age groups; and the juice by-product was investigated in a single dose.

Martin-Rincon et al. [40] investigated supplementation with mango leaf extract (Zynamite®) in combination with quercetin and found that the polyphenols enhanced performance and muscle pain recovery, regardless of oral contraceptive intake. However, this study had limitations, such as the small sample size. Additionally, the diet during the assessment was not standardized, and the polyphenol content of the participants' habitual diets had not been determined.

López-Ríos et al. [41] investigated the neurocognitive activity of mangiferin. The findings indicated significant spectral modifications in brain electrical activity. Although psychometric tests did not show significant modifications (except for reaction times across all groups), the study provides valuable evidence that mango leaf extract has no adverse effects on blood pressure, pulse, or heart rate variability. It is important to note that this study is limited in nature, being an exploratory investigation with a small number of participants.

Gelabert-Rebato et al. [42] showed the impact of mango leaf extract, rich in mangiferin, combined with luteolin on sprint exercise performance and muscle oxygen extraction capacity in active men. Although the results demonstrated promising effects of the extract on exercise performance, oxygen extraction, and cerebral oxygenation, the study was limited by its small sample size and the lack of an evaluation of oxidative stress biomarkers.

Buchwald-Werner et al. [43] conducted a study investigating the long-term consumption of a commercial powdered *M. indica* product (Careless™) on microcirculation and glucose metabolism. The researchers deemed these findings promising, indicating a modest beneficial effect of *M. indica* fruit preparation on microcirculation, endothelial function, and glucose metabolism by the end of the study. Some limitations can be cited for this study. The age range of the patients included was wide; the table with socio-demographic data does not make it clear how many men and women took part in the study.

Gerstgrasser et al. [44] observed the effects of Careless™ (mangiferin) on microcirculation. The results showed promising and beneficial effects in healthy women. These findings suggest that Careless™ may have potential therapeutic applications for individuals with microcirculation alterations and endothelial dysfunction. However, due to the small sample size, these findings cannot be extended to a larger number of patients.

In a study conducted by Na et al. [45], the effects of mangiferin on serum lipid profiles and free fatty acid concentrations were examined in overweight and dyslipidemic patients. Their findings suggest that mangiferin may improve lipid metabolism in individuals with these conditions. However, due to the broad age range included in the study, the findings may have limitations.

Patnaik [46] observed the effectiveness of mango leaves in controlling diabetes. Although this study is interesting, some limitations must be noted. Firstly, a questionnaire was used to evaluate the patients' anthropometric and biochemical parameters, limiting the reliability of the results. Furthermore, the sample was small (30 patients divided into three groups).

4.3. Antioxidant Effects of M. indica L.

Oxidative stress is related to the development or worsening of different metabolic conditions such as obesity, hyperglycemia, dyslipidemia, and cardiovascular disease. When considering *M. indica* L.'s *antioxidant properties*, it is essential to emphasize its phytochemical compounds [62]. The most relevant polyphenols with antioxidant activity in the mango fruit are the class of flavonoids (catechins, quercetin, kaempferol, rhamnetin, anthocyanins, and tannic acid) and the class of xanthones, which includes mangiferin (C2-b-d-glucopyranosyl-1,3,6,7-tetrahydroxyxanthone) [63]. Mangiferin can be obtained from the roots, bark, leaves, and fruits of *M. indica* L. and presents anticancer, antimicrobial, antiatherosclerotic, antiallergenic, anti-inflammatory, high iron-chelating, analgesic, and immunomodulatory properties as well as the capacity to reduce ROS production [62,64].

Moreover, the administration of mangiferin to diabetic nephropathy rats reduced serum levels of advanced glycation end products and enhanced antioxidant enzymes, including superoxide dismutase and glutathione peroxidase [65]. Also, another study involving the use of mangiferin in diabetic rats identified increased liver levels of the following antioxidant enzymes: superoxide dismutase, catalase, and glutathione peroxidase [66].

Furthermore, according to a study developed by Sferrazzo et al. [67], M. indica L. has proven to be a great example of an antioxidant source that stabilizes free radicals, contributing to minimizing oxidative stress. This experimental analysis using mango leaf extract (MLE) obtained from Sicilian mango (Lentini, Italy) has shown that the polyphenol derivatives of that extract present antioxidant activity. This activity was confirmed by using MLE treatment in humans, reducing the gene expression of pro-inflammatory enzymes and cytokines such as IL-1β, IL-6, TNFα, and COX2.

In rats, a study identified the antioxidant activity of the MLE on several oxidative stress systems induced by H_2O_2 related to antispasmodic activity spasmolytic. The results showed that chronic intake with MLE reduced lipid peroxidation in the small intestine and counteracted the ROS effects, protecting tissues from oxidative damage [68]. Figure 3 provides an illustration depicting the potential antioxidant effects of mangiferin. This bioactive compound may plausibly interact with Nrf2, as supported by prior research findings [69].

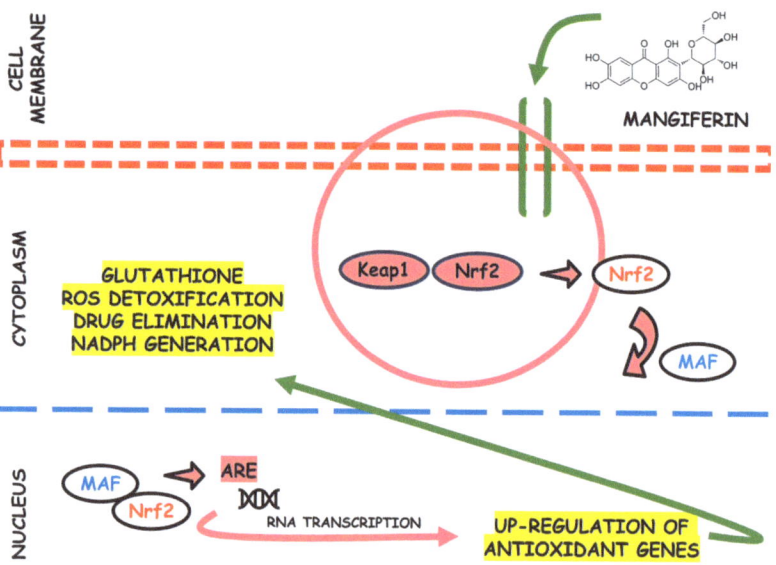

Figure 3. This scheme illustrates the potential antioxidant efficacy of mangiferin. This bioactive agent shows its effectiveness by potentially interacting with the nuclear factor erythroid 2-related factor 2 (Nrf2) pathway. This interaction initiates a series of events that activate cellular defense mechanisms against oxidative stress. The multifaceted actions of mangiferin in scavenging reactive oxygen species and enhancing endogenous antioxidant systems make it a compelling candidate for therapeutic interventions in oxidative stress-related disorders.

4.4. Anti-Inflammatory Effects of M. indica L.

As an oxidative stress, inflammation is related to obesity, hyperglycemia, dyslipidemia, metabolic syndrome, and cardiovascular diseases. As mango fruit and by-products possess many bioactive compounds, they can be considered to prevent inflammatory processes. A study developed by Sferrazzo et al. (2022) evaluated the anti-inflammatory activity of mango leaf extract (MLE) using macrophages treated with lipopolysaccharide (LPS) as an in vitro model. The use of LPS caused an increase in pro-inflammatory genes such as cyclooxygenase-2 (COX-2), interleukin-6 (IL6), interleukin-1β (IL-1β), and tumor necrosis factor-α (TNF-α), which was downregulated to regular levels by the use of MLE. The extract also caused a reduction in pro-inflammatory cytokines genes, IL-1β, IL-6, and COX-2 in hepatic stellate cells treated with LPS [67].

Mango pulp is an excellent source of polyphenols, most of them derived from gallic acid and galloyl-polyphenols such as mono-galloyl glucose and gallotannins (hexa-to nona-O-galloyl-glucoses). The anti-inflammatory properties of gallotannins and their associated metabolites (such as gallic acid and 4-O-methylgallic acid) are related to the reduction of pro-inflammatory cytokines, intracellular adhesion molecule 1 (ICAM-1), nuclear factor kappa B (NF-κB), and vascular cell adhesion molecule 1 (VCAM-1). Also, gallic acid can reduce inflammation responses in intestinal epithelial cells by reducing pro-inflammatory cytokines such as IL-1, IL-6, IL-17, and TNF-α and inducing anti-inflammatory cytokines such as IL-4 and IL-10. In addition, mango stem bark extract also showed anti-inflammatory activity by inhibiting TNF, prostaglandin E2 (PGE2), and nitric oxide (NO) [70].

Phospholipases A2 (PLA2) are a family of enzymes that regulate the arachidonic acid pathway; this, by the action of cyclooxygenase and lipoxygenase, liberates pro-inflammatory mediators such as leukotrienes, thromboxanes, prostacyclins, and prostaglandins. It has been proved that the aqueous stem bark extract of *M. indica* L. can suppress inflammation processes by inhibiting inflammatory PLA2 belonging to group IA, i.e., NN-XIa-PLA2 (a purified PLA2 enzyme obtained from *Naja naja* venom). The extract inhibited both in vitro and in situs PLA2 activity and dose-dependently inhibited the in vivo edema-inducing activity of NN-XIa-PLA2. Also, oral extract administration reduced mice ear edema induced by arachidonic acid and phorbol myristate acetate [71]. Figure 4 depicts the potential anti-inflammatory properties associated with mangiferin.

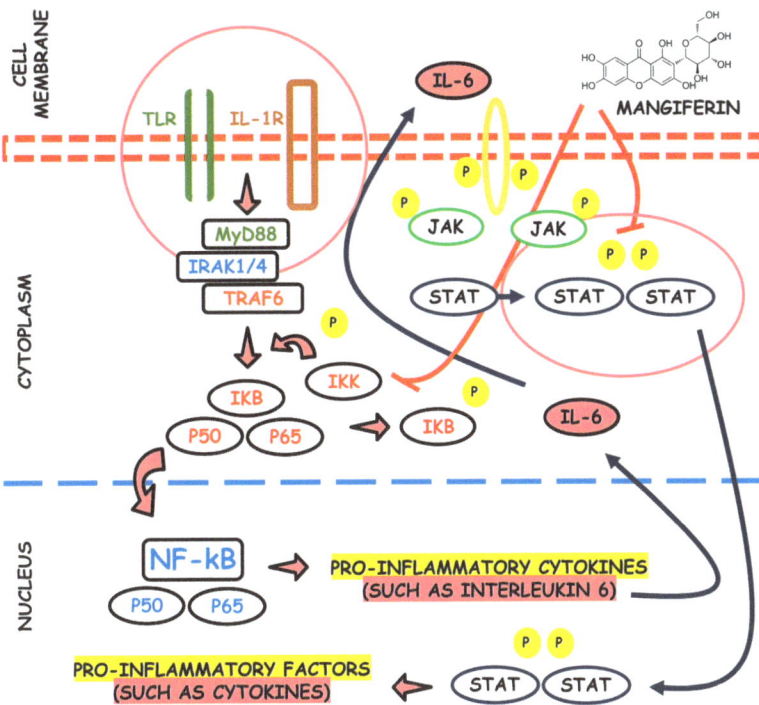

Figure 4. Visual representation of the potential anti-inflammatory attributes linked with the compound mangiferin.

4.5. Metabolic Effects of M. indica L.

Diabetes is considered one of the leading causes of death worldwide. Mango by-product consumption has potential value in treating risk factors for diabetes [21,25,72,73]. A study showed the hypoglycemic effect of the MLE in male normoglycemic and diabetic rats and observed a significant decrease in glycemia, surpassing the effects of glibenclamide. The extract maintained long-term hypoglycemic actions effectively and significantly improved insulin sensitivity in an animal model.

A study performed with human HepG2 and HL-7702 hepatocytes was evaluated under a high glucose level and submitted a mangiferin derivative compound: 1,3,6,7-tetrapropylene acyloxy-ketone (TPX). The results showed augmented glycogen synthesis and inhibition of gluconeogenesis. This evaluation of the article occurs due to the regulation of GSK3β, G6Pase, and PEPCK. Energy homeostasis and hepatic insulin resistance were also improved due to actions in AMPK and PI3K/AKT. These results are promising for the treatment of insulin-resistance-related diseases, including metabolic-associated fat liver diseases and diabetes [74].

Noh et al. [75] conducted a study examining the effects of mangiferin on glycemia and obesity both computationally and in C57BL/6 mice subjected to a high-fat diet. Their in silico model revealed mangiferin's capacity to bind with inflammatory biomarkers associated with macrophages and autophagy proteins. In vivo experiments demonstrated significant reductions in body weight and glucose and lipid metabolism improvements, as well as decreased insulin resistance in obese subjects. Furthermore, mangiferin decreased Kupffer cells and macrophages within adipose tissue and inhibited TNF-α and NF-κB expression in the same tissue. Additionally, it enhanced the liver's expression of fibroblast growth factor 21. In a separate investigation, Apontes et al. [76] administered 400 mg of mangiferin (derived from bark extract) per kilogram of diet to mice for five weeks. The outcomes demonstrated heightened glucose and pyruvate oxidation and increased ATP production without impacting fatty acid oxidation. This redirection of fuel utilization appeared to safeguard carbohydrates. The authors speculated that mangiferin has the potential to enhance carbohydrate metabolism. Another study by Saleem et al. [77] explored the effectiveness of mango leaf extract in diabetic Swiss albino mice. Their findings exhibited lowered postprandial glycemia and an improved lipid profile.

Irondi et al. [78] examined the impact of *M. indica* seed flour on diabetic Wistar rats subjected to a high-fat diet. Following this, the rats were provided diets enriched with flour or were administered metformin for 21 days. The inclusion of the flour led to enhancements in fasting glycemia, liver glycogen levels, glycosylated hemoglobin, and lipid profile, as well as reductions in hepatic and pancreatic malondialdehyde. Additionally, it resulted in favorable alterations in markers of liver function, including plasma aspartate aminotransferase, alanine aminotransferase, and alkaline phosphatase levels in diabetic rats.

Besides diabetes, the global prevalence of obesity has been a growing concern for public health officials worldwide. Obesity is a complex and multifactorial condition influenced by genetics, environment, lifestyle, and socioeconomic factors. The prevalence of obesity is indeed particularly high (>25% of adults) in several regions, including the Americas, the Middle East, and Pacific Island communities [79].

Some signaling pathways contribute to pro-obesity and anti-obesity mechanisms within the body. Pro-obesity mechanisms encompass insulin resistance, inflammation in adipose tissue, and the process of adipogenesis. Conversely, anti-obesity mechanisms involve signaling pathways and processes that aim to counteract these pro-obesity mechanisms [80,81]. In this way, Taing et al. [82] and Fang et al. [81] concluded that extracts from *M. indica* peel effectively hinder adipogenesis in vitro in a dose-dependent manner by suppressing mitotic clonal expansion, genes linked to mitochondrial biogenesis, and fatty acid oxidation in 3T3-L1 adipocytes via AMP-activated protein kinase (AMPK) signaling.

In addition to obesity, dyslipidemia frequently co-occurs, raising concerns about its potential implications, notably in terms of cardiovascular diseases, which continue to be the leading cause of global mortality [83]. Several studies have been conducted on mangoes to investigate their effects on cholesterol profiles and their potential relevance to cardiovascular health.

Various compounds present in *M. indica* leaves may have a twofold impact: firstly, by blocking enzymes responsible for the de novo synthesis of cholesterol, and secondly, by competing with cholesterol absorption [84]. Furthermore, in male Wistar rats, a standardized mango leaf extract demonstrated a more significant cholesterol-reducing effect compared to atorvastatin. This effect was attributed to three key molecules: iriflophenone 3-C-β-d-glucoside, mangiferin, and 3β-taraxerol [85]. In hyperlipidemic rats, a methanolic extract from mango leaves containing a significant amount of mangiferin with a bioaccessibility of 12% successfully decreased blood lipid levels [86].

In another study, Wistar rats underwent ischemia injury followed by reperfusion. The control group exhibited significant cardiac dysfunction, elevated serum cardiac injury markers, increased lipid peroxidation, and a notable decrease in tissue antioxidants. However, pretreatment with mangiferin effectively restored the balance between oxidants and antioxidants in the heart tissue, preserved cell membrane integrity, and reduced levels of pro-inflammatory cytokines, pro-apoptotic proteins, and transforming growth factor beta (TGF-β). Additionally, mangiferin significantly reduced the phosphorylation of p38 and c-Jun N-terminal protein kinase (JNK) while enhancing the phosphorylation of extracellular signal-regulated protein kinases 1 and 2 (ERK1/2), indicating its role in modulating the mitogen-activated protein kinase (MAPK) signaling pathway [87].

From the perspective of microcirculation, *M. indica* fruit preparation (Careless™) was administered in healthy women. It triggered the activation of evolutionarily conserved metabolic sensors, namely sirtuin 1 and adenosine monophosphate-activated protein kinase. These sensors have been recognized as pivotal in regulating microcirculation and maintaining endothelial function. At the same time, an in vitro test in primary human umbilical vein endothelial cells promoted endothelial nitric oxide synthase activation [44].

5. Conclusions

Incorporating mango into one's diet has been shown to enhance glycemic control, regulate plasma lipid levels, increase satiety, and improve endothelial function. The collective findings from the included studies indicate that the utilization of mango in various forms—ranging from fresh mango slices and mango puree to mango by-products, mango leaf extract, fruit powder, and mangiferin—yield many favorable effects. These encompass enhancements in glycemic control and improvements in plasma lipid profiles, including reductions in triglycerides, LDL cholesterol, and total cholesterol, alongside increased HDL cholesterol levels. Additionally, mango consumption is correlated with diminished appetite and reduced food intake, elevated mood scores, augmented physical performance during exercise, improved endothelial function, and a decreased incidence of respiratory tract infections. Despite the health benefits, further standardized clinical trials are imperative to establish the optimal administration, dosage, and intervention duration for mango and its derivative by-products. It is also crucial to investigate further the potential risks associated with their use.

The utilization of mango by-products aligns with the current trend and expanding market demand for improved, healthier products. Nutraceutical items could be more affordable, making them accessible to low-income populations. Additionally, this approach contributes to reducing environmental harm, a critical factor in ensuring humanity's sustainable presence on the Earth.

In light of this current landscape and alignment with the Sustainable Development Goals, there is a pressing need to promote the active development of nanomedicines derived from mangoes. This innovative approach holds tremendous potential for addressing various healthcare challenges. We can pioneer groundbreaking therapeutic approaches while safeguarding the environment by harnessing the distinctive attributes of mango-derived compounds at the nanoscale.

Author Contributions: Conceptualization, G.M., L.F.L., L.G.D., J.A.D., C.B.L., Y.C.N., M.D.B., E.B.J. and S.M.B.; methodology, G.M., L.F.L., N.M.M., L.G.D., E.L.G., A.C.A., C.B.L., Y.C.N., M.D.B. and S.M.B.; formal analysis, G.M., L.F.L., N.M.M., L.G.D., E.L.G., A.C.A., J.A.D., C.B.L., Y.C.N., M.D.B., E.B.J., F.B.G. and S.M.B.; investigation, G.M., L.F.L., N.M.M., L.G.D., E.L.G., A.C.A., J.A.D., C.B.L., Y.C.N., M.D.B., E.B.J., F.B.G. and S.M.B.; resources, G.M., L.F.L., N.M.M., L.G.D., E.L.G., A.C.A., J.A.D., C.B.L., Y.C.N., M.D.B., E.B.J., F.B.G. and S.M.B.; data curation, G.M., L.F.L., N.M.M., L.G.D., E.L.G., A.C.A., J.A.D., C.B.L., Y.C.N., M.D.B., E.B.J., F.B.G. and S.M.B.; writing—original draft preparation, G.M., L.F.L. and S.M.B.; writing—review and editing, G.M., L.F.L. and S.M.B.; visualization, G.M., L.F.L. and S.M.B.; supervision, G.M., L.F.L. and S.M.B.; project administration, G.M., L.F.L. and S.M.B.; funding acquisition, S.M.B.; software, G.M., L.F.L. and S.M.B.; validation, G.M., L.F.L. and S.M.B. All authors have read and agreed to the published version of the manuscript.

Funding: This research received no external funding.

Institutional Review Board Statement: Not applicable.

Informed Consent Statement: Not applicable.

Data Availability Statement: Not applicable.

Acknowledgments: The authors attribute to Smart Servier (https://smart.servier.com/, accessed on 10 October 2023) the scientific images that were used in this article under an attribution license of public copyrights (https://creativecommons.org/licenses/by/3.0/, accessed on 10 October 2023) and under a disclaimer of warranties. All Smart Servier's images were left unchanged when writing this article.

Conflicts of Interest: The authors declare no conflict of interest.

References

1. Alaiya, M.A.; Odeniyi, M.A. Utilisation of Mangifera indica plant extracts and parts in antimicrobial formulations and as a pharmaceutical excipient: A review. *Future J. Pharm. Sci.* **2023**, *9*, 29. [CrossRef]
2. Ahmad, R.; Alqathama, A.; Aldholmi, M.; Riaz, M.; Abdalla, A.N.; Aljishi, F.; Althomali, E.; Amir, M.; Abdullah, O.; Alamer, M.A.; et al. Antidiabetic and Anticancer Potentials of *Mangifera indica* L. from Different Geographical Origins. *Pharmaceuticals* **2023**, *16*, 350. [CrossRef] [PubMed]
3. Mirza, B.; Croley, C.R.; Ahmad, M.; Pumarol, J.; Das, N.; Sethi, G.; Bishayee, A. Mango (*Mangifera indica* L.): A magnificent plant with cancer preventive and anticancer therapeutic potential. *Crit. Rev. Food Sci. Nutr.* **2021**, *61*, 2125–2151. [CrossRef] [PubMed]
4. Shaban, N.Z.; Hegazy, W.A.; Abdel-Rahman, S.M.; Awed, O.M.; Khalil, S.A. Potential effect of Olea europea leaves, Sonchus oleraceus leaves and Mangifera indica peel extracts on aromatase activity in human placental microsomes and CYP19A1 expression in MCF-7 cell line: Comparative study. *Cell. Mol. Biol. (Noisy-Le-Grand Fr.)* **2016**, *62*, 11–19.
5. Lebaka, V.R.; Wee, Y.J.; Ye, W.; Korivi, M. Nutritional Composition and Bioactive Compounds in Three Different Parts of Mango Fruit. *Int. J. Environ. Res. Public. Health* **2021**, *18*, 741. [CrossRef] [PubMed]
6. Rokkam, M.P.; Gora, O.; Konda, M.R.; Koushik, A. A proprietary blend of Sphaeranthus indicus flower head and Mangifera indica bark extracts increases muscle strength and enhances endurance in young male volunteers: A randomized, double-blinded, placebo-controlled trial. *Food Nutr. Res.* **2023**, *67*, 8972. [CrossRef] [PubMed]
7. Yadav, S.P.; Paudel, P. The process standarzizing of mango (*Mangifera indica*) seed kernel for its value addition: A review. *Rev. Food Agric.* **2022**, *3*, 6–12. [CrossRef]
8. Pinneo, S.; O'Mealy, C.; Rosas, M., Jr.; Tsang, M.; Liu, C.; Kern, M.; Hooshmand, S.; Hong, M.Y. Fresh Mango Consumption Promotes Greater Satiety and Improves Postprandial Glucose and Insulin Responses in Healthy Overweight and Obese Adults. *J. Med. Food* **2022**, *25*, 381–388. [CrossRef]
9. Taing, M.-W.; Pierson, J.-T.; Shaw, P.N.; Dietzgen, R.G.; Roberts-Thomson, S.J.; Gidley, M.J.; Monteith, G.R. Mango fruit extracts differentially affect proliferation and intracellular calcium signalling in MCF-7 human breast cancer cells. *J. Chem.* **2015**, *2015*, 613268. [CrossRef]
10. Vilas-Franquesa, A.; Fryganas, C.; Casertano, M.; Montemurro, M.; Fogliano, V. Upcycling mango peels into a functional ingredient by combining fermentation and enzymatic-assisted extraction. *Food Chem.* **2023**, *434*, 137515. [CrossRef]
11. Zivković, J.; Kumar, K.A.; Rushendran, R.; Ilango, K.; Fahmy, N.M.; El-Nashar, H.A.S.; El-Shazly, M.; Ezzat, S.M.; Melgar-Lalanne, G.; Romero-Montero, A.; et al. Pharmacological properties of mangiferin: Bioavailability, mechanisms of action and clinical perspectives. *Naunyn-Schmiedeberg's Arch. Pharmacol.* **2023**. [CrossRef]
12. Kumar, M.; Saurabh, V.; Tomar, M.; Hasan, M.; Changan, S.; Sasi, M.; Maheshwari, C.; Prajapati, U.; Singh, S.; Prajapat, R.K.J.A. Mango (*Mangifera indica* L.) leaves: Nutritional composition, phytochemical profile, and health-promoting bioactivities. *Antioxidants* **2021**, *10*, 299. [CrossRef] [PubMed]

13. Khalid, M.; Alqarni, M.H.; Alsayari, A.; Foudah, A.I.; Aljarba, T.M.; Mukim, M.; Alamri, M.A.; Abullais, S.S.; Wahab, S.J.P. Anti-Diabetic Activity of Bioactive Compound Extracted from Spondias mangifera Fruit: In-Vitro and Molecular Docking Approaches. *Plants* **2022**, *11*, 562. [CrossRef] [PubMed]
14. Quintão, F.C.P. Competência Cultural na Atenção Primária à Saúde: Perspectiva dos Usuários da Comunidade da Mangueira. 2022. Available online: https://app.uff.br/riuff/handle/1/26778 (accessed on 25 October 2023).
15. Xavier, L.M.; Penha, T.A. O desempenho das exportações da manga no Brasil: Uma análise de constant market share. *Rev. Análise Econômica E Políticas Públicas-RAEPP* **2021**, *1*, 66–88.
16. Aggarwal, P.; Kaur, A.; Bhise, S. Value-added processing and utilization of mango by-products. In *Handbook of Mango Fruit: Production, Postharvest Science, Processing Technology and Nutrition*; John Wiley & Sons: Hoboken, NJ, USA, 2017; pp. 279–293.
17. Araújo, B.M.; Gonçalves, R.V.; Peluzio, M.D.; Leite, J.P.; dos Santos Chaves, G.; Lopes, S.O.; do Carmo Miranda, C.; de Queiroz, J.H. Uso do extrato de folhas de *Mangifera indica* L. e da mangiferina na lesão aterosclerótica em camundongos ApoE. *Biosci. J.* **2014**, *30*, 1873–1881.
18. Dorta, E.; González, M.; Lobo, M.G.; Sánchez-Moreno, C.; de Ancos, B.J.F.R.I. Screening of phenolic compounds in by-product extracts from mangoes (*Mangifera indica* L.) by HPLC-ESI-QTOF-MS and multivariate analysis for use as a food ingredient. *Food Res. Int.* **2014**, *57*, 51–60. [CrossRef]
19. Hu, K.; Dars, A.G.; Liu, Q.; Xie, B.; Sun, Z.J.F.c. Phytochemical profiling of the ripening of Chinese mango (*Mangifera indica* L.) cultivars by real-time monitoring using UPLC-ESI-QTOF-MS and its potential benefits as prebiotic ingredients. *Food Chem.* **2018**, *256*, 171–180. [CrossRef]
20. Souza, M.E. Potencial Antioxidante de Extratos da Casca de Manga (*Mangifera indica* L.) da Variedade Tommy Atkins Obtidos por Métodos a Baixa e a Alta Pressão e Dimensionamento de uma Coluna Para Extração Supercrítica. 2015. Available online: https://repositorio.ufsc.br/handle/123456789/135287 (accessed on 25 October 2023).
21. Caetano, M.M.M.; Toledo, R.C.L.; Brito, L.F.; Queiroz, J.H. Efeito da mangiferina e do extrato das folhas de manga Ubá (*Mangífera indica* L) na modulação da expressão do Receptor PPARα e do Fator de Transcrição NFκB no tecido cerebral de animais induzidos à Síndrome Metabólica. *São Paulo* **2015**, *1*, 7–8.
22. Lobo, F.A. Ação terapêutica da mangiferina como composto bioativo na modulação e prevenção da síndrome metabólica associada à obesidade. *Rev. Da JOPIC* **2022**, *7*.
23. Silva, F.M. Fracionamento do Extrato de Casca de Manga (*Mangifera indica* L.) CV. Tommy Atkinns Para Obtenção de Mangiferina por HPLC-UV/VIS Preparativo. 2019. Available online: https://repositorio.ufc.br/handle/riufc/43782 (accessed on 25 October 2023).
24. Nunezselles, A.; Delgadohernandez, R.; Garridogarrido, G.; Garciarivera, D.; Guevaragarcia, M.; Pardoandreu, G. The paradox of natural products as pharmaceuticals: Experimental evidences of a mango stem bark extract. *Pharmacol. Res.* **2007**, *55*, 351–358. [CrossRef]
25. Zarasvand, S.A.; Mullins, A.P.; Arjmandi, B.; Haley-Zitlin, V. Antidiabetic properties of mango in animal models and humans: A systematic review. *Nutr. Res.* **2023**, *111*, 73–89. [CrossRef] [PubMed]
26. Lum, P.T.; Sekar, M.; Gan, S.H.; Pandy, V.; Bonam, S.R. Protective effect of mangiferin on memory impairment: A systematic review. *Saudi J. Biol. Sci.* **2021**, *28*, 917–927. [CrossRef] [PubMed]
27. Arbos, K.A.; Stevani, P.C.; Castanha, R.D. Atividade antimicrobiana, antioxidante e teor de compostos fenólicos em casca e amêndoa de frutos de manga. *Rev. Ceres* **2013**, *60*, 161–165. [CrossRef]
28. dos Santos Borges, L. Compostos Fenólicos E Cumarinas em Três tipos de Frutas (Acerola, Maracujá e Manga): Uma Revisão da Literatura. Available online: https://doity.com.br/media/doity/submissoes/60900fba-5ffc-4d82-ac0e-2cc50a883292-resumo-expandidolivia-borges-finalpdf.pdf (accessed on 25 October 2023).
29. Ramos, S.A.; Silva, M.R.; Jacobino, A.R.; Damasceno, I.A.; Rodrigues, S.M.; Carlos, G.A.; Rocha, V.N.; Augusti, R.; Melo, J.O.; Capobiango, M. Caracterização físico-química, microbiológica e da atividade antioxidante de farinhas de casca e amêndoa de manga (*Mangifera indica*) e sua aplicação em brownie. *Res. Soc. Dev.* **2021**, *10*, e22310212436. [CrossRef]
30. Malherbe, C.J.; Willenburg, E.; de Beer, D.; Bonnet, S.L.; van der Westhuizen, J.H.; Joubert, E. Iriflophenone-3-C-glucoside from Cyclopia genistoides: Isolation and quantitative comparison of antioxidant capacity with mangiferin and isomangiferin using on-line HPLC antioxidant assays. *J. Chromatogr. B Anal. Technol. Biomed. Life Sci.* **2014**, *951–952*, 164–171. [CrossRef]
31. Henc, I.; Kokotkiewicz, A.; Łuczkiewicz, P.; Bryl, E.; Łuczkiewicz, M.; Witkowski, J.M. Naturally occurring xanthone and benzophenone derivatives exert significant anti-proliferative and proapoptotic effects in vitro on synovial fibroblasts and macrophages from rheumatoid arthritis patients. *Int. Immunopharmacol.* **2017**, *49*, 148–154. [CrossRef] [PubMed]
32. Wang, Y.; Julian McClements, D.; Chen, L.; Peng, X.; Xu, Z.; Meng, M.; Ji, H.; Zhi, C.; Ye, L.; Zhao, J.; et al. Progress on molecular modification and functional applications of anthocyanins. *Crit. Rev. Food Sci. Nutr.* **2023**, 1–19. [CrossRef] [PubMed]
33. Merecz-Sadowska, A.; Sitarek, P.; Kowalczyk, T.; Zajdel, K.; Jęcek, M.; Nowak, P.; Zajdel, R. Food Anthocyanins: Malvidin and Its Glycosides as Promising Antioxidant and Anti-Inflammatory Agents with Potential Health Benefits. *Nutrients* **2023**, *15*, 3016. [CrossRef]
34. Page, M.J.; McKenzie, J.E.; Bossuyt, P.M.; Boutron, I.; Hoffmann, T.C.; Mulrow, C.D.; Shamseer, L.; Tetzlaff, J.M.; Akl, E.A.; Brennan, S.E.; et al. The PRISMA 2020 statement: An updated guideline for reporting systematic reviews. *BMJ* **2021**, *372*, n71. [CrossRef]

35. Moher, D.; Liberati, A.; Tetzlaff, J.; Altman, D.G.; PRISMA Group. Preferred reporting items for systematic reviews and meta-analyses: The PRISMA statement. *Ann. Intern. Med.* **2009**, *151*, 264–269. [CrossRef]
36. Rosas, M., Jr.; Pinneo, S.; O'Mealy, C.; Tsang, M.; Liu, C.; Kern, M.; Hooshmand, S.; Hong, M.Y. Effects of fresh mango consumption on cardiometabolic risk factors in overweight and obese adults. *Nutr. Metab. Cardiovasc. Dis.* **2022**, *32*, 494–503. [CrossRef] [PubMed]
37. Arshad, F.; Umbreen, H.; Aslam, I.; Hameed, A.; Aftab, K.; Al-Qahtani, W.H.; Aslam, N.; Noreen, R. Therapeutic role of mango peels in management of dyslipidemia and oxidative stress in obese females. *BioMed Res. Int.* **2021**, *2021*, 3094571. [CrossRef] [PubMed]
38. Alkutbe, R.; Redfern, K.; Jarvis, M.; Rees, G. Nutrient Extraction Lowers Postprandial Glucose Response of Fruit in Adults with Obesity as well as Healthy Weight Adults. *Nutrients* **2020**, *12*, 766. [CrossRef] [PubMed]
39. Anaya-Loyola, M.A.; García-Marín, G.; García-Gutiérrez, D.G.; Castaño-Tostado, E.; Reynoso-Camacho, R.; López-Ramos, J.E.; Enciso-Moreno, J.A.; Pérez-Ramírez, I.F. A mango (*Mangifera indica* L.) juice by-product reduces gastrointestinal and upper respiratory tract infection symptoms in children. *Food Res. Int. (Ott. Ont.)* **2020**, *136*, 109492. [CrossRef]
40. Martin-Rincon, M.; Gelabert-Rebato, M.; Galvan-Alvarez, V.; Gallego-Selles, A.; Martinez-Canton, M.; Lopez-Rios, L.; Wiebe, J.C.; Martin-Rodriguez, S.; Arteaga-Ortiz, R.; Dorado, C.; et al. Supplementation with a Mango Leaf Extract (Zynamite®) in Combination with Quercetin Attenuates Muscle Damage and Pain and Accelerates Recovery after Strenuous Damaging Exercise. *Nutrients* **2020**, *12*, 614. [CrossRef]
41. López-Ríos, L.; Wiebe, J.C.; Vega-Morales, T.; Gericke, N. Central nervous system activities of extract *Mangifera indica* L. *J. Ethnopharmacol.* **2020**, *260*, 112996. [CrossRef]
42. Gelabert-Rebato, M.; Wiebe, J.C.; Martin-Rincon, M.; Galvan-Alvarez, V.; Curtelin, D.; Perez-Valera, M.; Juan Habib, J.; Pérez-López, A.; Vega, T.; Morales-Alamo, D.; et al. Enhancement of exercise performance by 48 hours, and 15-day supplementation with mangiferin and luteolin in men. *Nutrients* **2019**, *11*, 344. [CrossRef]
43. Buchwald-Werner, S.; Schön, C.; Frank, S.; Reule, C. Effects of *Mangifera indica* (Careless) on microcirculation and glucose metabolism in healthy volunteers. *Planta Medica* **2017**, *83*, 824–829. [CrossRef]
44. Gerstgrasser, A.; Röchter, S.; Dressler, D.; Schön, C.; Reule, C.; Buchwald-Werner, S. In Vitro Activation of eNOS by *Mangifera indica* (Careless™) and Determination of an Effective Dosage in a Randomized, Double-Blind, Human Pilot Study on Microcirculation. *Planta Medica* **2016**, *82*, 298–304. [CrossRef]
45. Na, L.; Zhang, Q.; Jiang, S.; Du, S.; Zhang, W.; Li, Y.; Sun, C.; Niu, Y. Mangiferin supplementation improves serum lipid profiles in overweight patients with hyperlipidemia: A double-blind randomized controlled trial. *Sci. Rep.* **2015**, *5*, 10344. [CrossRef]
46. Patnaik, R. Mango leaves in treating diabetes: A strategic study. *Int. J. Innov. Res. Dev.* **2014**, *3*, 432–441.
47. Pan, J.; Yi, X.; Zhang, S.; Cheng, J.; Wang, Y.; Liu, C.; He, X. Bioactive phenolics from mango leaves (*Mangifera indica* L.). *Ind. Crop. Prod.* **2018**, *111*, 400–406. [CrossRef]
48. Masibo, M.; He, Q. Mango bioactive compounds and related nutraceutical properties—A review. *Food Rev. Int.* **2009**, *25*, 346–370. [CrossRef]
49. Vazquez-Olivo, G.; Antunes-Ricardo, M.; Gutiérrez-Uribe, J.A.; Osuna-Enciso, T.; León-Félix, J.; Heredia, J.B. Cellular antioxidant activity and in vitro intestinal permeability of phenolic compounds from four varieties of mango bark (*Mangifera indica* L.). *J. Sci. Food Agric.* **2019**, *99*, 3481–3489. [CrossRef]
50. Barreto, J.C.; Trevisan, M.T.S.; Hull, W.E.; Erben, G.; de Brito, E.S.; Pfundstein, B.; Würtele, G.; Spiegelhalder, B.; Owen, R.W. Characterization and quantitation of polyphenolic compounds in bark, kernel, leaves, and peel of mango (*Mangifera indica* L.). *J. Agric. Food Chem.* **2008**, *56*, 5599–5610. [CrossRef]
51. Yap, K.M.; Sekar, M.; Seow, L.J.; Gan, S.H.; Bonam, S.R.; Rani, N.N.I.M.; Lum, P.T.; Subramaniyan, V.; Wu, Y.S.; Fuloria, N.K.; et al. *Mangifera indica* (Mango): A Promising Medicinal Plant for Breast Cancer Therapy and Understanding Its Potential Mechanisms of Action. *Breast Cancer Targets Ther.* **2021**, *13*, 471–503. [CrossRef] [PubMed]
52. Luo, F.; Lv, Q.; Zhao, Y.; Hu, G.; Huang, G.; Zhang, J.; Sun, C.; Li, X.; Chen, K. Quantification and purification of mangiferin from Chinese mango (*Mangifera indica* L.) cultivars and its protective effect on human umbilical vein endothelial cells under H2O2-induced stress. *Int. J. Mol. Sci.* **2012**, *13*, 11260–11274. [CrossRef]
53. Ajila, C.; Aalami, M.; Leelavathi, K.; Rao, U.P. Mango peel powder: A potential source of antioxidant and dietary fiber in macaroni preparations. *Innov. Food Sci. Emerg. Technol.* **2010**, *11*, 219–224. [CrossRef]
54. Oliver-Simancas, R.; Muñoz, R.; Díaz-Maroto, M.C.; Pérez-Coello, M.S.; Alañón, M.E. Mango by-products as a natural source of valuable odor-active compounds. *J. Sci. Food Agric.* **2020**, *100*, 4688–4695. [CrossRef]
55. Vergara-Valencia, N.; Granados-Pérez, E.; Agama-Acevedo, E.; Tovar, J.; Ruales, J.; Bello-Pérez, L.A. Fibre concentrate from mango fruit: Characterization, associated antioxidant capacity and application as a bakery product ingredient. *LWT-Food Sci. Technol.* **2007**, *40*, 722–729. [CrossRef]
56. Ribeiro, S.M.R.; Schieber, A. Bioactive compounds in mango (*Mangifera indica* L.). In *Bioactive Foods in Promoting Health*; Elsevier: Amsterdam, The Netherlands, 2010; pp. 507–523.
57. Masibo, M.; He, Q. Major mango polyphenols and their potential significance to human health. *Compr. Rev. Food Sci. Food Saf.* **2008**, *7*, 309–319. [CrossRef] [PubMed]
58. Marcal, S.; Pintado, M. Mango peels as food ingredient/additive: Nutritional value, processing, safety and applications. *Trends Food Sci. Technol.* **2021**, *114*, 472–489. [CrossRef]

59. Berardini, N.; Knödler, M.; Schieber, A.; Carle, R. Utilization of mango peels as a source of pectin and polyphenolics. *Innov. Food Sci. Emerg. Technol.* **2005**, *6*, 442–452. [CrossRef]
60. Jahurul, M.; Zaidul, I.; Norulaini, N.; Sahena, F.; Abedin, M.; Ghafoor, K.; Omar, A.M. Characterization of crystallization and melting profiles of blends of mango seed fat and palm oil mid-fraction as cocoa butter replacers using differential scanning calorimetry and pulse nuclear magnetic resonance. *Food Res. Int.* **2014**, *55*, 103–109. [CrossRef]
61. Asif, A.; Farooq, U.; Akram, K.; Hayat, Z.; Shafi, A.; Sarfraz, F.; Sidhu, M.A.I.; Rehman, H.-U.; Aftab, S. Therapeutic potentials of bioactive compounds from mango fruit wastes. *Trends Food Sci. Technol.* **2016**, *53*, 102–112. [CrossRef]
62. Maldonado-Celis, M.E.; Yahia, E.M.; Bedoya, R.; Landázuri, P.; Loango, N.; Aguillón, J.; Restrepo, B.; Guerrero Ospina, J.C. Chemical Composition of Mango (*Mangifera indica* L.) Fruit: Nutritional and Phytochemical Compounds. *Front. Plant Sci.* **2019**, *10*, 1073. [CrossRef]
63. Negi, J.S.; Bisht, V.K.; Singh, P.; Rawat, M.S.M.; Joshi, G.P. Naturally Occurring Xanthones: Chemistry and Biology. *J. Appl. Chem.* **2013**, *2013*, 621459. [CrossRef]
64. Imran, M.; Arshad, M.S.; Butt, M.S.; Kwon, J.H.; Arshad, M.U.; Sultan, M.T. Mangiferin: A natural miracle bioactive compound against lifestyle related disorders. *Lipids Health Dis.* **2017**, *16*, 84. [CrossRef]
65. Li, X.; Cui, X.; Sun, X.; Li, X.; Zhu, Q.; Li, W. Mangiferin prevents diabetic nephropathy progression in streptozotocin-induced diabetic rats. *Phytother. Res. PTR* **2010**, *24*, 893–899. [CrossRef]
66. Sellamuthu, P.S.; Arulselvan, P.; Muniappan, B.P.; Fakurazi, S.; Kandasamy, M. Mangiferin from Salacia chinensis prevents oxidative stress and protects pancreatic β-cells in streptozotocin-induced diabetic rats. *J. Med. Food* **2013**, *16*, 719–727. [CrossRef]
67. Sferrazzo, G.; Palmeri, R.; Restuccia, C.; Parafati, L.; Siracusa, L.; Spampinato, M.; Carota, G.; Distefano, A.; Di Rosa, M.; Tomasello, B.; et al. *Mangifera indica* L. Leaves as a Potential Food Source of Phenolic Compounds with Biological Activity. *Antioxidants* **2022**, *11*, 1313. [CrossRef] [PubMed]
68. Ybañez-Julca, R.O.; Asunción-Alvarez, D.; Quispe-Díaz, I.M.; Palacios, J.; Bórquez, J.; Simirgiotis, M.J.; Perveen, S.; Nwokocha, C.R.; Cifuentes, F.; Paredes, A. Metabolomic Profiling of Mango (*Mangifera indica* Linn) Leaf Extract and Its Intestinal Protective Effect and Antioxidant Activity in Different Biological Models. *Molecules* **2020**, *25*, 5149. [CrossRef] [PubMed]
69. Lauricella, M.; Emanuele, S.; Calvaruso, G.; Giuliano, M.; D'Anneo, A. Multifaceted Health Benefits of *Mangifera indica* L. (Mango): The Inestimable Value of Orchards Recently Planted in Sicilian Rural Areas. *Nutrients* **2017**, *9*, 525. [CrossRef] [PubMed]
70. Kim, H.; Castellon-Chicas, M.J.; Arbizu, S.; Talcott, S.T.; Drury, N.L.; Smith, S.; Mertens-Talcott, S.U. Mango (*Mangifera indica* L.) Polyphenols: Anti-Inflammatory Intestinal Microbial Health Benefits, and Associated Mechanisms of Actions. *Molecules* **2021**, *26*, 2732. [CrossRef] [PubMed]
71. Dhananjaya, B.L.; Shivalingaiah, S. The anti-inflammatory activity of standard aqueous stem bark extract of *Mangifera indica* L. as evident in inhibition of Group IA sPLA2. *An. Da Acad. Bras. De Cienc.* **2016**, *88*, 197–209. [CrossRef] [PubMed]
72. Wahab, S.; Khalid, M.; Alqarni, M.H.; Elagib, M.F.A.; Bahamdan, G.K.; Foudah, A.I.; Aljarba, T.M.; Mohamed, M.S.; Mohamed, N.S.; Arif, M. Antihyperglycemic Potential of Spondias mangifera Fruits via Inhibition of 11β-HSD Type 1 Enzyme: In Silico and In Vivo Approach. *J. Clin. Med.* **2023**, *12*, 2152. [CrossRef] [PubMed]
73. Sarkar, T.; Bharadwaj, K.K.; Salauddin, M.; Pati, S.; Chakraborty, R. Phytochemical Characterization, Antioxidant, Anti-inflammatory, Anti-diabetic properties, Molecular Docking, Pharmacokinetic Profiling, and Network Pharmacology Analysis of the Major Phytoconstituents of Raw and Differently Dried *Mangifera indica* (Himsagar cultivar): An In Vitro and In Silico Investigations. *Appl. Biochem. Biotechnol.* **2022**, *194*, 950–987. [CrossRef]
74. Fan, X.; Jiao, G.; Pang, T.; Wen, T.; He, Z.; Han, J.; Zhang, F.; Chen, W. Ameliorative effects of mangiferin derivative TPX on insulin resistance via PI3K/AKT and AMPK signaling pathways in human HepG2 and HL-7702 hepatocytes. *Phytomed. Int. J. Phytother. Phytopharm.* **2023**, *114*, 154740. [CrossRef]
75. Noh, J.W.; Lee, H.Y.; Lee, B.C. Mangiferin Ameliorates Obesity-Associated Inflammation and Autophagy in High-Fat-Diet-Fed Mice: In Silico and In Vivo Approaches. *Int. J. Mol. Sci.* **2022**, *23*, 15329. [CrossRef]
76. Apontes, P.; Liu, Z.; Su, K.; Benard, O.; Youn, D.Y.; Li, X.; Li, W.; Mirza, R.H.; Bastie, C.C.; Jelicks, L.A.; et al. Mangiferin stimulates carbohydrate oxidation and protects against metabolic disorders induced by high-fat diets. *Diabetes* **2014**, *63*, 3626–3636. [CrossRef]
77. Saleem, M.; Tanvir, M.; Akhtar, M.F.; Iqbal, M.; Saleem, A. Antidiabetic Potential of *Mangifera indica* L. cv. Anwar Ratol Leaves: Medicinal Application of Food Wastes. *Medicina* **2019**, *55*, 353. [CrossRef] [PubMed]
78. Irondi, E.A.; Oboh, G.; Akindahunsi, A.A. Antidiabetic effects of *Mangifera indica* Kernel Flour-supplemented diet in streptozotocin-induced type 2 diabetes in rats. *Food Sci. Nutr.* **2016**, *4*, 828–839. [CrossRef] [PubMed]
79. Lobstein, T.; Brinsden, H.; Neveux, M. World Obesity Atlas 2022. 2022. Available online: https://www.worldobesity.org/resources/resource-library/world-obesity-atlas-2022 (accessed on 25 October 2023).
80. Wen, X.; Zhang, B.; Wu, B.; Xiao, H.; Li, Z.; Li, R.; Xu, X.; Li, T. Signaling pathways in obesity: Mechanisms and therapeutic interventions. *Signal Transduct. Target. Ther.* **2022**, *7*, 298. [CrossRef] [PubMed]
81. Fang, C.; Kim, H.; Noratto, G.; Sun, Y.; Talcott, S.T.; Mertens-Talcott, S.U. Gallotannin derivatives from mango (*Mangifera indica* L.) suppress adipogenesis and increase thermogenesis in 3T3-L1 adipocytes in part through the AMPK pathway. *J. Funct. Foods* **2018**, *46*, 101–109. [CrossRef]
82. Taing, M.W.; Pierson, J.T.; Hoang, V.L.; Shaw, P.N.; Dietzgen, R.G.; Gidley, M.J.; Roberts-Thomson, S.J.; Monteith, G.R. Mango fruit peel and flesh extracts affect adipogenesis in 3T3-L1 cells. *Food Funct.* **2012**, *3*, 828–836. [CrossRef] [PubMed]

83. Kaminsky, L.A.; German, C.; Imboden, M.; Ozemek, C.; Peterman, J.E.; Brubaker, P.H. The importance of healthy lifestyle behaviors in the prevention of cardiovascular disease. *Prog. Cardiovasc. Dis.* **2022**, *70*, 8–15. [CrossRef] [PubMed]
84. Gururaja, G.; Mundkinajeddu, D.; Dethe, S.; Sangli, G.; Abhilash, K.; Agarwal, A. Cholesterol esterase inhibitory activity of bioactives from leaves of *Mangifera indica* L. *Pharmacogn. Res.* **2015**, *7*, 355. [CrossRef]
85. Dethe, S.M.; Gururaja, G.; Mundkinajeddu, D.; Kumar, A.S.; Allan, J.J.; Agarwal, A. Evaluation of cholesterol-lowering activity of standardized extract of *Mangifera indica* in albino Wistar rats. *Pharmacogn. Res.* **2017**, *9*, 21–26. [CrossRef]
86. Sandoval-Gallegos, E.M.; Ramírez-Moreno, E.; De Lucio, J.G.; Arias-Rico, J.; Cruz-Cansino, N.; Ortiz, M.I.; Cariño-Cortés, R. In vitro bioaccessibility and effect of *Mangifera indica* (Ataulfo) leaf extract on induced dyslipidemia. *J. Med. Food* **2018**, *21*, 47–56. [CrossRef]
87. Oliver-Simancas, R.; Labrador-Fernández, L.; Díaz-Maroto, M.C.; Pérez-Coello, M.S.; Alañón, M.E. Comprehensive research on mango by-products applications in food industry. *Trends Food Sci. Technol.* **2021**, *118*, 179–188. [CrossRef]

Disclaimer/Publisher's Note: The statements, opinions and data contained in all publications are solely those of the individual author(s) and contributor(s) and not of MDPI and/or the editor(s). MDPI and/or the editor(s) disclaim responsibility for any injury to people or property resulting from any ideas, methods, instructions or products referred to in the content.

Article

Virtual Insights into Natural Compounds as Potential 5α-Reductase Type II Inhibitors: A Structure-Based Screening and Molecular Dynamics Simulation Study

Sibhghatulla Shaikh [1,2,†], Shahid Ali [1,2,†], Jeong Ho Lim [1,2], Khurshid Ahmad [1,2], Ki Soo Han [3], Eun Ju Lee [1,2] and Inho Choi [1,2,*]

1 Department of Medical Biotechnology, Yeungnam University, Gyeongsan 38541, Republic of Korea; sibhghat.88@gmail.com (S.S.); ali.ali.md111@gmail.com (S.A.); lim2249@naver.com (J.H.L.); ahmadkhursheed2008@gmail.com (K.A.); gorapadoc0315@hanmail.net (E.J.L.)
2 Research Institute of Cell Culture, Yeungnam University, Gyeongsan 38541, Republic of Korea
3 Neo Cremar Co., Ltd., Seoul 05702, Republic of Korea; melong-h@cremar.co.kr
* Correspondence: inhochoi@ynu.ac.kr
† These authors contributed equally to this work.

Abstract: Androgenic alopecia (AGA) is a dermatological disease with psychosocial consequences for those who experience hair loss. AGA is linked to an increase in androgen levels caused by an excess of dihydrotestosterone in blood capillaries produced from testosterone by 5α-reductase type II (5αR2), which is expressed in scalp hair follicles; 5αR2 activity and dihydrotestosterone levels are elevated in balding scalps. The diverse health benefits of flavonoids have been widely reported in epidemiological studies, and research interest continues to increase. In this study, a virtual screening approach was used to identify compounds that interact with active site residues of 5αR2 by screening a library containing 241 flavonoid compounds. Here, we report two potent flavonoid compounds, eriocitrin and silymarin, that interacted strongly with 5αR2, with binding energies of −12.1 and −11.7 kcal/mol, respectively, which were more significant than those of the control, finasteride (−11.2 kcal/mol). Molecular dynamic simulations (200 ns) were used to optimize the interactions between compounds and 5αR2 and revealed that the interaction of eriocitrin and silymarin with 5αR2 was stable. The study shows that eriocitrin and silymarin provide developmental bases for novel 5αR2 inhibitors for the management of AGA.

Keywords: androgenic alopecia; androgen; dihydrotestosterone; 5α-reductase type II; natural compounds

Citation: Shaikh, S.; Ali, S.; Lim, J.H.; Ahmad, K.; Han, K.S.; Lee, E.J.; Choi, I. Virtual Insights into Natural Compounds as Potential 5α-Reductase Type II Inhibitors: A Structure-Based Screening and Molecular Dynamics Simulation Study. *Life* 2023, 13, 2152. https://doi.org/10.3390/life13112152

Academic Editors: Balazs Barna and Seung Ho Lee

Received: 22 August 2023
Revised: 30 October 2023
Accepted: 30 October 2023
Published: 1 November 2023

Copyright: © 2023 by the authors. Licensee MDPI, Basel, Switzerland. This article is an open access article distributed under the terms and conditions of the Creative Commons Attribution (CC BY) license (https://creativecommons.org/licenses/by/4.0/).

1. Introduction

Androgenic alopecia (AGA), also known as male pattern baldness, is a common type of hair loss. Hair is an essential bodily structure that protects the scalp, enhances human individuality, and serves a variety of purposes such as insulation, attractiveness, and tangibility [1]. AGA is caused by an androgen hormone imbalance, stress, hereditary diseases, malnourishment, 5α-reductase type II (5αR2) overactivity, thyroid dysfunction, drug addiction, and aging [2]. The testicles mainly produce androgens in the form of testosterone, which is then converted to dihydrotestosterone (DHT) by 5αR2. DHT interacts with androgen receptors in vulnerable scalp-based hair follicles and activates genes responsible for follicular shrinkage, thus causing AGA [3]. The absence of AGA in men with congenital impairment of 5αR2 demonstrates the relevance of DHT as an etiologic factor of this disorder [4]. Individuals with a genetic deficiency in the 5αR2 enzyme do not develop AGA. Furthermore, 5αR2 has been found in hair follicles on the scalp, with balding scalps exhibiting elevated levels of both 5αR2 activity and DHT concentrations. Overall, these findings support the rational use of inhibitors targeting 5αR2 as a therapeutic approach for treating AGA in men [5].

For many years, the treatment of AGA has been a point of contention in clinical dermatology. Many treatments for baldness are now available, including bioengineered hair transplants [6], hair follicle regeneration using rearranged stem cells [7], and medicinal treatment with the synthetic drugs minoxidil or finasteride [8,9]. Minoxidil stimulates hair growth by shortening the telogen phase and accelerating the transition to the anagen phase. It has also been demonstrated to expand hair follicles. In the treatment of androgenetic alopecia, minoxidil is still a significant advancement [10]. Finasteride is commonly used to treat androgen-dependent hair diseases such as androgenetic alopecia. This medicine is an orally given selective 5-alpha-reductase inhibitor used to treat androgenetic alopecia [11]. However, these drugs have been reported to have severe dermatological side effects, which include reduced libido, irritation, itching, erythema, and depression [12]. Hence, there is a pressing need for the development of a novel pharmaceutical agent that effectively stimulates hair growth while minimizing any potential adverse effects.

New drug development is one of the most important tasks of biomedical research from scientific and economic perspectives. However, despite the advancements made in informatics and computational biology, as well as parallel increases in drug development productivity, drug development has been sluggish due to heavy dependence on synthetic small molecules as a source of innovation. Computer-Aided Drug Design (CADD) has emerged as a highly effective technique for identifying promising lead compounds and advancing prospective pharmaceutical medicines addressing a wide range of disorders [13–15]. Currently, a variety of computational techniques are being used to identify interesting lead entities from large compound repositories. The application of CADD techniques in the context of drug development is progressing steadily. A prevalent trend in modern drug design revolves around the rational engineering of effective treatments with multi-targeting properties, improved efficacies, and reduced side effects, particularly with regard to toxicity considerations [16]. Natural products are a huge, diversified source of bioactive substances, and some have been utilized in traditional medicine for hundreds of years, which distinguishes them from synthetic small molecules [17]. The numerous health benefits of flavonoids, as described in epidemiological studies, have attracted the attention of the scientific community. These compounds are abundant in nature, particularly in fruits and vegetables, and possess diverse physical, chemical, and physiological properties. Several flavonoids have been well studied for their medicinal effects, which include antibacterial, hepatoprotective, antioxidant, anti-inflammatory, enzymatic activity modulation, anticancer, and antiviral activities [18]. Here, we aim to identify novel natural 5αR2 inhibitors using a computational approach, with the objective of finding a potential treatment for AGA.

2. Materials and Methods

2.1. Preparation of 3D Structure of 5αR2

The crystal structure of 5αR2 in complex with finasteride (PDB ID: 7BW1; resolution: 2.80 Å) was obtained from a protein data bank [19]. After removing water molecules and heteroatoms, including the co-crystallized ligand (finasteride), the clean structure of 5αR2 was prepared using the 'prepare protein' tool of the Discovery Studio 2021 (DS) and saved in monomer and .pdb format.

2.2. Retrieval and Preparation of Flavonoid Library

The unique collection of 241 flavonoid compounds was obtained from Selleck Chemicals (https://www.selleckchem.com (accessed on 13 December 2022)) in .sdf format. These compounds were minimized, prepared with DS, and converted to 'pdbqt' format for docking analysis.

2.3. Structure-Based Virtual Screening

The preeminent methodology used for the discovery of novel lead compounds in the field of drug development involves the physical evaluation of large chemical libraries against a specific biological target, also known as high-throughput screening. A complementary strategy, known as virtual screening (VS), entails the computational evaluation of large chemical repositories to identify molecules that exhibit complementarity to structurally characterized targets; compounds predicted to exhibit favorable binding characteristics are then subjected to experimental validation [20,21]. Receptor-based methodologies, also known as structure-based techniques, are intended to elucidate the interaction dynamics between a ligand and its receptor. Their primary goal is to distinguish between ligands with strong affinities for the target protein and those with weaker affinities. The possession of a three-dimensional target configuration is a critical requirement for carrying out a receptor-based VS initiative. This can take the form of a crystalline X-ray structure, an NMR-derived structure, or even a structure inferred through homology modeling. The dominance of receptor-based methodologies over ligand-based methodologies is growing, owing to the increasing availability of resolved three-dimensional structures of target proteins for research purposes. Structure-based VS anticipates the location and orientation of a ligand when it interacts with a protein [22]. In the present study, AutoDock Vina 1.1.2 [23] and AutoDock 4.2.5.1 [24] were used for VS and molecular docking studies to identify binding conformations with the lowest binding energies (BEs). Finasteride was used as a positive control for VS, and X, Y, and Z coordinates were set at -20.40, 17.50, and 45.54, respectively.

2.4. Physiochemical and Toxicity Prediction

The SwissADME web server was used to assess the predicted physiochemical and pharmacokinetic properties of the top two compounds (eriocitrin and silymarin) [25], and ProTox-II, an open-access web server, was used to predict their toxicities [26]. ProTox-II combines molecular similarity, pharmacophore, and machine learning models to predict toxicity endpoints, such as hepatotoxicity, immunotoxicity, carcinogenicity, mutagenicity, and cytotoxicity.

2.5. Molecular Dynamics (MD) Simulations

Studying the intramolecular dynamics of proteins can reveal hidden biological functions and intricate mechanisms. GROMACS2019.6 [27] was used to study the stabilities of 5αR2-eriocitrin, 5αR2-silymarin, and 5αR2-finasteride (control) complexes. MD simulation was carried out for 200 nanoseconds (ns) using the GROMOS96 43a1 force-field parameter [28]. The protein topology file was generated using the gmx tool, while the ligand topology file was created using the Swiss Param server. Subsequently, the TIP3 water model was employed, and a solvent box was generated at a distance of 10 Å. To achieve system equilibrium, counter ions such as Na^+ and Cl^- (5αR2-eriocitrin; Na^+ (0), Cl^- (10), 5αR2-silymarin; Na^+ (0), Cl^- (10), and 5αR2-finasteride; Na^+ (0), Cl^- (10)) were introduced, while preserving a salt concentration of 150 mM. The solvation of the protein was accomplished using the Simple Point Charge (spc216) water model. Using the energy-grps in the MDs parameters (mdp) file, the particle-mesh Ewald method was utilized to study interactions between 5αR2 and eriocitrin, silymarin, and finasteride. To achieve system equilibration, the NPT and NVT ensembles were employed at a temperature of 310 K and pressure of 1 bar. The GROMACS analysis module was utilized to examine trajectories by plotting graphs for Root Mean Square Deviation (RMSD), Root Mean Square Fluctuation (RMSF), Radius of Gyration (Rg), Solvent Accessible Surface Area (SASA), and H-bonding.

3. Results and Discussion

5αR2 is involved in the pathophysiology of AGA, and thus, 5αR2 inhibitors are considered crucial for the development of anti-baldness therapies. Accordingly, we screened 241 flavonoids against the active pocket of 5αR2 and identified 11 compounds whose BEs were superior to those of the finasteride (Table 1). These 11 compounds were further evaluated for their interactions with the active site residues of the 5αR2 using Pymol and DS in 2D and 3D view. Through a comprehensive examination of the interactions and thorough investigation of both two-dimensional and three-dimensional interactions, it was determined that eriocitrin and silymarin exhibited the most substantial interaction with 5αR2, as established by visual inspection and interaction analysis.

Table 1. BE of top-screened flavonoid compounds.

S. No.	Compound Name	Binding Energy (kcal/mol)
1.	Eriocitrin	−12.1
2.	Obacunone	−11.8
3.	Oroxin_B	−11.7
4.	Silymarin	−11.7
5.	Hesperidin	−11.7
6.	Baicalin	−11.6
7.	Diosmin	−11.6
8.	Scutellarin	−11.6
9.	Methyl-Hesperidin	−11.6
10.	Narirutin	−11.6
11.	Isosilybin	−11.6
12.	Finasteride (positive control)	−11.2

Figure 1A depicts finasteride, eriocitrin, and silymarin in the 5αR2 binding pocket. The low BE of eriocitrin and silymarin with 5αR2 provided the contributions of H-bonds and van der Waals interactions. Eriocitrin had a BE of −12.1 kcal/mol (Table 1) and interacted with several key residues of 5αR2, viz., Leu11, Tyr33, Gly34, Ala49, Trp53, Gln56, Glu57, His90, Tyr91, Arg94, Tyr98, Asn102, Arg103, Gly104, Tyr107, Arg114, Gly115, Phe118, Cys119, Phe129, Asn160, Asp164, Leu167, Asn193, Phe194, Glu197, Trp201, Phe216, Phe223, Leu224, and Arg227. Of these, the Gln56, Tyr91, Arg94, Asn102, Asn160, and Asp164 residues formed H-bonds with eriocitrin, while Tyr107, Leu111, Arg114, Phe118, Phe219, Cys119, Phe216, Trp201, His90, Glu197, Asn193, Tyr98, Leu167, Arg103, Gly104, Ala49, Arg227, Phe194, Tyr33, Trp53, Gly34, Leu224, Phe223, and Gly115 residues were involved in van der Waals interactions (Figure 1B).

Silymarin had a BE of −11.7 kcal/mol (Table 1) and interacted with the Tyr33, Lys35, Trp53, Glu57, Tyr91, Arg94, Tyr98, Arg103, Gly104, Tyr235, Pro181, Leu167, Asn160, Asp164, Arg168, Leu170, Arg171, Tyr178, Arg179, Asn193, Phe194, Glu197, Phe219, Ser220, Phe223, Leu224, and Arg227 residues of 5αR2. The Lys35, Glu57, Asn160, Asp164, Leu167, and Asn193 residues formed H-bonds with silymarin, while Ser220, Phe219, Tyr91, Arg94, Phe194, Tyr235, Pro181, Arg179, Leu170, Arg171, Arg168, Gly104, Arg103, Tyr33, and Trp53 residues were involved in van der Waals interactions (Figure 1C). The structures of eriocitrin, silymarin, and finasteride are shown in Table 2.

Finasteride, a 5αR2 inhibitor, reduces serum and scalp DHT levels by inhibiting testosterone to DHT conversion and is often used to treat AGA [29]. Clinical studies on men with alopecia revealed that finasteride administration reduced DHT levels in the scalp, promoting hair growth and confirming the role of DHT in the underlying pathophysiology of AGA [30]. In order to obtain a comprehensive understanding of the interacting residues between 5αR2 and the eriocitrin and silymarin, an interaction analysis was conducted on the co-crystallized ligand finasteride (PDB ID: 7BW1) and 5αR2 corresponding residues by redocking the finasteride with 5αR2. The results revealed that the Leu20, Leu23, Ala24, Ser31, Tyr33, Trp53, Gln56, Glu57, Tyr91, Arg94, Tyr98, Tyr107, Leu111, Arg114, Gly115, Phe118, Asn160, Asp164, Asn193, Phe194, Glu197, Trp201, Phe216, Phe219, Ser220, Fhe223, and Leu224 residues of 5αR2 were essential for the interaction with finasteride (Figure 1D). Remarkably, the Trp53, Glu57, Tyr91, Arg94, Tyr98, Asn160, Asp164, Asn193, Phe194, Glu197, Phe223, and Leu224 residues were identified as the common interacting residues of 5αR2 with eriocitrin and silymarin, as well as finasteride (Figure 1B–D), representing that these compounds bind at the same site of 5αR2 as finasteride.

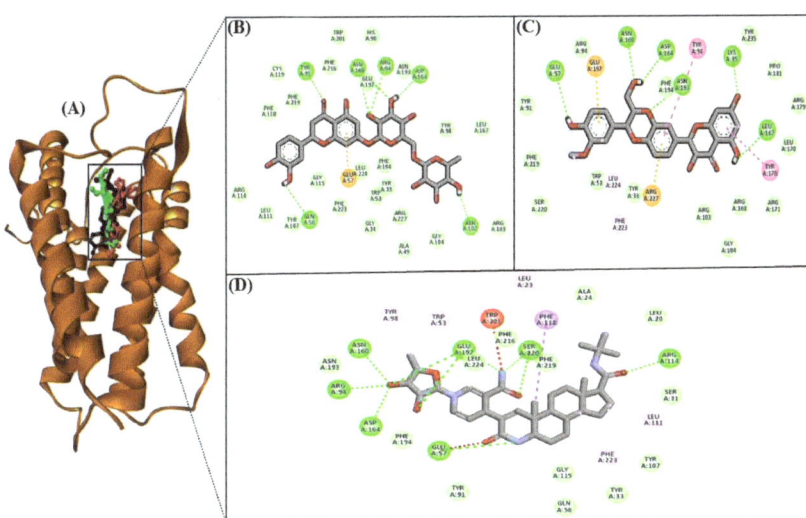

Figure 1. Visualization of finasteride (black), eriocitrin (red), and silymarin (cyan) in the 5αR2 binding pocket (**A**). 2D views of 5αR2 residues interacting with eriocitrin (**B**), silymarin (**C**), and finasteride (**D**).

Table 2. Structures of eriocitrin, silymarin, and finasteride and interacting residues of 5αR2.

Compounds	Structure	No. of H-Bond	H-Bonds Interacting Residues	Van der Waals Interactions
Eriocitrin	(structure)	6	Gln56, Tyr91, Arg94, Asn102, Asn160, and Asp164	Tyr107, Leu111, Arg114, Phe118, Phe219, Cys119, Phe216, Trp201, His90, Glu197, Asn193, Tyr98, Leu167, Arg103, Gly104, Ala49, Arg227, Phe194, Tyr33, Trp53, Gly34, Leu224, Phe223, and Gly115
Silymarin	(structure)	6	Lys35, Glu57, Asn160, Asp164, Leu167, and Asn193	Ser220, Phe219, Tyr91, Arg94, Phe194, Tyr235, Pro181, Arg179, Leu170, Arg171, Arg168, Gly104, Arg103, Tyr33, and Trp53
Finasteride	(structure)	7	Glu57, Arg94, Arg114, Asn160, Asp164, Glu197, and Ser220	Asn193, Leu224, Tyr91, Phe194, Phe216, Phe219, Ala24, Leu20, Ser31, Tyr107, Tyr33, Gly115, and Gln56

A higher negative BE of a compound with the target enzyme indicates a stronger interaction with its amino acid residues in the catalytic pocket, and the dissociation rate of such compounds from the target enzyme will be slower [31–33]. Interestingly, eriocitrin and silymarin had higher (negative) BEs than finasteride (control), revealing that these compounds have a strong interaction with 5αR2.

Silymarin is derived from *Silybum marianum* (L.) *gaernt* (the milk thistle), while *Citrus limon* is a rich source of eriocitrin. The pharmacological effects of these compounds have been well explored, especially their hepatoprotective, antioxidant, anticancer, anti-diabetic, anti-inflammatory, and cardioprotective activities [34,35], and accumulated evidence indicates these compounds are suitable therapeutics. We predicted the physicochemical properties and toxicities of eriocitrin and silymarin (Tables 3 and 4) and showed that both compounds possess acceptable selected parameters. The physicochemical parameters of the selected compounds were evaluated, including the number of heavy atoms, proportion Csp3, number of rotatable bonds, number of hydrogen bond donors and acceptors, molar refractivity, and TPSA. In addition, the lipophilicity, water solubility, and pharmacokinetics of these compounds were evaluated, and their estimated values are shown in Table 3.

Table 3. Estimated physicochemical properties of silymarin, eriocitrin, and finasteride.

Properties		Compound Name		
Physicochemical Properties		Eriocitrin	Silymarin	Finasteride (Control)
MW		596.53	482.44	372.54
Heavy atoms		42	35	27
Aromatic heavy atoms		12	18	0
Fraction Csp3		0.52	0.24	0.83
RB		6	4	3
HBA		15	10	2
HBD		9	5	2
Molar Refractivity		136.94	120.55	113.18
TPSA		245.29	155.14	58.20
		Lipophilicity		
iLOGP		1.95	2.79	3.32
XLOGP3		−1.35	1.9	3.03
WLOGP		−1.78	1.71	3.43
MLOGP		−3.24	−0.4	3.46
Silicos-IT Log P		−2.1	1.92	3.20
Consensus Log P		−1.3	1.59	3.29
		Water Solubility		
ESOL	Log S	−2.5	−4.14	−3.86
	Solubility (mg/mL)	1.87×10^0	3.46×10^{-2}	5.13×10^{-2}
	Class	Soluble	Moderately soluble	Soluble
Ali	Log S	−3.3	−4.78	−3.92
	Solubility (mg/mL)	2.98×10^{-1}	7.99×10^{-3}	4.50×10^{-2}
	Class	Soluble	Moderately soluble	Soluble
Silicos-IT	LogSw	0.1	−4.5	−4.54
	Solubility (mg/mL)	7.56×10^2	1.53×10^{-2}	1.07×10^{-2}
	class	Soluble	Moderately soluble	Moderately soluble

Table 3. Cont.

Properties		Compound Name		
Physicochemical Properties		Eriocitrin	Silymarin	Finasteride (Control)
		Pharmacokinetics		
GI absorption		Low	Low	High
BBB permeant		No	No	Yes
Pgp substrate		Yes	No	Yes
inhibitor	CYP1A2	No	No	No
	CYP2C19	No	No	No
	CYP2C9	No	No	No
	CYP2D6	No	No	No
	CYP3A4	No	Yes	No
log Kp (cm/s)		−10.9	−7.89	−6.42

Table 4. Toxicity predictions for silymarin, eriocitrin, and finasteride.

Classification	Target	Prediction			Probability		
		Silymarin	Eriocitrin	Finasteride (Control)	Silymarin	Eriocitrin	Finasteride (Control)
Organ toxicity	Hepatotoxicity	IA	IA	IA	0.78	0.8	0.98
Toxicity endpoints	Carcinogenicity	IA	IA	IA	0.72	0.91	0.61
	Immunotoxicity	A	A	A	0.97	0.99	0.99
	Mutagenicity	IA	IA	IA	0.69	0.88	0.81
	Cytotoxicity	IA	IA	IA	0.77	0.64	0.79
Tox21-Nuclear receptor signaling pathways	AhR	A	IA	IA	0.99	0.83	0.99
	Androgen Receptor (AR)	IA	IA	IA	0.95	0.98	0.87
	AR-LBD	IA	IA	IA	0.99	0.99	0.99
	Aromatase	IA	IA	IA	0.8	0.99	0.97
	Estrogen Receptor Alpha (ER)	IA	IA	IA	0.71	0.95	0.93
	ER-LBD	IA	IA	IA	0.96	0.99	0.98
	PPAR-Gamma	IA	IA	IA	0.97	0.98	0.98
Tox21-stress response pathways	nrf2/ARE	IA	IA	IA	0.92	0.99	0.97
	HSE	IA	IA	IA	0.92	0.99	0.97
	MMP	IA	IA	IA	0.73	0.97	0.93
	p53	IA	IA	IA	0.91	0.9	0.97
	ATAD5	IA	IA	IA	0.94	0.99	0.99

(IA = Inactive; A = Active).

Further, the toxicity assessment of the selected compounds, eriocitrin and silymarin, was assessed by ProTox-II. To estimate a wide range of toxicity endpoints, the ProTox-II methodology employs a comprehensive computational strategy that integrates molecular similarity, pharmacophores, fragment propensities, and machine-learning models. This platform predicts toxicity based on chemical compounds that have been confirmed using various experiments. The web server provides confidence levels for the results and permits similarity comparisons. The predicted lethal dose 50 (pLD50) of eriocitrin was determined to be 12,000 mg/kg, placing it in Toxicity Class six. This classification suggests that eriocitrin exhibits no toxicity. The average similarity between the predicted and actual toxicity statistics for eriocitrin was determined to be 98.6%, while the prediction accuracy was found to be 72.9%. On the other hand, the pLD50 of silymarin was estimated to be 2000 mg/kg, which placed it in the 'harmful if swallowed' category and Toxicity Class four. The average similarity between predicted and experimental toxicity was 76.44%, and the prediction accuracy was 69.26%. Toxicity endpoints for silymarin and eriocitrin, such as acute toxicity, hepatotoxicity, cytotoxicity, carcinogenicity, mutagenicity, and immunotoxicity, were within acceptable ranges (Table 4). Furthermore, there was no hepatotoxicity observed in both compounds.

Subsequently, MD simulations were conducted on the docked complexes of the aforementioned compounds with the 5αR2 enzyme, encompassing a duration of 200 ns. The primary objective of these simulations was to evaluate the stability of the docked complexes. MD simulations primarily aid in the understanding of conformational stability, a phenomenon that has a significant impact on the efficacy of therapeutic compounds in inhibiting target proteins. These simulations, on the other hand, demonstrate their utility by providing insights into interaction dynamics, including bonding events and stability patterns over time. RMSD is a metric utilized to evaluate protein stability, where lower RMSD deviations indicate greater stability. 5αR2-control, 5αR2-eriocitrin, and 5αR2-silymarin had RMSD average values of 0.45, 0.36, and 0.35 nm, respectively, and the RMSD plot revealed that 5αR2-eriocitrin and 5αR2-silymarin complexes had greater binding stability than the 5αR2-control complex. The bound structure of the 5αR2-control complex exhibited high deviation from its initial conformation, whereas the catalytic pocket of 5αR2 formed stable interactions with eriocitrin and silymarin. Furthermore, ligand RMSDs showed that 5αR2-eriocitrin exhibited a high deviation, whereas the 5αR2-control and 5αR2-silymarin complexes exhibited low deviations. It showed that the eriocitrin molecule did not tightly interact in the catalytic pocket of 5αR2 and therefore showed higher deviation in the pocket of the enzyme (Figure 2A–C).

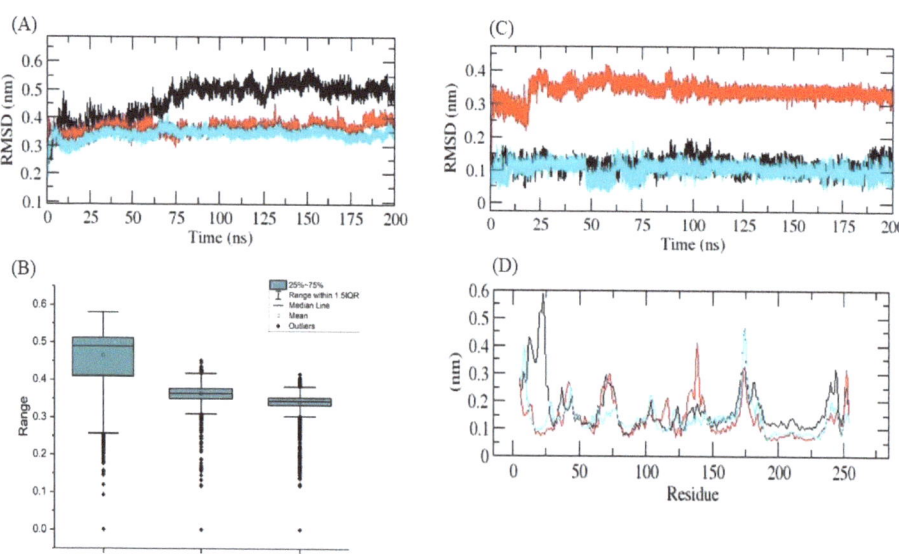

Figure 2. MD simulation studies of docked complexes. RMSD plot (**A**), average RMSD plot (**B**), RMSD plots of ligands in the 5αR2 catalytic pocket (**C**), and corresponding RMSF plots (**D**). Black, red, and cyan indicate finasteride, eriocitrin, and silymarin, respectively.

The average fluctuation of all residues throughout the simulation and the RMSD of 5αR2 during the binding of 5αR2-control, 5αR2-eriocitrin, and 5αR2-silymarin was plotted as a function of 5αR2 residue numbers. Consistent fluctuations in the catalytic pocket were observed in the backbones of 5αR2-silymarin and 5αR2-control, possibly due to orientation differences, while the 5αR2-eriocitrin complex exhibited high fluctuation in the 130–140 residue region of 5αR2 (Figure 2D), agreeing with the ligand RMSD which showed high deviation due to high fluctuations in the catalytic pocket of 5αR2. Notably, the 5αR2-silymarin complex demonstrated the least fluctuation overall.

Rg plots were used to obtain the compactness profiles of the complexes. The 5αR2-control, 5αR2-eriocitrin, and 5αR2-silymarin complexes had average Rg values of 1.79, 1.82, and 1.76 nm, respectively, and the Rg plots showed that 5αR2-eriocitrin and 5αR2-control complexes were less compact than the 5αR2-silymarin complex. These findings suggest that silymarin binding to 5αR2 increased enzyme stability, as evidenced by the reduced Rg compactness and little effect on the 5αR2 structure (Figure 3A), and that silymarin was more stable in the catalytic pocket of 5αR2. On the other hand, eriocitrin induced alteration in the conformational structure of the enzyme, and therefore, less compactness was shown by 5αR2.

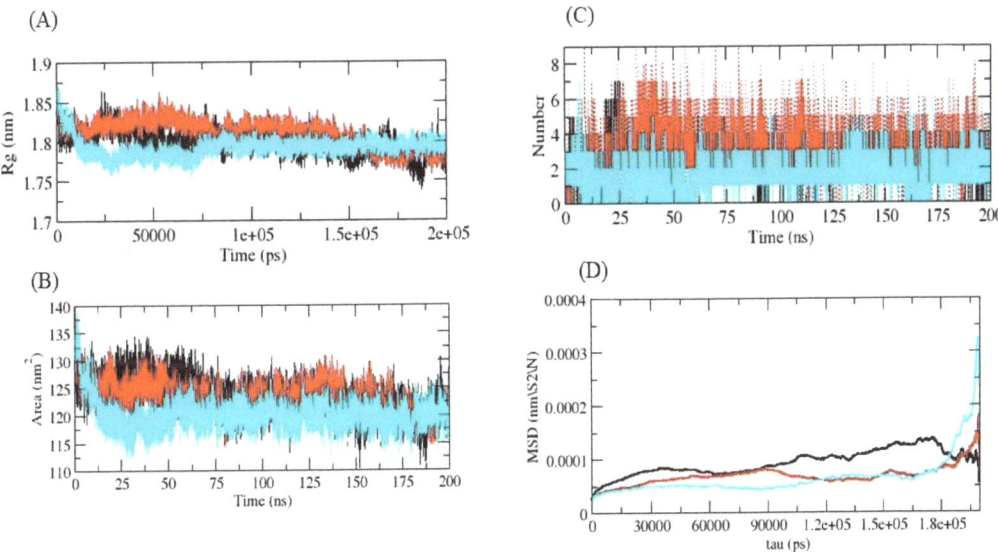

Figure 3. Rg plot of complexes (**A**), SASA plot (**B**), number of H-bonds formed with ligands (**C**), and MSD plot (**D**). Black, red, and cyan indicate finasteride, eriocitrin, and silymarin, respectively.

SASA provides a measure of the surface area of proteins which interact with solvent molecules. Average SASA values for 5αR2-control, 5αR2-eriocitrin, and 5αR2-silymarin complexes were plotted during the 200 ns simulation, and their SASA values were 128.50, 125.20, and 116.25 nm^2, respectively (Figure 3B). SASA analysis showed that surface exposure was reduced when silymarin and eriocitrin were bound, while the control compound increased the surface area of solvent accessibility. Thus, 5αR2-eriocitrin and 5αR2-silymarin were found to interact less with the solvent than the 5αR2-control.

H-bonds represent highly precise interactions between the inhibitor and the target. These interactions are crucial in determining the stability of the complex formed by the target and inhibitor. To evaluate the stabilities of the ligand–target complex, H-bond analysis was conducted during 200 ns simulations of 5αR2-control, 5αR2-eriocitrin, and 5αR2-silymarin in a solvent environment. Finasteride and silymarin were found to form two to four H-bonds with 5αR2, whereas eriocitrin formed two to six H-bonds (Figure 3C). In addition, the mean square displacement (MSD) of atoms from a set of original 5αR2 complex coordinates was computed (Figure 3D). The displacement of atoms from a set of beginning sites in the complexes 5αR2-control, 5αR2-eriocitrin, and 5αR2-silymarin was calculated, with 5αR2-silymarin having the greatest MSD value. The first two eigenvectors were projected in 2D. During the simulations, the 5αR2-control and 5αR2-eriocitrin exhibited a decreased diversity of conformation during simulation; however, the 5αR2-silymarin complex exhibited a greater diversity of conformations. This demonstrates that the 5αR2-silymarin complex was efficiently equilibrated and stable during the simulation (Figure 4A).

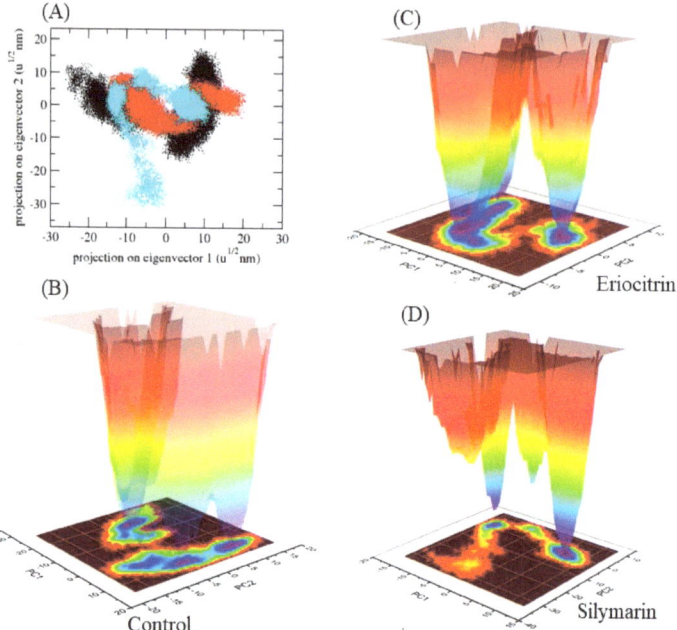

Figure 4. 2D projections of complexes (**A**) and Gibbs energy landscape plots (**B**–**D**).

Further, GROMACS analysis modules were used to calculate the Gibbs' free energy (GFE) landscape and project the respective first (PC1) and second (PC2) eigenvectors (darker blue shades indicate lower energy levels). During the simulations, ligand binding to 5αR2 caused fluctuations in the global minima of 5αR2, as observed in the GFE contour maps. 5αR2-control and 5αR2-silymarin had similar projections, whereas 5αR2-eriocitrin had different global minima, indicating that the global minima of eriocitrin had changed during simulation (Figure 4B–D).

Over the years, various potential 5αR2 inhibitors have been explored, and some have been synthesized. Furthermore, potent 5αR2 inhibitors have consistently been found to bind strongly to 5αR2 [36]. Finasteride and dutasteride are currently being used to treat AGA [37,38], but both have been associated with adverse effects like impotency and sexual dysfunction [38,39]. Thus, there is an urgent need to identify natural 5αR2 inhibitors with no side effects. Natural products and traditional medicines are extremely important, and their derivatives have long been recognized as valuable reservoirs of therapeutic agents and structural variability. A wide range of pharmaceutical agents currently on the market have their origins in natural reservoirs. Natural products have long been used to identify potential developmental leads [40–44], and in this study, we found that natural flavonoids, namely, eriocitrin and silymarin, stably interact with 5αR2, which implies their potential as therapeutic agents for treating AGA.

4. Conclusions

The involvement of 5αR2 in the pathophysiological mechanisms of AGA establishes the necessity of 5αR2 inhibitors in the advancement of therapeutic interventions for AGA. VS has been routinely employed to identify new drug leads. In the present study, in silico methodologies, viz., molecular docking-based VS, toxicity prediction, and MD simulation, were used to identify potential nontoxic 5αR2 inhibitors in a natural flavonoid library. Eriocitrin and silymarin were found to interact strongly with 5αR2 and form stable complexes and could be potential future anti-baldness drug candidates. However, further experimental research is needed to optimize them as 5αR2 inhibitors.

Author Contributions: Conceptualization, S.S. and I.C.; formal analysis, K.A. and I.C.; funding acquisition, I.C.; methodology, S.S., S.A. and K.A.; writing—original draft, S.S. and S.A.; writing—review and editing, J.H.L., K.A., K.S.H., E.J.L. and I.C. All authors have read and agreed to the published version of the manuscript.

Funding: This research was supported by the Basic Science Research Program through the National Research Foundation of Korea (NRF), funded by the Ministry of Education (2020R1A6A1A03044512).

Institutional Review Board Statement: Not applicable.

Informed Consent Statement: Not applicable.

Data Availability Statement: Not applicable.

Conflicts of Interest: The authors declare no conflict of interest.

References

1. Koch, S.L.; Tridico, S.R.; Bernard, B.A.; Shriver, M.D.; Jablonski, N.G. The biology of human hair: A multidisciplinary review. *Am. J. Hum. Biol.* **2020**, *32*, e23316. [CrossRef] [PubMed]
2. Richards, J.B.; Yuan, X.; Geller, F.; Waterworth, D.; Bataille, V.; Glass, D.; Song, K.; Waeber, G.; Vollenweider, P.; Aben, K.K.; et al. Male-pattern baldness susceptibility locus at 20p11. *Nat. Genet.* **2008**, *40*, 1282–1284. [CrossRef] [PubMed]
3. Adil, A.; Godwin, M. The effectiveness of treatments for androgenetic alopecia: A systematic review and meta-analysis. *J. Am. Acad. Dermatol.* **2017**, *77*, 136–141.e5. [CrossRef] [PubMed]
4. Sultan, C.; Lumbroso, S.; Poujol, N.; Boudon, C.; Georget, V.; Terouanne, B.; Belon, C.; Lobaccaro, J.M. Genetics and endocrinology of male sex differentiation: Application to molecular study of male pseudohermaphroditism. *C. R. Seances Soc. Biol. Fil.* **1995**, *189*, 713–740.
5. Trueb, R.M. Molecular mechanisms of androgenetic alopecia. *Exp. Gerontol.* **2002**, *37*, 981–990. [CrossRef]
6. Asakawa, K.; Toyoshima, K.E.; Ishibashi, N.; Tobe, H.; Iwadate, A.; Kanayama, T.; Hasegawa, T.; Nakao, K.; Toki, H.; Noguchi, S.; et al. Hair organ regeneration via the bioengineered hair follicular unit transplantation. *Sci. Rep.* **2012**, *2*, 424. [CrossRef]
7. Toyoshima, K.E.; Asakawa, K.; Ishibashi, N.; Toki, H.; Ogawa, M.; Hasegawa, T.; Irie, T.; Tachikawa, T.; Sato, A.; Takeda, A.; et al. Fully functional hair follicle regeneration through the rearrangement of stem cells and their niches. *Nat. Commun.* **2012**, *3*, 784. [CrossRef]
8. Gupta, A.K.; Carviel, J.; MacLeod, M.A.; Shear, N. Assessing finasteride-associated sexual dysfunction using the FAERS database. *J. Eur. Acad. Dermatol. Venereol.* **2017**, *31*, 1069–1075. [CrossRef]
9. Olsen, E.A.; Weiner, M.S.; Amara, I.A.; DeLong, E.R. Five-year follow-up of men with androgenetic alopecia treated with topical minoxidil. *J. Am. Acad. Dermatol.* **1990**, *22*, 643–646. [CrossRef]
10. Barbareschi, M. The use of minoxidil in the treatment of male and female androgenetic alopecia: A story of more than 30 years. *G. Ital. Dermatol. Venereol.* **2018**, *153*, 102–106. [CrossRef]
11. Mysore, V.; Shashikumar, B.M. Guidelines on the use of finasteride in androgenetic alopecia. *Indian J. Dermatol. Venereol. Leprol.* **2016**, *82*, 128–134. [CrossRef] [PubMed]
12. Yim, E.; Nole, K.L.; Tosti, A. 5alpha-Reductase inhibitors in androgenetic alopecia. *Curr. Opin. Endocrinol. Diabetes Obes.* **2014**, *21*, 493–498. [CrossRef] [PubMed]
13. Llorach-Pares, L.; Nonell-Canals, A.; Avila, C.; Sanchez-Martinez, M. Computer-Aided Drug Design (CADD) to De-Orphanize Marine Molecules: Finding Potential Therapeutic Agents for Neurodegenerative and Cardiovascular Diseases. *Mar. Drugs* **2022**, *20*, 53. [CrossRef] [PubMed]
14. Maghsoudi, S.; Taghavi Shahraki, B.; Rameh, F.; Nazarabi, M.; Fatahi, Y.; Akhavan, O.; Rabiee, M.; Mostafavi, E.; Lima, E.C.; Saeb, M.R.; et al. A review on computer-aided chemogenomics and drug repositioning for rational COVID-19 drug discovery. *Chem. Biol. Drug Des.* **2022**, *100*, 699–721. [CrossRef]

15. Cui, W.; Aouidate, A.; Wang, S.; Yu, Q.; Li, Y.; Yuan, S. Discovering Anti-Cancer Drugs via Computational Methods. *Front. Pharmacol.* **2020**, *11*, 733. [CrossRef]
16. Sabe, V.T.; Ntombela, T.; Jhamba, L.A.; Maguire, G.E.M.; Govender, T.; Naicker, T.; Kruger, H.G. Current trends in computer aided drug design and a highlight of drugs discovered via computational techniques: A review. *Eur. J. Med. Chem.* **2021**, *224*, 113705. [CrossRef]
17. Romano, J.D.; Tatonetti, N.P. Informatics and Computational Methods in Natural Product Drug Discovery: A Review and Perspectives. *Front. Genet.* **2019**, *10*, 368. [CrossRef]
18. Martins, B.T.; Correia da Silva, M.; Pinto, M.; Cidade, H.; Kijjoa, A. Marine natural flavonoids: Chemistry and biological activities. *Nat. Prod. Res.* **2019**, *33*, 3260–3272. [CrossRef]
19. Xiao, Q.; Wang, L.; Supekar, S.; Shen, T.; Liu, H.; Ye, F.; Huang, J.; Fan, H.; Wei, Z.; Zhang, C. Structure of human steroid 5alpha-reductase 2 with the anti-androgen drug finasteride. *Nat. Commun.* **2020**, *11*, 5430. [CrossRef]
20. Lionta, E.; Spyrou, G.; Vassilatis, D.K.; Cournia, Z. Structure-based virtual screening for drug discovery: Principles, applications and recent advances. *Curr. Top. Med. Chem.* **2014**, *14*, 1923–1938. [CrossRef]
21. Shoichet, B.K. Virtual screening of chemical libraries. *Nature* **2004**, *432*, 862–865. [CrossRef] [PubMed]
22. Ballante, F.; Kooistra, A.J.; Kampen, S.; de Graaf, C.; Carlsson, J. Structure-Based Virtual Screening for Ligands of G Protein-Coupled Receptors: What Can Molecular Docking Do for You? *Pharmacol. Rev.* **2021**, *73*, 527–565. [CrossRef] [PubMed]
23. Trott, O.; Olson, A.J. AutoDock Vina: Improving the speed and accuracy of docking with a new scoring function, efficient optimization, and multithreading. *J. Comput. Chem.* **2010**, *31*, 455–461. [CrossRef]
24. Morris, G.M.; Huey, R.; Lindstrom, W.; Sanner, M.F.; Belew, R.K.; Goodsell, D.S.; Olson, A.J. AutoDock4 and AutoDockTools4: Automated docking with selective receptor flexibility. *J. Comput. Chem.* **2009**, *30*, 2785–2791. [CrossRef] [PubMed]
25. Daina, A.; Michielin, O.; Zoete, V. SwissADME: A free web tool to evaluate pharmacokinetics, drug-likeness and medicinal chemistry friendliness of small molecules. *Sci. Rep.* **2017**, *7*, 42717. [CrossRef]
26. Banerjee, P.; Eckert, A.O.; Schrey, A.K.; Preissner, R. ProTox-II: A webserver for the prediction of toxicity of chemicals. *Nucleic Acids Res.* **2018**, *46*, W257–W263. [CrossRef]
27. Van Der Spoel, D.; Lindahl, E.; Hess, B.; Groenhof, G.; Mark, A.E.; Berendsen, H.J. GROMACS: Fast, flexible, and free. *J. Comput. Chem.* **2005**, *26*, 1701–1718. [CrossRef]
28. Pol-Fachin, L.; Fernandes, C.L.; Verli, H. GROMOS96 43a1 performance on the characterization of glycoprotein conformational ensembles through molecular dynamics simulations. *Carbohydr. Res.* **2009**, *344*, 491–500. [CrossRef]
29. Libecco, J.F.; Bergfeld, W.F. Finasteride in the treatment of alopecia. *Expert Opin. Pharmacother.* **2004**, *5*, 933–940. [CrossRef]
30. Kaufman, K.D.; Dawber, R.P. Finasteride, a Type 2 5alpha-reductase inhibitor, in the treatment of men with androgenetic alopecia. *Expert Opin. Investig. Drugs* **1999**, *8*, 403–415. [CrossRef]
31. Ahmad, S.S.; Khan, M.B.; Ahmad, K.; Lim, J.H.; Shaikh, S.; Lee, E.J.; Choi, I. Biocomputational Screening of Natural Compounds against Acetylcholinesterase. *Molecules* **2021**, *26*, 2641. [CrossRef] [PubMed]
32. Mojica, L.; de Mejia, E.G. Optimization of enzymatic production of anti-diabetic peptides from black bean (*Phaseolus vulgaris* L.) proteins, their characterization and biological potential. *Food Funct.* **2016**, *7*, 713–727. [CrossRef] [PubMed]
33. Ali, S.; Ahmad, K.; Shaikh, S.; Lim, J.H.; Chun, H.J.; Ahmad, S.S.; Lee, E.J.; Choi, I. Identification and Evaluation of Traditional Chinese Medicine Natural Compounds as Potential Myostatin Inhibitors: An In Silico Approach. *Molecules* **2022**, *27*, 4303. [CrossRef] [PubMed]
34. Yao, L.; Liu, W.; Bashir, M.; Nisar, M.F.; Wan, C.C. Eriocitrin: A review of pharmacological effects. *Biomed. Pharmacother.* **2022**, *154*, 113563. [CrossRef]
35. Karimi, G.; Vahabzadeh, M.; Lari, P.; Rashedinia, M.; Moshiri, M. "Silymarin", a promising pharmacological agent for treatment of diseases. *Iran. J. Basic Med. Sci.* **2011**, *14*, 308–317.
36. Aggarwal, S.; Thareja, S.; Verma, A.; Bhardwaj, T.R.; Kumar, M. An overview on 5alpha-reductase inhibitors. *Steroids* **2010**, *75*, 109–153. [CrossRef]
37. Iamsumang, W.; Leerunyakul, K.; Suchonwanit, P. Finasteride and Its Potential for the Treatment of Female Pattern Hair Loss: Evidence to Date. *Drug Des. Dev. Ther.* **2020**, *14*, 951–959. [CrossRef]
38. Dhurat, R.; Sharma, A.; Rudnicka, L.; Kroumpouzos, G.; Kassir, M.; Galadari, H.; Wollina, U.; Lotti, T.; Golubovic, M.; Binic, I.; et al. 5-Alpha reductase inhibitors in androgenetic alopecia: Shifting paradigms, current concepts, comparative efficacy, and safety. *Dermatol. Ther.* **2020**, *33*, e13379. [CrossRef]
39. Erdemir, F.; Harbin, A.; Hellstrom, W.J. 5-alpha reductase inhibitors and erectile dysfunction: The connection. *J. Sex. Med.* **2008**, *5*, 2917–2924. [CrossRef]
40. Mushtaq, S.; Abbasi, B.H.; Uzair, B.; Abbasi, R. Natural products as reservoirs of novel therapeutic agents. *EXCLI J.* **2018**, *17*, 420–451. [CrossRef]
41. Atanasov, A.G.; Zotchev, S.B.; Dirsch, V.M.; International Natural Product Sciences Taskforce; Supuran, C.T. Natural products in drug discovery: Advances and opportunities. *Nat. Rev. Drug Discov.* **2021**, *20*, 200–216. [CrossRef] [PubMed]
42. Shaikh, S.; Ali, S.; Lim, J.H.; Chun, H.J.; Ahmad, K.; Ahmad, S.S.; Hwang, Y.C.; Han, K.S.; Kim, N.R.; Lee, E.J.; et al. Dipeptidyl peptidase-4 inhibitory potentials of *Glycyrrhiza uralensis* and its bioactive compounds licochalcone A and licochalcone B: An in silico and in vitro study. *Front. Mol. Biosci.* **2022**, *9*, 1024764. [CrossRef] [PubMed]

43. Lee, E.J.; Shaikh, S.; Ahmad, K.; Ahmad, S.S.; Lim, J.H.; Park, S.; Yang, H.J.; Cho, W.K.; Park, S.J.; Lee, Y.H.; et al. Isolation and Characterization of Compounds from *Glycyrrhiza uralensis* as Therapeutic Agents for the Muscle Disorders. *Int. J. Mol. Sci.* **2021**, *22*, 876. [CrossRef] [PubMed]
44. Shaikh, S.; Lee, E.J.; Ahmad, K.; Ahmad, S.S.; Lim, J.H.; Choi, I. A Comprehensive Review and Perspective on Natural Sources as Dipeptidyl Peptidase-4 Inhibitors for Management of Diabetes. *Pharmaceuticals* **2021**, *14*, 591. [CrossRef] [PubMed]

Disclaimer/Publisher's Note: The statements, opinions and data contained in all publications are solely those of the individual author(s) and contributor(s) and not of MDPI and/or the editor(s). MDPI and/or the editor(s) disclaim responsibility for any injury to people or property resulting from any ideas, methods, instructions or products referred to in the content.

Review

Bowiea volubilis: From "Climbing Onion" to Therapeutic Treasure—Exploring Human Health Applications

Hlalanathi Gwanya, Sizwe Cawe, Ifeanyi Egbichi, Nomagugu Gxaba, Afika-Amazizi Mbuyiswa, Samkele Zonyane, Babalwa Mbolekwa and Madira C. Manganyi *

Department of Biological and Environmental Sciences, Botany Section, Walter Sisulu University, Nelson Mandela Drive, Mthatha Campus, Mthatha 5117, South Africa; hgwanya@wsu.ac.za (H.G.); scawe@wsu.ac.za (S.C.); iegbichi@wsu.ac.za (I.E.); ngxaba@wsu.ac.za (N.G.); ambuyiswa@wsu.ac.za (A.-A.M.); szonyane@wsu.ac.za (S.Z.); bmbolekwa@wsu.ac.za (B.M.)
* Correspondence: mmanganyi@wsu.ac.za; Tel.: +27-047-502-2361

Abstract: *Bowiea volubilis* subsp. *volubilis* is primarily used to address human respiratory infections, coughs, and colds due to its diverse pharmaceutical properties. Notably, the plant contains alkaloids that exhibit notable antifungal, antibacterial, and cytotoxic properties. Additionally, the presence of saponins, with recognized antioxidant and anticancer attributes, further contributes to its medicinal potential. Steroid compounds inherent to the plant have been associated with anti-inflammatory and anticancer activities. Moreover, the bulb of *B. volubilis* has been associated as a source of various cardiac glycosides. Despite these therapeutic prospects, *B. volubilis* remains inedible due to the presence of naturally occurring toxic substances that pose risks to both animals and humans. The review focuses on a comprehensive exploration concerning *B. volubilis* ethnobotanical applications, phytochemical properties, and diverse biological activities in relation to in vitro and in vivo applications for promoting human health and disease prevention. The aim of the study is to comprehensively investigate the phytochemical composition, bioactive compounds, and potential medicinal properties of Bowiea volubilis, with the ultimate goal of uncovering its therapeutic applications for human health. This review also highlights an evident gap in research, i.e., insufficient evidence-based research on toxicity data. This void in knowledge presents a promising avenue for future investigations, opening doors to expanded inquiries into the properties and potential applications of *B. volubilis* in the context of human diseases.

Keywords: *Bowiea volubilis*; climbing onion; ethnobotanical uses; human health applications

Citation: Gwanya, H.; Cawe, S.; Egbichi, I.; Gxaba, N.; Mbuyiswa, A.-A.; Zonyane, S.; Mbolekwa, B.; Manganyi, M.C. *Bowiea volubilis*: From "Climbing Onion" to Therapeutic Treasure—Exploring Human Health Applications. *Life* **2023**, *13*, 2081. https://doi.org/10.3390/life13102081

Academic Editor: Seung Ho Lee

Received: 21 September 2023
Revised: 14 October 2023
Accepted: 16 October 2023
Published: 19 October 2023

Copyright: © 2023 by the authors. Licensee MDPI, Basel, Switzerland. This article is an open access article distributed under the terms and conditions of the Creative Commons Attribution (CC BY) license (https://creativecommons.org/licenses/by/4.0/).

1. Introduction

The rise in antibiotic-resistant pathogens has led to higher mortality rates for infectious diseases [1]. In the quest to explore a safer and eco-friendly therapeutic treatment for human diseases and ailments, the utilization of plant-based remedies has been well-documented in historical records through the fabrics of cultural traditions and knowledge [2]. *Bowiea volubilis* stands as a suitable candidate in the ongoing battle against antibiotic resistance and emerging infectious diseases. Although *B. volubilis* is used for various purposes, its primary application is in traditional medicine for treating respiratory infections, coughs, and colds. The plant contains various phytochemical compounds that are responsible for its medicinal properties. For example, the plant contains alkaloids, which have been reported to have antifungal, antibacterial, and cytotoxic activities [3]. Other compounds include saponins, which have been reported to have antioxidant and anticancer properties [4], and steroid compounds, including sitosterol and stigmasterol, which have been reported to have anti-inflammatory and antitumor properties [5]. In addition, flavonoid compounds such as luteolin, quercetin, and kaempferol, as well as phenolic compounds such as gallic acid, caffeic acid, and chlorogenic acid, have been reported to have antioxidant, anti-inflammatory, and anticancer properties [6].

Bowiea volubilis Harv. ex Hook.f. subsp. *volubilis*, commonly known as climbing onion or sea onion, is a plant species belonging to the family Asparagaceae. This plant species is native to Southern Africa, specifically South Africa, Zimbabwe, Mozambique, Botswana, and Namibia. In South Africa, *B. volubilis* grows in rocky areas, on cliffs, and on the edges of dry forests with a dry climate at elevations ranging from sea level to around 1500 m in various parts of the country, such as the Namaqualand, Drakensberg Mountains, and Gauteng Province [7–10].

This perennial herbaceous bulb-like plant consists of several distinctive parts, each with its own specific characteristics and uses [11]. The bulbous part of the stem is the most used plant part as it is traditionally used as a medicinal plant in many African cultures. The plant has long, narrow, green leaves that grow in a spiral pattern around the stem. The stem of *B. volubilis* is long, thin, and twining and is used as a support structure for the plant, allowing it to climb and reach for sunlight. The plant produces small, greenish-white flowers that are arranged in an umbel, which are not widely used but play a role in the plant's pollination [12,13].

Environmental stressors pose a serious threat or detrimental effects on many plants [14]. *B. volubilis*, however, is an extremely adaptive plant that can tolerate a range of environmental stresses, including drought, heat, and poor soil conditions. Its ability to survive in harsh conditions is partly due to the specialized bulbous part of its stem, which stores water for extended periods. Studies have shown that the plant can mobilize various physiological mechanisms to cope with environmental stressors [15]. As such, in recent years, it has been used as a bioindicator of land restoration efforts in arid and semi-arid regions of Southern Africa [16].

While there is limited evidence-based research on its use for human consumption, there are studies suggesting the use of *B. volubilis* as a potential feed supplement for ruminants [17] due to its high content of saponins, which have been shown to have positive effects on rumen fermentation and nutrient utilization [18,19]. However, further investigation is necessary to establish the optimal dosage and potential hazards connected with administering this plant to domesticated animals.

The preparation of *B. volubilis* for treatment depends on the plant part used and the type of treatment required. These different preparation methods can ultimately affect the bioactive compounds' content and influence the therapeutic effects of the plant. For instance, the bulb of *B. volubilis* is commonly used to treat a wide range of illnesses, including respiratory infections, gastrointestinal disorders, and skin diseases, and can be prepared in different ways, including boiling, drying, and grinding into a fine powder [20,21]. The leaves and stems of *B. volubilis* are used to treat various ailments, including hypertension, fever, and arthritis. They can be prepared by crushing or boiling, while the stems are commonly used in decoctions or infusions [22].

Even though there are numerous modern therapeutic options available worldwide, most of the global population, particularly those in rural areas, still heavily rely on herbal medicines for their health and well-being. Over 4000 plant species in South Africa, both threatened and non-threatened species, are utilized for their medicinal benefits [23]. This paper seeks to focus on the plant's role in addressing human health issues, bridging the gap between traditional knowledge and modern scientific understanding by providing a comprehensive analysis of *B. volubilis* medicinal potential, drawing from ethnopharmacological, ethnobotanical, phytochemical, and toxicological perspectives. By achieving these aims, the research paper aims to contribute to the broader understanding of traditional herbal medicine's relevance in modern healthcare, especially in regions where herbal remedies continue to play a vital role in supporting human health.

2. Methodology

High-quality global search engines such as Google Scholar, Scopus, ScienceDirect, and PubMed were used to screen, collect, review, and analyse previous research information in order to compile this current review. Keywords such as *Bowiea volubilis*, climbing onion, ethnobotanical uses and human health applications, *Bowiea volubilis*, climbing onion, and toxicity were used in the search engines to discover relevant research papers. Afterward, the abstracts were pre-screened before studying the full documents (Figure 1). Finally, the results were analysed to provide new insights into the plant's impact on human health and its potential application in disease management.

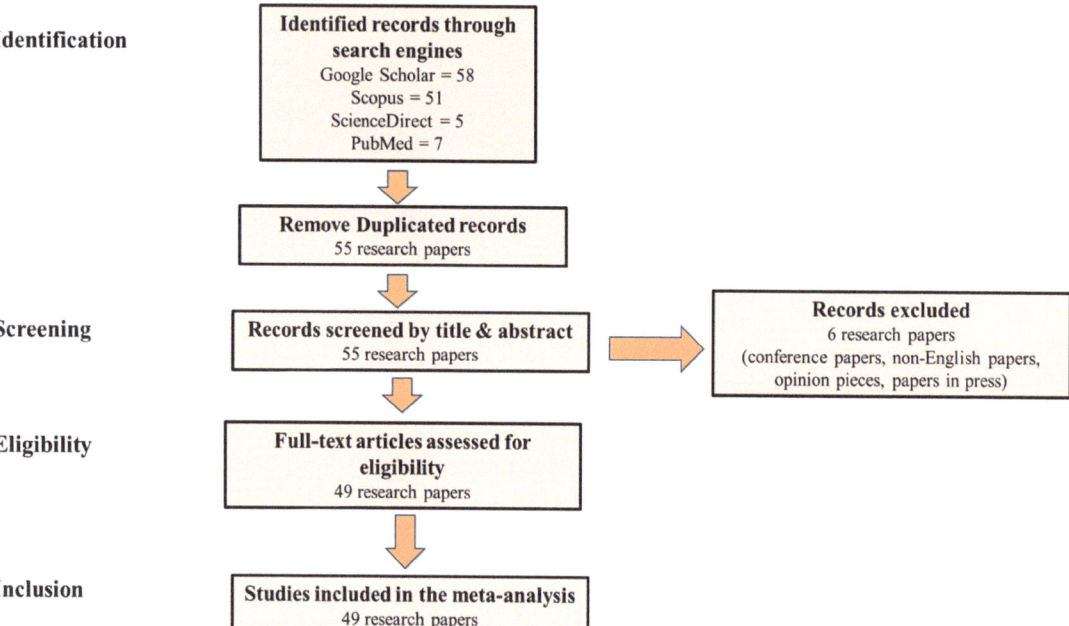

Figure 1. PRISMA Flowchart of studies through the systematic review process for *Bowiea volubilis* Health Applications.

3. Description, Distribution, and Habitat of *B. volubilis*

Bowiea vobulis is a succulent plant with a big, globose, green bulb that grows up to 15 cm in diameter, grows half buried in the ground, and sends up a twining, green-branched stem [8], as shown in Figure 2. The bulb is covered by thin branches that can wrap around anything for support [12]. It has six-petalled, half-inch wide, star-shaped, greenish-yellow flowers that develop from the top of the branches between January and February in its natural habitat and in summer in the Northern Hemisphere [13]. These green floral appendages are modified leaf petioles called cladophylls, which, like most true leaves, fall off with age and stress [8]. Since this plant does not have leaves, its photosynthesis is carried out by the stems [13]. The growth season of this plant is winter; during summer, *Bowiea volubilis* goes through a vegetative rest period [12].

Figure 2. *Bowiea volubilis* (**A**) onion-like bulb, (**B**) greenish-white flowers, and (**C**) whole plant [23].

B. volubilis is native to the grasslands and thickets of the eastern regions of South African Cape, such as the Eastern Cape, and it extends northwards all through sub-Sahara Africa (Zimbabwe, Zambia, Tanzania, and Uganda) to as far north as Kenya [8]. It has also been recorded in Mozambique, Malawi, and Angola. In South Africa, this plant is spread out in five (Eastern Cape, KwaZulu Natal, Gauteng, Mpumalanga, and North West) of the nine provinces [13]. In these regions, *B. volubilis* grows in low and medium altitudes along the mountain ranges, where it is hidden in thick river valleys, under bush clumps, and between boulder screes, where it is partially exposed to wet and dry conditions of summer [8].

4. Traditional Use of *B. volubilis*

The bulb of *B. volubilis* is extensively used medicinally as muthi (traditional medicine prescribed by herbalists or traditional healers (inyanga)), and native people of Southern Africa employ it for spiritual application, as they hold a profound place within the traditions of South Africa's traditional healers [24,25]. The indigenous people value its magical abilities as they can use them to make warriors brave and unstoppable, protect travellers, and find love [15,24,26–30]. In addition to its magical uses, the ingestion of medicine from this bulb is also used as an antidote to poison associated with sorcery [15,29]. The stems and leaves of *B. volubilis* are commonly used for decoction due to their bright green colour and deciduous climbing nature. Several studies conducted found that numerous tribes use it as a painkiller to treat backaches, headaches, muscle pain, and pelvic pain in women [10,15,29,31–39]. In addition, reports indicate that *B. volubilis* is used as a blood purifier in the Limpopo and Western Cape provinces of South Africa, respectively [40,41]. The bulb also treats cancer in the Limpopo Province of South Africa [40]. Several ethnobotanical surveys have shown that the *B. volubilis* bulb relieves gastrointestinal problems [42–45]. Cimi and Campbell [46] report that the plant is used to treat kidney problems in Makhanda (former Grahamstown). The use of this plant for urinary tract infections has been reported by Philander [23], Cock [47], and Coopoosamy and Naidoo [48]. In Eswatini (former Swaziland), the bulb is cut into pieces, boiled for five minutes, and the concoction is used to treat scabies [49]. Mixing roasted bulbs with water is also used as a purgative by the Bhaca, Mfengu, and Mpondo tribes in South Africa [37,50,51]. In Transkei, a decoction of the bulb is used to treat stomach-related problems [43]. Even sexually transmitted diseases are healed using *B. volubilis* [37,50,51]. Ramarumo et al. [40] report that the plant is used as an anthelmintic. Madikizela et al. [52] list *B. volubilis* as one of the plant species used to treat tuberculosis in Pondoland, South Africa. *B. volubilis* is used as a topical medication for various skin or mucous membrane diseases [23,38,40,45,48], as well as infection of the eye [23,36,47,48]. Furthermore, liver problems are managed using *B. volubilis* [36]. Certain problems associated with pregnancy and childbirth are treated with medicines made from the bulbs of *B. volubilis* [23,35,48,53,54]. With respect to reproductive health, *B. volubilis* is used to facilitate delivery, terminate pregnancies, and treat impotence in men [30,39,44,47]. It also shows that various inflammation-associated diseases are treated with *B. volubilis* [23,34].

5. Phytochemistry of *B. volubilis*

The bulb of *Bowiea volubilis* has long been known to be a source of several cardiac glycosides [42,55]. Cardiac glycosides are steroidal compounds that have proved to be fruitful in developing potential drugs for congestive heart failure [21]. Cardiac glycosides have long been isolated and characterized in *B. volubilis* [56–58]. These compounds consist of an aglycone or genin, which is bound to one or two sugar molecules [36,59]. Conversely, the aglycone contains an unsaturated lactone ring with either a 5-membered ring known as cardenalide or a 6-membered ring known as bufadienolide [60]. The cardiac glycoside has two classes of compounds that differ in the structure of the aglycone bovogenin A and structurally related bufadienolides [55]. A number of bufadienolides glycosides that are specific to *B. volubilis* have been isolated [61]. These include bovoruboside, sciliburoside, sciliguacoside, scillicyanoside, scilliphaeoside, bovuside A, glucobovuside, bovogenin E, and bowieasubstanz G [42]. Figure 3 presents the six phytochemical compounds of *Bowiea volubilis*. The bufadienolides present in the bulb were fractioned and characterized by LC-MS. The other bufadienolides were identified by means of thin-layer chromatography (TLC) [60], FAB-MS, NMR, and C-NMR [15].

The sugar moieties of cardiac glycosides often contain unusual 2-deoxy sugar that influences their structure, pharmacological properties, and side effects [21]. In the case of cardiac glycoside ingestion, enzymes in the body hydrolyse the glycosidic bonds, which result in the release of bioactive steroidal compounds and sugar moieties [58]. The primary pharmacological action of the cardiac glycoside is to inhibit the Na^+/K^+ ATPase pump and increase the intracellular Ca^{2+} levels pumped out of the cell by Na^+/Ca^{2+} exchanger during diastole [62]. As a consequence, the intracellular Ca concentration rises, thereby inducing positive inotropy [36,42,58].

Alkaloids, which are prominently present within *B. volubilis*, have been extensively investigated for their diverse biological activities [63]. Notably, they have demonstrated efficacy in combating fungal and bacterial infections, as well as exhibiting cytotoxic effects [64]. Saponins, a class of glycosides that are known for their multifaceted health-related attributes [65], are also notably present in *B. volubilis*. These compounds, with their antioxidant and anticancer properties, contribute significantly to the plant's therapeutic potential [66]. Furthermore, the inclusion of steroids, such as sitosterol and stigmasterol, further enhances *B. volubilis* medicinal repertoire [67]. These steroids are associated with noteworthy anti-inflammatory and antitumor capabilities, reinforcing the plant's prospective health benefits [68]. Additionally, the presence of flavonoids, a group of phenolic compounds acclaimed for their biological significance, adds another dimension to *B. volubilis* potential therapeutic prowess. Notably, flavonoids are recognized for their antioxidant and anti-inflammatory attributes, both of which contribute to the plant's overall health-promoting effects [69]. Within *B. volubilis*, the occurrence of specific flavonoid compounds such as luteolin, quercetin, and kaempferol further contributes to its diversified therapeutic potential [70]. Moreover, phenolic compounds, including constituents like gallic acid, caffeic acid, and chlorogenic acid, further enhance *B. volubilis* therapeutic appeal [71]. Renowned for their antioxidant, anti-inflammatory, and anticancer activities [72], these compounds provide a robust foundation for the plant's potential role in mitigating various health-related concerns.

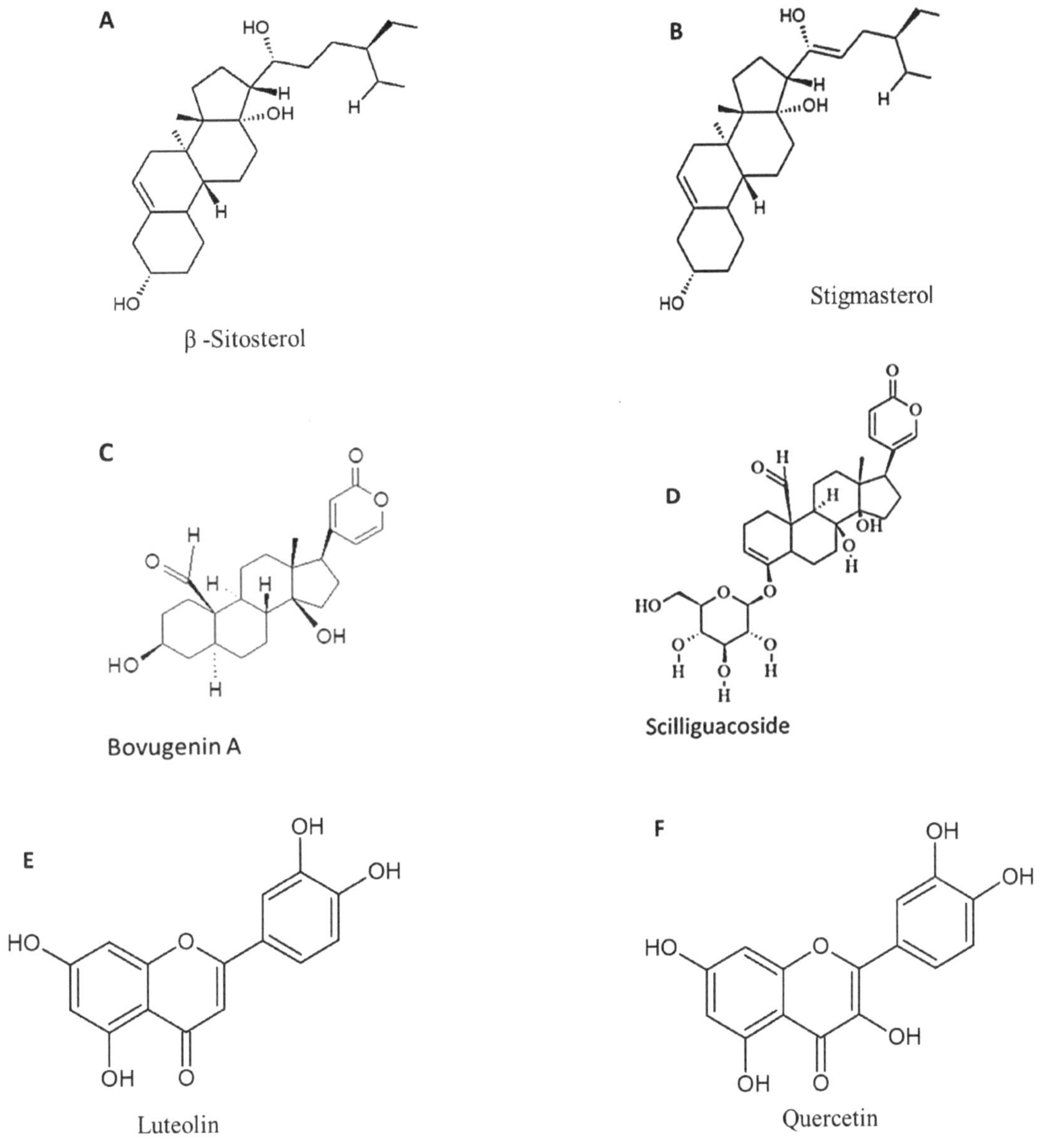

Figure 3. Phytochemical compounds of Bowiea volubilis, (**A,B**) steroids, (**C,D**) cardiac glycosides, and (**E,F**) flavonoids.

6. Biological Activity of *B. volubilis*

6.1. Antibacterial Activity of *B. volubilis*

Evidence exists on the extensive use of *B. volubilis* to traditionally treat and cure various ailments caused by pathogenic bacteria [16,50,73]. The frequent use of *B. volubilis* to treat pelvic pain, rash, liver infections, jaundice, and sexually transmitted infections has been recorded, leading one to assume that the plant has high anti-pathogenic activity [36,37,42,50]. However, studies have shown that ethanol, dichloromethane (DCM), ethyl acetate, water, and n-hexane extracts of *B. volubilis* perform poorly against bacterial pathogenic activity [16,37,50,55]. The activity of *B. volubilis* against bacteria such as *Bacillus subtilis, Escherichia coli, Klebsiella pneumoniae, Staphylococcus aureus, Oligella ureolytica, Ureaplasma urealyticum, Neisseria gonorrhoeae,* and *Gardnerella vaginalis*, which are implicated in the development of skin infections and on rare occasions pneumonia and meningitis as well as urogenital infections, is documented [74,75]. Masondo et al. [55] investigated the antimicrobial activity of botanically grown and muthi market-sourced *B. volubilis* against *B. subtilis, S. aureus, K. pneumoniae,* and *E. coli*. The results reported the Minimum Inhibitory Concentration (MIC) values, with the highest MIC observed in ethanol extracts against *S. aureus* as well as DCM extracts against *K. pneumoniae* and *E. coli*. In another study, Stafford et al. [16] indicated that MIC showed the highest MIC of 6.25 mg/mL against *Bacillus subtilis*. Greater activities were reported by Buwa and Van Staden [50], with a MIC value of 12.5 mg/mL for ethanol extracts in all strains, while the water solvent exhibited 3.125 mg/mL and the ethyl acetate extract showed no activity. Un-remarkable antibacterial activity of *B. volubilis* bulb and leaf tissue was also reported by Van Vuuren and co-workers [37]. Methanol, water, and DCM extracts against *Oligella ureolytica, Ureaplasma urealyticum, Neisseria gonorrhoeae,* and *Gardnerella vaginalis* showed MIC values ranging between 1.5 and 4.0 mg/mL for the CH_2Cl_2:MeOH (1:1 DCM and methanol) extracts and greater than 16.0 for the water extract.

6.2. Antifungal Activity of B. volubilis

In this current research, many studies have been reviewed on screening *B. volubilis* plants for their antifungal activity. *B. volubilis* water extracts exhibited strong inhibitory effects with a MIC value of 6.25 mg/mL against *Candida albicans* [58]. In another study, a water extract of a muthi market-sourced (MM) bulb tested against *C. albicans* showed a MIC result of 1.56 mg/mL, which was the best compared to the rest of the MIC values of other extracts [55]. Aremu et al. [42] demonstrated the activity of *B. volubilis* leaf water extracts against *C. albicans*. The results showed a distinguished MIC value of 0.5 mg/mL [42]. Discovering that bulbs can be substituted with leaves was a good indication that this plant will be available in the future. It is worth noting that, for an extract to be considered a good antifungal drug, there should be minimal drug resistance, low toxicity or minimal side effects, stability, good bioavailability, and most importantly, broad spectrum and efficacy. In addition, it has been stated that it is better for the extract to be fungicidal rather than fungistatic [9,73]. It is a well-established fact that *B. volubilis* plant extracts are significantly more effective against plant-pathogenic fungi compared to bacteria, as reported by multiple studies. Even in the early years of research, a study was conducted on 13 extracts, and only 5 extracts suppressed fungal growth, proving that, indeed, plant-pathogenic fungi are more resilient to plant extracts than plant-pathogenic bacteria [19,50,76].

6.3. Anti-Inflammatory Activity of B. volubilis

Several studies have proven that medicinal plants, including *B. volubilis* are an excellent source of anti-inflammatory agents [55,57]. A study conducted by Stafford et al. [16] revealed that *B. volubilis* water extracts exhibited greater performance compared to ethanol extracts when assessing the anti-inflammatory potential using cyclooxygenase (COX-1 and -2) inhibitory assays. The non-polar solvent extracts of both botanical garden-grown (BG) and muthi market-sourced (MM) *B. volubilis* bulbs showed significant inhibitory activity of greater than 70%. In the same study, it was further proven that the majority (75%) of BG extracts showed a higher percentage of inhibition compared to the MM (50%) extracts with regards to COX-1 inhibition. The inhibitory activity of the water extracts of both BG and MM bulbs against COX-2 enzymes was too small to show any activity at all, and it was found that the MM water extracts had far better COX-1 inhibitory activity compared to the BG bulbs. Results from Masondo et al. [55] showed that there was a higher COX-2 inhibition compared to COX-1 when focusing on the MM ethanol extract. An assessment of the effectiveness of various *Bowiea volubilis* bulbs extracts on the inhibition of cyclooxygenase (COX) was conducted, and it was proven that these extracts showed a high success in vitro COX assays as compared to other anti-inflammatory related enzymes, such as 5-lipoxygenase [42]. The effectiveness of *B. volubilis* was then confirmed against pain and anti-inflammatory-related illnesses as the above information was considered as further proof [44,55]. The variation of the results in which weak activity can be obtained may be due to the fact that the active compound(s) in certain extracts may be present in inadequate quantities [57,77].

6.4. Antiviral Activity of B. volubilis

Among various biological activities documented for *B. volubilis*, there is accumulating evidence of this plant species' antiviral activities. In one study, the methanolic extract of *B. volubilis* bulbs was investigated for its antiviral activity against herpes simplex virus type 1 (HSV-1) [78]. The study found that the extract exhibited notable antiviral activity against HSV-1 in vitro, exhibiting an IC_{50} value of 0.34 mg/mL [78]. Another in vitro study also recorded the antiviral activity of *B. volubilis* crude extract against dengue virus type 2 (DENV-2) [79]. This study found that the extract exhibited antiviral activity at an EC_{50} value of 64.4 µg/mL against the test strain. However, the study also reported that the extract was cytotoxic at high concentrations, indicating that further research is needed to determine the safety and efficacy of the extract for use as an antiviral agent [79]. The crude extract of *B. volubilis* has also been evaluated for antiviral activity against the HIV-1 strain in another study [80]. Using a peripheral blood mononuclear cell (PBMC)-based assay, the study found that the extract exhibited significant antiviral activity against HIV-1 in vitro, reducing viral replication by up to 70% at a concentration of 50 µg/mL [80]. On the other hand, Feng et al. [81] investigated the antiviral activity of extracts from the bulb of *B. volubilis* against the respiratory syncytial virus (RSV). The study found that the extracts exhibited significant antiviral activity against RSV, with the most active extract showing an IC_{50} value of 0.13 µg/mL [81].

While *B. volubilis* has been traditionally used in some cultures for medicinal purposes, there is insufficient scientific research specifically focused on its antiviral properties. Nevertheless, the findings documented thus far are promising and suggest that this species can be utilized, amongst other products, as an antiviral agent. Viral infection outbreaks in humans are becoming formidable pandemic threats [82], so there is an essential need for novel and natural antiviral agents. However, it is important to note that the antiviral activity of natural products is a complex and dynamic field of research, and the efficacy and safety of using *B. volubilis* or its extracts for antiviral purposes have not been fully established [78–81]. Further research, including in vitro and in vivo studies, is needed to adequately assess the antiviral properties of *B. volubilis* and its mechanism of action against specific viruses. Figure 4 and Table 1 summarizes bio-compounds that are associated with specific activities. Scientific validation of its diverse uses in traditional medicine has been demonstrated via antimicrobial, anti-inflammatory, and toxicity assays. The anti-inflammatory activity is promising; however, the available studies reveal usually low antibacterial activity, especially with bulb extracts. Bowiea volubilis includes cardiac glycosides and related chemicals, according to phytochemical screens; information on additional types of compounds is not yet available.

Table 1. Biological properties of *B. volubilis* with its associated bio-compounds and activity levels.

Plant Part	Extraction Solvent	Bioactive Compounds	Biological Properties	Activity Level	Ref.
Bulb	Water	N/A	Antifungal	>25 mg/mL	[12]
Bulb	Water	N/A	Antibacterial	>16.0 mg/mL	[37]
Bulb	Methanol	N/A	Antibacterial	1.4–4.0 mg/mL	[37]
Bulb, leaves		Cardiac glycosides	Anti-inflammatory		[38]
Leaf	Petrolium ether	Glycosides of bovogenin A	Antifungal	0.5 mg/mL	[42]
Bulb	Water	N/A	Antibacterial	>12.5 mg/mL	[50]

Table 1. Cont.

Plant Part	Extraction Solvent	Bioactive Compounds	Biological Properties	Activity Level	Ref.
Bulb	Ethanol	N/A	Antibacterial	3.125 mg/mL	[50]
Bulb	Ethyl acetate	N/A	Antibacterial	No value	[50]
Bulb	Water extract	Cardiac glycoside	Antibacterial		[54]
Bulb	Ethanol	Cardiac glycoside	Antifungal	3.13 mg/mL (BG) and 12.50 mg/mL (MM)	[55]
Bulb	Ethanol	Cardiac glycoside	Antibacterial	1.56–6.25 mg/mL	[72]
Bulb	Petroleum ether	Cardiac glycoside	Antifungal	12.50 mg/mL (BG and MM)	[73]
Bulb	Dichloromethene (DCM)	Cardiac glycoside	Antifungal	12.50 mg/mL (BG and MM)	[73]
Bulb	Water	Cardiac glycoside	Antifungal	3.13 mg/mL (BG) and 1.56 mg/mL (MM)	[73]
Bulb	Petroleum ether	Cardiotoxic glycosides of the bufadienolide group	Anti-inflammatory	COX-1 = 100% (MM)	[73]
Bulb	Petroleum ether	Cardiotoxic glycosides of the bufadienolide group	Anti-inflammatory	COX-2 = 100% (BG)	[73]
Bulb	Methanol extract	N/A	Antiviral	IC_{50} = 0.34 mg/mL	[78]
Bulb	Aqueous extracts	N/A	Antiviral	IC_{50} = 0.13 µg/mL	[81]
Bulb	Ethanol	Prostaglandin	Anti-inflammatory	COX-1 = 100%	[83]
Bulb	Water	Cardiac glycosides	Anti-inflammatory	COX-1 = 45%	[83]

BG: botanical garden-grown, MM: muthi market-sourced.

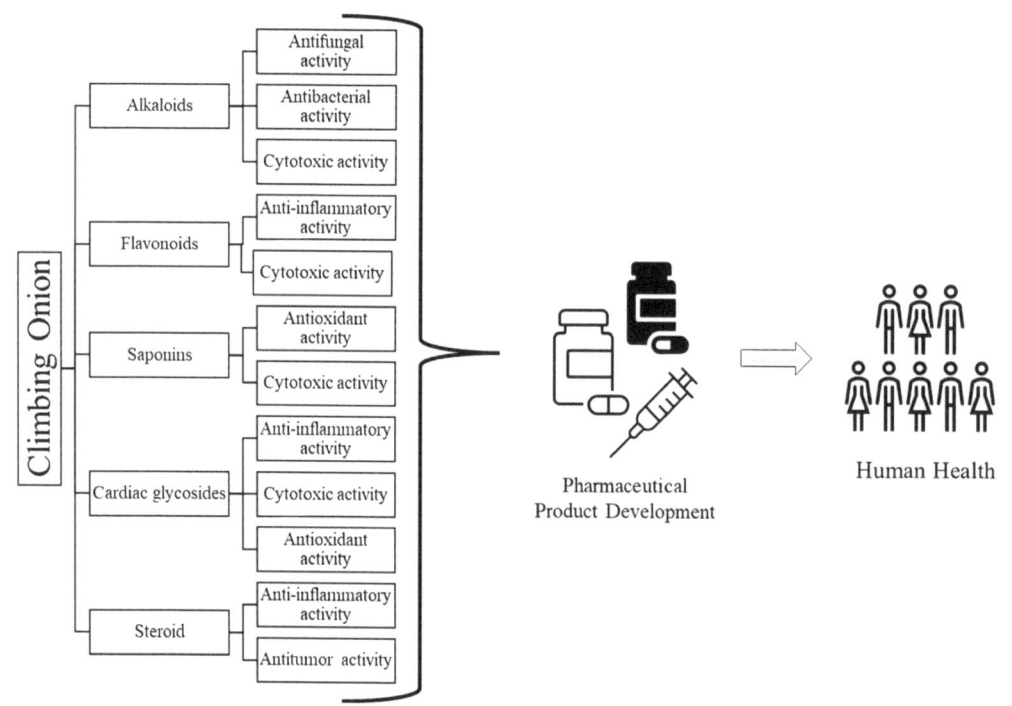

Figure 4. A flow chart representing various bio-compounds isolated from *Bowiea volubilis* with their activities.

7. Toxicological Data of *B. volubilis*

The administration of traditional medicine is not well-documented, and it is roughly passed on by word of mouth from experienced healers to trainers or parents to children. This has led to the healthcare system ignoring this form of medication. *Bowiea volubilis* is rated as the top-selling traditional plant with a conservation status that is now vulnerable [55], which demonstrates the intensive use of the indigenous plant for medical relief. It is important to evaluate the safe levels of classified chemical responses to characterize exposure's toxic effects and health. Generally, plants respond by defensive modification against herbivores and microorganisms by producing toxins, cytotoxins, and metabolic toxins that affect the central nervous system, brain, kidney, liver, heart, and lungs in humans and animals [73]. *B. volubilis* is highly poisonous and toxic to humans and animals due to the production of cardiac glycoside, a phytochemical aglycone (steroid) or polycyclic steroid compound linked to one or more sugar molecules by a glycosidic bond [42,61]. The poisonous effects of the climbing onion were reported as early as 1915, and human and animal post-mortems suggested death due to toxicity caused by cardiac glycoside compounds.

The toxicity of medicinal plants may be indirect (consumption by mistake, or incorrect selection or recommendation) or direct (misuse, overdose, incorrect preparations) [6]. Numerous studies suggest that cardiac glycoside compounds are the main contributor to toxicity at certain levels, with accumulative effects over time [55]. Ndhlala et al. [73] further emphasized that the growth stage and/or part of the plant, route and amount of administration, solubility of the compound, frequency of intoxication, and age and susceptibility of the victim all influence the severity of the toxicity. Cardenolides and bufadienolides are two compounds of cardiac glycoside listed by [61], while [42,55] argue that the cardiac glycoside compound groups are bovogenin A and bufadienolides. The varying cardiac glycosides, bovogenin A, bufadienolides, and cardenolides are associated with the toxicity effect caused by *B. volubilis* consumption. The cardiac glycoside inhibits the Na^+/K^+ ATPase pump, which increases intracellular Na^+ concentration and, in turn, increases the intracellular Ca^+ level, resulting in a positive inotropic effect [42,61]. The acquisition of toxicological information on *B. volubilis* is critical to our society, especially to our traditional society that depends on and trusts in medicinal plants. This study assesses the safety administration of medical plants by dosage evaluation and, most importantly, the risk assessment of the medicinal plant.

In the study conducted by Emamzadeh-Yazdi et al. [84], in vitro cytotoxicity assay (XTT colourimetric) using Vero cells (kidney epithelial cells of African monkey) and HEK 293 cells (human embryonic kidney) exposed to fresh and dry *B. volubilis* ethanol extract both showed toxicity (quite toxic). The fresh and dry ethanol extract exhibited toxicity on Vero cells with toxicity activity at IC_{50} of 50 µg/mL (quite toxic), while on HEK 293 cells, the dry ethanol extract yielded IC_{50} = 23.34 µg/mL (IC_{50} < 50 µg/mL; quite toxic) and the fresh ethanol extract exhibited IC_{50} = 28.83 µg/mL (IC_{50} < 50 µg/mL; quite toxic). Another in vitro cytotoxicity study conducted by Fasinu et al. [85] investigated the effect of *B. volubilis* aqueous extract on metabolic enzyme activity in HLM cells (Human liver microsomes). The results showed toxic activity. The in vitro cytotoxicity assay revealed that the extract on enzyme CYP1A2 exhibited toxicity activity at IC_{50} of 92.5 µg/mL (IC_{50} < 50 µg/mL quite toxic), while the extract on enzyme CYP2C9 was non-toxic, and the extract on enzyme CYP2C19 displayed toxicity activity at an IC_{50} of >1000 µg/mL (non-toxic). However, the extract on enzyme CYP3A4 revealed toxicity activity at an IC_{50} of 8.1 µg/mL of $IC_{50} \leq 20$ µg/mL highly toxic). More in vivo studies are necessary to evaluate the toxicity activity of *B. volubilis*; it is still of utmost importance to generate sufficient toxicological data that will give an overview in terms of risk assessment, safety, and dosage evaluation. Studies report anecdotal reports on the toxicity of a plant, but no substantial evidence-based research is available. It is essential to acknowledge anecdotal reports but also to stress the lack of scientific evidence when discussing the toxicity of a plant like *B. volubilis*. Despite this, *B. volubilis* is ranked among the top 10 medicinal plants sold in South Africa and has displayed potential antiviral effects that might be used to treat human diseases. Natural compounds isolated from B. *volubilis* can show promising antiviral activity in laboratory studies; translating these findings into effective treatments for human diseases is a complex process. Rigorous scientific research, including preclinical and clinical trials, is necessary to establish the safety and efficacy of any potential treatments; nonetheless, it is a suitable candidate.

8. *Bowiea volubilis* as a Potential Therapeutic Drug: Addressing the Major Challenges toward Human Diseases

Respiratory infections encompass a range of diseases that affect the respiratory system, which includes the lungs, airways, and other related structures. These infections arise from diverse pathogens, encompassing viruses, bacteria, and fungi [86]. The manifestation of symptoms and associated effects vary depending on the precise infectious agent, with certain respiratory infections posing considerable challenges for effective treatment. Respiratory infections can lead to various complications, especially if not properly managed or treated. In the realm of medical intervention, challenges in treating respiratory ailments are notable due to multifaceted factors that encompass various dimensions. The rise of antibiotic-resistant strains of bacteria complicates the efficacy of conventional treatment strategies [87]. This phenomenon is particularly relevant in bacterial respiratory infections like pneumonia, where traditional antibiotics might exhibit diminished effectiveness. In underdeveloped countries, access to proper healthcare remains a challenge [88]. The presence of alkaloids and scillaren-type cardiac glycosides has been reported across all parts of *B. volubilis*, as indicated by research conducted by Mulholland et al. [89]. Cardiac glycosides, in particular, are known to act by selectively and effectively inhibiting Na^+/K^+-ATPase. The observed inhibitory effect on cytochrome P450 (CYP) enzymes *in B. volubilis* can be attributed to its alkaloid content, which serves as substrates for human CYPs. A separate study by Salminen et al. [90] highlighted the inhibitory potential of structurally similar alkaloids from plants on major human drug-metabolizing CYPs, including CYP3A4, CYP2D6, and CYP2C19.

The review conducted by Smith et al. [91] offers a comprehensive analysis of regionally relevant herbal medicine utilization, with a specific focus on remedies that have been advocated for COVID-19 treatment. Notably, within this context, *Bowiea volubilis* emerges as a viable and promising candidate. The study's findings shed light on the potential role of *B. volubilis* within the framework of herbal medicine's response to the ongoing COVID-19 pandemic. The identification of *B. volubilis* as a suitable candidate underscores its significance and merits further exploration in the pursuit of effective and holistic approaches to address respiratory ailments, including those associated with COVID-19. The utilization of *Bowiea volubilis* in the context of respiratory infections holds significance due to its potential therapeutic properties that have been traditionally recognized and are increasingly being explored through scientific investigation. While the scientific evidence is still evolving, the recognition of *Bowiea volubilis* as a potential therapeutic agent for respiratory infections underscores the need for continued research and exploration to establish its efficacy, safety, and potential integration into healthcare practices.

Coughs and colds are prevalent respiratory illnesses characterized by symptoms such as nasal congestion, sore throat, sneezing, and coughing. While primarily caused by viral infections, bacterial infections, and environmental factors can also contribute to their onset. These conditions often result in discomfort, impaired daily functioning, and increased healthcare utilization [92,93]. The management of coughs and colds revolves around alleviating symptoms and preventing complications. Traditional medicinal plants have been explored for their potential to provide relief from these ailments. *Bowiea volubilis*, known for its ethnobotanical uses, has drawn attention due to its bioactive compounds with potential therapeutic properties.

Gastrointestinal disorders encompass a diverse range of conditions that affect the digestive tract, including the stomach, intestines, liver, gallbladder, and pancreas. These disorders can manifest with symptoms such as abdominal pain, bloating, diarrhea, constipation, nausea, and vomiting. The causes of gastrointestinal disorders vary and can include infections, inflammation, dietary factors, genetic predisposition, and lifestyle choices [94,95]. Managing these disorders requires a multifaceted approach that addresses both symptom relief and underlying causes. Traditional medicinal plants like *Bowiea volubilis* have been explored for their potential to alleviate gastrointestinal symptoms and promote digestive health [96]. Research into the effects of *B. volubilis* on gastrointestinal disorders is still emerging. Investigating its potential impact on inflammation, gut motility, and microbial balance could provide insights into its suitability as a complementary or alternative therapeutic option [97]. While there is a potential role for *Bowiea volubilis* in gastrointestinal health, it is important to emphasize the need for comprehensive research to establish its effectiveness, safety, and appropriate usage.

Cancer is a complex and multifaceted disease characterized by uncontrolled cell growth and the potential to invade other tissues and organs [98,99]. Despite significant progress in cancer research and the development of various anti-cancer drugs, the effectiveness can be limited due to factors such as drug resistance, adverse side effects, and incomplete tumor eradication. Additionally, cancer cells can evolve by developing mechanisms to evade the effectiveness of drugs, leading to treatment resistance and disease recurrence [100]. In this context, the exploration of medicinal plants as alternative or complementary treatments for cancer has gained attention. *Bowiea volubilis*, with its historical use in traditional medicine, presents a unique opportunity for investigation. The plant's bioactive compounds, including alkaloids, saponins, and cardiac glycosides, have shown potential in various therapeutic contexts, including anti-inflammatory, antioxidant, and anti-cancer activities [101]. Research into the potential of *B. volubilis* as an alternative or adjunctive treatment for cancer is still in its early stages. Studies on its cytotoxic effects, potential mechanisms of action, and interactions with existing anti-cancer drugs could shed light on its role in cancer management. However, it is important to note that rigorous scientific investigation, including preclinical and clinical trials, is needed to establish its safety, efficacy, and appropriate usage.

Skin conditions impact a substantial portion of the global population, ranging from 30% to 70%, making them a prevalent cause for seeking medical attention in general medical practice [102]. Over 3000 distinct skin diseases, encompassing both short-term and long-lasting forms, affect people across various age groups and social backgrounds [103]. Moreover, skin diseases can have a significant impact on quality of life due to visible symptoms, discomfort, and social stigma. Furthermore, skin disorders encompass a wide spectrum of ailments that impact the integrity, visual attributes, and operational capabilities of the skin. These afflictions can stem from genetic predispositions, environmental catalysts, immune reactions, infections, and lifestyle influences. Gaining insight into the scientific facets of skin diseases entails delving into their fundamental origins, manifestations, and intricacies [104]. In line with this, inflammation serves as a pervasive characteristic shared by numerous dermatological disorders. Conditions such as eczema, psoriasis, and acne are marked by immune reactions that precipitate inflammatory processes, giving rise to manifestations of redness, irritation, and pruritus [105]. The skin microbiome, consisting of diverse microorganisms, plays a role in skin health. The implications for wound healing and protection against potential pathogens or environmental conditions highlight the crucial role of skin homeostasis. Imbalances in the microbiome can contribute to conditions such as acne [106]. *B. volubilis* is reported to contain various bioactive compounds, including alkaloids, saponins, and flavonoids, that could contribute to its anti-inflammatory properties. This potential anti-inflammatory effect could have implications for managing various inflammatory skin conditions, as well as other disorders characterized by inflammation. The potential of *Bowiea volubilis* in combatting skin pathogens is of interest due to its reported bioactive constituents that could exhibit antimicrobial properties. These properties could make *B. volubilis* a candidate for addressing skin infections caused by various pathogens, including bacteria, fungi, and other microorganisms. However, it is important to note that scientific research on its effectiveness and safety for treating skin diseases is limited.

Sexually transmitted diseases (STDs), also referred to as sexually transmitted infections (STIs) encompass a group of infections caused by various pathogens that are typically transmitted through sexual activity. These infections result in overgrowth of opportunistic bacterial microflora causing pelvic pain, vaginal discharge, penile discharge, genital ulcers, and other symptoms and indicators of STIs, in some cases, infertility [107,108]. As of 2020, the World Health Organization (WHO) approximated a total of 374 million new infections attributed to the four most prevalent STIs: chlamydia (129 million), gonorrhea (82 million), syphilis (7.1 million), and trichomoniasis (156 million) [109]. Recent models indicate that sub-Saharan Africa and the Western/Eastern Pacific regions bear a disproportionate burden of 75% of global STI control costs [109]. Despite the fact that antimicrobial resistance is a global public health problem, front-line practitioners often underestimate the effect of antibiotic-resistant STIs [110]. Medicinal plants have garnered attention as potential sources of alternative or complementary treatments for STDs. These plants contain bioactive compounds that exhibit antimicrobial, anti-inflammatory, and immunomodulatory properties, which could contribute to their effectiveness against STD-causing pathogens [111]. The potential role of *B. volubilis* in managing STDs is an area that requires thorough scientific investigation. Traditional knowledge might suggest its historical use in addressing STDs such as *B. volubilis* as a suitable alternative [112]. Drawing from the findings, it is shown that *B. volubilis* exhibits potential antifungal, antibacterial, and antiviral properties that might be useful in the treatment of STDs as well as other health complications (Table 2).

Table 2. An overall summary of the health benefits of *B. volubilis*.

Respiratory infections	Lungs, airways, and other related structures Pneumonia COVID-19 Coughs and colds, nasal congestion, sore throat, sneezing, and coughing.
Gastrointestinal disorders	Digestive tract, including the stomach, intestines, liver, gallbladder, and pancreas Abdominal pain, bloating, diarrhoea, constipation, nausea, and vomiting
Cancer	
Skin conditions	Eczema, psoriasis, and acne
Sexually transmitted diseases (STDs)	Antifungal, antibacterial, and antiviral properties

9. Conservation Statues of *B. volubilis*

In South Africa, most of the medicinal plants, including *B. volubilis*, are collected from the wild, and they are decreasing at an alarming rate because of extensive exploitation [12,53]. According to herbalists, *B. vobulis* is rated one of the top six medicinal species to have become scarce because of over-exploitation. This is particularly worrying if the harvestable part of the plant is a non-renewable part, such as the bulb, rhizome, and bark. In *B. volubilis*, the most used part is the bulb [28]. Studies have shown that bulbous medicinal plants, including *B. volubilis*, are at risk of going extinct because of threats like over-exploitation, habitat destruction, human settlement, and agricultural expansion [40]. This has led to this plant being categorized as a vulnerable species in the International Union for Conservation of Nature's (IUCN) Plant Red Data list [38]. A vulnerable species is a species whose population has declined by 30 to 50% and the cause of its decline is known [26]. It has been estimated that the population of this plant has declined by 30% in the last 30 years, and the number of individual bulbs in the muthi market has decreased tremendously [40]. Together with *Siphonochilus aethiopicus* (Schweinf.), B.L. Burtt, and *Eucomis autumnalis* (Mill.) Chitt.,

Bowiea volubilis is among the top three traded medicinal plants in South Africa assigned as being rare [53].

Various conservation strategies have been described for medicinal plants. These include in situ and ex situ conservation strategies [12]. In situ conservation is described as the conservation of the threatened species in the plant's natural habitats, with the aim of maintaining and recovering a viable population of that species in the natural environment [38]. Ex situ, on the other hand, is concerned with the conservation of a threatened species outside the plant's natural habitat [33]. With this strategy, the threatened species is cultivated and naturalized to ensure their continued survival and sometimes to produce large quantities of planting material for use in drug development [12]. The ex situ conservation is the one that has been proposed for *B. volubilis* [53]. To assist with the ex situ conservation of *B. volubilis*, it has been proposed that these plants be cultivated from seeds with seed coats that are acid scarified [100]. Other researchers proposed that *B volubilis* be grown from vegetative propagules using bulb scales [12]. However, it has been reported that seed and bulb scale propagation are both too slow at multiplying the needed plant material [40]. They turned to the micropropagation technique with tissue culture and found that it had saved the population of *B. volubilis* [56]. Through this technique, thousands of plantlets that can be used in the cultivation of this species have been produced [12]. Although the cultivation of medicinal plants is recognized as a conservation strategy that can provide additional or alternative stocks, concerns have been raised about the potency of their active ingredients [53]. Moreover, traditional healers believe that cultivated medicinal plants are less potent than the ones collected from the wild [40].

10. Conclusions

Human diseases pose a significant and ongoing concern for global public health. The diverse range of diseases that affect individuals and populations worldwide can have far-reaching implications on various aspects of society, including healthcare systems, economies, and quality of life. Medicinal plants play a significant role in the management of various human diseases due to their diverse array of bioactive compounds and therapeutic properties. These plants have been used for centuries across cultures and traditions to alleviate symptoms, promote healing, and support overall well-being. *Bowiea volubilis* holds a significant place as a recognized and widely traded medicinal plant in Southern Africa. The findings of this investigation underscore its substantial potential, encompassing antifungal, anti-inflammatory, antibacterial, and cytotoxic attributes. Consequently, the therapeutic spectrum of *B. volubilis* positions it as a promising contender for addressing conditions linked to pain, microbial infections, and inflammation-driven ailments. Notably, the scientific landscape also indicates its historical application in treating conditions such as infertility, skin disorders, cystitis, headaches, and sexually transmitted diseases. *B. volubilis* has been of interest due to its potential applications in addressing various aspects of human health and disease. Our findings outline various human diseases such as respiratory infections, cough and colds, gastrointestinal disorders, cancer, skin conditions, and sexually transmitted diseases. In regard to this, scientific research is essential to validate its effectiveness and safety in treating specific diseases. In conclusion, this study serves as a catalyst for new avenues of drug development aimed at addressing the challenges posed by human diseases and improving overall health outcomes. With the plant being threatened in the wild, conservation strategies aimed at continuously making this plant available for future use are only limited to cultivation of the plant ex situ conservation. Therefore, further research is needed to explain the specific conservation measures that can be taken to protect the *B. volubilis* population, especially in light of its potential contributions to medicine and our understanding of human diseases. The unique chemical compounds found within *B. volubilis* have demonstrated promising pharmacological properties, and their potential applications in treating or preventing various human diseases remain largely untapped.

Author Contributions: Writing—original draft preparation, review and editing, project management, H.G.; writing—original draft preparation, review and editing, S.C., I.E., N.G., A.-A.M., S.Z. and B.M.; conceptualization, methodology, writing and editing, M.C.M. All authors have read and agreed to the published version of the manuscript.

Funding: This research received no external funding.

Institutional Review Board Statement: Not applicable.

Informed Consent Statement: Not applicable.

Data Availability Statement: Not applicable.

Acknowledgments: H.G, S.C., I.E., N.G., A-A.M., S.Z., B.M. and M.C.M.would like to appreciate Walter Sisulu University.

Conflicts of Interest: The authors declare no conflict of interest.

References

1. Kaptchouang Tchatchouang, C.D.; Fri, J.; Montso, P.K.; Amagliani, G.; Schiavano, G.F.; Manganyi, M.C.; Baldelli, G.; Brandi, G.; Ateba, C.N. Evidence of Virulent Multi-Drug Resistant and Biofilm-Forming Listeria Species Isolated from Various Sources in South Africa. *Pathogens* **2022**, *11*, 843. [CrossRef] [PubMed]
2. Birhan, Y.S.; Kitaw, S.L.; Alemayehu, Y.A.; Mengesha, N.M. Medicinal plants with traditional healthcare importance to manage human and livestock ailments in Enemay District, Amhara Region, Ethiopia. *Acta Ecol. Sin.* **2023**, *43*, 382–399. [CrossRef]
3. Van Jaarsveld, E.J. *Bowiea volubilis*-an overlooked South African ornamental plant. *Veld Flora* **1992**, *78*, 63–64.
4. Stafford, G.I.; Jäger, A.K.; Van Staden, J. Effect of storage on the chemical composition and biological activity of several popular South African medicinal plants. *J. Ethnopharmacol.* **2005**, *97*, 107–115. [CrossRef] [PubMed]
5. Asres, K.; Bucar, F.; Kartnig, T.; Witvrouw, M. Antiviral activity against human immunodeficiency virus type 1 (HIV-1) and type 2 (HIV-2) of ethnobotanically selected Ethiopian medicinal plants. *Phytother. Res.* **2005**, *19*, 999–1006. [CrossRef]
6. Shokoohinia, Y.; Sadeghi-aliabadi, H.; Mosaddegh, M.; Abdollahi, M. Review on the potential therapeutic roles of *Bowiea volubilis*. *J. Evid.-Based Complement. Altern. Med.* **2018**, *23*, 662–669.
7. Zavala, M.A.; Hulme, P.E. The diversity and biogeography of alien plants in novel ecosystems: Insights from a global quantitative analysis. *Plant Syst. Evo.* **2015**, *17*, 588–597.
8. Manning, J.; Goldblatt, P. *Plants of the Greater Cape Floristic Region 1: The Core Cape Flora*. Strelitzia; South African National Biodiversity Institute: Pretoria, South Africa, 2012; Volume 29.
9. Manning, J.C.; Goldblatt, P. *Plants of the Greater Cape Floristic Region 2: The Extra Cape Flora*. Strelitzia; South African National Biodiversity Institute: Pretoria, South Africa, 2013; Volume 30, pp. 91–94.
10. Hulley, P.E.; Van Wyk, B.E. *Plants of the Klein Karoo*; Umdaus Press: Pretoria, South Africa, 2011.
11. Ndlovu, M.; Masika, P.J. The genus Bowiea: Review of ethnobotanical, phytochemical and pharmacological properties. *J. Med. Plants Res.* **2011**, *5*, 3981–3988.
12. Pourakbari, R.; Taher, S.M.; Mosayyebi, B.; Ayoubi-Joshaghani, M.H.; Ahmadi, H.; Aghebati-Maleki, L. Implications for glycosylated compounds and their anti-cancer effects. *Int. J. Biol. Macromol.* **2020**, *163*, 1323–1332. [CrossRef]
13. Zacchino, S.; Yunes, R.; Cechinel, V.; Enriz, R.; Kouznetsov, V.; Ribas, J.C. The need for new antifungal drugs: Screening for antifungal compounds with a selective mode of action with emphasis on the inhibitors of the fungal cell wall. In *Plant Derived Antimycotics*; Haworth Press: New York, NY, USA, 2003; pp. 1–47.
14. Rasensky, J.; Jonak, C. Drought, salt, and temperature stress-induced metabolic rearrangements and regulatory networks. *J. Exp. Bot.* **2012**, *63*, 1593–1608. [CrossRef]
15. Mir, R.A.; Mir, M.U. Plant osmolytes: Potential for crop improvement under adverse conditions. *Inter. J. Agric. Environ. Biotechnol.* **2015**, *8*, 719–728.
16. Mugivhisa, L.L.; Hackleton, C.M. *Bowiea volubilis* Harv. Ex Hook, F. as a potential candidate for land restoration in South Africa. *Afr. J. Range Forage Sci.* **2017**, *34*, 147–153.
17. Ndlovu, L.R.; Chimonyo, M.; Okoh, A.I.; Muchenje, V. Nutritional value of *Bowiea volubilis* as a potential feed supplement for small ruminants: Rumen fermentation and in vitro digestibility. *Trop. Anim. Health Prod.* **2008**, *40*, 1–9.
18. Xego, S.; Kambizi, L.; Nchu, F. Threatened medicinal plants of South Africa: Case of the family. *Afr. J. Tradit. Complement. Altern. Med.* **2016**, *13*. [CrossRef]
19. Koorbanally, C.; Crouch, N.R.; Mulholland, D.A. The phytochemistry and ethnobotany of the southern African genus Eucomis (Hyacinthaceae: Hyacinthoideae). In *Phytochemistry: Advances in Research*; Research Signpost: Kerala, India, 2006; pp. 69–85, ISBN 81-308-0034-9.
20. Afolayan, A.J.; Adebola, P.O. In vitro propagation: A biotechnological tool capable of solving the problem of medicinal plants decimation in South Africa. *Afr. J. Biotech.* **2004**, *3*, 683–687.

21. Afzal, S.; Singh, N.K.; Singh, N.; Sohrab, S.; Rani, M.; Mishra, S.K.; Agarwal, S.C. Effect of metals and metalloids on the physiology and biochemistry of medicinal and aquatic plants. In *Metals Metalloids Soil Plant Water Systems*; Academic Press: Cambridge, MA, USA, 2022; pp. 199–216.
22. Patel, S. Plant-derived cardiac glycosides: Role in heart ailments and cancer management. *Biomed. Pharmacother.* **2016**, *84*, 1036–1041. [CrossRef]
23. Mander, M.; Wynberg, R.; Schroeder, D. The socio-economic contribution of wild harvested plants in South Africa: A case study of the rooibos industry. *J. Environ. Manag.* **2020**, *264*, 110456.
24. Philander, L.A. An ethnobotany of Western Cape Rasta bush medicine. *J. Ethnopharmacol.* **2011**, *138*, 578–594. [CrossRef]
25. Hutchings, A. *Zulu Medicinal Plants: An Inventory*; University of Natal Press: Pietermaritzburg, South Africa, 1996.
26. Symmonds, R.; Bircher, C.; Crouch, N. Bulb scaling and seed success with *Bowiea volubilis*. *Plant Life* **1997**, *17*, 25–26.
27. Pooley, E. *A Field Guide to Wildflowers: KwaZulu-Natal and the Eastern Region*; Natal Flora Publications Trust: Durban, South Africa, 1998.
28. Raimondo, D.; Von Staden, L.; Foden, W.; Victor, J.E.; Helme, N.A.; Tuner, R.C.; Kamundi, D.A. (Eds.) *Red List of South African Plants. Strelitzia 25*; South African National Biodiversity Institute: Pretoria, South Africa, 2009.
29. Ördögh, M.; Farkas, D. The Effect of Different Substrates on Morphological Characteristics of Acclimatized *Bowiea volubilis*. *Rev. Agri. Rural Dev.* **2021**, *10*, 9–15.
30. Rasethe, M.T.; Semenya, S.S.; Maroyi, A. Medicinal plants traded in informal herbal medicine markets of the Limpopo Province, South Africa. *Evid.-Based Complement. Altern. Med.* **2019**, *11*, 2609532. [CrossRef]
31. Bisi-Johnson, M.A.; Obi, C.L.; Kambizi, L.; Nkomo, M. A survey of indigenous herbal diarrhoeal remedies of OR Tambo district, Eastern Cape Province, South Africa. *Afr. J. Biotechnol.* **2010**, *9*. [CrossRef]
32. Hutchings, A.; Van Staden, J. Plants used for stress-related ailments in traditional Zulu, Xhosa and Sotho medicine. Part 1: Plants used for headaches. *J. Ethnopharmacol.* **1994**, *43*, 89–124. [CrossRef]
33. Hannweg, K.F. Development of Micropropagation Protocols for Selected Indigenous Plant Species. Doctoral Dissertation, University of Kwazulu-Natal, Pietermaritzburg, South Africa, 1995.
34. Maroyi, A. Diversity of use and local knowledge of wild and cultivated plants in the Eastern Cape province, South Africa. *J. Ethnobiol. Ethnomed.* **2017**, *13*, 43. [CrossRef]
35. Ndawonde, B.G.; Zobolo, A.M.; Dlamini, E.T.; Siebert, S.J. A survey of plants sold by traders at Zululand muthi markets, with a view to selecting popular plant species for propagation in communal gardens. *Afr. J. Range Forage Sci.* **2007**, *24*, 103–107. [CrossRef]
36. Van Der Bijl, P.; Van Der Bijl, P. Cardiotoxicity of plants in South Africa. *Cardiovasc. J. Afr.* **2012**, *23*, 476–477.
37. Van Vuuren, S.F.; Naidoo, D. An antimicrobial investigation of plants used traditionally in Southern Africa to treat sexually transmitted infections. *J. Ethnopharmacol.* **2010**, *130*, 552–558. [CrossRef]
38. Sagbo, I.; Mbeng, W. Plants used for cosmetics in the Eastern Cape Province of South Africa: A case study of skin care. *Pharmacogn. Rev.* **2018**, *12*, 139–156. [CrossRef]
39. Husen, A. (Ed.) *Environmental Pollution and Medicinal Plants*; CRC Press: Boca Raton, FL, USA, 2022.
40. Ramarumo, L.; Maroyi, A.; Tshisikhawe, M.P. *Bowiea volubilis* Harv. ex Hook.f. subsp. volubilis: A therapeutic plant species used by the traditional healers in the Soutpansberg Region, Vhembe Biosphere Reserve, Limpopo Province, South Africa. *J. Pharm. Sci. Res.* **2019**, *11*, 2538–2542.
41. Njoroge, G.N.; Bussmann, R.W. Herbal usage and informant consensus in ethnoveterinary management of cattle diseases among the Kikuyus (Central Kenya). *J. Ethnopharmacol.* **2008**, *108*, 332–339. [CrossRef]
42. Aremu, A.O.; Moyo, M.; Amoo, S.O.; Van Staden, J. Ethnobotany, therapeutic value, phytochemistry and conservation status of *Bowiea volubilis*: A widely used bulbous plant in southern Africa. *J. Ethnopharmacol.* **2015**, *174*, 308–316. [CrossRef] [PubMed]
43. Bhat, R.B. Plants of Xhosa people in the Transkei region of Eastern Cape (South Africa) with major pharmacological and therapeutic properties. *J. Med. Plants Res.* **2013**, *7*, 1474–1480.
44. Street, R.A. Heavy Metals in South African Medicinal Plants. Doctoral Dissertation, University of KwaZulu-Natal, Pietermaritzburg, South Africa, 2008.
45. Asare, G.A.; Ongong'a, R.O.; Anang, Y.; Asmah, R.H.; Rahman, H. Effect of a Benign Prostatic Hyperplasia (BPH) Xenobiotic-Croton membranaceus Müll. Arg. Root Extract on CYP1A2, CYP3A4, CYP2D6, and GSTM1 Drug Metabolizing Enzymes in Rat Model. 2020; preprint.
46. Cimi, P.V.; Campbell, E.E. An investigation of medicinal and cultural use of plants by Grahamstown community members in the Eastern Cape Province of South Africa. *S. Afr. Mus. Assoc.* **2017**, *39*, 1–9.
47. Cock, I.; Mavuso, N.; Van Vuuren, S. A review of Plant-Based therapies for the treatment of urinary tract infections in traditional Southern African Medicines. *Evid. Based Compleme. Altern. Med.* **2021**, *2021*, 7341124. [CrossRef]
48. Coopoosamy, R.; Naidoo, K. An ethnobotanical study of medicinal plants used by traditional healers in Durban, South Africa. *Afr. J. Pharm. Pharmacol.* **2012**, *6*, 818–823. [CrossRef]
49. Amusan, O.O.G.; Dlamini, P.S.; Msonthi, L.P.; Makhubu, L.P. Some herbal remedies from Manzini region of Swaziland. *J. Ethnopharmacol.* **2002**, *79*, 109–112. [CrossRef]
50. Buwa, L.V.; Van Staden, J. Antibacterial and antifungal activity of traditional medicinal plants used against venereal diseases in South Africa. *J. Ethnopharmacol.* **2006**, *103*, 139–142. [CrossRef]

51. Mongalo, N.I. Petlophorum africanum Sond [Mesetlha]: A review of its ethnomedicinal uses, toxicology, phytochemistry and pharmacological activities. *J. Med. Plants Res.* **2013**, *7*, 3484–3491.
52. Madikizela, L.M.; Tavengwa, N.T.; Chimuka, L. Status of pharmaceuticals in African water bodies: Occurrence, removal and analytical methods. *J. Environ. Manag.* **2017**, *193*, 211–220. [CrossRef]
53. Steenkamp, V. Traditional herbal remedies used by South African women for gynaecological complaints. *J. Ethnopharmacol.* **2003**, *86*, 97–108. [CrossRef]
54. Veale, D.J.; Furman, K.I.; Oliver, D.W. South African traditional herbal medicines used during pregnancy and childbirth. *J. Ethnopharmacol.* **1992**, *36*, 185–191. [CrossRef]
55. Masondo, N.A.; Ndhlala, A.R.; Aremu, A.O.; Van Staden, J.; Finnie, J.F. A comparison of the pharmacological properties of garden cultivated and muthi market-sold *Bowiea volubilis*. *S. Afr. J. Bot.* **2013**, *86*, 135–138. [CrossRef]
56. Finnie, J.F.; Drewes, F.E.; Van Staden, J. Bowiea volubilis Harv. ex hook. f.(sea onion): In vitro culture and the production of cardiac glycosides. In *Medicinal and Aromatic Plants VII*; Springer: Berlin/Heidelberg, Germany, 1994; pp. 84–97.
57. Taylor, J.L.S.; Rabe, T.; Mcgaw, L.J.; Jäger, A.K.; Van Staden, J. Towards the scientific validation of traditional medicinal plants. *Plant Growth Regul.* **2001**, *34*, 23–37. [CrossRef]
58. Aremu, A.O.; Van Staden, J.; Finnie, J.F. Does micropropagation influence the antimicrobial properties of *Bowiea volubilis*? *S. Afr. J. Bot.* **2013**, *86*, 157–158.
59. Ewhea, A.S.; Morah, F.; Obeten, A.U. Anti-microbial and anthelminthic activities of Spilanthes filicaulis (Schum. & Thonn.) CD Adams. *World Sci. News* **2023**, *175*, 1–12.
60. Jäger, A.K.; Hutchings, A.; Van Staden, J. Screening of Zulu medicinal plants for prostaglandin-synthesis inhibitors. *J. Ethnopharmacol.* **1996**, *52*, 95–100. [CrossRef]
61. Steyn, P.S.; Van Heerden, F.R. Bufadienolides of plant and animal origin. *Nat. Prod. Rep.* **1998**, *15*, 397–413. [CrossRef]
62. Winnicka, K.A.; Bielawski, K.R.; Bielawska, A.N. Cardiac glycosides in cancer research and cancer therapy. *Acta Pol. Pharm.* **2006**, *63*, 109–115.
63. Eshun, R.I.; Atangwho, D.G.; Asiedu-Gyekye, R.N. Ethnobotanical, Phytochemical, and Pharmacological Profile of *Bowiea volubilis* Harv. ex Hook.f. *J. Complement. Integr. Med.* **2014**, *11*, 253–261.
64. Salim, K.P.; Anto, R.J.; Chetty, A.S.; Sen, S. Anti-inflammatory effect of the saponin isolated from *Bowiea volubilis*—A sea onion. *J. Ethnopharmacol.* **2009**, *124*, 569–572.
65. Cheok, C.Y.; Salman, H.A.K.; Sulaiman, R. Extraction and quantification of saponins: A review. *Food Res. Inter.* **2014**, *59*, 16–40. [CrossRef]
66. Boatwright, J.S.; De Lange, M.S.; Van der Merwe, A.L. The Genus Bowiea (Hyacinthaceae: Urgineoideae) in Southern Africa: Morphology, Taxonomy, and Conservation. *Taxon.* **2008**, *57*, 1255–1264.
67. Phambu, N.M.; Nthambeleni, C.M.; Happi, E.E.E. An overview of the medicinal importance of *Bowiea volubilis*. *J. Ethnopharmacol.* **2020**, *246*, 112202.
68. Guleria, P.; Tiku, V.N.; Singh, R. Phytochemical analysis and antimicrobial activity of *Bowiea volubilis* Harv. ex Hook.f. *J. Appl. Pharm. Sci.* **2013**, *3*, 120–124.
69. Singh, B.; Singh, J.P.; Singh, N.; Kaur, A. Saponins in pulses and their health promoting activities: A review. *Food Chem.* **2017**, *233*, 540–549. [CrossRef]
70. Bakrim, S.; Benkhaira, N.; Bourais, I.; Benali, T.; Lee, L.H.; El Omari, N.; Sheikh, R.A.; Goh, K.W.; Ming, L.C.; Bouyahya, A. Health benefits and pharmacological properties of stigmasterol. *Antioxidants* **2022**, *11*, 1912. [CrossRef]
71. Brodowska, K.M. Natural flavonoids: Classification, potential role, and application of flavonoid analogues. *Eur. J. Bio. Res.* **2017**, *7*, 108–123.
72. Oueslati, S.; Ksouri, R.; Falleh, H.; Pichette, A.; Abdelly, C.; Legault, J. Phenolic content, antioxidant, anti-inflammatory and anticancer activities of the edible halophyte *Suaeda fruticosa* Forssk. *Food Chem.* **2012**, *132*, 943–947. [CrossRef]
73. Ndhlala, A.; Ncube, B.; Okem, A.; Mulaudzi, R.; van Staden, J. Toxicology of some important medicinal plants in southern Africa. *Food Chem. Toxicol.* **2013**, *62*, 609–621. [CrossRef]
74. Van Vuuren, S.F.; Motlhatlego, K.E.; Netshia, V. Traditionally used polyherbals in a southern African therapeutic context. *J. Ethnopharmacol.* **2022**, *288*, 114977. [CrossRef]
75. Tong, S.Y.; Davis, J.S.; Eichenberger, E.; Holland, T.L.; Fowler, V.G., Jr. Staphylococcus aureus infections: Epidemiology, pathophysiology, clinical manifestations, and management. *Clin. Microbiol. Rev.* **2015**, *28*, 603–661. [CrossRef] [PubMed]
76. Heisey, R.; Gorham, B.K. Antimicrobial effects of plant extracts on Streptococcus mutans, Candida albicans, Trichophyton rubrum and other micro-organisms. *Lett. Appl. Microbiol.* **1992**, *14*, 136–139. [CrossRef]
77. Dzoyem, J.P.; Eloff, J.N. Anti-inflammatory, anticholinesterase and antioxidant activity of leaf extracts of twelve plants used traditionally to alleviate pain and inflammation in South Africa. *J. Ethnopharmacol.* **2015**, *160*, 194–201. [CrossRef] [PubMed]
78. Adeleke, E.I.; Adeyemi, O.; Adebiyi, O.; Adeniji, O.O.; Abolaji, A.; Akinyemi, J.O. Antiviral Activity of *Bowiea volubilis* Against Herpes Simplex Virus Type 1. *J. Pure Appl. Microbiol.* **2018**, *12*, 2287–2294.
79. Tran, N.-L.; Park, J.-H.; Lee, B.-H.; Tung, T.T.; Yang, E.-J.; Oh, W.K. Antiviral Activity of *Bowiea volubilis* Extract and Compounds against Dengue Virus. *J. Nat. Prod.* **2017**, *80*, 2215–2220.
80. Barrientos, A.G.; Juma, J.O.; Okinda Owuor, E.M.; Anzala, J.O.; Holman, M.J. Evaluation of the antiviral activity of a crude extract from *Bowiea volubilis* Harv. against HIV-1 using a PBMC-based assay. *Afr. J. Tradit. Complement. Altern. Med.* **2010**, *7*, 287–292.

81. Feng, R.; Yu, H.; Wang, X.-L.; Hou, W. Antiviral activity of extracts from the bulb of *Bowiea volubilis* against respiratory syncytial virus (RSV). *J. Ethnopharmacol.* **2014**, *153*, 101–107.
82. Ali, S.I.; Sheikh, W.M.; Rather, M.A.; Venkatesalu, V.; Bashir, S.M.; Nabi, S.U. Medicinal plants: Treasure for antiviral drug discovery. *Phytother. Res.* **2021**, *35*, 3447–3483. [CrossRef]
83. Iwalewa, E.O.; Mcgaw, L.J.; Naidoo, V.; Eloff, J.N. Inflammation: The foundation of diseases and disorders. A review of phytomedicines of South African origin used to treat pain and inflammatory conditions. *Afr. J. Biotechnol.* **2007**, *6*, 2868–2885.
84. Emamzadeh-Yazdi, S. Antiviral, Antibacterial, and Cytotoxic Activities of South African Plants Containing Cardiac Glycosides. Masters Dissertation, University of Pretoria, Plant Science, Pretoria, South Africa, 2013.
85. Fasinu, P.; Bouic, P.J.; Rosenkranz, B. The Inhibitory activity of the extract of popular medicinal herbs on CYP1A2, 2C9, 2C19 and 2A4 and the implications for herb-drug interaction. *Afr. J. Tradit. Complement. Altern. Med.* **2014**, *11*, 54–61. [CrossRef]
86. Kuek, L.E.; Lee, R.J. First contact: The role of respiratory cilia in host-pathogen interactions in the airways. *Am. J. Physiol. Lung Cell. Mol. Physiol.* **2020**, *319*, L603–L619. [CrossRef] [PubMed]
87. Morris, S.; Cerceo, E. Trends, epidemiology, and management of multi-drug resistant gram-negative bacterial infections in the hospitalized setting. *Antibiotics* **2020**, *9*, 196. [CrossRef] [PubMed]
88. Meawed, T.E.; Ahmed, S.M.; Mowafy, S.M.; Samir, G.M.; Anis, R.H. Bacterial and fungal ventilator associated pneumonia in critically ill COVID-19 patients during the second wave. *J. Infect. Public Health* **2021**, *14*, 1375–1380. [CrossRef]
89. Mulholland, D.A.; Nuzillard, J.M.; Stermitz, F.R. Cardenolides from *Bowiea volubilis*. *Phytochem.* **2013**, *96*, 295–301.
90. Salminen, K.A.; Meyer, A.; Bernal, M.R.; Schuster, D.; Karonen, M. Alkaloids as inhibitors of human cytochrome P450 3A4, 2D6, and 2C9: Implications in drug development. *Front. Pharmacol.* **2011**, *2*, 28.
91. Smith, D.J.; Bi, H.; Hamman, J.; Ma, X.; Mitchell, C.; Nyirenda, K.; Monera-Penduka, T.; Oketch-Rabah, H.; Paine, M.F.; Pettit, S.; et al. Potential pharmacokinetic interactions with concurrent use of herbal medicines and a ritonavir-boosted COVID-19 protease inhibitor in low and middle-income countries. *Front. Pharmacol.* **2023**, *14*, 1210579. [CrossRef] [PubMed]
92. Eccles, R. Understanding the symptoms of the common cold and influenza. *Lancet Infect. Infect. Dis.* **2005**, *5*, 718–725. [CrossRef]
93. Moghadami, M. A narrative review of influenza: A seasonal and pandemic disease. *Iran. J. Med. Sci.* **2017**, *42*, 2.
94. Rustamovich, T.D.; Alisherovna, K.M.; Nizamitdinovich, K.S.; Djamshedovna, K.D. Gastrointestinal Conditions in Rheumatoid Arthritis Patients. *Texas J. Med. Sci.* **2022**, *15*, 68–72.
95. Parker, C.H.; Naliboff, B.D.; Shih, W.; Presson, A.P.; Kilpatrick, L.; Gupta, A.; Liu, C.; Keefer, L.A.; Sauk, J.S.; Hirten, R.; et al. The role of resilience in irritable bowel syndrome, other chronic gastrointestinal conditions, and the general population. *J. Gastroenterol. Hepatol.* **2021**, *19*, 2541–2550. [CrossRef]
96. Mangoale, R.M.; Afolayan, A.J. Comparative phytochemical constituents and antioxidant activity of wild and cultivated Alepidea amatymbica Eckl & Zeyh. *BioMed Res. Inter.* **2020**, *2020*, 5808624.
97. Casado-Bedmar, M.; Viennois, E. MicroRNA and gut microbiota: Tiny but mighty—Novel insights into their cross-talk in inflammatory bowel disease pathogenesis and therapeutics. *J. Crohn's Colitis* **2022**, *16*, 992–1005. [CrossRef] [PubMed]
98. Ansari, M.A.; Chung, I.M.; Rajakumar, G.; Alzohairy, M.A.; Alomary, M.N.; Thiruvengadam, M.; Pottoo, F.H.; Ahmad, N. Current nanoparticle approaches in nose to brain drug delivery and anticancer therapy—A review. *Curr. Pharm. Des.* **2020**, *26*, 1128–1137. [CrossRef] [PubMed]
99. Aqeel, R.; Srivastava, N.; Kushwaha, P. Micelles in Cancer Therapy: An Update on Preclinical and Clinical Status. *Recent Pat. Nanotechnol.* **2022**, *16*, 283–294.
100. Duan, C.; Yu, M.; Xu, J.; Li, B.Y.; Zhao, Y.; Kankala, R.K. Overcoming Cancer Multi-drug Resistance (MDR): Reasons, mechanisms, nanotherapeutic solutions, and challenges. *Biomed. Pharmacother.* **2023**, *162*, 114643. [CrossRef]
101. Mofokeng, M.M.; du Plooy, C.P.; Araya, H.T.; Amoo, S.O.; Mokgehle, S.N.; Pofu, K.M.; Mashela, P.W. Medicinal plant cultivation for sustainable use and commercialisation of high-value crops. *S. Afr. J. Sci.* **2022**, *118*, 1–7. [CrossRef]
102. Rani, H.; Srivastava, A.K. Phytomedicines and Their Prospects in Treatment of Common Skin Diseases. *Adv. Pharm. Biotechnol. Recent Prog. Future Appl.* **2020**, 289–315. [CrossRef]
103. Richard, M.A.; Paul, C.; Nijsten, T.; Gisondi, P.; Salavastru, C.; Taieb, C.; Trakatelli, M.; Puig, L.; Stratigos, A. EADV Burden of Skin Diseases Project Team, Prevalence of most common skin diseases in Europe: A population-based study. *J. Eur. Acad. Dermatol. Venereol.* **2022**, *36*, 1088–1096. [CrossRef]
104. De Pessemier, B.; Grine, L.; Debaere, M.; Maes, A.; Paetzold, B.; Callewaert, C. Gut-skin axis: Current knowledge of the interrelationship between microbial dysbiosis and skin conditions. *Microorganisms* **2021**, *9*, 353. [CrossRef]
105. Gendrisch, F.; Esser, P.R.; Schempp, C.M.; Wölfle, U. Luteolin as a modulator of skin aging and inflammation. *Biofactors* **2021**, *47*, 70–180. [CrossRef]
106. Boxberger, M.; Cenizo, V.; Cassir, N.; La Scola, B. Challenges in exploring and manipulating the human skin microbiome. *Microbiome* **2021**, *9*, 1–14. [CrossRef]
107. Shaskolskiy, B.; Dementieva, E.; Leinsoo, A.; Runina, A.; Vorobyev, D.; Plakhova, X.; Kubanov, A.; Deryabin, D.; Gryadunov, D. Drug resistance mechanisms in bacteria causing sexually transmitted diseases and associated with vaginosis. *Front. Microbiol.* **2016**, *7*, 747. [CrossRef] [PubMed]
108. World Health Organization (WHO). News-Room. Key Facts. Sexually Transmitted Infections (STIs). 2023. Available online: https://www.who.int/news-room/fact-sheets/detail/sexually-transmitted-infections-(stis)#:~:text=Scope%20of%20the%20problem,and%20trichomoniasis%20(156%20million) (accessed on 16 August 2023).

109. Haese, E.C.; Thai, V.C.; Kahler, C.M. Vaccine candidates for the control and prevention of the sexually transmitted disease gonorrhea. *Vaccines* **2021**, *9*, 804. [CrossRef] [PubMed]
110. Tien, V.; Punjabi, C.; Holubar, M.K. Antimicrobial resistance in sexually transmitted infections. *J. Travel Med.* **2020**, *27*, taz101. [CrossRef] [PubMed]
111. Kacholi, D.S.; Mvungi, H.A. Plants Used by Nyamwezi Traditional Health Practitioners To Remedy Sexually Transmitted Infections in Sikonge, Tanzania. *J. Educ. Humanit. Sci.* **2021**, *10*, 89–101.
112. Hashemi, N.; Ommi, D.; Kheyri, P.; Khamesipour, F.; Setzer, W.N.; Benchimol, M. A review study on the anti-trichomonas activities of medicinal plants. *Int. J. Parasitol. Drugs Drug Resist.* **2021**, *15*, 92–104. [CrossRef]

Disclaimer/Publisher's Note: The statements, opinions and data contained in all publications are solely those of the individual author(s) and contributor(s) and not of MDPI and/or the editor(s). MDPI and/or the editor(s) disclaim responsibility for any injury to people or property resulting from any ideas, methods, instructions or products referred to in the content.

Article

Anti-Yeasts, Antioxidant and Healing Properties of Henna Pre-Treated by Moist Heat and Molecular Docking of Its Major Constituents, Chlorogenic and Ellagic Acids, with *Candida albicans* and *Geotrichum candidum* Proteins

Sulaiman A. Alsalamah [1], Mohammed Ibrahim Alghonaim [1], Mohammed Jusstaniah [2] and Tarek M. Abdelghany [3,*]

1. Department of Biology, College of Science, Imam Mohammad Ibn Saud Islamic University, Riyadh 11623, Saudi Arabia; saalsalamah@imamu.edu.sa (S.A.A.); mialghonaim@imamu.edu.sa (M.I.A.)
2. University Medical Service Center, Building 70, King Abdulaziz University, Jeddah 21589, Saudi Arabia; mjusstanih@kau.edu.sa
3. Botany and Microbiology Department, Faculty of Science, Al-Azhar University, Cairo 11725, Egypt
* Correspondence: tabdelghany.201@azhar.edu.eg or tabdelghany@yahoo.com

Citation: Alsalamah, S.A.; Alghonaim, M.I.; Jusstaniah, M.; Abdelghany, T.M. Anti-Yeasts, Antioxidant and Healing Properties of Henna Pre-Treated by Moist Heat and Molecular Docking of Its Major Constituents, Chlorogenic and Ellagic Acids, with *Candida albicans* and *Geotrichum candidum* Proteins. *Life* 2023, 13, 1839. https://doi.org/10.3390/life13091839

Academic Editor: Seung Ho Lee

Received: 25 July 2023
Revised: 24 August 2023
Accepted: 28 August 2023
Published: 30 August 2023

Copyright: © 2023 by the authors. Licensee MDPI, Basel, Switzerland. This article is an open access article distributed under the terms and conditions of the Creative Commons Attribution (CC BY) license (https://creativecommons.org/licenses/by/4.0/).

Abstract: *Lawsonia inermis*, known as henna, has traditionally been utilized in cosmetics and folk medicine because of their valuable health effects. A lack of information about the processes that increase or decrease release, as well as the biological activities of constituents of natural origin, is an important pharmacological problem. This investigation evaluates the influence of moist heat on the flavonoid and phenolic contents of henna powder and their biological activities. HPLC analysis reflected the existence of 20 and 19 compounds of flavonoids and phenolics in the extract of unpre-treated henna by moist heat (UPMH) and pre-treated henna by moist heat (PMH). Several compounds such as chlorogenic acid, ellagic acid, rutin, rosmarinic acid, kaempferol, and pyrocatechol occurred with high concentrations of 57,017.33, 25,821.09, 15,059.88, 6345.08, 1248.42, and 819.19 µg/mL UPMH while occurred with low concentrations of 44,286.51, 17,914.26, 3809.85, 5760.05, 49.01, and 0.0 µg/mL, respectively in PMH. *C. albicans*, *C. tropicalis*, and *G. candidum* were more affected by UPMH with inhibition zones of 30.17 ± 0.29, 27 ± 0.5, and 29 ± 1.5 mm than PMH with inhibition zones of 29 ± 0.5, 25.33 ± 0.58, and 24.17 ± 0.29 mm, respectively. UPMH henna exhibited less MIC and MFC against the tested yeasts than PMH. Moreover, UPMH henna showed good wound healing, where the rat of migration, wound closure %, and area difference % were 14.806 um, 74.938 um^2, and 710.667% compared with PMH henna 11.360 um, 59.083 um^2, 545.333%, respectively. Antioxidant activity of UPMH and PMH henna. Promising antioxidant activity was recorded for both UPMH or PMH henna with IC$_{50}$ 5.46 µg/mL and 7.46 µg/mL, respectively. The docking interaction of chlorogenic acid and ellagic acid with the crystal structures of *G. candidum* (4ZZT) and *C. albicans* (4YDE) was examined. The biological screening demonstrated that the compounds had favorable docking results with particular proteins. Chlorogenic acid had robust behavior in the *G. candidum* (4ZZT) active pocket and displayed a docking score of −7.84379 Kcal/mol, higher than ellagic acid's −6.18615 Kcal/mol.

Keywords: anti-yeast; moist heat; *Lawsonia inermis*; healing; phenolic; flavonoid; molecular docking

1. Introduction

Lawsonia inermis (*Lythraceae* family) is commonly recognized as henna and is native to subtropical areas of North Africa and Asia. Traditionally, it has been utilized as a dandruff-fighting and a controller for fungi when functional to the hair, feet, and hands, besides coloring of skin, hair, and nails was attributed to henna [1,2]. Several pharmacological properties were associated with henna extract, such as alleviating and ameliorating wound healing, antifungal, antibacterial, antioxidant, nootropic, hepatoprotective, anti-ulcer, anti-cancer, anti-inflammatory, and anti-cancer activity. However, the leaves of this plant represent the most valuable part, but roots, stem, and bark have been utilized in ethno medicine conventional medicine for over nine centuries [3,4].

Species of *Candida* are pathogenic yeast forming mucocutaneous and systemic complaints in humans, particularly in diabetes patients, transplant recipients, xerostomia, malignancy, malnutrition, and lowly oral hygiene (immunocompromised patients) [5,6]. Several species of *Candida* become resistant to antifungal compounds [7]. Therefore, there is a requisite for novel compounds to fight pathogenic yeasts with greater efficiency and low toxicity. Yiğit [3] tested the paste of *L. inermis* extract against several clinical *Candida* isolates including *Candida albicans*, *C. parapsilosis*, *C. glabrata*, *C. tropicalis*, *C. krusei*, *C. kefyr* with different levels of inhibition, where the inhibition zone was more than 20 mm against 35.4% of the isolates, and was up to 15 mm against 38.0%, while the rest of isolates were resistance to the extract of *L. inermis*. As reported by Samadi et al. [8], henna extract can be applied to treat oral cavity infections resulting from *C. albicans* because of the promising anti-candidal activity of the extract, with 2.8 mg/mL as the minimum inhibitory concentration. Besides *Candida* spp., the henna leaves extract reflected a fungicidal effect against filamentous fungi, including *Penicillium ochrochloron*, *P. funiculosum*, *Aspergillus flavus*, *A. ochraceus* due to the presence of apigenin 5-glucoside in the extract [9]. *Trichophyton* spp. *Curvularia* spp. and *Geotrichum* spp. were affected by henna leaf extract because of the existence of terpenes, aliphatic compounds, and flavonoids [10].

Different extraction solvents (chloroform, ethanol, and methanol) and methods (hot sequential extraction) were applied to evaluate the biological activities of *L. inermis* [11]. Despite the use of natural materials since the beginning of humanity, and even the demand for them is increasing daily, there is a big gap between how to apply, the correct extraction methods, and pre-treatments to evaluate its effects on the type and quantity of active ingredients. In the present investigation, moist heat was applied to the powder of henna before the extraction process and its biological activities.

Some plants, such as *Laurus nobilis*, were subjected to microwave, and high temperature of oven up to 120 °C [12]. Also, Al-Rajhi et al. [13] studied the effect of moist heat on the phenolic and flavonoid contents of *L. nobilis*, where moist heat induced the release of constituents and increased its biological activities. To date, numerous studies have reported the biological activities of henna using several extraction solvents, but few, if any, investigations have assessed the phytochemical characterization and biological effect of pre-treated henna by moist heat. Therefore, the current investigation aimed to study the effect of moist heat on henna before extraction, anti-yeast and antioxidant activity, and healing properties. Also, the molecular docking interaction of the most detected constituents of henna extract with some tested yeasts.

2. Materials and Methods

2.1. Chemical Used

Methanol, acetonitrile, and 2,2-diphenyl-1-picrylhydrazyl (DPPH) were obtained from Sigma-Aldrich (Steinheim, Germany). Potato dextrose agar (PDA) medium was obtained from Oxoid Ltd., Basingstoke, Hampshire, UK.

2.2. Henna Source and Its Pre-Treated with Moist Heat

Dried henna leaves were obtained from the company of Abnaa Sayed Elobied Agro-Export, P.O. Box 10725, Khartoum, Sudan. The plant was validated by Prof. Marei A. Hamed, Prof. of the plant. The sample of leaves was kept under herbarium number SH 4325 in the Faculty of Science, Al-Azhar University, Egypt. The leaves were ground by a mill, and then passed via a 40-mesh sieve. The powder of Sudanian Henna was used in the current investigation. Henna powder (250 g) was autoclaved for 10 min at 100 °C, then cooled at room temperature (25 °C) and became pretreated by moist heat (PMH). At the same time, 250 g of henna was kept without pre-treatment by moist heat (UPMH) at

30 °C for 10 min with a humidity of 32%. The PMH and UPMH henna powders were extracted by mixing with 600 mL of methanol on the magnetic stirrer for 12 h. For removing any remains of the powders, the mixture at 5000 rpm for 10 min was centrifuged. Via rotary evaporator, the supernatant was concentrated to get a known weight, followed by re-dissolved in 0.5 mL of dimethyl sulfoxide (DMSO) [13].

2.3. Assessment of Phenolic and Flavonoid Constituents by HPLC

UPMH and PMH henna extract were subjected to HPLC (Agilent 1260 series) for phenolic and flavonoid constituents' detection. The separation process was performed via Zorbax Eclipse Plus C8 column (4.6 mm × 250 mm i.d., 5 µm). The flow rate of the mobile phase (MP) of water (W) and acetonitrile containing 0.05% trifluoroacetic acid (A) was 0.9 mL/min. The MP was automated sequentially in a linear gradient in the flowing order: 0 min (82% W); 82% W from 0–1 min; 75% W from 1–11 min; 60% W from 11–18 min; 82% W from 18–24 min. The ultraviolet (UV) detector was adopted at 280 nm and 330 nm for phenolic and flavonoid constituents' detection, respectively. The solution of tested samples was injected in volume 5 µL with the column maintained at 40 °C. The input data of standard molecules of phenolic and flavonoids was used for the quantitative determination of the extract's compounds [14].

2.4. Anti-Yeast Activity of UPMH and PMH Henna Extracts

The anti-yeast activity of the henna extracts was assessed according to Al-Rajhi et al. [15] with some modification via cup-plate agar diffusion technique against *Candida albicans* (ATCC 10231), *Geotrichum candidum* (RCMB 027016), and *Candida tropicalis*. The tested yeasts were standardized (Corresponding to 0.5 McFarland scale), sowed in molten sterile Sabouraud dextrose agar medium, and poured into petri dishes. After solidification, via sterile cork borer (6 mm radius), four cups were cut and removed. Via automatic microlitre pipette, 100 µL of 20 µg/mL of each extract was injected in each cup, and then kept in the refrigerator at 4 °C for 30 min to allow the extract to diffuse through the agar layer. Followed by the incubation at 35 °C for 48 h. The visualized inhibition zones were recorded using a calibrated ruler in millimeters.

2.5. Evaluation of Minimum Inhibitory Concentration of UPMH and PMH Henna Extracts

The extract of henna, including the unpre-treated and pre-treated henna powders, was tested to detect minimum inhibitory concentration against tested yeasts. According to the CLSI M27-A3 standard manner. For each species of yeasts, a dilution was prepared in an equivalent to 0.5 McFarland. A dilution of each extract was prepared (1 mg/mL). Using 96-well plates, 100 µL of RPMI 1640 broth adjusted at pH 7 using a buffer of MOPS were transferred into each well. 100 µL of each extract was mixed with RPMI in wells of 1st column, and then serial dilution was performed. From the suspensions of tested yeast, 100 µL (0.5 McFarland) were added to each well, and wells without yeast cell suspensions were used as a negative control. Followed by incubation for one day at 34 ± 2 °C. The lowest concentration of extract that gave 50% declined growth was defined as MIC compared to the controls. The antifungal nystatin was used as a positive control, while the solvent of the extraction (methanol) was used as a negative control, respectively.

2.6. Estimation of Minimum Fetal Concentration (MFC) of UPMH and PMH Henna Extracts

To assess the MFC, 50 µL of the clear homogenized well suspension (devoid of visual growth) was cultivated using Sabouraud Dextrose Agar plates, followed by incubation for 48 h at 35 °C. The lowest dose of the extract affected growth inhibition (99.9%) compared to growth at control (without treatment) was MFC. The count of each yeast colony (CFU/mL) at different doses was compared with the count of each yeast colony at control (without treatment).

2.7. Antioxidant Activity of UPMH and PMH Henna Extracts

The 1,1-diphenyl-2-picrylhydrazyl free radical (DPPH), developed by Elansary et al. [9] with minor modifications, was used to study the antioxidant scavenging activity. Various dilutions (1 mL) containing different concentrations of the plant extracts were combined with 1 mL of 0.2 mM DPPH dissolved in methanolic. Using a Helios spectrophotometer (Unicam, Cambridge, UK), the absorbance at 520 nm was measured following a 30-min incubation time at 25 °C. The same process was used in a blank experiment to create a solution devoid of the tested plant extract, and the absorbance was recorded. The percent inhibition of each solution's free radical-scavenging activity was then determined using the following equation:

$$\% \text{ inhibition} = \left(\frac{\text{Absorbance of blank} - \text{Absorbance of extract}}{\text{Absorbance of blank}}\right) \times 100$$

Antioxidant activity was itemized as IC_{50}, which is the amount of the tested extracts needed to result in a 50% drop in the initial DPPH concentration.

2.8. Healing Properties of UPMH and PMH Henna Extracts

A multiwell plate was used for scratch wound examination. The plate was coated with an extracellular matrix substrate of 10 µg/mL fibronectin. Followed by incubation at 37 °C for 2 h. Then, the unbound extracellular matrix was removed and washed with phosphate-buffered saline. The growing cells from a dish containing tissue culture were detached with trypsin. The cells were developed on the scratch wound assay plate, followed by incubation to permit cells to spread and to obtain a confluent monolayer. The monolayer cell, including the confluent monolayer, was scraped using a pipette tip. Once scratched, slightly wash the monolayer of cells to remove separated cells. Then, replace with fresh medium containing tested extracts. The plate was incubated at 37 °C in the incubator of cell culture for 24–48 h. After the end of the incubation period, the cell monolayer was washed using phosphate-buffered saline. Then, the cells were fixed for 15 minutes using 3.7% paraformaldehyde. The cells were stained for 10 min using crystal violet (1% in ethanol). Then, the cell culture was examined using a phase-contrast microscope [16]. The following analysis was calculated according to the following equations:

$$\text{Rat of migration (RM)} = \frac{Wi - Wf}{t} \times 100$$

where, Wi = average of initial wound width (um), Wf = average of final wound width (um), t = time span of the assay in hours

$$\text{Wound clouser \%} = \frac{A_{t0} - A_{t\Delta t}}{A_{t0}} \times 100$$

where, A_{t0} = intial wound area, $A_{t\Delta t}$ = wound area after n hours

$$\text{Area difference \%} = \text{intial area} - \text{final area}$$

2.9. Molecular Docking Investigation

Docking studies were carried out using the MOE (Molecular Operating Environment) software 2019.0102 program.

Ligand preparation: Chemical structures of substrate molecules (chlorogenic acid and ellagic acid) were drawn using ChemDraw Ultra 15.0, and this structure was saved as MDL files (".sdf") for MOE to show. These structures were optimized by adding hydrogens, and energies were minimized with parameters (gradient: 0.05, Force Field: MMFF94X).

Preparation of receptor structure: The G. candidum and C. albicans models were predicted through homology modeling. The best model was selected for docking analysis. This model is subjected to 3D protonation and energy minimization using parameters (gradient:

0.05, Force Field: MMFF94X + Solvation). The minimized structure was used as the receptor protein for Docking. The protein molecules utilised throughout our investigation were obtained from Protein Data Bank (http://www.rcsb.org/pdb accessed on 23 September 2015) using PDB codes (4ZZT) and (4YDE), respectively for *G. candidum* and *C. albicans*.

Docking Run: The MOE docking program with default parameters was used to bind the selected ligands with receptor proteins and to find the correct conformation of the substrate. Free energy of binding of the ligand from a given pose was estimated by MOE London dG scoring function. The top five poses were determined using hydrogen bonds with lengths under 3.5 Å and binding free energies (S, kcal/mol) between substances and amino acids that are part of proteins. Additionally, the RMSD and RMSD-refine fields were used to compare the results pose-with-pose in the co-crystal ligand position and before and after amendment, respectively.

2.10. Statistical Analysis

Standard deviation was calculated from the average three replicates of the obtained results via Microsoft programs of Excel version 365 and SPSS v.25. For variance analysis, one-way ANOVA, besides the test of post hoc Tukey, were applied to values analysis with a parametric distribution. The confidence interval was set to 95%, and the border of the accepted error was set up to 5%.

3. Results and Discussion

3.1. Flavonoid and Phenolic Contents of Unpre-Treated and Pre-Treated Henna by Moist Heat

The effect of moist heat was evaluated on the contents of flavonoid and phenolic in Henna extract, as well as its anti-yeast properties and other biological activities were performed (Figure 1). From HPLC analysis, the extract of unpre-treated henna by moist heat (UPMH) reflected the existence of 20 compounds, while the extract of pre-treated henna by moist heat (PMH) reflected the existence of 19 compounds of flavonoids and phenolics with different retention times, area, area %, and concentrations (Figures 2 and 3 and Table 1). Chlorogenic acid, ellagic acid, gallic acid, rosmarinic acid, and rutin represent the highest concentrations in both UPMH and PMH henna extracts. The results indicated that the concentrations of the most detected compounds decreased in PMH henna extract compared to UPMH. For example, the concentrations of chlorogenic acid, ellagic acid, rutin, rosmarinic acid, kaempferol, and pyrocatechol were 57,017.33, 25,821.09, 15,059.88, 6345.08, 1248.42, and 819.19 µg/mL in UPMH, while become 44,286.51, 17,914.26, 3809.85, 5760.05, 49.01, and 0.0 µg/mL in PMH henna extract On the other hand, quercetin, naringenin, gallic acid, and coumaric acid concentrations were 96.76, 133.45, 9349.90, and 270.56 µg/mL in UPMH while becoming 1269.47, 2146.89, 32,349.91, and 402.02 µg/mL in PMH henna extract, respectively. Some studies reported that the effect of moist heat induced the discharge of extract constituents, unlike the current study. For example, Juániz et al. [17] found that phenolic constituents of vegetables liberated more if they were pretreated with heat due to cell wall destruction by heat. The decrease of most compounds in pre-treated henna extract or missing compound (pyrocatechol) indicated that these compounds are unstable at high temperatures or heat-sensitive and, at the same time, may easily oxidize and transform into other compounds. This explanation may match with Khoddami et al. [14]. While the concentration of some compounds increased, it may be due to high temperature causing the destruction of plant cell walls, causing the release of internal compounds. Another clarification of this phenomenon is that some phenolic and flavonoid constituents exist in an insoluble form. Heat may break these constituents, leading to the discharge of these bound constituents.

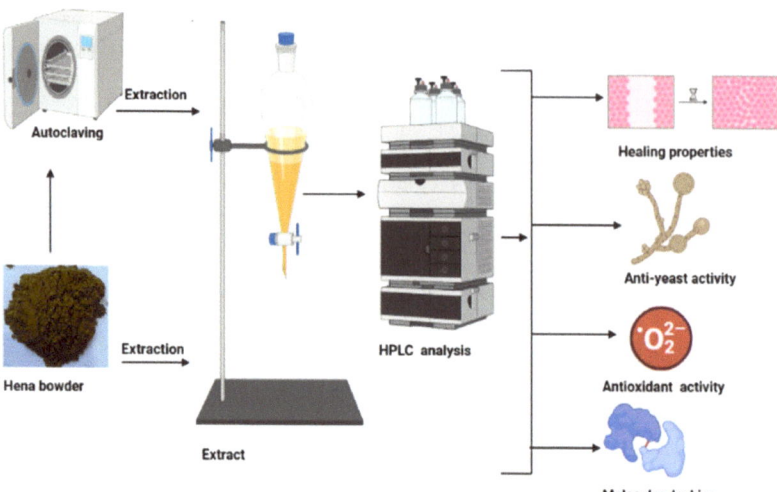

Figure 1. Designed tests for unpre-treated (UPMH) and pre-treated henna powder by moist heat (PMH) in an autoclave with different assessments, including HPLC analysis, healing properties, anti-yeast activity, antioxidant activity, and molecular docking of the main constituents of henna extract.

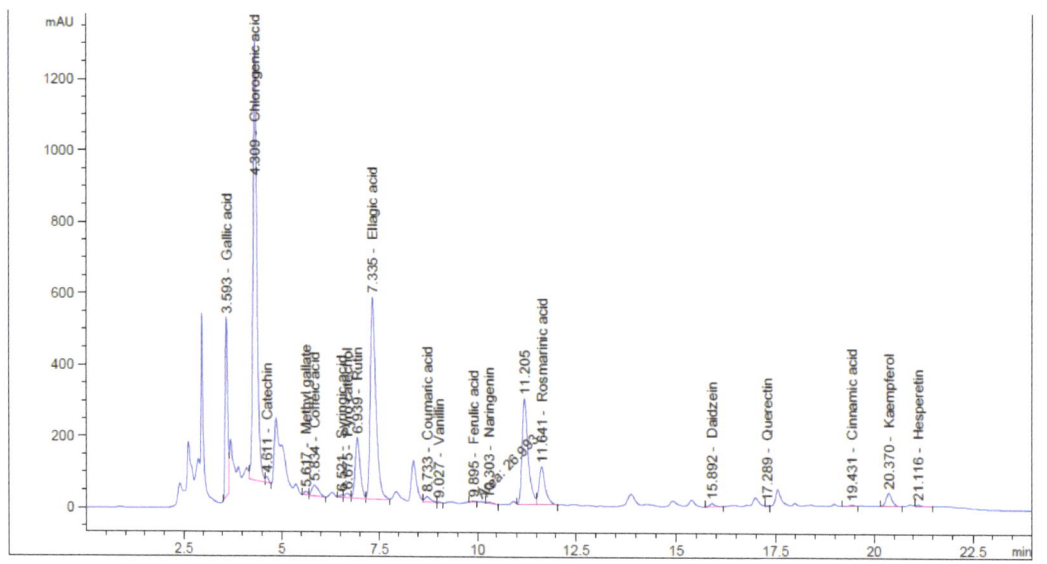

Figure 2. Phenolic and flavonoid compounds detection in UPMH henna extract by moist heat indicated by HPLC chromatogram.

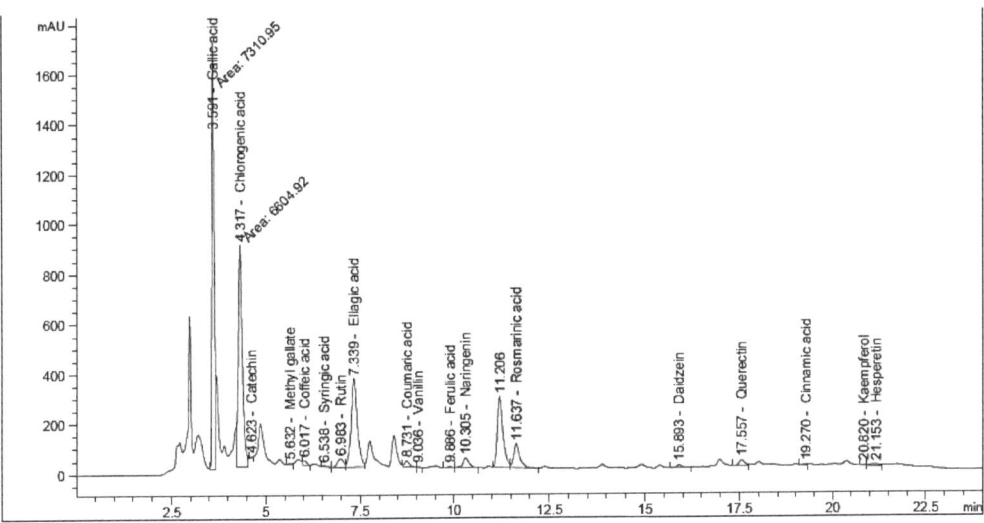

Figure 3. Phenolic and flavonoid compounds detection in PMH henna extract by moist heat indicated by HPLC chromatogram.

Table 1. Identified phenolic and flavonoid compounds in unpre-treated (UPMH) and pre-treated (PMH) henna extracts by moist heat.

Compound	UPMH Henna Extract				PMH Henna Extract			
	Retention Time	Area	Area (%)	Conc. (µg/mL)	Retention Time	Area	Area (%)	Conc. (µg/mL)
Gallic acid	3.593	2113.04	8.9652	9349.90	3.591	7310.95	30.9495	32,349.91
Chlorogenic acid	4.309	8503.60	36.0792	57,017.33	4.317	6604.92	27.9606	44,286.51
Catechin	4.611	91.37	0.3877	1049.75	4.623	36.99	0.1566	424.95
Methyl gallate	5.617	67.04	0.2844	173.53	5.632	28.98	0.1227	75.00
Caffeic acid	5.834	358.26	1.5200	1469.54	6.017	103.97	0.4401	426.45
Syringic acid	6.521	35.46	0.1504	138.64	6.538	33.00	0.1397	129.02
Pyrocatechol	6.675	100.78	0.4276	819.19	6.652	0.00	0.00	0.00
Rutin	6.939	1493.20	6.3354	15,059.88	6.983	377.75	1.5991	3809.85
Ellagic acid	7.335	5712.41	24.2367	25,821.09	7.339	3963.18	16.7774	17,914.26
Coumaric acid	8.733	145.80	145.7976	270.56	8.731	216.64	0.9171	402.02
Vanillin	9.027	4.98	4.97855	9.41	9.036	4.57	0.0193	8.63
Ferulic acid	9.895	26.99	0.1145	82.29	9.886	41.57	0.1760	126.72
Naringenin	10.303	27.96	0.1186	133.45	10.305	449.88	1.9045	2146.89
Rosmarinic acid	11.641	1155.13	4.9010	6345.08	11.637	1048.63	4.4392	5760.05
Daidzein	15.892	86.45	86.4480	251.57	15.893	113.65	0.4811	330.74
Quercetin	17.289	15.82	0.0671	96.76	17.557	207.59	0.8788	1269.47
Cinnamic acid	19.431	38.59	0.1637	35.39	19.270	41.77	0.1768	35.30
Kaempferol	20.370	373.40	1.5843	1248.42	20.820	14.66	0.0621	49.01
Hesperetin	21.116	44.49	0.1888	118.00	21.153	122.14	0.5171	323.97

Previously, Routray and Orsat [18] revealed that some phenolic ingredients were degraded when the natural plant extracts were exposed to the high power of the microwave. Dezashibi et al. [19] found that ultrasound of henna extract increments the phenolic constituents. Moreover, the content of total phenolic was affected by storage temperature and period. Other reports indicated that rises in temperature from 40 to 80 °C caused rises in total phenolic content [20]. Effects of boiling, microwaving, and boiling were reported on the content of ascorbic acid, β-carotene, and vitamin E in some plants, where visualized changes were observed [21].

3.2. Anti-Yeast Activity of Unpre-Treated and Pre-Treated Henna by Moist Heat

From the anti-yeast activity experiment, the extract of UPMH henna reflected significantly more inhibitory action with inhibition zones, 30.17 ± 0.29, 27 ± 0.5, and 29 ± 1.5 mm against *C. albicans*, *C. tropicalis*, and *G. candidum* compared with extract of PMH henna with inhibition zones, 29 ± 0.5, 25.33 ± 0.58, and 24.17 ± 0.29 mm, respectively (Table 2 and Figure 4). Moreover, the MIC and MFC of the extract of UPMH henna were less than the extract of PMH henna against all tested yeasts but with different degrees of sensitivity. Both types of extracts exhibited the highest anti-yeast activity than standard antifungal agents. According to the MIC and MFC, *G. candidum* was more sensitive, followed by *C. albicans* and *C. tropicalis*. The calculated index of MFC/MIC (≤ 2) indicated the cidal properties of both extracts. *L. inermis* showed anti-yeast activity but with different levels of inhibition depending on several factors, such as solvent extract, as mentioned previously by Suleiman and Mohamed [22], where the extraction by ethanol exhibited similar activity to nystatin, but the petroleum ether extract reflected more activity. Also, differences between the activities may be due to the region, the plant's development, climatic changes, soil fertilizers, and the tested fungal population. Petroleum ether and ethanol extract of henna at 10 mg/mL demonstrated 26.3- and 25.3-mm inhibition zones against *Saccharomyces cerevisiae* 22.7-and-17 mm inhibition zones, respectively, against *C. albicans*. The inhibitory potential of the extract by the two types of solvents was more than that observed by the nystatin but did not give antifungal action against *Pichia fabianii* [22]. Kouadri [23] recorded strong anti-*C. albicans* activity of Saudi henna with an inhibition zone of 26 mm with a low MIC of 3.12 mg/mL due to several biologically active constituents. In vivo study, the henna (4%) was formulated as a vaginal cream, which showed promising management for infection caused by *C. albicans* in female rats. Moreover, it gave a similar effect to the antifungal agent (clotrimazole) [24].

Table 2. Anti-yeast activity, MIC, MFC, and MFC/MIC index of UPMH, PMH henna extracts, positive control (Nystatin) and negative control (solvent used).

	Mean Inhibition Zone (mm)				MIC (µg/mL)		MFC (µg/mL)		MFC/MIC Index	
	UPMH	PMH	+ve C	−ve C	UPMH	PMH	UPMH	PMH	UPMH	PMH
C. albicans	30.17 ± 0.29 [a]	29.00 ± 0.50 [b]	26.0 ± 1.32 [c]	0.0	15.63 ± 0.09 [a]	15.64 ± 0.07 [b]	31.23 ± 0.03 [a]	31.25 ± 0.25 [a]	1.99	1.99
C. tropicalis	27.0 ± 0.50 [a]	25.33 ± 0.58 [b]	25.17 ± 0.76 [b]	0.0	62.50 ± 1.00 [a]	125.33 ± 1.53 [b]	125 ± 3.0 [a]	249.67 ± 1.53 [b]	2.0	1.99
G. candidum	29.0 ± 1.50 [a]	24.17 ± 0.29 [b]	27.0 ± 1.00 [c]	0.0	7.83 ± 0.35 [a]	15.62 ± 0.02 [b]	7.8 ± 0.20 [a]	15.62 ± 0.04 [b]	0.99	1.0

Different higher letters for each species within a row (between UMH, MH, and +ve C in case inhibition zone or between UMH and MH in case MIC or between UMH and MH in case MFC) reveal significant differences ($p \leq 0.05$).

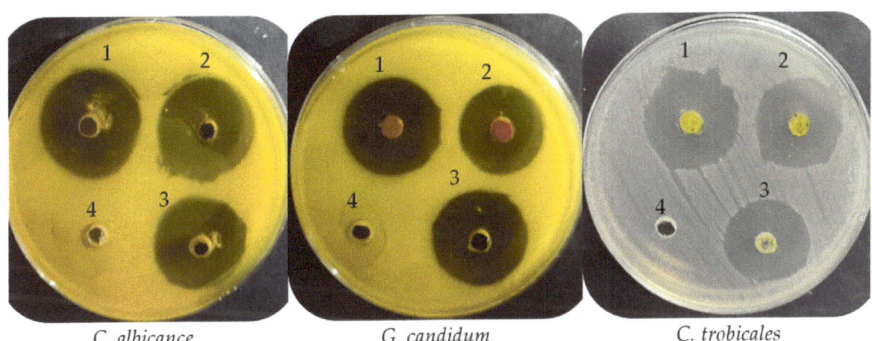

| | C. albicance | G. candidum | C. trobicales |

Figure 4. Anti-yeast activity of UPMH (1), PMH (2) Henna extracts, positive control (3) and negative control (4) against *C. albicance*, *G. candidum* and *C. trobicales*.

3.3. Healing Properties of Unpre-Treated and Pre-Treated Henna by Moist Heat

There is a difference between the healing properties of the extract of UPMH and PMH henna, where the extract of UPMH henna provides reliable healing compared to the extract of PMH henna and control (cells without treatment) (Table 3 and Figure 5). The greatest common data resulting from the assessment of wound healing is the gap closure rate, which determines the rapidity of the cells' collective motion. Rat of migration (RM), wound closure % and area difference % were 14.806 um, 74.938 um^2, and 710.667% using extract of UPMH henna; 11.360 um, 59.083 um^2, 545.333% using extract of PMH henna, compared to control cells, 11.554 um, 58.903 um^2, and 554.667%, respectively with significant differs (Table 3). Daemi et al. [25] demonstrated that the healing process was accelerated by henna via minimizing tissue inflammation and increasing the uptake of glucose. An ointment containing henna extract shows a promising effect in managing episiotomy wounds [26]. El Massoudi et al. [27] explained the healing properties of henna. They mentioned that henna is richneed with different active molecules such as flavonoids, saponins, polyphenols, and others, which are vital in lowering oxidative stress and accelerating wound healing.

Table 3. Healing properties of the extract of UPMH and PMH henna extract.

Treatment	At 0 h		At 24 h		At 48 h		RM um	Wound Closure % um^2	Area Difference %
	Area	Width	Area	Width	Area	Width			
Control (without treatment)	885	884.081	737	736.024	381	380.021	11.554 [a]	58.903 [a]	554.667 [a]
	937	936.009	737	736.000	377	376.021			
	959	958.052	741	740.219	361	360.355			
	945	944.008	837	836.038	337	336.095			
	959	958.000	849	848.021	413	412.000			
	965	964.000	843	842.086	453	452.004			
Mean									
	941.667	940.692	790.667	789.731	387	386.083			
Extract of UPMH henna	931	930.002	813	812.089	249	248.008	14.806 [b]	74.938 [b]	710.667 [b]
	953	952.034	819	818.002	281	280.007			
	945	944.172	823	822.01	249	248.129			

Table 3. Cont.

Treatment	At 0 h		At 24 h		At 48 h		RM um	Wound Closure % um²	Area Difference %
	Area	Width	Area	Width	Area	Width			
Extract of UPMH henna	943	942.034	837	836.117	183	182.176	14.806 [b]	74.938 [b]	710.667 [b]
	971	970.132	749	748	229	228.000			
	947	946.008	763	762.042	235	234.034			
	Mean								
	948.333	947.397	800.667	799.71	237.667	236.726			
Extract of PMH henna	913	912.020	845	844.009	305	304.105	11.360 [c]	59.083 [c]	545.333 [c]
	923	922.020	887	886.009	255	254.031			
	919	918.035	839	838.01	337	336.381			
	947	946.002	885	884.274	447	446.000			
	917	916.020	885	884.081	479	478.004			
	919	918.002	847	846.002	443	442.018			
	Mean								
	923	922.017	864.667	863.731	377.667	376.757			

Higher letters for the treatments within a column reveal significant differences ($p \leq 0.05$).

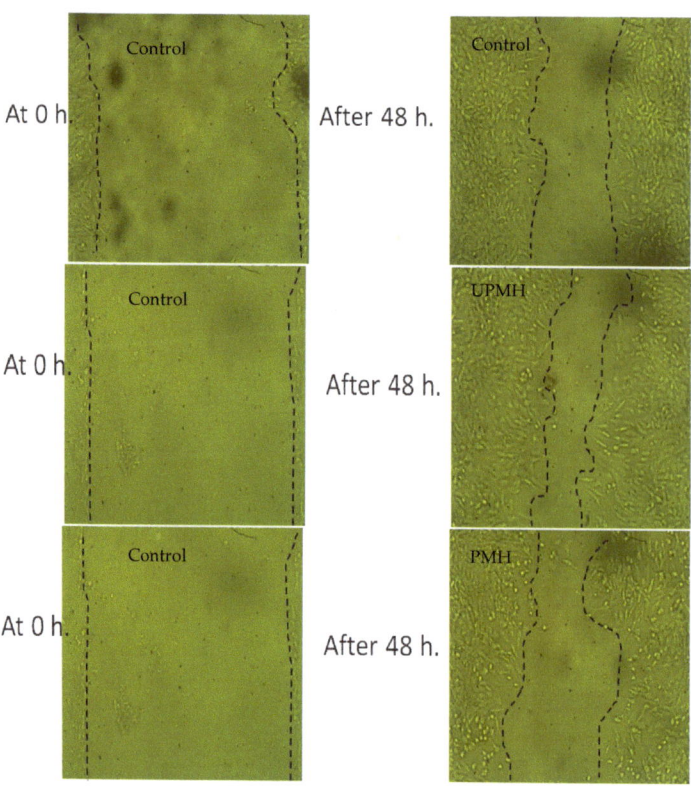

Figure 5. Images of scratch test illustrated the effect of UPMH and PMH henna extract by moist heat on the wounding area at 0 and 48 h.

3.4. Antioxidant Properties of Unpre-Treated and Pre-Treated Henna by Moist Heat

The antioxidant activity of UPMH and PMH henna by moist heat was recorded in Table 4, compared with standard (Ascorbic acid). Generally, henna extract of either UPMH or PMH exhibited promising DPPH scavenging %. However, the extract of UPMH henna reflected more activity than the extract of PMH henna with IC_{50} 5.46 µg/mL and 7.46 µg/mL, respectively, compared with the IC_{50} value of ascorbic acid 2.52 µg/mL. The low activity of antioxidants for the PMH henna may be due to a low concentration of active compounds, as mentioned in HPLC analysis, due to exposure to moist heat. From Table 4, the antioxidant potential increments with the increasing dosage of the tested extracts with concentration-dependent liner. Réblov [28] studied the antioxidant activity of several phenolic acids under stress of temperature. For instance, vanillic acid was effective at 90 °C, while gallic and caffeic acids presented antioxidant potential at 150 °C. In vivo investigation, henna protected them from oxidative stress and possessed hepatoprotective properties in Wistar rats [29]. Çubukçu et al. [30] reported that treatment by heat had harmful effects on the antioxidant activities of some plants, such as garlic and onion, while freezing enhanced the antioxidant properties of garlic and had negative effects on onion.

Table 4. Antioxidant activity of unpre-treated (UPMH), pre-treated (PMH) henna extract by moist heat and ascorbic acid.

Concentration (µg/mL)	DPPH Scavenging %		
	UPMH	PMH	Ascorbic Acid
1000	98.5 [a]	97.7 [a]	99.3 [ab]
500	95.0 [a]	94.5 [a]	96.3 [b]
250	90.8 [a]	90.1 [a]	94.8 [b]
125	83.2 [a]	82.6 [a]	91.9 [b]
62.50	75.6 [a]	74.4 [a]	84.2 [b]
31.25	69.1 [a]	66.5 [b]	76.1 [c]
15.63	61.7 [a]	59.3 [b]	67.6 [c]
7.81	53.3 [a]	50.9 [b]	60.4 [c]
3.90	45.5 [a]	41.5 [b]	52.1 [c]
1.95	37.8 [a]	33.2 [b]	43.7 [c]
0.0	0.0	0.0	0.0
IC_{50}	5.46 µg/mL	7.46 µg/mL	2.52 µg/mL

Different higher letters at each concentration within a row (between UPMH, PMH, and ascorbic acid) reveal significant differences ($p \leq 0.05$).

3.5. Molecular Docking of Chlorogenic Acid and Ellagic Acid with 4ZZT Protein of G. candidum and 4YDE Protein of C. albicans

In the current decade, molecular docking has attracted the attention of several investigators in drug design, development, and discovery. Structure-based computer modeling of ligand-receptor interactions is widely used in modern drug development. To identify conformational changes that vary with the environment and to characterize the interaction of the molecule with the protein with which it interacts inside the body, molecular docking calculations are commonly used in structure-based drug design investigations.

Chlorogenic and ellagic acids were docked using MOE (Molecular Operating Environment) in a vital trial to get insight into the potential pathways through which these compounds exert their antibacterial action. Both target proteins interact effectively with inhibitor compounds. The docking results of compounds' interaction with the crystal structures of G. candidum (4ZZT) and C. albicans (4YDE) (4YDE code of the farnesyltransferase enzyme, and 4ZZT code of the enzyme cellobiohydrolase enzyme, which represent one of the virulence factors in yeast cells) revealed that chlorogenic acid was found the most promising than ellagic acid. Chlorogenic acid revealed substantial binding interactions within the (4ZZT) active site with a binding score of -7.84379 Kcal mol^{-1} according to five conventional hydrogen bonds between GLU 217 and O 19 with 2.80 Å, GLN 175 and O 17 with 2.97 Å, HIS 228 and O 19 with 3.24 Å, TRP 380 and C 1 with 3.19 Å, and TRP 371 and 6-ring with 3.61 Å. Also, ellagic acid outlined four significant interactions with active site residues of *G. candidum* (4ZZT) between GLU 217 and O 20 with 2.67 Å, THR 226 and O 22 with 3.20 Å, ARG 399 and O 24 with 3.08 Å, and ARG 251 and 6-ring with 4.09 Å.

Similarly, ranked the inhibitor compounds with *C. albicans* (4YDE) showed similar behavior and adopted direct H-donor and H-acceptor bonds with active site residues. Chlorogenic acid highlighted a binding score of -5.69876 Kcal mol^{-1}, and exhibited several key interactions with (4YDE) with ASP 527, ASP 569, and LYS 512 via C 26, O 40, and O 23, respectively. While ellagic acid had a binding score of -4.5145 Kcal mol^{-1}, and only one donor interaction between the O 20 atom and the ASP 527 amino acid residue, it demonstrated reduced effectiveness to (4YDE). The 2D and 3D docking interactions are given in Figures 6–9, and the obtained results are shown in Tables 5–8. Several reports confirmed the biological activities of drugs theoretically via molecular docking interaction [31–36]. The most potent constituent in henna (3a, 4a-Dihydroxy-a-tetralone) was docked with sterol 14-demethylase protein of *Staphylococcus aureus*, which reflected a docking score of -7.196 kcal mol^{-1} [37]. Alteration of yeast cells from the yeast shape to filamentous shape is linked with pathogenicity [38]. Therefore, the inhibitors of these enzymes (4YDE and 4ZZT) are potentially therapeutic against infection caused by yeasts.

Table 5. Docking scores and energies of chlorogenic acid and ellagic acid with the crystal structure of *C. albicans* (4YDE).

Mol	S	Rmsd_Refine	E_Conf	E_Place	E_Score1	E_Refine	E_Score2
Chlorogenic acid	−5.69876	1.6749115	−2.57228	−41.8448	−11.4368	−30.2837	−5.69876
Chlorogenic acid	−5.65288	1.729309	−2.45688	−57.55	−12.6225	−28.5364	−5.65288
Chlorogenic acid	−5.58586	1.9682481	14.59859	−48.7804	−12.7552	−28.076	−5.58586
Chlorogenic acid	−5.44998	1.4506954	9.080463	−45.4938	−11.5779	−26.7653	−5.44998
Chlorogenic acid	−5.41043	2.7580004	3.998731	−46.4482	−12.3195	−27.2703	−5.41043
Ellagic acid	−4.5145	4.868453	14.66806	−16.8034	−6.74256	−25.2636	−4.5145
Ellagic acid	−4.23481	3.9006364	16.43244	−48.9339	−7.72468	−19.8999	−4.23481
Ellagic acid	−4.2133	4.5235729	14.63135	7.640516	−6.31675	−19.1073	−4.2133
Ellagic acid	−4.11208	3.4813149	14.97426	−26.417	−8.96398	−18.4696	−4.11208
Ellagic acid	−4.08451	2.8578191	14.7638	−21.7358	−6.67745	−16.6032	−4.08451

Table 6. Docking scores and energies of chlorogenic acid and ellagic acid with the crystal structure of *G. candidum* (4ZZT).

Mol	S	Rmsd_Refine	E_Conf	E_Place	E_Score1	E_Refine	E_Score2
Chlorogenic acid	−7.84379	1.9669101	0.067377	−99.6069	−12.9229	−50.3995	−7.84379
Chlorogenic acid	−7.25803	1.8288656	12.89607	−99.626	−13.0726	−46.7111	−7.25803
Chlorogenic acid	−7.23898	2.5565467	5.357535	−91.2855	−12.8967	−46.316	−7.23898
Chlorogenic acid	−7.23589	2.2799456	14.8198	−101.412	−13.962	−45.1973	−7.23589
Chlorogenic acid	−7.12938	1.8402419	12.1299	−83.0073	−13.2714	−44.543	−7.12938
Ellagic acid	−6.18615	0.88734156	15.62339	−92.791	−13.3693	−39.0664	−6.18615
Ellagic acid	−6.08951	1.0905997	17.58312	−94.3595	−13.0066	−37.9829	−6.08951
Ellagic acid	−6.08205	1.4268098	15.37853	−85.0092	−12.7316	−33.7174	−6.08205
Ellagic acid	−6.07779	1.5727613	16.8008	−94.8967	−12.7112	−34.4453	−6.07779
Ellagic acid	−6.07731	2.3918681	15.11539	−88.7238	−13.13	−34.386	−6.07731

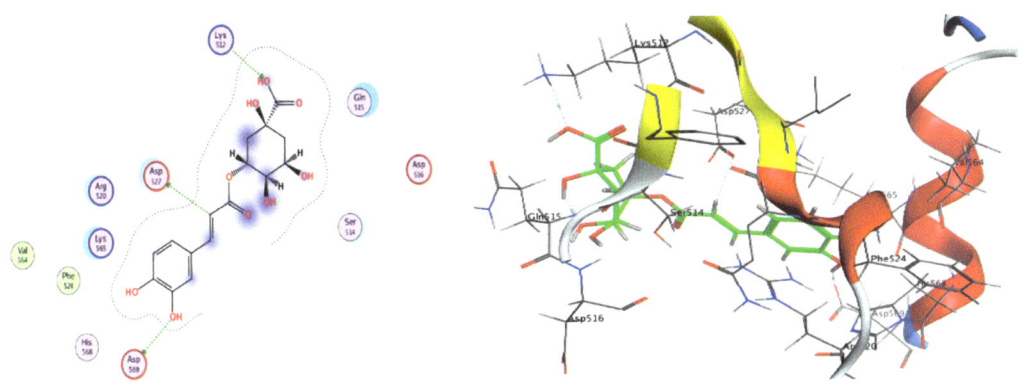

| The interaction between chlorogenic acid and active sites of 4YDE protein | The most likely binding conformation of chlorogenic acid and the corresponding intermolecular interactions are identified |

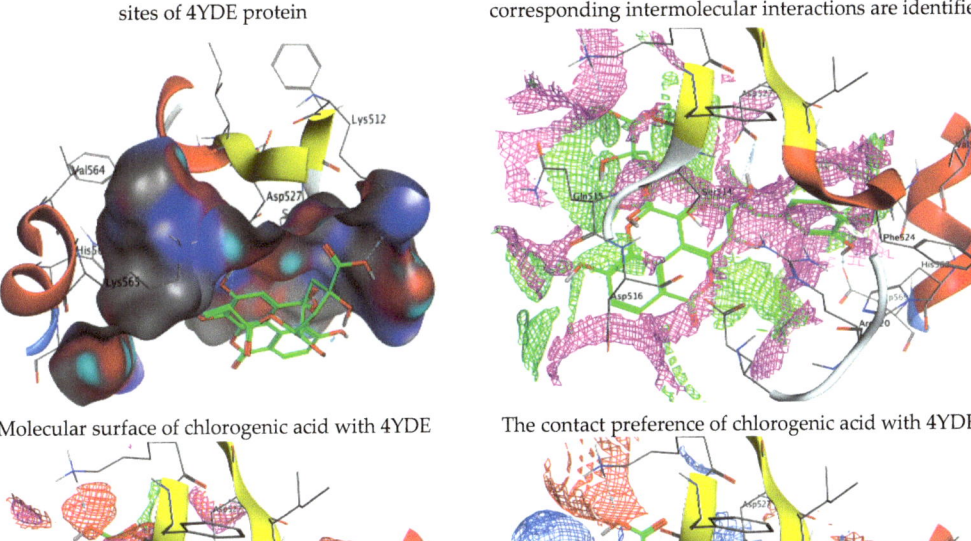

| Molecular surface of chlorogenic acid with 4YDE | The contact preference of chlorogenic acid with 4YDE |
| Interaction potential of chlorogenic acid with 4YDE | The electrostatic map of chlorogenic acid with 4YDE |

Figure 6. Molecular docking process of chlorogenic acid with the crystal structure of *C. albicans* (4YDE).

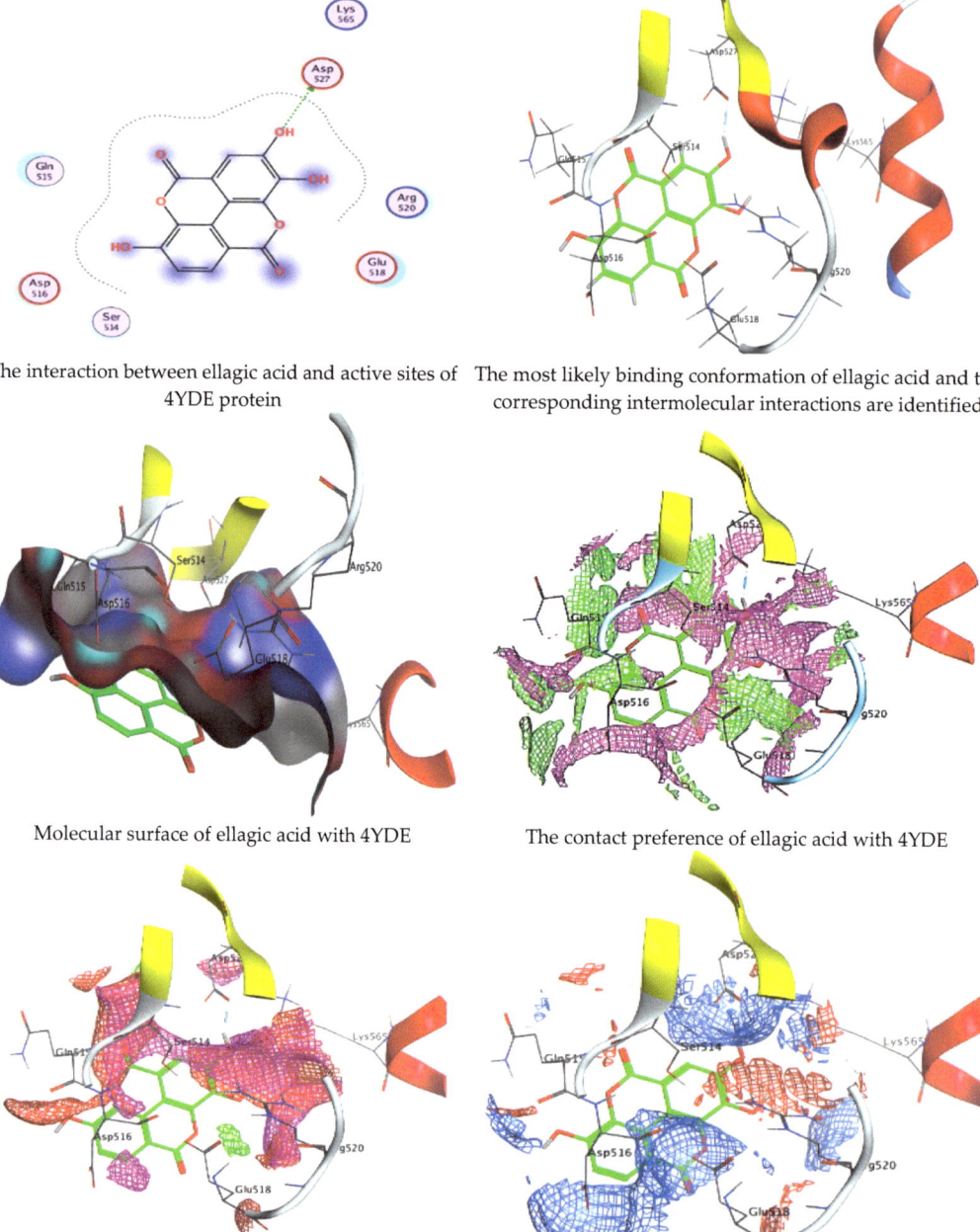

Figure 7. Molecular docking process of ellagic acid with the crystal structure of *C. albicans* (4YDE).

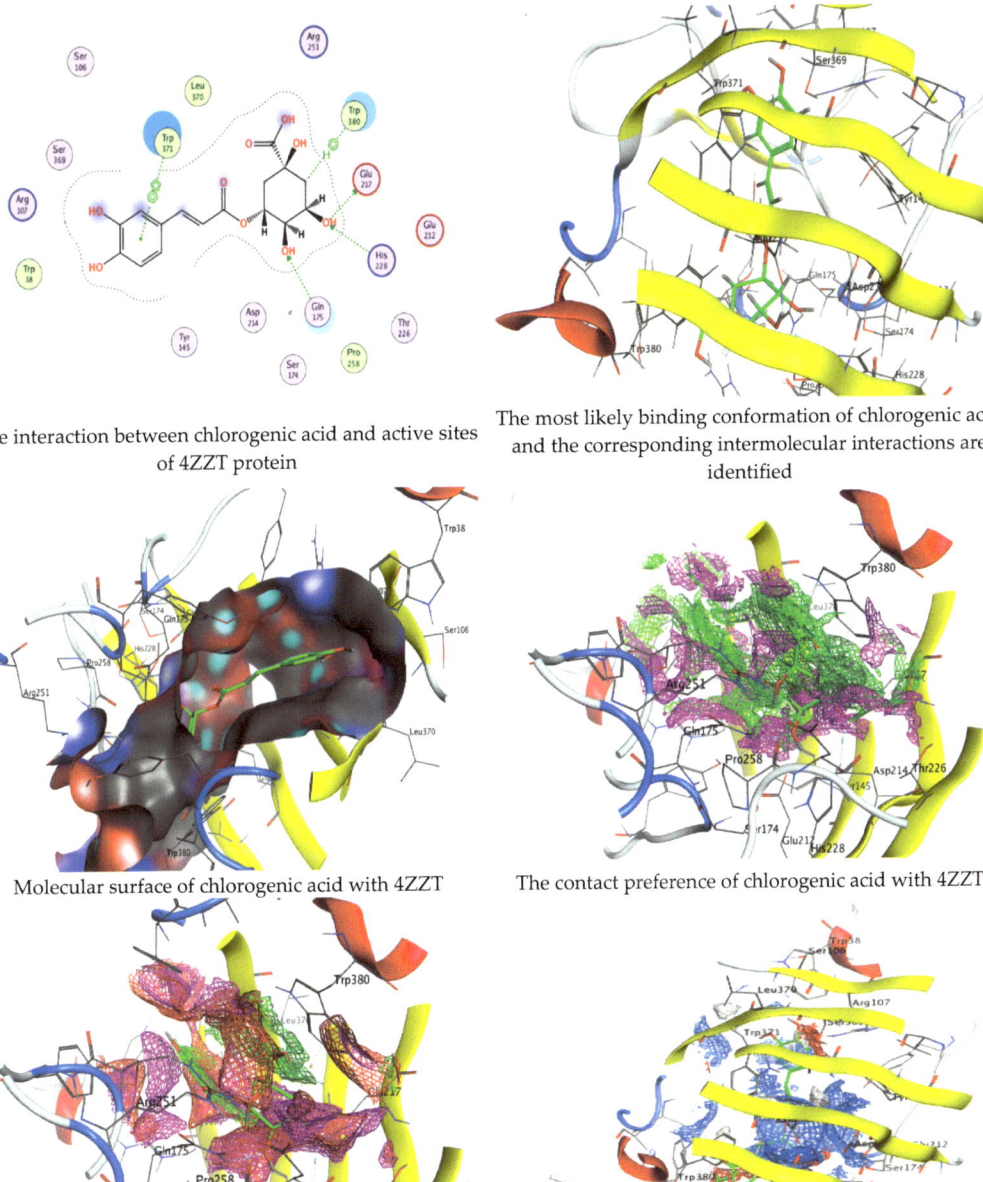

Figure 8. Molecular docking process of chlorogenic acid with the crystal structure of *G. candidum* (4ZZT).

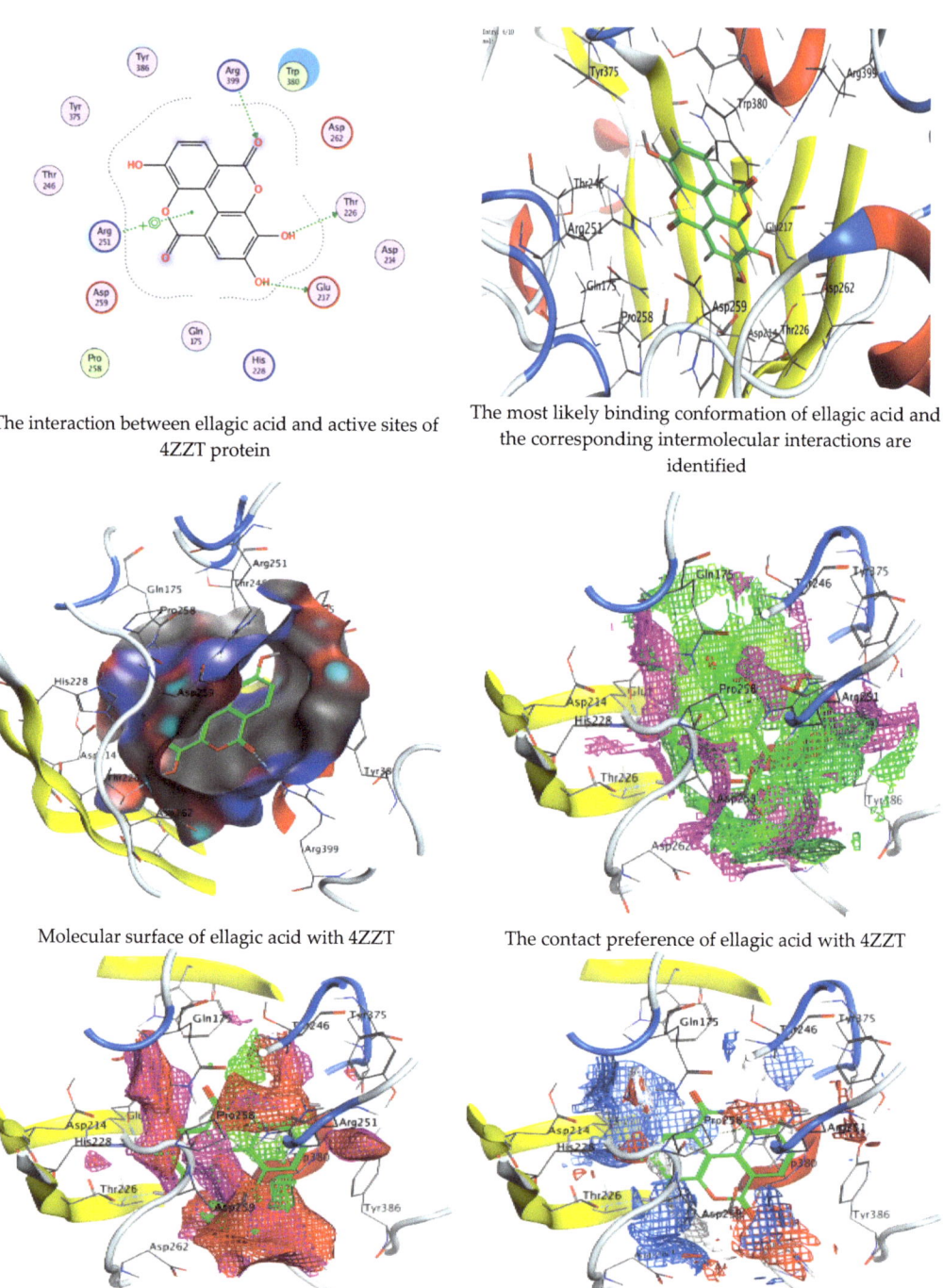

Figure 9. Molecular docking process of ellagic acid with the crystal structure of *G. candidum* (4ZZT).

Table 7. Interaction of chlorogenic acid and ellagic acid with the crystal structure of C. albicans (4YDE).

Mol	Ligand			Receptor			Interaction	Distance	E (kcal/mol)
Chlorogenic acid	C	26	OD1	ASP	527	(B)	H-donor	3.48	−0.5
	O	40	OD2	ASP	569	(B)	H-donor	2.94	−2.0
	O	23	NZ	LYS	512	(B)	H-acceptor	3.18	−0.8
Ellagic acid	O	20	OD1	ASP	527	(B)	H-donor	2.84	−5.0

Table 8. Interaction of chlorogenic acid and ellagic acid with the crystal structure of G. candidum (4ZZT).

Mol	Ligand			Receptor			Interaction	Distance	E (kcal/mol)
Chlorogenic acid	O	19	OE2	GLU	217	(A)	H-donor	2.80	−3.8
	O	17	NE2	GLN	175	(A)	H-acceptor	2.97	−1.1
	O	19	NE2	HIS	228	(A)	H-acceptor	3.24	−1.7
	C	1	6-ring	TRP	380	(A)	H-pi	3.91	−0.6
	6-ring		5-ring	TRP	371	(A)	pi-pi	3.61	−0.0
Ellagic acid	O	20	OE2	GLU	217	(A)	H-donor	2.67	−6.2
	O	22	OG1	THR	226	(A)	H-donor	3.20	−1.2
	O	24	NH1	ARG	399	(A)	H-acceptor	3.08	−2.4
	6-ring		NH2	ARG	251	(A)	pi-cation	4.09	−3.2

4. Conclusions

The obtained results indicated that pre-treated henna powder with moist heat provided undesirable findings, where several phenolic and flavonoid compounds were decreased as exposure to moist heat. Moreover, the anti-yeast, antioxidant, and healing properties of un-pretreated henna were acceptable compared to pretreated henna by moist heat. In the search for new bioactive analogues, chlorogenic acid and ellagic acid exhibited good potential for G. candidum and C. albicans inhibition. According to docking data, chlorogenic acid performed the best interaction with the crystal structure of G. candidum (4ZZT), and the results can be utilized to guide future experimental studies. This investigation recommended the application of henna without any cocked process.

Author Contributions: Investigation, methodology, supervision, S.A.A.; formal analysis, Writing—review and editing, M.I.A.; investigation, resources, M.J.; investigation, writing—original draft preparation, conceptualization, T.M.A. All authors have read and agreed to the published version of the manuscript.

Funding: This work was supported and funded by the Deanship of Scientific Research at Imam Mohammad Ibn Saud Islamic University (IMSIU) (grant number IMSIU-RP23038), Riyadh, Saudi Arabia.

Institutional Review Board Statement: Not applicable.

Informed Consent Statement: Not applicable.

Data Availability Statement: Not applicable.

Acknowledgments: The authors wish to appreciate the Deanship of Scientific Research at Imam Mohammad Ibn Saud Islamic University (IMSIU) for supporting and funding the current study (grant number IMSIU-RP23038).

Conflicts of Interest: The authors declare no conflict of interest.

References

1. Jothiprakasam, V.I.; Ramesh, S.A.; Rajasekharan, S.K. Preliminary phytochemical screening and antibacterial activity of *Lawsonia inermis* Linn (Henna) leaf extracts against reference bacterial strains and clinically important AMPC beta-lactamases producing Proteus mirabilis. *Int. J. Pharm. Pharm. Sci.* **2013**, *5*, 219–222.
2. Rahmoun, N.; Boucherit-Otmani, Z.; Boucherit, K.; Benabdallah, M.; Choukchou-Braham, N. Antifungal activity of the Algerian *Lawsonia inermis* (henna). *Pharm. Biol.* **2013**, *51*, 131–135. [CrossRef]
3. Yiğit, D. Antifungal activity of *Lawsonia inermis* L. (Henna) against clinical Candida isolates. *Erzincan Univ. J. Sci. Technol.* **2017**, *10*, 196–202.
4. Supian, F.N.A.; Osman, N.I. Phytochemical and Pharmacological Activities of Natural Dye Plant, *Lawsonia inermis* L. (Henna). *J. Young Pharm.* **2023**, *15*, 201–211. [CrossRef]
5. Coronado-Castellote, L.; Jimenez-Soriano, Y. Clinical and microbiological diagnosis of oral candidiasis. *J. Clin. Exp. Dent.* **2013**, *5*, e279–e286. [CrossRef] [PubMed]
6. Hemaid, A.S.S.; Abdelghany, M.M.E.; Abdelghany, T.M. Isolation and identification of Candida spp. from immunocompromised patients. *Bull. Natl. Res. Cent.* **2021**, *45*, 163. [CrossRef]
7. Badiee, P.; Alborzi, A.; Shakiba, E.; Farshad, S.; Japoni, A. Susceptibility of *Candida* species isolated from immunocompromised patients to antifungal agents. *East. Mediterr. Health J.* **2011**, *17*, 425–430. [CrossRef]
8. Samadi, F.M.; Suhail, S.; Sonam, M.; Sharma, N.; Singh, S.; Gupta, S.; Dobhal, A.; Pradhan, H. Antifungal efficacy of herbs. *J. Oral. Biol. Craniofacial Res.* **2019**, *9*, 28–32. [CrossRef]
9. Elansary, H.O.; Szopa, A.; Kubica, P.; Ekiert, H.; Al-Mana, F.A.; Al-Yafrsi, M.A. Antioxidant and biological activities of *Acacia saligna* and *Lawsonia inermis* natural populations. *Plants* **2020**, *9*, 908. [CrossRef]
10. Abirami, S.; Raj, B.E.; Soundarya, T.; Kannan, M.; Sugapriya, D.; Al-Dayan, N.; Mohammed, A.A. Exploring antifungal activities of acetone extract of selected Indian medicinal plants against human dermal fungal pathogens. *Saudi J. Biol. Sci.* **2021**, *28*, 2180–2187. [CrossRef]
11. Chowdhury, M.M.H.; Kubra, K.; Ahmed, S.R. Antimicrobial, Phytochemical and toxicological evaluation of *Lawsonia inermis* extracts against clinical isolates of pathogenic bacteria. *Res. J. Med. Plant* **2014**, *8*, 187–195. [CrossRef]
12. Khodja, Y.K.; Dahmoune, F.; Bey, M.B.; Madani, K.; Khettal, B. Conventional method and microwave drying kinetics of *Laurus nobilis* leaves: Effects on phenolic compounds and antioxidant activity. *Braz. J. Food Technol.* **2020**, *23*, e2019214. [CrossRef]
13. Al-Rajhi, A.M.H.; Qanash, H.; Almashjary, M.N.; Hazzazi, M.S.; Felemban, H.R.; Abdelghany, T.M. Anti-*Helicobacter pylori*, Antioxidant, Antidiabetic, and Anti-Alzheimer's Activities of Laurel Leaf Extract Treated by Moist Heat and Molecular Docking of Its Flavonoid Constituent, Naringenin, against Acetylcholinesterase and Butyrylcholinesterase. *Life* **2023**, *13*, 1512. [CrossRef] [PubMed]
14. Khoddami, A.; Wilkes, M.A.; Roberts, T.H. Techniques for Analysis of Plant Phenolic Compounds. *Molecules* **2013**, *18*, 2328–2375. [CrossRef] [PubMed]
15. Al-Rajhi, A.M.H.; Qanash, H.; Bazaid, A.S.; Binsaleh, N.K.; Abdelghany, T.M. Pharmacological Evaluation of *Acacia nilotica* Flower Extract against *Helicobacter pylori* and Human Hepatocellular Carcinoma In Vitro and In Silico. *J. Funct. Biomater.* **2023**, *14*, 237. [CrossRef] [PubMed]
16. Martinotti, S.; Ranzato, E. Scratch Wound Healing Assay. *Methods Mol. Biol.* **2020**, *2109*, 225–229. [CrossRef] [PubMed]
17. Juániz, I.; Ludwig, I.A.; Huarte, E.; Pereira-Caro, G.; Moreno-Rojas, J.M.; Cid, C.; De Peña, M.-P. Influence of heat treatment on antioxidant capacity and (poly)phenolic compounds of selected vegetables. *Food Chem.* **2016**, *197*, 466–473. [CrossRef]
18. Routray, W.; Orsat, V. Microwave-Assisted Extraction of Flavonoids: A Review. *Food Bioprocess. Technol.* **2012**, *5*, 409–424. [CrossRef]
19. Dezashibi, Z.; Samarin, A.M.; Hematyar, N.; Khodaparast, M.H. Phenolics in Henna: Extraction and stability. *Eur. J. Exp. Biol.* **2013**, *3*, 38–41.
20. Dobroslavić, E.; Repajić, M.; Dragović-Uzelac, V.; Garofulić, I.E. Isolation of *Laurus nobilis* Leaf Polyphenols: A Review on Current Techniques and Future Perspectives. *Foods* **2022**, *11*, 235. [CrossRef]
21. Lee, S.; Choi, Y.; Jeong, H.S.; Lee, J.; Sung, J. Effect of different cooking methods on the content of vitamins and true retention in selected vegetables. *Food Sci. Biotechnol.* **2017**, *27*, 333–342. [CrossRef] [PubMed]
22. Suleiman, E.A.; Mohamed, E.A. In vitro activity of *Lawsonia inermis* (Henna) on some pathogenic fungi. *J. Mycol.* **2014**, *2014*, 375932.
23. Soliman, S.S.M.; Semreen, M.H.; El-Keblawy, A.A.; Abdullah, A.; Uppuluri, P.; Ibrahim, A.S. Assessment of herbal drugs for promising anti-Candida activity. *BMC Complement. Altern. Med.* **2017**, *17*, 257. [CrossRef] [PubMed]
24. Kouadri, F. In vitro antibacterial and antifungal activities of the Saudi *Lawsonia inermis* extracts against some nosocomial infection pathogens. *J. Pure Appl. Microbiol.* **2018**, *12*, 281–286. [CrossRef]
25. Yaralizadeh, M.; Abedi, P.; Namjoyan, F.; Fatahinia, M.; Chegini, S.N. A comparison of the effects of *Lawsonia inermis* (Iranian henna) and clotrimazole on *Candida albicans* in rats. *J. Med. Mycol.* **2018**, *28*, 419–423. [CrossRef]
26. Daemi, A.; Farahpour, M.R.; Oryan, A.; Karimzadeh, S.; Tajer, E. Topical administration of hydroethanolic extract of *Lawsonia inermis* (henna) accelerates excisional wound healing process by reducing tissue inflammation and amplifying glucose uptake. *Kaohsiung J. Med. Sci.* **2019**, *35*, 24–32. [CrossRef] [PubMed]

27. Miraj, S.; Zibanejad, S.; Kopaei, M.R. Healing effect of *Quercus persica* and *Lawsonia inermis* ointment on episiotomy wounds in primiparous women. *J. Res. Med. Sci.* **2020**, *25*, 11. [CrossRef] [PubMed]
28. El Massoudi, S.; Zinedine, A.; Rocha, J.M.; Benidir, M.; Najjari, I.; El Ghadraoui, L.; Benjelloun, M.; Errachidi, F. Phenolic Composition and Wound Healing Potential Assessment of Moroccan Henna (*Lawsonia inermis*) Aqueous Extracts. *Cosmetics* **2023**, *10*, 92. [CrossRef]
29. Réblová, Z. Effect of temperature on the antioxidant activity of phenolic acids. *Czech J. Food Sci.* **2012**, *30*, 171–175. [CrossRef]
30. Kumar, M.; Kaur, P.; Chandel, M.; Singh, A.P.; Jain, A.; Kaur, S. Antioxidant and hepatoprotective potential of *Lawsonia inermis* L. leaves against 2-acetylaminofluorene induced hepatic damage in male Wistar rats. *BMC Complement. Altern. Med.* **2017**, *17*, 56. [CrossRef]
31. Çubukçu, H.C.; Kılıçaslan, N.S.D.; Durak, I. Different effects of heating and freezing treatments on the antioxidant properties of broccoli, cauliflower, garlic and onion. An experimental in vitro study. *Sao Paulo Med. J.* **2019**, *137*, 407–413. [CrossRef] [PubMed]
32. Qanash, H.; Alotaibi, K.; Aldarhami, A.; Bazaid, A.S.; Ganash, M.; Saeedi, N.H.; Ghany, T.A. Effectiveness of oil-based nanoemulsions with molecular docking of its antimicrobial potential. *BioResources* **2023**, *18*, 1554–1576. [CrossRef]
33. Dehghan, M.; Fathinejad, F.; Farzaei, M.H.; Barzegari, E. In silico unraveling of molecular anti-neurodegenerative profile of Citrus medica flavonoids against novel pharmaceutical targets. *Chem. Pap.* **2023**, *77*, 595–610. [CrossRef]
34. Al-Rajhi, A.M.H.; Yahya, R.; Abdelghany, T.M.; Fareid, M.A.; Mohamed, A.M.; Amin, B.H.; Masrahi, A.S. Anticancer, Anticoagulant, Antioxidant and Antimicrobial Activities of *Thevetia peruviana* Latex with Molecular Docking of Antimicrobial and Anticancer Activities. *Molecules* **2022**, *27*, 3165. [CrossRef]
35. Al-Rajhi, A.M.H.; Ghany, T.M.A. Nanoemulsions of some edible oils and their antimicrobial, antioxidant, and anti-hemolytic activities. *BioResources* **2023**, *18*, 1465–1481. [CrossRef]
36. Qanash, H.; Bazaid, A.S.; Aldarhami, A.; Alharbi, B.; Almashjary, M.N.; Hazzazi, M.S.; Felemban, H.R.; Abdelghany, T.M. Phytochemical Characterization and Efficacy of *Artemisia judaica* Extract Loaded Chitosan Nanoparticles as Inhibitors of Cancer Proliferation and Microbial Growth. *Polymers* **2023**, *15*, 391. [CrossRef] [PubMed]
37. Kavepour, N.; Bayati, M.; Rahimi, M.; Aliahmadi, A.; Ebrahimi, S.N. Optimization of aqueous extraction of henna leaves (*Lawsonia inermis* L.) and evaluation of biological activity by HPLC-based profiling and molecular docking techniques. *Chem. Eng. Res. Des.* **2023**, *195*, 332–343. [CrossRef]
38. McGeady, P.; Logan, D.A.; Wansley, D.L. A protein-farnesyl transferase inhibitor interferes with the serum-induced conversion of *Candida albicans* from a cellular yeast form to a filamentous form. *FEMS Microbiol. Lett.* **2002**, *213*, 41–44. [CrossRef]

Disclaimer/Publisher's Note: The statements, opinions and data contained in all publications are solely those of the individual author(s) and contributor(s) and not of MDPI and/or the editor(s). MDPI and/or the editor(s) disclaim responsibility for any injury to people or property resulting from any ideas, methods, instructions or products referred to in the content.

Review

Comprehensive Overview of the Effects of *Amaranthus* and *Abelmoschus esculentus* on Markers of Oxidative Stress in Diabetes Mellitus

Wendy N. Phoswa *[ID] and Kabelo Mokgalaboni [ID]

Department of Life and Consumer Sciences, University of South Africa (UNISA), Science Campus, Private Bag X6, Florida, Roodepoort 1710, South Africa; mokgak@unisa.ac.za
* Correspondence: phoswwn@unisa.ac.za

Abstract: The use of medicinal plants in the management of diabetes mellitus (DM) is extensively reported. However, there is still very limited information on the role of these plants as markers of oxidative stress in DM. This current review evaluated the effect of *Amaranthus spinosus*, *Amaranthus hybridus*, and *Abelmoschus esculentus* on markers of oxidative stress in rodent models of DM. Current findings indicate that these plants have the potential to reduce prominent markers of oxidative stress, such as serum malondialdehyde and thiobarbituric acid-reactive substances, while increasing enzymes that act as antioxidants, such as superoxide dismutase, catalase, glutathione, and glutathione peroxidase. This may reduce reactive oxygen species and further ameliorate oxidative stress in DM. Although the potential benefits of these plants are acknowledged in rodent models, there is still a lack of evidence showing their efficacy against oxidative stress in diabetic patients. Therefore, we recommend future clinical studies in DM populations, particularly in Africa, to evaluate the potential effects of these plants. Such studies would contribute to enhancing our understanding of the significance of incorporating these plants into dietary practices for the prevention and management of DM.

Keywords: antioxidants; diabetes mellitus; *Amaranthus*; *Abelmoschus esculentus*; oxidative stress

Citation: Phoswa, W.N.; Mokgalaboni, K. Comprehensive Overview of the Effects of *Amaranthus* and *Abelmoschus esculentus* on Markers of Oxidative Stress in Diabetes Mellitus. *Life* **2023**, *13*, 1830. https://doi.org/10.3390/life13091830

Academic Editor: Seung Ho Lee

Received: 1 August 2023
Revised: 25 August 2023
Accepted: 28 August 2023
Published: 29 August 2023

Copyright: © 2023 by the authors. Licensee MDPI, Basel, Switzerland. This article is an open access article distributed under the terms and conditions of the Creative Commons Attribution (CC BY) license (https://creativecommons.org/licenses/by/4.0/).

1. Introduction

Diabetes mellitus (DM) is a chronic, life-threatening disease that has caused more than 6.7 million deaths worldwide [1]. This condition affects approximately 537 million adults (20–79 years old) [1]. According to the International Diabetes Federation, the number of people living with diabetes is predicted to reach 643 million by 2030 [1]. There are three commonly known types of DM: type 1 diabetes mellitus (T1DM), type 2 diabetes mellitus (T2DM), and gestational diabetes mellitus (GDM). T2DM is considered the most common, and it affects [2] at least 95% of the diabetic population [3]. In 2021, the prevalence of T2DM was reported to be 10.5% [4], and most of the population is from low- and middle-income countries (LMICs) [1].

Diabetic patients in LMICs face many challenges, which include a lack of awareness and knowledge about the disease, difficulty accessing health care systems, including medications, and inadequate diabetes management strategies, which likely result from a poor socio-economic background [5]. Biological risk factors associated with DM include older age, increased body mass index (BMI), obesity, stress, physical inactivity, and chronic inflammation due to other infectious diseases [6–8]. DM is associated with health complications such as cardiovascular diseases, kidney diseases, vision impairment, and neurological conditions [9]. Oxidative stress has been reported to play a major role in the pathophysiology of DM-related complications [10].

In 2021, it was documented that DM caused at least 966 billion USD in health expenditure, with 9% of total spending on adults [1]. Low- and middle-income countries already have overwhelming health burdens resulting from other common diseases such as tuberculosis and the human immunodeficiency virus [11]. Therefore, healthcare systems in place must implement efficacious medicines that are less toxic and cost-effective in the management of DM.

While medical and pharmacological drugs are currently available for managing DM, these are still associated with severe side effects in different individuals, increased complications, and a rising mortality rate in DM. For instance, using sodium-glucose cotransporter 2 inhibitors increases the risk of hypotension, diabetic ketoacidosis, kidney injury, and bone fractures [12]. Regrettably, prolonged use of glucophage is linked to cobalamin deficiency, increasing the risk of additional complications, including anemia, in T2DM patients [13]. Given the mentioned drawbacks of pharmaceutical medications, there has been a burgeoning interest in the utilization of functional foods and herbal remedies for the treatment and management of DM. This interest is partly attributable to their inherent properties. Numerous studies have highlighted the antioxidant characteristics of medicinal plants in the treatment and management of DM.

For instance, our team recently found the potential beneficial effects of *Corchorus olitorius* and curcumin in a rodent model of obesity and DM and T2DM, respectively [14,15].

Interestingly, functional fruits have also demonstrated potential benefits, especially on oxidative stress in DM [16]. Although there are rising calls for more research support for medicinal plant use in the treatment and management of DM [11], there is still limited clinical evidence to support their efficacy, especially in DM. Moreover, evidence from previous preclinical studies has not focused on common markers of oxidative stress. In this study, we aim to gather evidence from preclinical studies evaluating the effect of *Amaranthus hybridus*, *Amaranthus spinosus*, and *Abelmoschus esculentus* in DM primarily due to their beneficial properties and safety profile [17], with the main focus on various biomarkers of oxidative stress. Therefore, this review will highlight and document the potential benefits of these selected medicinal plants in DM.

2. Oxidative Stress and Diabetes Mellitus

Oxidative stress occurs as a result of an imbalance between the production and clearance of reactive oxygen species (ROS) [18] and contributes to the pathogenesis and pathophysiology of DM [19,20]. DM is a metabolic disorder characterized by increased blood glucose levels resulting from insulin resistance and impaired insulin secretion [21].

In DM, several factors contribute to oxidative stress, including hyperglycemia, dyslipidemia, insulin resistance, and inflammation. Hyperglycemia and hyperlipidemia can lead to increased cellular oxidative stress through mitochondrial electron leak or incomplete fatty acid oxidation, the formation of advanced glycation end products (AGEs), lipid hydroperoxides, and induced free fatty acids (FFA), diacylglycerol (DAG), and ceramides. Lipid peroxidation has been identified as one factor that leads to DM development [22,23]. This occurs when there is uncontrolled high blood glucose and free fatty acids, which in turn activate DAG and protein kinase C (PKC) [24–26]. Activation of PKC induces the inflammatory response by promoting the secretion of endothelin 1 (ET-1), vascular cell adhesion molecule (VCAM-1), intercellular adhesion molecule (ICAM-1), nuclear factor kappa-light chain enhancer of activated β cells (NF-κβ), and NADPH oxidase [27–29]. Notably, NADPH oxidase activation mediates ROS generation through superoxide [30,31]. Excessive ROS production damages cells, resulting in a pronounced inflammatory response [32]. Hence, it is comprehensible why diabetes is frequently linked to inflammation [32–35]. Conversely, catalase (CAT), an active enzyme, functions as an antioxidant by catalyzing the conversion of hydrogen peroxide into water and oxygen [36]. Nevertheless, diminished CAT activity results in oxidative stress in the pancreatic beta cells, which contain numerous mitochondria. This excess production of reactive oxygen species (ROS) ultimately leads to dysfunction in β-cells and the onset of diabetes [18]. Therefore, this would subject the cells or organs to oxidative stress by allowing the accumulation of harmful oxidants and free radicals. Reactive oxygen species can also increase insulin resistance, leading to further hyperglycemia [37–41]. This occurs when caloric intake exceeds energy expenditure, thereby causing an increase in citric acid cycle activity. This subsequently leads to excess mitochondrial NADH (mNADH) and ROS [42]. Inflammation, which is common in DM, can also contribute to oxidative stress by activating immune cells that produce ROS [43]. Several studies have indicated that oxidative stress can contribute to the development of diabetic complications such as neuropathy, retinopathy, and nephropathy [43–46]. Therefore, controlling oxidative stress may be an important DM management strategy [44]. This can be achieved through lifestyle modifications such as regular exercise, a healthy diet, and smoking cessation [47]. Antioxidant supplements such as vitamin C, vitamin E, and alpha-lipoic acid may also be beneficial in reducing oxidative stress [48]. Some medicinal plants, such as *Amaranthus spinosus, Amaranthus hybridus,* and *Abelmoschus esculentus,* have been shown to have antioxidant effects.

3. *Amaranthus* Species

Amaranthus, a herbaceous plant native to Central America, has been cultivated for centuries due to its valuable properties [49]. It has spread to various nations, successfully established itself, and naturalized in numerous regions across the globe [50]. In Africa, it is esteemed as a traditional food plant, thus providing a valuable source of nutrition. All parts of the plant, including seeds, roots, leaves, and stems, are recognized for their edible and medicinal attributes [51]. In addition, *Amaranthus* is affordable and cost-effective [52], making it significant in improving nutrition, ensuring food security, and alleviating poverty, especially in low- and middle-income countries [53]. *Amaranthus* has been recognized as a superfood, making it an interesting plant for further exploration in research [54]. Among the various *Amaranthus* species, the common ones include *Amaranthus thunbergii, A. greazican, A. deflexus, A. hypochondriacus, A. viridis, A. spinosus,* and *A. hybridus*. This pivotal plant possesses an abundance of vital nutrients, making it a rich source of essential nutrients, including vitamins and minerals [54]. These micronutrients have been extensively studied for their crucial role in promoting optimal well-being. More interestingly, *Amaranthus* has been reported to possess various compounds, including amino acids such as lysine, arginine, histidine, leucine, cysteine, phenylalanine, isoleucine, valine, threonine, and methionine [55].

The *Amaranthus* plant also contains important active compounds that promote its activities [56–58] (Figure 1). The use of *Amaranthus* in the public interest extends beyond its nutritional benefits to its therapeutic properties, especially in the management of cholesterol and blood glucose levels in DM [59,60]. While many studies reporting on the therapeutic properties of *Amaranthus* in DM have focused on inflammatory markers [61–63], only a limited number of studies have explored its benefits in alleviating oxidative stress in DM. The present study will review studies reporting the potential benefits of *Amaranthus spinosus* and *hybridus* on markers of oxidative stress in DM. This will help to improve current knowledge on the importance of consuming these plants, especially in diabetic individuals, in order to manage the condition and in non-diabetic individuals to prevent the risk of developing DM.

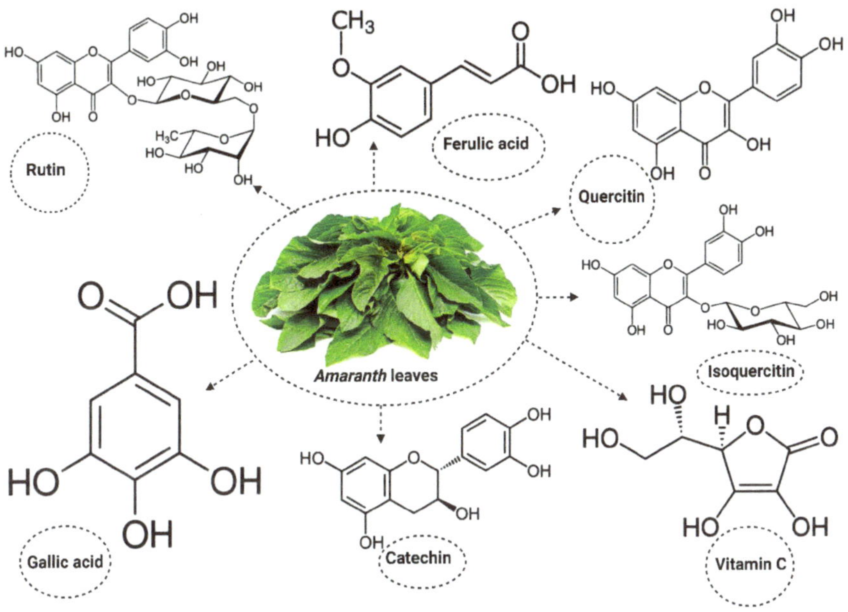

Figure 1. Some of the active compounds present in *Amaranthus*.

3.1. Amaranthus spinosus

As presented in Table 1, *Amaranthus spinosus* has been shown to alleviate oxidative stress. *Amaranthus spinosus* has been shown to reduce hyperglycemia-associated oxidative stress significantly in rodent models of obesity [64]. A study by Kumar et al. (2011) reported that methanol extract of *Amaranthus spinosus* at a dose of 200 and 400 mg/kg for 15 days had antioxidant effects against DM [65]. This was demonstrated by a reduction in the serum level of malondialdehyde (MDA) concomitant with the increased activity of enzymes that act as antioxidants, such as glutathione (GSH) and catalase (CAT), in alloxan-induced diabetic rats. Furthermore, the same study showed an increase in total thiols. Similar findings were observed by Mishra et al. (2012), whereby *Amaranthus spinosus* leaf extract (ASEt) at doses of 250 and 500 mg/kg for 21 days reduced oxidative stress and improved pancreatic cell function in diabetic rats [66]. These positive impacts were demonstrated by a significant increase in superoxide dismutase (SOD), CAT, GSH, and glutathione peroxidase (GPx). An increase in antioxidant enzymes observed after *Amaranthus* treatment shows its potential as an antioxidant remedy (Figure 2). Therefore, this suggests that this plant could play an important role in alleviating oxidative stress.

Table 1. Overview of studies evaluating the antioxidant effect of *Amaranthus* rodents in a model of diabetes mellitus.

Experimental Model	Treatment and Duration	Experimental Outcomes	Country	Reference
Alloxan-induced diabetic albino Wistar rats.	Methanol extract of *Amaranthus spinosus* (MEAS) leaves. Oral administration of MEAS (200 and 400 mg/kg) for 15 days.	*Amaranthus spinosus* in diabetic rats significantly reduced malondialdehyde (MDA) and increased glutathione (GSH), catalase (CAT), and total thiols (TT) compared to diabetic control.	India	[65]
Streptozotocin (STZ)-induced diabetes in Wistar rats.	Ethanolic extract of *Amaranthus spinosus* leaves. Intraperitoneal injection of *Amaranthus* at doses (250 and 500 mg/kg) for 21 days.	*Amaranthus spinosus* in diabetic rats significantly increased GSH, superoxide dismutase (SOD), CAT, and glutathione peroxidase (GPx) compared to the diabetic control.	India	[66]
STZ-induced diabetic albino Wistar rats.	Ethanol extract of *Amaranthus hybridus* leaves. Oral administration of AHELE at doses (200 and 400 mg/kg) for 14 days.	*Amaranthus hybridus* in diabetic rats significantly reduced thiobarbituric acid reactive substances (TBARS) and increased SOD and CAT activity compared to diabetic control.	India	[67]

MEAS: methanol extract of *Amaranthus spinosus*, STZ: streptozotocin, SOD: superoxide dismutase, MDA: malondialdehyde, CAT: catalase, GSH: glutathione, GPx: glutathione peroxidase, IC50: inhibitory concentration 50, TT: total thiols, TBARS: thiobarbituric acid reactive substances, AHELE: *Amaranthus hybridus* ethanol leaf extract.

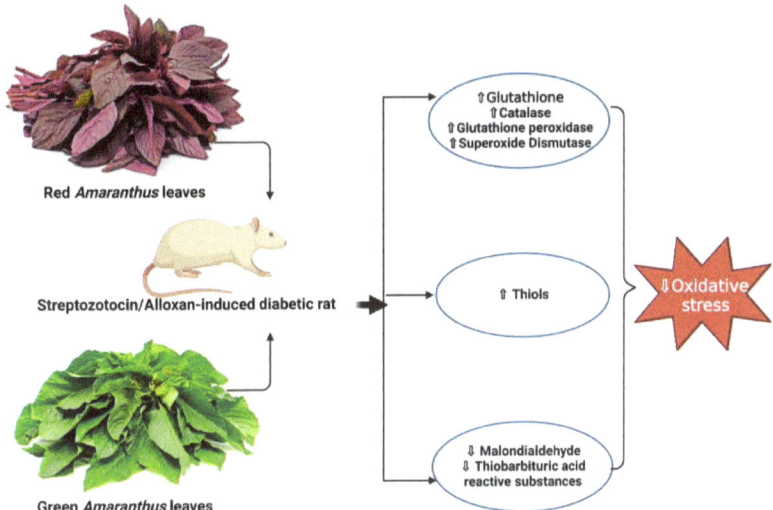

↑ = increase and ↓ = decrease

Figure 2. Overview showing the impact of *Amaranthus* on oxidative stress in diabetic rats. Administration of *Amaranthus* in diabetic rats ameliorates oxidative stress by reducing malondialdehydes and thiobarbituric acid while increasing the activity of antioxidant enzymes such as glutathione, glutathione peroxidase, catalase, and superoxide dismutase. (https://www.seeds-gallery.eu/9135-large_default/amaranth-red-garnet-seeds-Amaranthus-tricolor.jpg (accessed on 25 July 2023), https://specialtyproduce.com/produce/Green_Amaranth_12831.php (accessed on 25 July 2023).

3.2. Amaranthus hybridus

Similarly, *Amaranthus hybridus* has been reported to have anti-oxidative stress effects in DM [67]. *Amaranthus hybridus* ethanol leaf extract (AHELE) has been identified to have a nephron protective effect against oxidative damage in streptozotocin (STZ)-induced diabetic rats [68]. In a study by Balasubramanian et al. (2016), this was indicated by a significant reduction in the marker of lipid peroxidation, thiobarbituric acid reactive substances (TBARS) ($p < 0.001$), and a significant increase in antioxidant markers such as superoxide dismutase (SOD) ($p < 0.001$) and CAT ($p < 0.01$). In support of their findings, they also discovered that AHELE possessed both nephroprotective and hepatoprotective effects by reducing the levels of MDA in the liver and kidneys of STZ-induced diabetic rats [68]. Therefore, the evidence from rat models of diabetes induced by STZ or alloxan indicates the potential of the *Amaranthus* plant as an anti-oxidative stress agent (Figure 2, Table 1). MDA is an end-product of fatty acid peroxidation [69], and its high level is an indication of lipid peroxidation [70]. During lipid peroxidation, lipids are degraded in the cell membrane, thus leading to cell damage [69,71]. Lipid peroxidation has been identified as one of the factors leading to the development of DM [22,23].

4. *Abelmoschus esculentus*

Abelmoschus esculentus L. is also known as okra and belongs to the Malvaceae plant family. Although this plant is found in Africa, it is widely distributed in Asia, America, and Southern Europe [72]. The fruit, seeds, roots, leaves, flowers, and pods of *Abelmoschus esculentus* contain vital bioactive chemicals that contribute to this plant's beneficial effects [73,74] (Figure 3). For example, the seed contains oligomeric catechins and flavonol derivatives [74–76]. The root contains carbohydrates and flavonol glycosides [77], while the leaves contain minerals, tannins, and flavonol glycosides [78]. The pod contains carotene, folic acid, thiamine, riboflavin, protein, fiber, calcium, iron, zinc, niacin, vitamin C, oxalic acid, and amino acids [79–81]. These compounds mediate the various functions that *Abelmoschus esculentus* possesses. Such potential benefits include but are not limited to anti-hyperglycemic, anti-inflammatory, and antioxidant effects [82–84].

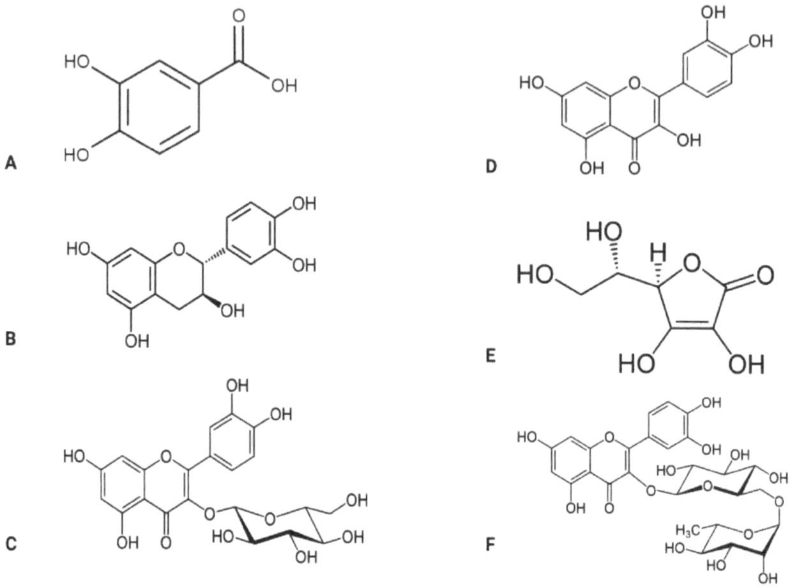

Figure 3. Active compounds found in the *Abelmoschus esculentus* plant. (**A**) protocatechuic acid, (**B**,**C**) catechin, (**D**) quercetin, (**E**) vitamin C, (**F**) rutin.

4.1. Antioxidant Effect of Abelmoschus esculentus

The antioxidant properties of *Abelmoschus esculentus* have been revealed by previous research [83–86], and this is attributable mainly to its active compounds, such as polyphenols and flavonoids [87] (Figure 3). Polyphenols mediate antioxidant activity by reducing MDA and increasing SOD, GPx, and catalase activity [77,78,82,84–93]. Some of the active polyphenol compounds include isoquercetin, quercetin, quercetin-3-O-gentiobioside, quercetin-3-O-glucoside, protocatechuic acid, and rutin. All these phenolic compounds exhibit free radical scavenging and ferric-reducing properties [92] and further inhibit the activities of α-glucosidase and α-amylase [94]. Specifically, the *Abelmoschus esculentus* seeds are excellent sources of phenols, including procyanidin B1 and B2, which facilitate the free radical scavenging activities of 1,1-diphenyl-2-picrylhydrazyl (DPPH) and 2,2′-casino-bis (3-ethylbenzothi azoline-6-sulfonic acid (ABTS) [90,95,96].

4.1.1. Effect of *Abelmoschus esculentus* on Oxidative Stress in Animal Models of Diabetes Mellitus

Mice and rats have been used for decades to mimic DM observed in humans, primarily to explore the beneficial effects, toxicity, and desirable doses of different compounds against various metabolic disorders [97]. Oxidative stress is implicated in the progression of insulin resistance into DM due to the increased production of free radical molecules. These ROS molecules (hydrogen peroxide, superoxide anion, and hydroxyl radicals) are generated by the partial reduction of oxygen molecules [98]. However, when these molecules are excessively produced in the body, they cause damage to cellular proteins, membrane lipids, and nucleic acids and reduce lifespan [99,100].

Several biomarkers have been widely considered predictors of oxidative stress; these include SOD, MDA, CAT, GPx, and GSH. The existing research suggests that *Abelmoschus esculentus* has antioxidant potential, partly due to its high phenolic and flavonoid content. For instance, two groups recently demonstrated a very high content of phenols and flavonoids in *Abelmoschus esculentus* mucilage and seed peel, respectively [101,102]. In contrast, other studies showed an increased half-maximal inhibitory concentration (IC50) in *Abelmoschus esculentus* extracts and mucilage compared to vitamin C [101,103]. This suggests the reduced antioxidant capability of *Abelmoschus esculentus*. IC50 refers to the number of antioxidant compounds necessary to scavenge 50% of the initial DPPH radicals. The compound's increased effectiveness in scavenging DPPH radicals results in a reduced IC50 value, indicating an ideal level of antioxidant activity for the compound [104]. The summarized effect of *Abelmoschus esculentus* in rodent models of diabetes is presented in Table 2. Although the potential benefits are acknowledged, different studies have contradictory findings, with others showing negative results on oxidative stress markers. Below, we outline the effects of *Abelmoschus esculentus* on various markers of oxidative stress.

4.1.2. *Abelmoschus esculentus* on Oxidative Stress with a Focus on ROS

In general, the evidence supports using *Abelmoschus esculentus* treatment as a possible antioxidant in at least three preclinical studies (Table 2). These studies revealed a significant ($p < 0.05$) reduction in the levels of ROS in a rodent model of DM and, therefore, an attenuation of oxidative stress [105–107]. Elevated ROS levels are associated with oxidative stress and organ damage [18]. Therefore, the potential of *Abelmoschus esculentus* to reduce these excess ROS may be of importance in reducing organ and tissue damage in diabetes mellitus. *Abelmoschus esculentus* potential to reduce ROS is associated with its high polyphenols, flavonoids, and vitamin C content as they scavenge free radical molecules, thus alleviating oxidative stress [79–81]. These compounds accomplish this activity by transferring hydrogen atoms to unstable ROS molecules, stabilizing them, and subsequently preventing any cell, tissue, or organ damage [108,109]. Similarly, *Abelmoschus esculentus* is rich in quercetin and catechin, which have antioxidant properties [74–76]. These compounds reduce NADPH oxidase activity, reducing ROS production and further oxidative stress [110,111].

4.1.3. *Abelmoschus esculentus* on Oxidative Stress with a Focus on SOD

SOD is an antioxidant enzyme that protects cells against ROS if it is upregulated in the body [112]. In a model of DM, the central feature is oxidative stress, which exacerbates the condition. The administration of antioxidants, however, seems important as they reduce oxidative stress by increasing SOD levels. For example, as presented in Table 2, *Abelmoschus esculentus* is significantly ($p < 0.05$) associated with an increased SOD [96,106,113,114]. SOD is a crucial antioxidant that defends the body's organs and cells from oxidative damage. Therefore, an increase in SOD in the body is important to help break down potentially harmful oxygen molecules in cells. An improvement of antioxidant status by *Abelmoschus esculentus* extract, as demonstrated by a significant increase in SOD, is commendable and thus may be relevant to attenuating oxidative stress (Figure 4). Such an effect of okra may further lead to ameliorating secondary complications associated with oxidative stress. However, other studies showed a significant ($p < 0.05$) decrease in SOD following the administration of *Abelmoschus esculentus* [91,102,115,116]. This suggests the limitation of *Abelmoschus esculentus* as an antioxidant; reduction of SOD in hyperglycemia or overproduction of ROS may subject the cells to damage and subsequently to apoptosis.

4.1.4. *Abelmoschus esculentus* on Oxidative Stress with a Focus on CAT Activity

CAT is an active enzyme involved in the catalysis of hydrogen peroxide into water and oxygen [36]. However, due to reduced CAT activity, beta cells of the pancreas that contain many mitochondria undergo oxidative stress by producing excess ROS that leads to β-cells dysfunction and, ultimately, diabetes [18]. Interestingly, evidence presented in Table 2 showed that *Abelmoschus esculentus* treatment in rodent models of diabetes significantly increases CAT activity [91,95,105,106,113,115,117]. This suggests that *Abelmoschus esculentus* may ameliorate oxidative stress, further reduce complications of DM, or prevent DM in non-diabetics (Figure 4). Although the potential benefits of *Abelmoschus esculentus* on CAT activity have been noted in DM rodent models, another group of researchers has reported contradictory findings, as shown by significantly reduced CAT activity [116]. This, disappointingly, suggests a limited beneficial impact of *Abelmoschus esculentus* in improving the activity of CAT enzymes in diabetic models and, thus, its limited efficacy in reducing oxidative stress.

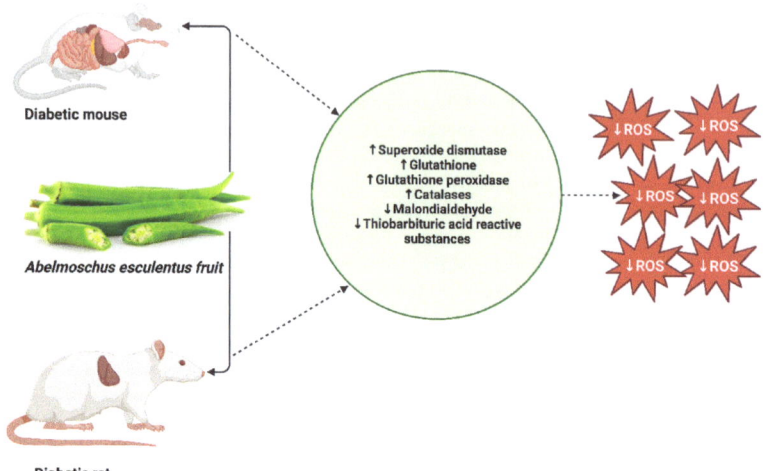

Figure 4. Potential mode of action of *Abelmoschus esculentus* in rodent models of diabetes. In a nutshell, oral administration of *Abelmoschus esculentus* ameliorates oxidative stress in rodent models of diabetes induced by either alloxan monohydrate or a high-fat diet coupled with streptozotocin. ROS: reactive oxygen species (https://prove.es/en/unknown-but-really-healthy-vegetable-called-okra/ (accessed on 25 July 2023)).

4.1.5. Effects of *Abelmoschus esculentus* on Oxidative Stress: Focusing on MDA and TBARS

Additional oxidative stress markers include MDA, TBARS, lipid hydroperoxides (LH), and 4-hydroxy-2-Nonenal (4-HNE), the end products of lipid peroxidation. However, among these markers, MDA is widely studied and is regarded as an ideal marker of oxidative stress [118]. MDA seems to be elevated in DM, thus increasing the likelihood of developing complications associated with oxidative stress. It is assumed that the reduction of these markers can substantially alleviate oxidative stress and associated complications. Interestingly, preclinical evidence gathered in Table 2 showed *Abelmoschus esculentus* promising potential in reducing MDA amongst rodent models of DM [91,95,105,106,113,115,117]. Disappointingly, evidence from a preclinical model of GDM reported by Tian [113] showed different findings, which suggest the limitation of *Abelmoschus esculentus* in alleviating oxidative stress in GDM. Although the results were unfavorable, we believe this is due to insulin resistance occurring in the late stage of pregnancy and different pathophysiological mechanisms between GDM and DM, thus limiting the antioxidant potential of *Abelmoschus esculentus* [119]. Consistently, TBARS, a derivative of thiobarbituric acid and MDA, increases in response to oxidative stress [120]. Notably, an increased level of TBARS is observed in T1DM and T2DM, signifying oxalate toxicity induced by lipid peroxidation [121,122]. However, existing evidence showed a significant decrease in TBARS levels following *Abelmoschus esculentus* treatment in rodent models of DM [95,107]. Therefore, this reduction would suggest *Abelmoschus esculentus* potential for attenuating oxalate toxicity and a further reduction in oxidative stress.

Table 2. Overview of studies evaluating the antioxidant effect of *Abelmoschus esculentus* in a rodent model of diabetes mellitus.

Experimental Model	Treatment and Duration	Experimental Outcomes	Country	Reference
Alloxan monohydrate induced diabetes in Swiss albino female and male mice.	The suspension was prepared by dissolving the powdered peel seed (PPS) and *Abelmoschus esculentus* mucilage (PM) of *Abelmoschus esculentus* into distilled water. PPS and PM were administered orally at 150 and 200 mg/kg body weight for three weeks.	Total flavonoid content was higher in PPS than in PM. At the same time, the total phenol content was higher in PM than PPS. *Abelmoschus esculentus* had reduced antioxidant capabilities, as shown by a higher IC50 value in *Abelmoschus esculentus* PM and PPS compared to vitamin C.	Bangladesh	[101]
High-fat diet (HFD)-streptozotocin (STZ)-induced diabetes in SPF-grade C57BL/6 male mice.	*Abelmoschus esculentus* powder was isolated using distilled water and 80% ethanol precipitation from *Abelmoschus esculentus* and orally administered at 200 and 400 mg/kg of *Abelmoschus esculentus* powder for eight weeks.	*Abelmoschus esculentus* in diabetic mice reduced the level of reactive oxygen species (ROS) in diabetic mice compared to mice in the control group. A dose of 400 *Abelmoschus esculentus* powder in diabetic mice significantly increased superoxide dismutase (SOD), glutathione (GSH), and catalase (CAT) and decreased malondialdehyde (MDA) in the kidney when compared to the control group. The same pattern was observed at 200; however, this was not significantly different from diabetic controls.	China	[106]
HFD-fed-specific pathogen-free (SPF)-grade C57BL/6 male mice administered with STZ to induce diabetes.	*Abelmoschus esculentus* powder was dissolved in distilled water, and 200 or 400 mg/kg of body weight was orally administered for eight weeks.	*Abelmoschus esculentus* powder in diabetic mice decreased ROS and MDA and increased SOD, glutathione peroxidase (GPx), and CAT in the liver compared to the diabetic control. Nuclear factor erythroid 2–related factor 2 (Nrf2), heme oxygenase-1 (HO-1), and superoxide dismutase 2 (SOD2) expression were significantly upregulated by *Abelmoschus esculentus* powder.	China	[105]
Streptozotocin-induced diabetes in male Wistar albino rats.	*Abelmoschus esculentus* peel powder (AEPP) and *Abelmoschus esculentus* seed powder (AESP) were mixed with distilled water and administered orally at a dose of (100 or 200 mg/kg bw) for 28 days.	Administration of both doses of AEPP and AESP in diabetic rats significantly increased liver, kidney, and pancreatic SOD, CAT, GPx, and GSH levels compared to the diabetic controls.	India	[95]
STZ-induced diabetes in male Wistar rats.	*Abelmoschus esculentus* mucilage extract and *Abelmoschus esculentus* seed aqueous solution were made by dissolving the powder in ethanol. The extract was orally administered at 150 and 200 mg/kg of body weight for 30 days.	*Abelmoschus esculentus* in diabetic rats significantly decreased MDA and increased catalase, SOD, and GSH activity compared to diabetic controls.	Saudi Arabia	[115]

Table 2. Cont.

Experimental Model	Treatment and Duration	Experimental Outcomes	Country	Reference
STZ-induced diabetes in male Wistar rats.	About 50 g of *Abelmoschus esculentus* powder was used to make the following extracts: (ethanolic extract 75%, ethanolic extract 90%, aqueous extract, and ethyl acetate extract). *Abelmoschus esculentus* extracts were orally administered at 200 and 400 mg/kg doses for eight weeks.	*Abelmoschus esculentus* extracts had reduced antioxidant properties, as revealed by an increased half-maximal inhibitory concentration (IC50) level compared to vitamin C and quercetin.	Iran	[103]
STZ-induced diabetes mellitus in adult Wistar rats.	*Abelmoschus esculentus* powder was mixed with food and given as a food pellet to the rats.	*Abelmoschus esculentus*-mixed diets in diabetic rats significantly reduced SOD, CAT, GSH, MDA, and α-amylase. However, GPx in *Abelmoschus esculentus* was not significantly different from diabetic controls.	Nigeria	[116]
Alloxan-induced diabetes in female and male Wistar rats.	Whole *Abelmoschus esculentus* (WAE), *Abelmoschus esculentus* peel (AEP), and *Abelmoschus esculentus* seed (AES). *Abelmoschus esculentus* samples (WAE, AEP, and AES) were administered at 100, 200, and 300 mg/kg by single forced oral feeding once daily for 21 days.	*Abelmoschus esculentus* in diabetic rats significantly increased CAT activity and GSH and decreased MDA compared to diabetic controls.	Nigeria	[117]
HFD fed Sprague–Dawley male rats were injected with STZ to induce diabetes.	*Abelmoschus esculentus* subfractions (F1, F2, and FR), the ethanol-extracted subfraction, and distilled water residue of *Abelmoschus esculentus* oral feed for 12 weeks.	*Abelmoschus esculentus* in diabetic rats has benefits in regulating dipeptidyl peptidase-4 (DPP-4) and the glucagon-like peptide 1 receptor (GLP-1R), thus reducing oxidative stress. *Abelmoschus esculentus* in diabetic rats significantly decreased serum and kidney TBARS compared to diabetic controls. All extracts lowered peroxidation except for fraction 1.	Taiwan	[107]
Alloxan-induced diabetes in male Wistar strain rats.	*Abelmoschus esculentus* powder (2 g/kg rat body weight) was mixed well in 0.2% carboxy methyl cellulose (CMC) and fed to animals by gavage technique at 2 g/kg body weight for 35 days.	*Abelmoschus esculentus* in diabetic rats significantly increased lipid peroxidation in erythrocytes, GSH, and decreased MDA in the kidney compared to diabetic controls.	India	[123]
Alloxan monohydrate induced diabetes in male and female Wistar rats.	The animal diet was prepared by mixing 33 g of the *Abelmoschus esculentus* powder with 67 g of normal rat feed, and 66 g of *Abelmoschus esculentus* was mixed with 34 g of normal rat feed to obtain the 33% and 66% supplement ratios and fed to rats for 16 days.	*Abelmoschus esculentus* powder in diabetic rats significantly decreased SOD and MDA levels and increased GSH and CAT activity compared to diabetic controls.	Nigeria	[91]

Table 2. *Cont.*

Experimental Model	Treatment and Duration	Experimental Outcomes	Country	Reference
STZ-induced diabetes in male Sprague–Dawley rats.	The *Abelmoschus esculentus* powder samples (250 g) were subjected to extraction procedures using ten volumes of either 75% ethanol or distilled water. 100 mL/kg of solution (aqueous extract, ethanol extract, and aqueous extract of Indole Acetic Acid (IAA) from *Abelmoschus esculentus*) was administered orally daily for six weeks.	*Abelmoschus esculentus* powder significantly decreased SOD in diabetic rats that received ethanol extract compared to diabetic controls. Diabetic rats that received an aqueous extract of *Abelmoschus esculentus* at 100 mg/kg had significantly increased liver total phenolic content.	Canada	[102]
STZ-induced hyperglycemia in male Wistar rats.	Three fresh *Abelmoschus esculentus* pods were sliced and infused in 250 mL of 3.6 mL *Abelmoschus esculentus* infusion water for 28 days.	*Abelmoschus esculentus* in diabetic rats significantly increased SOD when compared to diabetic controls.	Bangladesh	[114]
STZ-induced gestational diabetes (GDM) in Sprague–Dawley rats.	The intervention group was administered orally a solution containing 200 mg/kg of *Abelmoschus esculentus* extract.	*Abelmoschus esculentus* in GDM rats significantly increased SOD, MDA, CAT, GSH, and GPx in the liver and pancreas compared to GDM controls.	China	[113]

SOD: superoxide dismutase, MDA: malondialdehyde, CAT: catalase, GSH: glutathione, GPx: glutathione peroxidase, CMC: carboxy methyl cellulose, IAA-AE: Indole Acetic Acid-*Abelmoschus esculentus*, AEPP: *Abelmoschus esculentus* peel powder, AESP: *Abelmoschus esculentus* seed powder, GDM: gestational diabetes, F1: fraction 1, FR: fractional residue, DPP-4: dipeptidyl peptidase-4, GLP-1R: glucagon-like peptide 1 receptor, WAE: whole *Abelmoschus esculentus*, AEP: *Abelmoschus esculentus* peel, AES: *Abelmoschus esculentus* seed, SPF: specific pathogen-free, IC50: half maximal inhibitory concentration, TT: total thiols, TBARS: thiobarbituric acid reactive substances, Nrf2: Nuclear factor erythroid 2–related factor 2.

4.1.6. *Abelmoschus esculentus* on Glutathione and Glutathione Peroxidase

GSH is a group of enzymes that protect the body from oxidative stress by reducing lipid hydroperoxides and free hydrogen peroxide [124,125]. The general overview of the effect of *Abelmoschus esculentus* on markers of GPx and GSH is outlined in Table 2.

However, a decrease in its activity subjects the cells or organs to oxidative stress by allowing the accumulation of harmful oxidants and free radicals. Therefore, antioxidant compounds that increase the activity of GSH and GPx can be used as an alternative therapy to ameliorate oxidative stress and protect the cells from oxidative damage (Figure 4). Our current review found contradictory reports from various rodent models of DM induced by STZ or alloxan monohydrate. Of interest was that several studies reported a significant increase in the activity of GSH [91,95,105,106,115,123]. Additionally, Tian et al. [113] and Aleisa et al. [115] have reported a noteworthy augmentation in the activity of GPx, further substantiating the potential of *Abelmoschus esculentus* and its constituents as antioxidant agents.

In contrast to the aforementioned encouraging discoveries, other researchers [91,95,105,106,115,116,123] have recently identified a notable decline in GSH levels after administering *Abelmoschus esculentus* treatment to rodent models with diabetes. This once again highlights a potential limitation of *Abelmoschus esculentus* as an antioxidant. *Abelmoschus esculentus* ability to regulate oxidative stress seems to be attributable to its high content of phenols, flavonoids, and associated minerals, as presented in Figure 3.

5. Limitations

The current review has several limitations; for instance, the evidence gathered here is primarily from preclinical studies, with mainly mice and rats used for such experimentation. Additionally, the majority of models of diabetes presented here were primarily developed by using the administration of either STZ or alloxan monohydrate, with a few using HFD to induce diabetes. It is commonly known that these two drugs cause pancreatic damage, and the model developed through their administration mimics that of T1DM, while HFD resembles that of T2DM seen in humans. Since the pathogenesis of these conditions differs, the interpretation may be skewed toward T1DM, as most studies have induced diabetes through drug intervention. While evidence from the preclinical studies supports the use of both *Amaranthus* and *Abelmoschus esculentus* as herbal treatments for DM against oxidative stress, it is still not clear as to which part of these plants is more beneficial, as some studies used leaves, seeds, and pods.

The exact dose or form by which these plants are administered is also not specified, as a powder has been used to prepare the extract or given to rodents as a food-powder mixture. Although other clinical studies have been conducted on *Abelmoschus esculentus* in diabetic populations, ranging from randomized controlled trials to quasi-experiments, such studies did not focus on oxidative stress or related markers [126]. Moreover, regarding *Amaranthus*, only one trial has been conducted in diabetic patients, and the results are promising [127]. However, since then, clinical evidence has been scarce. Lastly, Amaranthus, and *Abelmoschus esculentus* have been proven effective in reducing diabetes. However, to the best of our knowledge, no studies have been conducted using a combined treatment of both of these plants in diabetic models or clinical trials. As the evidence reviewed here is derived from preclinical studies, it is important to note that in some cases, the evidence from experimental studies is not fully translatable to humans due to different physiological systems and functions.

6. Conclusions and Recommendations

The study discussed the effects of different parts of *Amaranthus* and *Abelmoschus esculentus*, highlighting their potential as alternative remedies to attenuate oxidative stress in diabetes models. Current evidence on the recommended dosages for these plants has proven these plants to be considered safe in the management of DM. Both plants are rich in carbohydrates, proteins, fatty acids, vitamins, fiber, minerals, and other bioactive phytochemicals that promote good health and are less expensive, making them economically affordable natural antioxidants. Evidence exploring the efficacy and safety of *Abelmoschus esculentus* in diabetes on other markers exists [17,101], but studies in this population still lack exploration of oxidative stress. While the potential benefits of *Amaranthus* have been extensively explored in diabetic models, there is limited evidence in clinical trials, especially on oxidative stress. The preclinical evidence gathered in this review revealed that *Amaranthus* treatment in diabetes could ameliorate hyperglycemic-associated oxidative stress (Figure 2). Remarkably, our summarized evidence also demonstrated that the administration of *Abelmoschus esculentus* to diabetic rodents also attenuates oxidative stress (Figure 4). Although there have been limited clinical trials involving *Abelmoschus esculentus* in diabetic populations [126], its potential benefits are widely recognized within the broader research community, and its safety has been confirmed. However, since there are currently no trials investigating the effects of these plants on oxidative stress, it is recommended that future trials be conducted, especially in African countries where the prevalence of DM is high. Therefore, we plan to explore oxidative stress with well-designed and adequately powered clinical studies. Therefore, future studies may be necessary to fully understand its efficacy, optimal dosage, and impact on oxidative stress. These studies should be well-designed and adequately powered, focusing on diabetic patients.

Author Contributions: Conceptualization, W.N.P. and K.M.; methodology, W.N.P. and K.M.; validation, W.N.P. and K.M.; investigation, W.N.P. and K.M.; data curation, W.N.P. and K.M.; writing—original draft preparation, W.N.P. and K.M.; writing—review and editing, W.N.P. and K.M.; visualization, W.N.P. and K.M. All authors have read and agreed to the published version of the manuscript.

Funding: This research is part of a bigger study entitled "The use and functional properties of African indigenous fruits and vegetables in alleviating household food and nutrition insecurity for local communities" which is funded by the University of South Africa Women in Research (WiR-2023). Opinions expressed, and conclusions arrived at, are those of the author and not necessarily to be attributed to the (WiR-2023).

Institutional Review Board Statement: Not applicable.

Informed Consent Statement: Not applicable.

Data Availability Statement: Not applicable.

Acknowledgments: K.M. is partially funded by Research Development Grants for nGAP Scholars (NGAP23022780506) and the Research Excellence Award for Next Generation Researchers (NONF230515106418). However, the content presented in this manuscript is the authors' sole responsibility and does not necessarily represent the official views of the NRF nGAP or the funders.

Conflicts of Interest: The authors declare no conflict of interest.

Abbreviations

DM: Diabetes mellitus; T2DM: Type 2 diabetes mellitus; T1DM: Type 1 diabetes mellitus; AHELE: *Amaranthus hybridus* ethanol leaf extract; CAT: Catalase; GSH: Glutathione; GPx: Glutathione peroxidase; SOD: Superoxide dismutase; MDA: Malondialdehyde; TBARS: Thiobarbituric acid reactive substances; and STZ: Streptozotocin; IAA: Indole Acetic Acid; PM: Powdered nucillage; PPS: powdered peel seed; ROS: Reactive oxygen species; AEP: *Abelmoschus esculentus* powder; HFD: High-fat diet; AEPP: *Abelmoschus esculentus* Peel powder; AESP: *Abelmoschus esculentus* seed powder; DPPH: 1,1-diphenyl-2-picrylhydrazyl; ABTS: 2,2′-casino-bis-3-ethylbenzothiazoline-6-sulfonic acid; LMICs: low- and middle-income countries; FFA: Free fatty acids; AGEs: advanced glycation end products; mNADH: mitochondrial NADH; DAG: diacylglycerol; PKC: Protein kinase C; ET-1: endothelin 1; VCAM-1: vascular cell adhesion molecule; ICAM-1: intercellular adhesion molecule; NF-κβ: nuclear factor kappa-light-chain-enhancer of activated β cell; CMC: Carboxy methyl cellulose; GDM: gestational diabetes; F1: Fraction 1; FR: Fractional residue; GLP-1R: Glucagon like peptide 1 receptor; WAE: Whole *Abelmoschus esculentus*; SPF: Specific pathogen-free.

References

1. Ogurtsova, K.; Guariguata, L.; Barengo, N.C.; Ruiz, P.L.-D.; Sacre, J.W.; Karuranga, S.; Sun, H.; Boyko, E.J.; Magliano, D.J. IDF Diabetes Atlas: Global Estimates of Undiagnosed Diabetes in Adults for 2021. *Diabetes Res. Clin. Pract.* **2022**, *183*, 109118. [CrossRef] [PubMed]
2. DeFronzo, R.A.; Ferrannini, E.; Groop, L.; Henry, R.R.; Herman, W.H.; Holst, J.J.; Hu, F.B.; Kahn, C.R.; Raz, I.; Shulman, G.I.; et al. Type 2 Diabetes Mellitus. *Nat. Rev. Dis. Primers* **2015**, *1*, 15019. [CrossRef] [PubMed]
3. Gojka Roglic WHO Global Report on Diabetes: A Summary. *Int. J. Non.-Commun. Dis.* **2016**, *1*, 3–8. [CrossRef]
4. Sun, H.; Saeedi, P.; Karuranga, S.; Pinkepank, M.; Ogurtsova, K.; Duncan, B.B.; Stein, C.; Basit, A.; Chan, J.C.N.; Mbanya, J.C.; et al. IDF Diabetes Atlas: Global, Regional and Country-Level Diabetes Prevalence Estimates for 2021 and Projections for 2045. *Diabetes Res. Clin. Pract.* **2022**, *183*, 109119. [CrossRef] [PubMed]
5. Karachaliou, F.; Simatos, G.; Simatou, A. The Challenges in the Development of Diabetes Prevention and Care Models in Low-Income Settings. *Front. Endocrinol.* **2020**, *11*, 518. [CrossRef]
6. Harris, M.L.; Oldmeadow, C.; Hure, A.; Luu, J.; Loxton, D.; Attia, J. Stress Increases the Risk of Type 2 Diabetes Onset in Women: A 12-Year Longitudinal Study Using Causal Modelling. *PLoS ONE* **2017**, *12*, e0172126. [CrossRef]
7. Nguyen, C.T.; Pham, N.M.; Lee, A.H.; Binns, C.W. Prevalence of and Risk Factors for Type 2 Diabetes Mellitus in Vietnam: A Systematic Review. *Asia Pac. J. Public Health* **2015**, *27*, 588–600. [CrossRef]
8. Sanada, H.; Yokokawa, H.; Yoneda, M.; Yatabe, J.; Yatabe, M.S.; Williams, S.M.; Felder, R.A.; Jose, P.A. High Body Mass Index Is an Important Risk Factor for the Development of Type 2 Diabetes. *Intern. Med.* **2012**, *51*, 1821–1826. [CrossRef]
9. Deshpande, A.D.; Harris-Hayes, M.; Schootman, M. Epidemiology of Diabetes and Diabetes-Related Complications Diabetes Special Issue. *Phys. Ther.* **2008**, *88*, 1254–1264. [CrossRef]
10. Giacco, F.; Brownlee, M. Oxidative Stress and Diabetic Complications. *Circ. Res.* **2010**, *107*, 1058–1070. [CrossRef]
11. Shao, Y.; Williamson, C. The HIV-1 Epidemic: Low- to Middle-Income Countries. *Cold Spring Harb. Perspect. Med.* **2012**, *2*, a007187. [CrossRef] [PubMed]
12. Pittampalli, S.; Upadyayula, S.; Hema, H.M.; Mekala, M.; Lippmann, S. Risks vs Benefits for SGLT2 Inhibitor Medications. *Fed. Pract.* **2018**, *35*, 45–48. [PubMed]
13. Shurrab, N.T.; Arafa, E.-S.A. Metformin: A Review of Its Therapeutic Efficacy and Adverse Effects. *Obes. Med.* **2020**, *17*, 100186. [CrossRef]
14. Mokgalaboni, K.; Phoswa, W.N. Corchorus Olitorius Extract Exhibit Anti-Hyperglycemic and Anti-Inflammatory Properties in Rodent Models of Obesity and Diabetes Mellitus. *Front. Nutr.* **2023**, *10*, 1099880. [CrossRef]
15. Mokgalaboni, K.; Ntamo, Y.; Ziqubu, K.; Nyambuya, T.M.; Nkambule, B.B.; Mazibuko-Mbeje, S.E.; Gabuza, K.B.; Chellan, N.; Tiano, L.; Dludla, P.V. Curcumin Supplementation Improves Biomarkers of Oxidative Stress and Inflammation in Conditions of Obesity, Type 2 Diabetes and NAFLD: Updating the Status of Clinical Evidence. *Food Funct.* **2021**, *12*, 12235–12249. [CrossRef]
16. Mokgalaboni, K.; Dlamini, S.; Phoswa, W.N.; Modjadji, P.; Lebelo, S.L. The Impact of Punica Granatum Linn and Its Derivatives on Oxidative Stress, Inflammation, and Endothelial Function in Diabetes Mellitus: Evidence from Preclinical and Clinical Studies. *Antioxidants* **2023**, *12*, 1566. [CrossRef]
17. Tavakolizadeh, M.; Peyrovi, S.; Ghasemi-Moghaddam, H.; Bahadori, A.; Mohkami, Z.; Sotoudeh, M.; Ziaee, M. Clinical Efficacy and Safety of Okra (*Abelmoschus esculentus* (L.) Moench) in Type 2 Diabetic Patients: A Randomized, Double-Blind, Placebo-Controlled, Clinical Trial. *Acta Diabetol.* **2023**. [CrossRef]
18. Incalza, M.A.; D'Oria, R.; Natalicchio, A.; Perrini, S.; Laviola, L.; Giorgino, F. Oxidative Stress and Reactive Oxygen Species in Endothelial Dysfunction Associated with Cardiovascular and Metabolic Diseases. *Vasc. Pharmacol.* **2018**, *100*, 1–19.
19. Rehman, K.; Akash, M.S.H. Mechanism of Generation of Oxidative Stress and Pathophysiology of Type 2 Diabetes Mellitus: How Are They Interlinked? *J. Cell Biochem.* **2017**, *118*, 3577–3585. [CrossRef]
20. Kuyvenhoven, J.P.; Meinders, A.E. Oxidative Stress and Diabetes Mellitus Pathogenesis of Long-Term Complications. *Eur. J. Intern. Med.* **1999**, *10*, 9–19. [CrossRef]
21. American Diabetes Association Diagnosis and Classification of Diabetes Mellitus. *Diabetes Care* **2014**, *37*, S81–S90. [CrossRef] [PubMed]
22. de Souza Bastos, A.; Graves, D.T.; de Melo Loureiro, A.P.; Júnior, C.R.; Corbi, S.C.T.; Frizzera, F.; Scarel-Caminaga, R.M.; Câmara, N.O.; Andriankaja, O.M.; Hiyane, M.I.; et al. Diabetes and Increased Lipid Peroxidation Are Associated with Systemic Inflammation Even in Well-Controlled Patients. *J. Diabetes Complicat.* **2016**, *30*, 1593–1599. [CrossRef] [PubMed]
23. Augustine, J.; Troendle, E.P.; Barabas, P.; McAleese, C.A.; Friedel, T.; Stitt, A.W.; Curtis, T.M. The Role of Lipoxidation in the Pathogenesis of Diabetic Retinopathy. *Front. Endocrinol.* **2021**, *11*, 621938. [CrossRef]
24. Geraldes, P.; King, G.L. Activation of Protein Kinase C Isoforms and Its Impact on Diabetic Complications. *Circ. Res.* **2010**, *106*, 1319–1331. [CrossRef] [PubMed]
25. Zabielski, P.; Chacinska, M.; Charkiewicz, K.; Baranowski, M.; Gorski, J.; Blachnio-Zabielska, A.U. Effect of Metformin on Bioactive Lipid Metabolism in Insulin-Resistant Muscle. *J. Endocrinol.* **2017**, *233*, 329–340. [CrossRef] [PubMed]
26. Kolczynska, K.; Loza-Valdes, A.; Hawro, I.; Sumara, G. Diacylglycerol-Evoked Activation of PKC and PKD Isoforms in Regulation of Glucose and Lipid Metabolism: A Review. *Lipids Health Dis.* **2020**, *19*, 113. [CrossRef]

27. Abdala-Valencia, H.; Berdnikovs, S.; Cook-Mills, J.M. Mechanisms for Vascular Cell Adhesion Molecule-1 Activation of ERK1/2 during Leukocyte Transendothelial Migration. *PLoS ONE* **2011**, *6*, e26706. [CrossRef]
28. Kawanami, D.; Maemura, K.; Takeda, N.; Harada, T.; Nojiri, T.; Saito, T.; Manabe, I.; Imai, Y.; Nagai, R. C-Reactive Protein Induces VCAM-1 Gene Expression through NF-KB Activation in Vascular Endothelial Cells. *Atherosclerosis* **2006**, *185*, 39–46. [CrossRef]
29. Park, J.Y.; Kim, Y.M.; Song, H.S.; Park, K.Y.; Kim, Y.M.; Kim, M.S.; Pak, Y.K.; Lee, I.K.; Lee, J.D.; Park, S.J.; et al. Oleic Acid Induces Endothelin-1 Expression through Activation of Protein Kinase C and NF-KB. *Biochem. Biophys. Res. Commun.* **2003**, *303*, 891–895. [CrossRef]
30. Sedeek, M.; Nasrallah, R.; Touyz, R.M.; Hébert, R.L. NADPH Oxidases, Reactive Oxygen Species, and the Kidney: Friend and Foe. *J. Am. Soc. Nephrol.* **2013**, *24*, 1512–1518. [CrossRef]
31. Sinenko, S.A.; Starkova, T.Y.; Kuzmin, A.A.; Tomilin, A.N. Physiological Signaling Functions of Reactive Oxygen Species in Stem Cells: From Flies to Man. *Front. Cell Dev. Biol.* **2021**, *9*, 714370. [CrossRef]
32. Volpe, C.M.O.; Villar-Delfino, P.H.; Dos Anjos, P.M.F.; Nogueira-Machado, J.A. Cellular Death, Reactive Oxygen Species (ROS) and Diabetic Complications Review-Article. *Cell Death Dis.* **2018**, *9*, 119. [CrossRef]
33. Zhang, E.; Miramini, S.; Patel, M.; Richardson, M.; Ebeling, P.; Zhang, L. Role of TNF-α in Early-Stage Fracture Healing under Normal and Diabetic Conditions. *Comput. Methods Programs Biomed.* **2022**, *213*, 106536. [CrossRef]
34. Adams, A.P. Is Diabetic Retinopathy an inflammatory Disease? *Br. J. Ophthalmol.* **2000**, *86*, 363–365. [CrossRef] [PubMed]
35. Hameed, I.; Masoodi, S.R.; Mir, S.A.; Nabi, M.; Ghazanfar, K.; Ganai, B.A. Type 2 Diabetes Mellitus: From a Metabolic Disorder to an Inflammatory Condition. *World J. Diabetes* **2015**, *6*, 598. [CrossRef] [PubMed]
36. Mahomoodally, M.F.; Désiré, A.-L.D.; Rosette, M.A.-L.E. Chapter2.2 Catalase. In *Antioxidants Effects in Health*; Nabavi, S.M., Silva, A.S., Eds.; Elsevier: Amsterdam, The Netherlands, 2022; pp. 81–90. ISBN 978-0-12-819096-8.
37. Anderson, E.J.; Lustig, M.E.; Boyle, K.E.; Woodlief, T.L.; Kane, D.A.; Lin, C.T.; Price, J.W.; Kang, L.; Rabinovitch, P.S.; Szeto, H.H.; et al. Mitochondrial H_2O_2 Emission and Cellular Redox State Link Excess Fat Intake to Insulin Resistance in Both Rodents and Humans. *J. Clin. Investig.* **2009**, *119*, 573–581. [CrossRef] [PubMed]
38. Fisher-Wellman, K.H.; Neufer, P.D. Linking Mitochondrial Bioenergetics to Insulin Resistance via Redox Biology. *Trends Endocrinol. Metab.* **2012**, *23*, 142–153. [CrossRef]
39. Iwakami, S.; Misu, H.; Takeda, T.; Sugimori, M.; Matsugo, S.; Kaneko, S.; Takamura, T. Concentration-Dependent Dual Effects of Hydrogen Peroxide on Insulin Signal Transduction in H4IIEC Hepatocytes. *PLoS ONE* **2011**, *6*, e27401. [CrossRef]
40. Berdichevsky, A.; Guarente, L.; Bose, A. Acute Oxidative Stress Can Reverse Insulin Resistance by Inactivation of Cytoplasmic JNK. *J. Biol. Chem.* **2010**, *285*, 21581–21589. [CrossRef]
41. Tucker, P.S.; Fisher-Wellman, K.; Bloomer, R.J. Can Exercise Minimize Postprandial Oxidative Stress in Patients with Type 2 Diabetes? *Curr. Diabetes Rev.* **2008**, *4*, 309–319. [CrossRef]
42. Maddux, B.A.; See, W.; Lawrence, J.C.; Goldfine, A.L.; Goldfine, I.D.; Evans, J.L. Protection against oxidative stress-induced insulin resistance in rat l6 muscle cells by micromolar concentrations of-lipoic acid. *Diabetes* **2001**, *50*, 404–410. [CrossRef] [PubMed]
43. Chatterjee, S. Oxidative Stress, Inflammation, and Disease. In *Oxidative Stress and Biomaterials*; Elsevier: Amsterdam, The Netherlands, 2016; pp. 35–58. ISBN 9780128032701.
44. Folli, F.; Corradi, D.; Fanti, P.; Davalli, A.; Paez, A.; Giaccari, A.; Perego, C.; Muscogiuri, G. The role of oxidative stress in the pathogenesis of type 2 diabetes mellitus micro-and macrovascular complications: Avenues for a mechanistic-based therapeutic approach. *Curr. Diabetes Rev.* **2011**, *7*, 313–324. [CrossRef] [PubMed]
45. Kitada, M.; Zhang, Z.; Mima, A.; King, G.L. Molecular Mechanisms of Diabetic Vascular Complications. *J. Diabetes Investig.* **2010**, *1*, 77–89. [CrossRef] [PubMed]
46. Vikram, A.; Tripathi, D.N.; Kumar, A.; Singh, S. Oxidative Stress and Inflammation in Diabetic Complications. *Int. J. Endocrinol.* **2014**, *2014*, 679754. [CrossRef]
47. Psaltopoulou, T.; Ilias, I.; Alevizaki, M. The Role of Diet and Lifestyle in Primary, Secondary, and Tertiary Diabetes Prevention: A Review of Meta-Analyses. *Rev. Diabet. Stud.* **2010**, *7*, 26–35. [CrossRef]
48. El-Senousey, H.K.; Chen, B.; Wang, J.Y.; Atta, A.M.; Mohamed, F.R.; Nie, Q.H. Effects of Dietary Vitamin C, Vitamin E, and Alpha-Lipoic Acid Supplementation on the Antioxidant Defense System and Immune-Related Gene Expression in Broilers Exposed to Oxidative Stress by Dexamethasone. *Poult. Sci.* **2018**, *97*, 30–38. [CrossRef]
49. Thapa, R.; Edwards, M.; Blair, M.W. Relationship of Cultivated Grain Amaranth Species and Wild Relative Accessions. *Genes* **2021**, *12*, 1849. [CrossRef]
50. Das, S. Distribution and Maintenance of Amaranth Germplasm Worldwide. In *Amaranthus: A Promising Crop of Future*; Das, S., Ed.; Springer: Singapore, 2016; pp. 99–106. ISBN 978-981-10-1469-7.
51. Baraniak, J.; Kania-Dobrowolska, M. The Dual Nature of Amaranth—Functional Food and Potential Medicine. *Foods* **2022**, *11*, 618. [CrossRef]
52. Manyelo, T.G.; Sebola, N.A.; van Rensburg, E.J.; Mabelebele, M. The Probable Use of Genus Amaranthus as Feed Material for Monogastric Animals. *Animals* **2020**, *10*, 1504. [CrossRef]
53. Yadav, K.; Bhatia, A.L.; Sisodia, R. Modulation of Radiation Induced Biochemical Changes in Testis of Swiss Albino Mice by Amaranthus Paniculatus Linn. *Asian J. Exp. Sci.* **2004**, *18*, 63–74.

54. Ruth, O.N.; Unathi, K.; Nomali, N.; Chinsamy, M. Underutilization versus Nutritional-Nutraceutical Potential of the Amaranthus Food Plant: A Mini-Review. *Appl. Sci.* **2021**, *11*, 6879. [CrossRef]
55. Reyad-ul-Ferdous, M.D. Present Biological Status of Potential Medicinal Plant of Amaranthus Viridis: A Comprehensive Review. *Am. J. Clin. Exp. Med.* **2015**, *3*, 12. [CrossRef]
56. Adegbola, P.I.; Adetutu, A.; Olaniyi, T.D. Antioxidant Activity of Amaranthus Species from the Amaranthaceae Family A Review. *South. Afr. J. Bot.* **2020**, *133*, 111–117. [CrossRef]
57. Mishra, A.; Jha, S.K.; Ojha, P. Study on Zero Energy Cool Chamber (ZECC) for Storage of Vegetables. *Int. J. Sci. Res. Publ. (IJSRP)* **2020**, *10*, p9767. [CrossRef]
58. Gins, V.K.; Motyleva, S.M.; Kulikov, I.M.; Tumanyan, A.F.; Romanova, E.V.; Baikov, A.A.; Gins, E.M.; Terekhin, A.A.; Gins, M.S. Antioxidant Profile of *Amaranthus Paniculatus* L. Of the Pamyat of Kovas Variety. *IOP Conf. Ser. Earth Environ. Sci.* **2021**, *624*, 012152. [CrossRef]
59. Chaturvedi, A.; Sarojini, G.; Devi, N.L. Hypocholesterolemic Effect of Amaranth Seeds (*Amaranthus Esculantus*). *Plant Foods Hum. Nutr.* **1993**, *44*, 63–70. [CrossRef]
60. Jia, W.; Gaoz, W.; Tang, L. Antidiabetic Herbal Drugs Officially Approved in China. *Phytother. Res.* **2003**, *17*, 1127–1134. [CrossRef]
61. Kar, A.; Bhattacharjee, S. Bioactive Polyphenolic Compounds, Water-Soluble Vitamins, in Vitro Anti-Inflammatory, Anti-Diabetic and Free Radical Scavenging Properties of Underutilized Alternate Crop *Amaranthus Spinosus* L. from Gangetic Plain of West Bengal. *Food Biosci.* **2022**, *50*, 102072. [CrossRef]
62. Schröter, D.; Neugart, S.; Schreiner, M.; Grune, T.; Rohn, S.; Ott, C. Amaranth's 2-Caffeoylisocitric Acid—An Anti-Inflammatory Caffeic Acid Derivative That Impairs NF-KB Signaling in LPS-Challenged RAW 264.7 Macrophages. *Nutrients* **2019**, *11*, 571. [CrossRef]
63. Nawale, R.B.; Mate, G.S.; Wakure, B.S. Ethanolic Extract of *Amaranthus Paniculatus* Linn. Ameliorates Diabetes-Associated Complications in Alloxan-Induced Diabetic Rats. *Integr. Med. Res.* **2017**, *6*, 41–46. [CrossRef]
64. Prince, M.R.U.; Zihad, S.M.N.K.; Ghosh, P.; Sifat, N.; Rouf, R.; Al Shajib, G.M.; Alam, M.A.; Shilpi, J.A.; Uddin, S.J. Amaranthus Spinosus Attenuated Obesity-Induced Metabolic Disorders in High-Carbohydrate-High-Fat Diet-Fed Obese Rats. *Front. Nutr.* **2021**, *8*, 653918. [CrossRef] [PubMed]
65. Ashok Kumar, B.S.; Lakshman, K.; Nandeesh, R.; Arun Kumar, P.A.; Manoj, B.; Kumar, V.; Sheshadri Shekar, D. In Vitro Alpha-Amylase Inhibition and in Vivo Antioxidant Potential of Amaranthus Spinosus in Alloxan-Induced Oxidative Stress in Diabetic Rats. *Saudi J. Biol. Sci.* **2011**, *18*, 1–5. [CrossRef] [PubMed]
66. Mishra, S.B.; Verma, A.; Mukerjee, A.; Vijayakumar, M. *Amaranthus Spinosus* L. (Amaranthaceae) Leaf Extract Attenuates Streptozotocin-Nicotinamide Induced Diabetes and Oxidative Stress in Albino Rats: A Histopathological Analysis. *Asian Pac. J. Trop. Biomed.* **2012**, *2*, S1647–S1652. [CrossRef]
67. Balasubramanian, T.; Karthikeyan, M.; Muhammed Anees, K.P.; Kadeeja, C.P.; Jaseela, K. Antidiabetic and Antioxidant Potentials of *Amaranthus hybridus* in Streptozotocin-Induced Diabetic Rats. *J. Diet. Suppl.* **2017**, *14*, 395–410. [CrossRef] [PubMed]
68. Balasubramanian, T.; Karthikeyan, M. Therapeutic Effect of *Amaranthus hybridus* on Diabetic Nephropathy. *J. Dev. Drugs* **2015**, *5*, 1000147. [CrossRef]
69. Ayala, A.; Muñoz, M.F.; Argüelles, S. Lipid Peroxidation: Production, Metabolism, and Signaling Mechanisms of Malondialdehyde and 4-Hydroxy-2-Nonenal. *Oxid. Med. Cell. Longev.* **2014**, *2014*, 360438. [CrossRef]
70. Ghonimi, N.A.M.; Elsharkawi, K.A.; Khyal, D.S.M.; Abdelghani, A.A. Serum Malondialdehyde as a Lipid Peroxidation Marker in Multiple Sclerosis Patients and Its Relation to Disease Characteristics. *Mult. Scler. Relat. Disord.* **2021**, *51*, 102941. [CrossRef] [PubMed]
71. Ademowo, O.S.; Dias, H.K.I.; Burton, D.G.A.; Griffiths, H.R. Lipid (per) Oxidation in Mitochondria: An Emerging Target in the Ageing Process? *Biogerontology* **2017**, *18*, 859–879. [CrossRef]
72. Chowdhury, N.S.; Jamaly, S.; Farjana, F.; Begum, N.; Zenat, E.A. A Review on Ethnomedicinal, Pharmacological, Phytochemical and Pharmaceutical Profile of Lady's Finger (*Abelmoschus esculentus* L.) Plant. *Pharmacol. Pharm.* **2019**, *10*, 94–108. [CrossRef]
73. Esmaeilzadeh, D.; Razavi, B.M.; Hosseinzadeh, H. Effect of *Abelmoschus esculentus* (Okra) on Metabolic Syndrome: A Review. *Phytother. Res.* **2020**, *34*, 2192–2202. [CrossRef]
74. Elkhalifa, A.E.O.; Alshammari, E.; Adnan, M.; Alcantara, J.C.; Awadelkareem, A.M.; Eltoum, N.E.; Mehmood, K.; Panda, B.P.; Ashraf, S.A. Okra (*Abelmoschus esculentus*) as a Potential Dietary Medicine with Nutraceutical Importance for Sustainable Health Applications. *Molecules* **2021**, *26*, 696. [CrossRef] [PubMed]
75. Arapitsas, P. Identification and Quantification of Polyphenolic Compounds from Okra Seeds and Skins. *Food Chem.* **2008**, *110*, 1041–1045. [CrossRef] [PubMed]
76. Nwankwo, C.I.; Chinomso, N.J.; Ndubuisi, N.S.; Obinna, A.; Ogochukwu, A.P.; Nnaemeka, U.E.; Gift, P.O.; Esther, O.; Virginus, U. Phytochemical And Proximate Composition of Igbo Okra (*Abelmoschus esculentus*) Seeds. *World J. Pharm. Life Sci.* **2021**, *7*, 20–24.
77. Sunilson, J.; Anbu, J.; Jayaraj, P.; Mohan, M.S.; Anita, A.; Kumari, G.; Varatharajan, R. Antioxidant and Hepatoprotective Effect of the Roots of Hibiscus Esculentus Linn. *Int. J. Green Pharm.* **2008**, 200–203.
78. Maria, E.E.C.; Luciana, M.P.d.S.; Elba, d.S.F.; Amanda, P.d.F.; Carlos, A.d.A.G.; Jailane, d.S.A.; Tatiane, S.-G. Nutritional, Antinutritional and Phytochemical Status of Okra Leaves (*Abelmoschus esculentus*) Subjected to Different Processes. *Afr. J. Biotechnol.* **2015**, *14*, 683–687. [CrossRef]

79. Petropoulos, S.; Fernandes, Â.; Barros, L.; Ferreira, I.C.F.R. Chemical Composition, Nutritional Value and Antioxidant Properties of Mediterranean Okra Genotypes in Relation to Harvest Stage. *Food Chem.* **2018**, *242*, 466–474. [CrossRef] [PubMed]
80. Gemede, H.F.; Haki, G.D.; Beyene, F.; Woldegiorgis, A.Z.; Rakshit, S.K. Proximate, Mineral, and Antinutrient Compositions of Indigenous Okra (*Abelmoschus esculentus*) Pod Accessions: Implications for Mineral Bioavailability. *Food Sci. Nutr.* **2016**, *4*, 223–233. [CrossRef]
81. Romdhane, M.H.; Chahdoura, H.; Barros, L.; Dias, M.I.; Corrêa, R.C.G.; Morales, P.; Ciudad-Mulero, M.; Flamini, G.; Majdoub, H.; Ferreira, I.C.F.R. Chemical Composition, Nutritional Value, and Biological Evaluation of Tunisian Okra Pods (*Abelmoschus Esculentus* L. Moench). *Molecules* **2020**, *25*, 4739. [CrossRef]
82. Majd, N.E.; Tabandeh, M.R.; Shahriari, A.; Soleimani, Z. Okra (Abelmoscus Esculentus) Improved Islets Structure, and Down-Regulated PPARs Gene Expression in Pancreas of High-Fat Diet and Streptozotocin-Induced Diabetic Rats. *Cell J.* **2018**, *20*, 31–40. [CrossRef]
83. Fabianová, J.; Šlosár, M.; Kopta, T.; Vargová, A.; Timoracká, M.; Mezeyová, I.; Andrejiová, A. Yield, Antioxidant Activity and Total Polyphenol Content of Okra Fruits Grown in Slovak Republic. *Horticulturae* **2022**, *8*, 966. [CrossRef]
84. Puji, S.; Wahyuningsih, A.; Winarni, D.; Pramudya, M.; Setianingsih, N.; Ayubu Mwendolwa, A.; Nindyasari, F. Antioxidant Potential of Red Okra Pods (*Abelmoschus esculentus* Moench). *EPIC Biol. Sci.* **2021**, *1*, 158–163.
85. Gemede, H.F.; Haki, G.D.; Beyene, F.; Rakshit, S.K.; Woldegiorgis, A.Z. Indigenous Ethiopian Okra (*Abelmoschus esculentus*) Mucilage: A Novel Ingredient with Functional and Antioxidant Properties. *Food Sci. Nutr.* **2018**, *6*, 563–571. [CrossRef] [PubMed]
86. Hu, L.; Yu, W.; Li, Y.; Prasad, N.; Tang, Z. Antioxidant Activity of Extract and Its Major Constituents from Okra Seed on Rat Hepatocytes Injured by Carbon Tetrachloride. *Biomed. Res. Int.* **2014**, *2014*, 1–9. [CrossRef] [PubMed]
87. Umeno, A.; Horie, M.; Murotomi, K.; Nakajima, Y.; Yoshida, Y. Antioxidative and Antidiabetic Effects of Natural Polyphenols and Isoflavones. *Molecules* **2016**, *21*, 708. [CrossRef]
88. Wahyuningsih, S.P.A.; Savira, N.I.I.; Anggraini, D.W.; Winarni, D.; Suhargo, L.; Kusuma, B.W.A.; Nindyasari, F.; Setianingsih, N.; Mwendolwa, A.A. Antioxidant and Nephroprotective Effects of Okra Pods Extract (*Abelmoschus esculentus* L.) against Lead Acetate-Induced Toxicity in Mice. *Scientifica* **2020**, *2020*, 4237205-10. [CrossRef]
89. Mousavi, A.; Pourakbar, L.; Siavash Moghaddam, S. Effects of Malic Acid and EDTA on Oxidative Stress and Antioxidant Enzymes of Okra (*Abelmoschus esculentus* L.) Exposed to Cadmium Stress. *Ecotoxicol. Environ. Saf.* **2022**, *248*, 114320. [CrossRef]
90. Wahyuningsih, S.P.A.; Fachrisa, A.; Nisaâ€ᵀᴹ, N.; Kusuma, B.W.A.; Shoukat, N.; Ahmar, R.F.; Alifiyah, N.I. Potential of Red Okra Extract (*Abelmoschus esculentus* L. Moench) to Restore Kidney Damage Due to Sodium Nitrite. *Biosaintifika J. Biol. Biol. Educ.* **2021**, *13*, 84–91. [CrossRef]
91. Tanko, Y.; Idris, N.; Om, A.; Nm, G.; Muhammad, A.; Ka, M.; Yusuf, R. Evaluation of the Effect of Okra (*Abelmoschus esculentus*) Supplement on Blood Glucose Levels and Antioxidant Biomarkers on Alloxan Induced Diabetic Wistar Rats. *J. Biomed. Sci.* **2016**, *1*, 55–62.
92. Adetuyi, F.O.; Ibrahim, T.A. Effect of Fermentation Time on the Phenolic, Flavonoid and Vitamin C Contents and Antioxidant Activities of Okra (*Abelmoschus esculentus*) Seeds. *Niger. Food J.* **2014**, *32*, 128–137. [CrossRef]
93. Wang, K.; Li, M.; Wen, X.; Chen, X.; He, Z.; Ni, Y. Optimization of Ultrasound-Assisted Extraction of Okra (*Abelmoschus esculentus* (L.) Moench) Polysaccharides Based on Response Surface Methodology and Antioxidant Activity. *Int. J. Biol. Macromol.* **2018**, *114*, 1056–1063. [CrossRef]
94. Wu, D.T.; Nie, X.R.; Shen, D.D.; Li, H.Y.; Zhao, L.; Zhang, Q.; Lin, D.R.; Qin, W. Phenolic Compounds, Antioxidant Activities, and Inhibitory Effects on Digestive Enzymes of Different Cultivars of Okra (*Abelmoschus esculentus*). *Molecules* **2020**, *25*, 1276. [CrossRef] [PubMed]
95. Sabitha, V.; Ramachandran, S.; Naveen, K.R.; Panneerselvam, K. Investigation of in Vivo Antioxidant Property of *Abelmoschus esculentus* (L) Moench. Fruit Seed and Peel Powders in Streptozotocin-Induced Diabetic Rats. *J. Ayurveda Integr. Med.* **2012**, *3*, 188–193. [CrossRef] [PubMed]
96. Fekadu Gemede, H.; Desse Haki, G.; Beyene, F.; Woldegiorgis, A.Z.; Kumar Rakshit, S. Phenolic Profiles and Antioxidant of Ethiopian Indigenous Okra (*Abelmoschus esculentus*) Pod And Seed Accessions: A New Source of Natural Antioxidants. *Ann. Food Sci. Technol.* **2019**, *20*, 809–819.
97. Kottaisamy, C.P.D.; Raj, D.S.; Prasanth Kumar, V.; Sankaran, U. Experimental Animal Models for Diabetes and Its Related Complications—A Review. *Lab. Anim. Res.* **2021**, *37*, 1–14. [CrossRef]
98. Sies, H.; Jones, D.P. Reactive Oxygen Species (ROS) as Pleiotropic Physiological Signalling Agents. *Nat. Rev. Mol. Cell Biol.* **2020**, *21*, 363–383. [CrossRef]
99. Martemucci, G.; Costagliola, C.; Mariano, M.; D'andrea, L.; Napolitano, P.; D'Alessandro, A.G. Free Radical Properties, Source and Targets, Antioxidant Consumption and Health. *Oxygen* **2022**, *2*, 48–78. [CrossRef]
100. Shields, H.J.; Traa, A.; Van Raamsdonk, J.M. Beneficial and Detrimental Effects of Reactive Oxygen Species on Lifespan: A Comprehensive Review of Comparative and Experimental Studies. *Front. Cell Dev. Biol.* **2021**, *9*, 23. [CrossRef]
101. Uddin Zim, A.F.M.I.; Khatun, J.; Khan, M.F.; Hossain, M.A.; Haque, M.M. Evaluation of in Vitro Antioxidant Activity of Okra Mucilage and Its Antidiabetic and Antihyperlipidemic Effect in Alloxan-Induced Diabetic Mice. *Food Sci. Nutr.* **2021**, *9*, 6854–6865. [CrossRef]

102. Adewale, M.E.; Masisi, K.; Le, K.; Olaiya, C.O.; Esan, A.M.; Aluko, R.E.; Moghadasian, M.H. The Effects of Okra (*Abelmoschus esculentus* (L) Moench) Fruit Extracts on Diabetes Markers in Streptozotocin-Induced Diabetic Rats. *Arch. Diabetes Obes.* **2021**, *3*, 296–304.
103. Nasrollahi, Z.; ShahaniPour, K.; Monajemi, R.; Ahadi, A.M. Effect of Quercetin and *Abelmoschus esculentus* (L.) Moench on Lipids Metabolism and Blood Glucose through AMPK-α in Diabetic Rats (HFD/STZ). *J. Food Biochem.* **2022**, *46*, e14506. [CrossRef]
104. Martinez-Morales, F.; Alonso-Castro, A.J.; Zapata-Morales, J.R.; Carranza-Álvarez, C.; Aragon-Martinez, O.H. Use of Standardized Units for a Correct Interpretation of IC50 Values Obtained from the Inhibition of the DPPH Radical by Natural Antioxidants. *Chem. Pap.* **2020**, *74*, 3325–3334. [CrossRef]
105. Liao, Z.; Zhang, J.; Liu, B.; Yan, T.; Xu, F.; Xiao, F.; Wu, B.; Bi, K.; Jia, Y. Polysaccharide from Okra (*Abelmoschus esculentus* (L.) Moench) Improves Antioxidant Capacity via PI3K/AKT Pathways and Nrf2 Translocation in a Type 2 Diabetes Model. *Molecules* **2019**, *24*, 1906. [CrossRef] [PubMed]
106. Liao, Z.; Zhang, J.; Wang, J.; Yan, T.; Xu, F.; Wu, B.; Xiao, F.; Bi, K.; Niu, J.; Jia, Y. The Anti-Nephritic Activity of a Polysaccharide from Okra (*Abelmoschus esculentus* (L.) Moench) via Modulation of AMPK-Sirt1-PGC-1α Signaling Axis Mediated Anti-Oxidative in Type 2 Diabetes Model Mice. *Int. J. Biol. Macromol.* **2019**, *140*, 568–576. [CrossRef] [PubMed]
107. Peng, C.H.; Lin, H.C.; Lin, C.L.; Wang, C.J.; Huang, C.N. Abelmoschus Esculentus Subfractions Improved Nephropathy with Regulating Dipeptidyl Peptidase-4 and Type 1 Glucagon-like Peptide Receptor in Type 2 Diabetic Rats. *J. Food Drug Anal.* **2019**, *27*, 135–144. [CrossRef] [PubMed]
108. Liu, Y.; Liu, C.; Li, J. Comparison of Vitamin c and Its Derivative Antioxidant Activity: Evaluated by Using Density Functional Theory. *ACS Omega* **2020**, 25467–25475. [CrossRef] [PubMed]
109. Kaźmierczak-Barańska, J.; Boguszewska, K.; Adamus-Grabicka, A.; Karwowski, B.T. Two Faces of Vitamin c—Antioxidative and pro-Oxidative Agent. *Nutrients* **2020**, *12*, 1501. [CrossRef]
110. Mhya, D.H.; Mohammed, A.; Dawus, T.T. Investigation of NADPH-Oxidase's Binding Subunit(s) for Catechin Compounds Induce Inhibition. *Eur. J. Adv. Chem. Res.* **2023**, *4*, 10–18. [CrossRef]
111. Chen, X.; Touyz, R.M.; Park, J.B.; Schiffrin, E.L. Antioxidant Effects of Vitamins C and E Are Associated with Altered Activation of Vascular NADPH Oxidase and Superoxide Dismutase in Stroke-Prone SHR. *Hypertension* **2001**, *38*, 606–611. [CrossRef]
112. Karmakar, A.; Das, A.K.; Ghosh, N.; Sil, P.C. Chapter2.7 Superoxide Dismutase. In *Antioxidants Effects in Health*; Nabavi, S.M., Silva, A.S., Eds.; Elsevier: Amsterdam, The Netherlands, 2022; pp. 139–166. ISBN 978-0-12-819096-8.
113. Tian, Z.-H.; Miao, F.-T.; Zhang, X.; Wang, Q.-H.; Lei, N.; Guo, L.-C. Therapeutic Effect of Okra Extract on Gestational Diabetes Mellitus Rats Induced by Streptozotocin. *Asian Pac. J. Trop. Med.* **2015**, *8*, 1038–1042. [CrossRef]
114. Tyagita, N.; Utami, K.P.; Zulkarnain, F.H.; Rossandini, S.M.; Pertiwi, N.P.; Rifki, M.A.; Safitri, A.H. Okra Infusion Water Improving Stress Oxidative and Inflammatory Markers on Hyperglycemic Rats. *Bangladesh J. Med. Sci.* **2019**, *18*, 748–752. [CrossRef]
115. Aleissa, M.S.; AL-Zharani, M.; Alneghery, L.M.; Hasnain, M.S.; Almutairi, B.; Ali, D.; Alarifi, S.; Alkahtani, S. Comparative Study of the Anti-Diabetic Effect of Mucilage and Seed Extract of Abelmoschus Esculentus against Streptozotocin-Induced Diabetes in Rat Model. *J. King Saud. Univ. Sci.* **2022**, *34*. [CrossRef]
116. Uadia, P.O.; Imagbovowman, I.O.; Oriakhi, K.; Eze, I.G. Effect of Abelmoschus Esculentus (Okra)-Based Diet on Streptozotocin-Induced Diabetes Mellitus in Adult Wistar Rats. *Trop. J. Pharm. Res.* **2020**, *19*, 1737–1743. [CrossRef]
117. Abbas, A.Y.; Muhammad, I.; Abdulrahman, M.B.; Bilbis, L.S. Antioxidant Effect of Ex-Maradi Okra Fruit Variety (*Abelmuscus esculentus*) on Alloxan-Induced Diabetic Rats. *Trop. J. Nat. Prod. Res.* **2020**, *4*, 105–112. [CrossRef]
118. Mas-Bargues, C.; Escrivá, C.; Dromant, M.; Borrás, C.; Viña, J. Lipid Peroxidation as Measured by Chromatographic Determination of Malondialdehyde. Human Plasma Reference Values in Health and Disease. *Arch. Biochem. Biophys.* **2021**, *709*, 108941. [CrossRef] [PubMed]
119. Ruszała, M.; Pilszyk, A.; Niebrzydowska, M.; Kimber-trojnar, Ż.; Trojnar, M.; Leszczyńska-gorzelak, B. Novel Biomolecules in the Pathogenesis of Gestational Diabetes Mellitus 2.0. *Int. J. Mol. Sci.* **2022**, *23*, 4364. [CrossRef]
120. Aguilar Diaz De Leon, J.; Borges, C.R. Evaluation of Oxidative Stress in Biological Samples Using the Thiobarbituric Acid Reactive Substances Assay. *J. Vis. Exp.* **2020**, *2020*, e61122. [CrossRef]
121. Strom, A.; Kaul, K.; Brüggemann, J.; Ziegler, I.; Rokitta, I.; Püttgen, S.; Szendroedi, J.; Müssig, K.; Roden, M.; Ziegler, D. Lower Serum Extracellular Superoxide Dismutase Levels Are Associated with Polyneuropathy in Recent-Onset Diabetes. *Exp. Mol. Med.* **2017**, *49*, e394. [CrossRef]
122. Puntel, R.L.; Roos, D.H.; Paixão, M.W.; Braga, A.L.; Zeni, G.; Nogueira, C.W.; Rocha, J.B.T. Oxalate Modulates Thiobarbituric Acid Reactive Species (TBARS) Production in Supernatants of Homogenates from Rat Brain, Liver and Kidney: Effect of Diphenyl Diselenide and Diphenyl Ditelluride. *Chem. Biol. Interact.* **2007**, *165*, 87–98. [CrossRef]
123. Mishra, N.; Kumar, D.; Rizvi, S.I. Protective Effect of Abelmoschus Esculentus Against Alloxan-Induced Diabetes in Wistar Strain Rats. *J. Diet. Suppl.* **2016**, *13*, 634–646. [CrossRef]
124. Hatai, B.; Ganguly, A.; Bandopadhyay, S.; Banerjee, S.; Hatai, J. Impact of Glutathione Peroxidase Activity (GPX) as Oxidative-Stress Marker and Its Role on Inflammation with Osteoarthritis Patients. *Int. J. Adv. Res.* **2017**, *5*, 1288–1294. [CrossRef]
125. Hisalkar, P.J.; Patne, A.B.; Fawade, M.M.; Karnik, A.C. Evaluation of Plasma Superoxide Dismutase and Glutathione Peroxidase in Type 2 Diabetic Patients c Patients. *Res. Artic. Biol. Med.* **2012**, *4*, 65–72.

126. Mokgalaboni, K.; Lebelo, L.S.; Modjadji, P.; Ghaffary, S. Okra Ameliorates Hyperglycaemia in Pre-Diabetic and Type 2 Diabetic Patients: A Systematic Review and Meta-Analysis of the Clinical Evidence. *Front. Pharmacol* **2023**, *14*, 1132650. [CrossRef] [PubMed]
127. Yelisyeyeva, O.; Semen, K.; Zarkovic, N.; Kaminskyy, D.; Lutsyk, O.; Rybalchenko, V. Activation of Aerobic Metabolism by Amaranth Oil Improves Heart Rate Variability Both in Athletes and Patients with Type 2 Diabetes Mellitus. *Arch. Physiol. Biochem.* **2012**, *118*, 47–57. [CrossRef] [PubMed]

Disclaimer/Publisher's Note: The statements, opinions and data contained in all publications are solely those of the individual author(s) and contributor(s) and not of MDPI and/or the editor(s). MDPI and/or the editor(s) disclaim responsibility for any injury to people or property resulting from any ideas, methods, instructions or products referred to in the content.

Article

Antiviral Potential of Specially Selected Bulgarian Propolis Extracts: In Vitro Activity against Structurally Different Viruses

Neli Milenova Vilhelmova-Ilieva [1,*], Ivanka Nikolova Nikolova [1], Nadya Yordanova Nikolova [1], Zdravka Dimitrova Petrova [1,2], Madlena Stephanova Trepechova [1], Dora Ilieva Holechek [1], Mina Mihaylova Todorova [3], Mariyana Georgieva Topuzova [4], Ivan Georgiev Ivanov [4] and Yulian Dimitrov Tumbarski [5,*]

[1] Department of Virology, The Stephan Angeloff Institute of Microbiology, Bulgarian Academy of Sciences, 26 Georgi Bonchev Str., 1113 Sofia, Bulgaria; inikolova@microbio.bas.bg (I.N.N.); nadyanik@yahoo.com (N.Y.N.); zdr.z1971@abv.bg (Z.D.P.); madi_trepechova@yahoo.com (M.S.T.); doraholechek@yahoo.com (D.I.H.)

[2] Institute of Morphology, Pathology and Anthropology with Museum, Bulgarian Academy of Sciences, 25 Georgi Bonchev, 1113 Sofia, Bulgaria

[3] Department of Organic Chemistry, Paisii Hilendarski University of Plovdiv, 24 Tsar Asen Str., 4000 Plovdiv, Bulgaria; mm_todorova@abv.bg

[4] Department of Organic Chemistry and Inorganic Chemistry, University of Food Technologies, 26 Maritsa blvd., 4002 Plovdiv, Bulgaria; marianagt@mail.bg (M.G.T.); ivanov_ivan.1979@yahoo.com (I.G.I.)

[5] Department of Microbiology, University of Food Technologies, 26 Maritsa blvd., 4002 Plovdiv, Bulgaria

* Correspondence: nelivili@gmail.com (N.M.V.-I.); tumbarski@abv.bg (Y.D.T.)

Citation: Vilhelmova-Ilieva, N.M.; Nikolova, I.N.; Nikolova, N.Y.; Petrova, Z.D.; Trepechova, M.S.; Holechek, D.I.; Todorova, M.M.; Topuzova, M.G.; Ivanov, I.G.; Tumbarski, Y.D. Antiviral Potential of Specially Selected Bulgarian Propolis Extracts: In Vitro Activity against Structurally Different Viruses. *Life* 2023, *13*, 1611. https://doi.org/10.3390/life13071611

Academic Editor: Seung Ho Lee

Received: 15 June 2023
Revised: 19 July 2023
Accepted: 21 July 2023
Published: 23 July 2023

Copyright: © 2023 by the authors. Licensee MDPI, Basel, Switzerland. This article is an open access article distributed under the terms and conditions of the Creative Commons Attribution (CC BY) license (https://creativecommons.org/licenses/by/4.0/).

Abstract: Propolis is a natural mixture of resins, wax, and pollen from plant buds and flowers, enriched with enzymes and bee saliva. It also contains various essential oils, vitamins, mineral salts, trace elements, hormones, and ferments. It has been found that propolis possesses antimicrobial, antiviral, and anti-inflammatory properties. We have studied the antiviral activity of six extracts of Bulgarian propolis collected from six districts of Bulgaria. The study was conducted against structurally different viruses: human coronavirus strain OC-43 (HCoV OC-43) and human respiratory syncytial virus type 2 (HRSV-2) (enveloped RNA viruses), human herpes simplex virus type 1 (HSV-1) (enveloped DNA virus), human rhinovirus type 14 (HRV-14) (non-enveloped RNA virus) and human adenovirus type 5 (HadV-5) (non-enveloped DNA virus). The influence of the extracts on the internal replicative cycle of viruses was determined using the cytopathic effect (CPE) inhibition test. The virucidal activity, its impact on the stage of viral adsorption to the host cell, and its protective effect on healthy cells were evaluated using the final dilution method, making them the focal points of interest. The change in viral infectivity under the action of propolis extracts was compared with untreated controls, and Δlgs were determined. Most propolis samples administered during the viral replicative cycle demonstrated the strongest activity against HCoV OC-43 replication. The influence of propolis extracts on the viability of extracellular virions was expressed to a different degree in the various viruses studied, and the effect was significantly stronger in those with an envelope. Almost all extracts significantly inhibited the adsorption step of the herpes virus and, to a less extent, of the coronavirus to the host cell, and some of them applied before viral infection demonstrated a protective effect on healthy cells. Our results enlarge the knowledge about the action of propolis and could open new perspectives for its application in viral infection treatment.

Keywords: propolis extracts; antiviral activity; virucidal activity; viral adsorption; human coronavirus; human respiratory syncytial virus; herpes simplex virus; human rhinovirus; human adenovirus

1. Introduction

Propolis (bee glue) is a hydrophobic substance with sticky consistency produced by European honeybees (*Apis mellifera* L.), serving as a building and defensive material in their hives. Bees use propolis to smooth the internal walls of the hive, fill up cracks, repair and seal up the cells of the honeycomb, and embalm dead invaders inside the hive, thus removing the unpleasant smell and the microflora accompanying their decomposition and protecting the bee colony from infections [1,2]. In order to produce propolis, the worker bees collect resins from flowers and leaf buds of various plant species. Thereafter, they transport the material to the hive and mix it with beeswax and saliva secreted by their salivary glands. Some recent studies have revealed that the number of chemical compounds identified in propolis has reached 850. They form a heterogenous mixture, which includes mainly polyphenolic compounds like flavonoids (quercetin, galangin, chrysin), aromatic acids and esters, aliphatic acids and esters, volatile compounds, waxy acids, carbohydrates, alcohols, aldehydes, ketones, steroids, enzymes, micro- and macronutrients, amino acids, vitamins, essential oils, pollen, and organic matter [3]. The physicochemical properties, phytochemical composition, and biological activities of propolis vary widely, which depends on the botanical source and the climatic characteristics of the geographic region from which it originates [4].

Propolis is one of the most valuable bee products, which has been used by humans for thousands of years as a remedy in folk and traditional medicine due to the lack of toxicity and the remarkable therapeutic properties. It has been found that the unique phytochemical composition of propolis determines its diverse pharmacological activities, such as antibacterial, antifungal, antiviral, antioxidant, immunomodulatory, antiparasitic, anti-allergic, anti-inflammatory, anticarcinogenic, anaesthetic, hepatoprotective, gastroprotective, anti-ulcerogenic, antidiabetic, astringent and other health beneficial effects [5,6].

Viral outbreaks are widely spread and represent an important problem for the health sector. The application of non-toxic natural products without adverse effects as alternatives of chemotherapeutics is essential for the therapy of viral infections. In this respect, propolis has shown great promise to be used as a potentially effective antiviral agent [7]. It has been reported to possess inhibitory activity against both DNA and RNA viruses. Amoros et al. (1992) [8] investigated the in vitro antiviral effect of propolis on herpes simplex virus type 1 and 2 (HSV-1 and HSV-2), adenovirus type 2, vesicular stomatitis virus and poliovirus type 2. The obtained results demonstrated that propolis had remarkable activity against poliovirus and herpes viruses, while vesicular stomatitis virus and adenovirus were less susceptible. Serkedjieva et al. (1992) [9] evaluated the in vitro antiviral activity of Bulgarian propolis on H3N2 and H1N1 influenza viruses and stated that propolis inhibited viral replication. Years later, the anti-influenza virus activity was confirmed by Kujumgiev et al. (1999) [10], who investigated the inhibitory effect of propolis extracts on avian influenza virus A/strain Weybridge (H7N7). Other studies reveal the antiviral potential of propolis against herpes simplex virus (HSV-1 and HSV-2) [11], coronavirus 2 (SARS-CoV-2) [12], varicella zoster virus [13], infectious bursal disease virus (IBDV), Reovirus [14] and canine distemper virus [15].

In the present study, the antiviral activity against structurally different viruses of six Bulgarian propolis extracts collected from six districts of Bulgaria was determined.

2. Materials and Methods

2.1. Host Cell Lines

The human colon carcinoma (HCT-8) cells were obtained from the American Type Culture Collection (ATCC) based in Manassas, VA, USA. The HCT-8 [HRT-18] cell line (ATCC-CCL-244, LGC Standards) was cultured in RPMI 1640 growth medium (ATCC-30-2001), supplemented with 10% horse serum (ATCC-30-2021), 0.3 g/L L-glutamine (Sigma-Aldrich, Darmstadt, Germany), and an antibiotic mixture of 100 IU penicillin and 0.1 mg streptomycin/mL (both from Sigma-Aldrich) at 37 °C with a constant supply of 5% CO_2.

Madin–Darbey bovine kidney (MDBK) cells were provided by the National Bank for Industrial Microorganisms and Cell Cultures in Sofia, Bulgaria. These cells were grown in DMEM growth medium (Gibco, Grand Island, NY, USA) with 10% fetal bovine serum

(Gibco BRL, USA), 10 mM HEPES buffer (AppliChem GmbH, Darmstadt, Germany), and antibiotics (100 IU/mL penicillin, 100 µg/mL streptomycin). The incubation occurred in a HERA cell 150 incubator (Heraeus, Hanau, Germany) at 37 °C with a 5% CO_2 atmosphere.

The human epithelial type 2 (HEp-2) cells, derived from human laryngeal carcinoma, were obtained from the National Bank for Industrial Microorganisms and Cell Cultures in Sofia, Bulgaria. These cells were cultured in DMEM growth medium (Gibco, Grand Island, NY, USA) with 10% fetal bovine serum (Gibco, BRL, USA), 10 mM HEPES buffer (AppliChem GmbH, Darmstadt, Germany), and antibiotics (100 IU/mL penicillin, 100 µg/mL streptomycin). They were maintained in a HERA cell 150 incubator (Heraeus, Hanau, Germany) at 37 °C with a humidified atmosphere of 5% CO_2.

Lastly, the human cervical epithelioid carcinoma cells (HeLa Ohio-I) were generously provided by Dr D. Barnard from Utah State University, Logan, UT, USA. These cells were cultured in DMEM growth medium (Gibco, Grand Island, NY, USA) with 10% fetal bovine serum (Gibco, BRL, USA), 10 mM HEPES buffer (AppliChem GmbH, Darmstadt, Germany), and antibiotics (100 IU/mL penicillin, 100 µg/mL streptomycin) at 37 °C with a 5% CO_2 atmosphere in a HERA cell 150 incubator (Heraeus, Hanau, Germany).

2.2. Viruses

Human coronavirus OC-43 (HCoV-OC43) (ATCC: VR-1558) strain was cultured in HCT-8 cells using RPMI 1640 medium supplemented with 2% horse serum, 100 U/mL penicillin, and 100 µg/mL streptomycin. After 5 days of infection, cell lysis was performed through two freeze and thaw cycles, and the virus was titrated following the Reed and Muench formula. Both virus and mock aliquots were stored at −80 °C, as outlined in our previous study [16]. The infectious titer of the stock virus was found to be $10^{6.5}$ $CCID_{50}$/mL.

Herpes simplex virus type 1, Victoria strain (HSV-1), obtained from Prof. S. Dundarov at the National Center of Infectious and Parasitic Diseases in Sofia, was replicated in confluent monolayers of MDBK cells using a maintenance solution Dulbecco's modified Eagle medium (DMEM) from Gibco BRL, Paisley, Scotland, UK, supplemented with 0.5% fetal bovine serum (Gibco BRL, Scotland, UK), and antibiotics (100 IU/mL penicillin, 100 µg/mL streptomycin). Following incubation at 37 °C in a 5% CO_2 incubator, the viral yield was frozen at −80 °C [16]. The infectious titer of the stock virus was determined to be $10^{8.5}$ $CCID_{50}$/mL.

Human rhinovirus type 14 (strain 1059) (HRV-14) used for the experiments was purchased from the American Type Culture Collection (Manassas, VA, USA). HRV-14 stocks were prepared in HeLa Ohio-I cells using a maintenance DMEM medium with 2% fetal bovine serum and antibiotics (100 IU/mL penicillin, 100 µg/mL streptomycin). After incubation at 33 °C in a 5% CO_2 incubator, the recovered virus was frozen at −80 °C. The stock virus titer was determined to be $10^{3.5}$ $CCID_{50}$/mL.

Human respiratory syncytial virus type 2 (Long; HRSV-2), kindly provided by the Regional Center for Hygiene and Epidemiology, Plovdiv, Bulgaria, was grown in HEp-2 cells using DMEM maintenance medium (Gibco, BRL) containing 10 mmol/l HEPES buffer (Gibco, BRL), 0.5% fetal calf serum (Gibco BRL), and antibiotics (100 IU/mL penicillin, 100 µg/mL streptomycin). Following incubation at 37 °C in a 5% CO_2 incubator, the viral yield was frozen at −80 °C. The infectious viral titer was determined to be $10^{4.5}$ $CCID_{50}$/mL.

Human adenovirus type 5 (HadV-5), kindly provided by the District Center for Hygiene and Epidemiology, Plovdiv, Bulgaria, was replicated in HEp-2 cells in the presence of DMEM (Gibco, BRL) maintenance medium containing 10 mmol/L HEPES buffer (Gibco, BRL), 0.5% fetal calf serum (Gibco BRL), and antibiotics (100 IU/mL penicillin, 100 µg/mL streptomycin). The resulting amount of virus was frozen at −80 °C. The infectious viral titer was determined to be $10^{5.0}$ $CCID_{50}$/mL.

2.3. Raw Propolis Material

Six fresh propolis samples collected from beekeepers at the end of the active beekeeping season (August–October) in six locations in Bulgaria were used in the study (Table 1). The samples were stored in plastic containers at room temperature in darkness until analysis.

Table 1. Origin of the propolis samples.

Propolis Sample (PS)	Town/Village	Municipality	District	GPS Coordinates
PS1	Silistra	Silistra	Silistra	44°07′ N 27°17′ E
PS2	Simitli	Simitli	Blagoevgrad	41°54′ N 23°08′ E
PS3	Gorna Malina	Gorna Malina	Sofia	42°41′ N 23°42′ E
PS4	Shumen	Shumen	Shumen	43°16′ N 26°55′ E
PS5	Vladimir	Radomir	Pernik	42°26′ N 23°05′ E
PS6	Cherven breg	Dupnitsa	Kyustendil	42°18′ N 23°10′ E

2.4. Reference Compound

Remdesivir (GS-5734, RDV, REM, Veklury®) (Gilead Science Ireland UC) was initially dissolved in double distilled water to a concentration of 150 mg/mL and then diluted in RPMI nutrient medium to the required concentrations.

Acyclovir {ACV, [9-(2-hydroxyethoxymethyl)-guanine]} was kindly provided by the Deutsches Kresforschung Zentrum, Heidelberg, with a stock concentration of 3 mM solution in DMSO. Then, falling dilutions were made in DMEM medium to the required concentration.

Ribavirin (1-(β-D-ribofuranosyl)-1H-1,2,4-triazole-3-carboxamide), kindly provided by Prof. R. W. Sidwell, Utah State University, Logan, USA, was dissolved directly into the DMEM medium.

2.5. Preparation of Propolis Extracts

The raw propolis samples were finely ground using a blender (Bosch, Germany). The propolis extracts were prepared by weighing 1 g of sample and pouring 10 mL of 70% ethanol (Sigma-Aldrich, Merck, Germany) in a plastic tube. Next, the samples were left at room temperature for 72 h in darkness and periodically shaken on vortex V-1 (Biosan, Latvia) during the extraction period. The obtained extracts were filtered through filter paper and then stored at 4 °C for further analyses [17].

2.6. Total Phenolic Content

The total phenolic content (TPC) was determined by the standard method using a Folin–Ciocalteu reagent (Sigma-Aldrich, Merck), 1 mL of which was mixed with 0.8 mL of 7.5% sodium carbonate (Sigma-Aldrich, Merck) and 0.2 mL of the tested propolis extract. Then, the mixture was kept at room temperature for 20 min (in darkness), and the absorbance was measured at 765 nm (Camspec M107, Spectronic-Camspec Ltd., UK) against a blank (distilled water). The results were presented as mg equivalent of gallic acid (GAE)/g propolis [18].

2.7. Total Flavonoid Content

The total flavonoid content (TFC) was determined according to the standard procedure [18]. An aliquot of 1 mL of the tested propolis extract was mixed with 0.1 mL of 10% $Al(NO_3)_3$, 0.1 mL of 1 M CH_3COOK, and 3.8 mL of distilled water. The sample was left at room temperature for 40 min, and then the absorbance was measured at 415 nm using quercetin as a standard. The results are expressed as mg quercetin equivalents (QE)/g propolis.

2.8. Antioxidant activity

DPPH radical scavenging assay. The reaction mixture containing 2.85 mL of DPPH reagent (2,2-diphenyl-1-picrylhydrazyl) and 0.15 mL of the tested propolis extract was incubated at 37 °C for 15 min. The reduction of absorbance was measured at 517 nm against a blank (methanol). The antioxidant activity was expressed as mM Trolox® equivalents (TE)/g propolis [18].

Ferric-reducing antioxidant power (FRAP) assay. The FRAP reagent was freshly prepared with 300 mM acetate buffer with pH 3.6, 10 mM 2,4,6-Tris(2-pyridyl)-s-triazine (TPTZ) in 40 mM hydrochloric acid, and 20 mM Iron (III) chloride hexahydrate in distilled water in a ratio of 10:1:1. The reaction mixture (3 mL of FRAP reagent and 0.1 mL of the propolis extract) was incubated at 37 °C for 10 min, in darkness. The absorbance was measured at 593 nm against a blank (distilled water). The antioxidant activity was expressed as mM TE/g propolis [18].

2.9. Cytotoxicity Assay

A confluent monolayer of cell culture in 96-well plates (Costar®, Corning Inc., Kennebunk, ME, USA) was subjected to treatment with 0.1 mL/well of the support medium, either without the tested propolis extracts or with varying decreasing concentrations of the extracts. The cells were then incubated under specific conditions similar to those used for subsequent virus experiments: 33 °C with 5% CO_2 for 5 days (for HCT-8), 33 °C with 5% CO_2 for 2 days (for HeLa Ohio-I), and 37 °C with 5% CO_2 for 2 days (for MDBK and HEp-2 cells). After the designated incubation period, the propolis extracts were removed, and the cells were washed before being incubated with neutral red (NR) dye at 37 °C for 3 h. The concentration of the test sample that reduced cell viability by 50% compared to untreated controls was defined as the 50% cytotoxic concentration (CC_{50}). Each sample was tested in triplicate, with four wells per replicate.

The maximally tolerated concentration (MTC) of the extracts, which is the concentration at which they do not affect the cell monolayer, was also determined. The methodology is described in more detail in our previous study [16].

2.10. Antiviral Activity Assay

To assess the antiviral activity of propolis extracts, the cytopathic effect inhibition (CPE) test was employed. A 96-well plate with a confluent cell monolayer was infected with 100 cell culture infectious doses of 50% ($CCID_{50}$) in 0.1 mL. After 2 h of adsorption at 33 °C (for HCoV OC-43 and HRV-14), 2 h of adsorption at 37 °C (for HRSV-2) and 1 h of adsorption at 37 °C (for HSV-1 and HadV-5) unattached virus was removed, and the tested extract was added at different concentrations, and the cells were incubated for 5 days at 33 °C (for HCoV OC-43); 2 days at 33 °C (for HRV-14) or 2 days at 37 °C (for HRSV-2, HSV-1 and HadV-5) and in the presence of 5% CO_2. The cytopathic effect was determined using a neutral red uptake assay, and the percentage of CPE inhibition for each test sample concentration was calculated using the following formula:

$$\% \text{ CPE} = [\text{OD test sample} - \text{OD virus control}]/[\text{OD toxicity control} - \text{OD virus control}] \times 100$$

where ODtest sample is the mean of the ODs of the wells inoculated with the virus and treated with the test sample at the corresponding concentration, ODs virus control is the mean of the ODs of the virus control wells (no compound in the medium). OD control for toxicity is the mean of the ODs of the wells not inoculated with the virus but treated with the corresponding concentration of the test compound. The 50% inhibitory concentration (IC_{50}) is defined as the concentration of the test substance that inhibits 50% of viral replication compared to the viral control. The selectivity index (SI) is calculated from the CC_{50}/IC_{50} ratio [16].

2.11. Virucidal Assay

Preparations were made with a total volume of 1 mL containing virus (10^4 $CCID_{50}$) and the tested propolis extract at its maximum permissible concentration (MTC) in a 1:1 ratio. In parallel, a sample was created with untreated virus diluted 1:1 with DMEM medium. Both the control and experimental samples were incubated at room temperature for various time intervals (15, 30, 60, 90, and 120 min). Using the endpoint dilution method of Reed and Muench (1938) [19], the residual infectious virus content in each sample and the Δlgs compared to untreated controls were then determined.

2.12. Effect on Viral Adsorption

To initiate the experiments, twenty-four well plates were chilled to 4 °C and then inoculated with 10^4 $CCID_{50}$ of either HCoV OC-43 or HSV-1, depending on whether HCT-8 or MDBK monolayers were used, respectively. Concurrently, the monolayer was treated with the tested propolis extracts at their maximum permissible concentration (MTC) and kept at 4 °C during the virus adsorption time. The virus and propolis extract were removed at different time intervals, which varied for the two types of viruses (15, 30, 45, and 60 min for HSV-1, and 15, 30, 60, 90, and 120 min for HCoV OC-43). Subsequently, the cells were washed with PBS and covered with maintenance medium before being incubated at 37 °C (for HSV-1) or 33 °C (for HCoV OC-43) in the presence of 5% CO_2 for 24 h. After three cycles of freezing and thawing, the infectious viral titer of each sample was determined and compared to the viral titer of the control for the respective time interval. The Δlgs (logarithmic differences) were then calculated. Each sample was prepared in quadruplicate for reliable data analysis.

2.13. Pre-Treatment of Healthy Cells

Previously grown monolayers of MDBK or HCT-8 cells in 24-well cell culture plates (CELLSTAR, Greiner Bio-One) were exposed to the propolis extracts at their maximum permissible concentration (MTC). The samples were then incubated at 37 °C for different time intervals of 15, 30, 60, 90, and 120 min. After the designated time, the extracts were removed, and the cells were washed with PBS before being inoculated with the respective virus strain (1000 $CCID_{50}$ in 1 mL/well). For HCoV OC-43, the virus adsorption lasted for 120 min, and for HSV-1, it was 60 min. Afterwards, any unadsorbed virus was removed, and the cells were covered with a support medium. Subsequently, the samples were incubated at 33 °C (for HCoV OC-43) or 37 °C (for HSV-1) in the presence of 5% CO_2 for 24 h. Following this incubation period, the samples were subjected to triplicate freezing and thawing, and the infectious virus titers were determined. Δlg (logarithmic differences) were calculated by comparing the viral titer of the treated samples to the viral titer of the control (untreated with extract) for the respective time interval. Each sample was prepared in quadruplicate for accurate and reliable analysis.

2.14. Statistical Analysis

Data on cytotoxicity and antiviral effects were analysed statistically. The values of CC_{50} and IC_{50} were presented as means ± SD. The differences' significance between the cytotoxicity values of propolis extracts and the reference substances, as well as between the effects of the test products on the viral replication, was performed by Student's *t*-test, with *p*-values of <0.05 were considered significant. The final data sets were analysed with the Graph Pad Prism 4 software.

3. Results

From the studies carried out so far on the composition of various types of propolis, it has been established that its composition includes about 850 ingredients. The main compounds that contribute to its biological activities are polyphenols and especially flavonoids [3]. Therefore, we focused our attention on the ingredients contained in the propolis extracts we studied. The propolis extracts presented in this manuscript were selected after an initial screening selection. Eighty propolis extracts from different territories of Bulgaria were studied. Of all the extracts, these six showed the highest content of polyphenols and flavonoids, as well as the most distinct antioxidant activity. Table 2 presents the results of total phenolic content (TPC) and total flavonoid content (TFC) of the six propolis extracts. Of the studied samples, PS6 showed the highest amount of TPC

(256.1 ± 0.56 mg GAE/g propolis), and PS5 demonstrated the lowest (151.7 ± 0.32 mg GAE/g propolis). Comparing the obtained data for TFC, the highest value was reported for PS2 (124.1 ± 2.23 mg QE/g propolis) and the lowest for PS6 (74.6 ± 0.15 mg QE/g propolis).

Table 2. Total phenolic content (TPC), total flavonoid content (TFC) and antioxidant activity of the ethanolic propolis extracts.

Propolis Sample	TPC, mg GAE/g	TFC, mg QE/g	Antioxidant Activity	
			DPPH, mM TE/g	FRAP, mM TE/g
PS1	216.9 ± 0.28	78.3 ± 0.38	873.2 ± 14.50	867.5 ± 60.00
PS2 *	168.0 ± 0.63	124.1 ± 2.23	1201.4 ± 26.23	715.3 ± 6.12
PS3	243.1 ± 0.28	87.9 ± 0.11	1106.9 ± 43.50	1000.1 ± 37.50
PS4	230.8 ± 0.30	77.0 ± 0.50	1083.3 ± 34.80	978.9 ± 17.50
PS5 *	151.7 ± 0.32	87.0 ± 2.01	1016.3 ± 22.20	645.3 ± 11.25
PS6	256.1 ± 0.56	74.6 ± 0.15	1133.5 ± 23.22	1085.0 ± 22.50

* The result is presented in our previous study [20].

Antioxidant activity results were evaluated by two methods. The DPPH assay showed the highest antioxidant activity in PS2, while PS1 had the lowest antioxidant potential of the studied propolis extracts. According to the FRAP method, PS6 possessed the highest antioxidant activity, and PS5 showed the lowest value by the same method (Table 2).

In order to avoid the negative influence of toxic concentrations of the propolis extracts when conducting the antiviral experiments, the cytotoxicity they exert on the cells was determined in advance. The effect of the extracts on the cells was determined against the four cell lines on which the antiviral experiments were carried out in the next step (HCT-8, MDBK, HEp-2 and HeLa Ohio cells). Against MDBK, HEp-2 and HeLa Ohio cell lines, cytotoxicity was measured after two days of incubation with the extracts, as this is the time interval at which antiviral experiments are determined for the respective viruses. To achieve a good cytopathic effect with HCoV OC-43, a longer time of 5 days was required. Therefore, cytotoxicity with the HCT-8 cell line was also measured on day 5.

From the experiments performed, it can be seen that, in general, the highest cytotoxicity was reported for the HCT-8 cell line, most likely because the exposure time was the longest. From the rest of the cell lines, where the effect was measured for the same time interval, the extracts showed the weakest cytotoxicity against the HeLa Ohio cell line, close but slightly higher toxicity in MDBK cells, and the most sensitive to the action of propolis extracts from the cells tested turned out to be HEp-2. The weakest cytotoxicity was shown by the extracts PS4, PS3 and PS1 against HeLa Ohio cells, as well as PS1 and PS3 against the MDBK cell line. As a general effect on the four cell lines, PS6 showed the highest toxicity. When comparing the cytotoxicity of the tested samples to the reference substances used, it is noticed that they are less toxic than Ribavirin but demonstrate several times higher cytotoxicity compared to Acyclovir and Remdesivir (Table 3).

Once the non-toxic concentration range of the propolis samples was identified, their impact on the internal replicative cycle of the virus was studied at concentrations below the CC_{50}. In general, the influence of the extracts is strongest for the herpes virus and to a slightly lesser extent for the coronavirus and the rhinovirus. The effect was weakest with adenovirus, and almost in the same range was the inhibition reported with RSV-2. The highest selective index (SI) of all propolis samples showed PS4 against HSV-1 (SI = 45.3) and HCoV OC-43 (SI = 43.3). PS2 also showed significant activity against HSV-1 replication (SI = 32.9). The distinct activity was also demonstrated by PS1 (SI = 28.7) applied to the replication of HRV-14 and PS2 (SI = 26.6), affecting HCoV OC-43. The substances PS4 (SI = 22.3) for HRV-14 and PS6 (SI = 21.7) for HSV-1 have similar activity. Significantly lower was the influence of PS2 (SI = 10.0) on HRV-14; PS5 (SI = 8.9) in HSV-1; and PS 1 and PS2 with SI between 7.5 and 8.6 for RSV-2 and HAdV-5 (Table 4).

Table 3. In vitro assessment of cytotoxicity of the propolis extracts.

Propolis Sample	Cytotoxicity (µg/mL)							
	HCT-8		MDBK		HEp-2		HeLa Ohio	
	CC_{50}	MTC	CC_{50}	MTC	CC_{50}	MTC	CC_{50}	MTC
PS1	58.0 ± 8.2 **	10.0	120.0 ± 6.7 **	32.0	56.5 ± 2.1 *	30.0	158.0 ± 7.2 **	70.0
PS2	48.0 ± 4.5 **	10.0	72.5 ± 3.8 **	10.0	68.2 ± 3.3 *	30.0	77.7 ± 2.9 *	35.0
PS3	62.6 ± 5.2 **	10.0	104.7 ± 7.3 **	32.0	57.0 ± 2.5 *	30.0	174.0 ± 6.5 **	70.0
PS4	52.0 ± 3.6 **	10.0	73.5 ± 3.2 **	10.0	67.7 ± 2.9 *	30.0	190.0 ± 7.0 **	70.0
PS5	59.8 ± 7.3 **	10.0	71.8 ± 4.7 **	10.0	62.5 ± 2.4 *	30.0	77.0 ± 2.5 *	35.0
PS6	57.0 ± 6.5 **	10.0	66.2 ± 5.7 **	10.0	59.4 ± 1.5 *	20.0	58.0 ± 2.0 *	30.0
Acyclovir	nd	nd	291.0 ± 9.4 **	nd	nd	nd	nd	nd
Remdesivir	250.0 ± 4.3	nd	nd	nd	nd	nd	nd	nd
Ribavirin	nd	nd	nd	nd	14.0 ± 0.5	nd	34.0 ± 0.5	nd

nd—no data; * $p < 0.05$; when comparing the value of each propolis extract with the corresponding reference substance for the given cell line; ** $p < 0.001$ when comparing the value of each propolis extract with the corresponding reference substance for the given cell line.

Table 4. In vitro antiviral activity of the propolis extracts.

Propolis Sample	Antivirus Activity (µg/mL)									
	HCoV OC-43		HSV-1		HAdV-5		RSV-2		HRV-14	
	IC_{50} (µg/mL)	SI	IC_{50} (µg/mL)	SI	IC_{50} (µg/mL)	SI	IC_{50} (µg/mL)	SI	IC_{50} (µg/mL)	SI
PS1	-	-	-	-	7.5 ± 0.2 **	7.5	6.7 ± 0.3 **	8.4	5.5 ± 0.2 **	28.7
PS2	1.8 ± 0.3 *	26.6	2.2 ± 0.3 *	32.9	8.2 ± 0.3 **	8.3	7.9 ± 0.3 **	8.6	7.8 ± 0.2 **	10.0
PS3	-	-	-	-	-	-	47.0 ± 2.2 **	1.2	7.0 ± 0.1 **	2.8
PS4	1.2 ± 0.1 *	43.3	1.4 ± 0.02 *	45.3	-	-	44.0 ± 2.4 **	1.5	8.5 ± 0.3 **	22.3
PS5	10.2 ± 0.9	5.8	8.2 ± 1.2 **	8.9	-	-	46.0 ± 1.8 **	1.4	-	-
PS6	10.8 ± 0.7	5.2	3.3 ± 0.8 **	21.7	-	-	-	-	-	-
Acyclovir	nd	nd	0.33 ± 0.03	881.8	nd	nd	nd	nd	nd	nd
Remdesivir	12.5 ± 0.9	200.0	nd	nd	nd	nd	nd	nd	nd	nd
Ribavirin	nd	nd	nd	nd	0.2 ± 0.01	70.0	0.3 ± 0.01	46.6	0.5 ± 0.02	68.0

-, lack of inhibition of viral replication; nd—no data; * $p < 0.05$; when comparing the value of each propolis extract with the corresponding reference substance for the given virus strain; ** $p < 0.001$, when comparing the value of each propolis extract with the corresponding reference substance for the given virus strain.

Having assessed the effect of the investigated propolis samples on the replication of structurally diverse viruses, the subsequent phase of our research involved examining the impact of the extracts on the vitality of extracellular virions. The results from the experiments showed a stronger effect on enveloped viruses compared to non-enveloped ones. The effect was monitored at different time intervals, and, in general, a dependence of the effect on the exposure time was noticed. With a longer exposure, the inhibition of virus particles increased. The most significant was the effect of PS5 (Δlg = 2.25) at 90 and 120 min on HCoV OC-43 virions, with a similar effect on HSV-1 and, to a less extent, on HRSV-2. A distinct effect of Δlg = 2.0 in HCoV OC-43 was also demonstrated by PS4 and PS6, whose influence on HRSV-2 was less pronounced, respectively, Δlg = 1.6 and

Δlg = 2.0 at the longest interval of 120 min. PS6 was the only one of the six investigated extracts which exhibited a significant effect (Δlg = 1 = 8) at 120 min on HRV-14. PS1 also exerted a suppressive effect on extracellular HSV-1, even at a contact duration interval of 30 min; compared to HCoV OC-43 virions, the effect was weaker (Δlg = 1.75) (Tables 5 and 6).

Table 5. Virucidal activity of enveloped viruses.

Propolis Sample	Δlg														
	HCoV OC-43					HRSV-2					HSV-1				
	15 min	30 min	60 min	90 min	120 min	15 min	30 min	60 min	90 min	120 min	15 min	30 min	60 min	90 min	120 min
PS1	1.25	1.25	1.25	1.75	1.75	0.0	0.0	0.5	1.0	1.0	1.0	1.75	1.75	2.0	2.0
PS2	1.0	1.25	1.5	1.5	1.5	0.2	0.2	0.6	1.5	1.5	1.0	1.0	1.5	1.5	1.5
PS3	0.5	0.5	1.0	1.5	1.5	0.2	0.2	0.7	1.2	1.6	1.0	1.0	1.25	1.25	1.25
PS4	0.5	1.0	1.5	2.0	2.0	0.3	0.3	0.7	1.4	1.6	1.0	1.0	1.5	1.5	1.5
PS5	1.25	1.25	1.25	2.25	2.25	0.3	0.3	0.9	1.5	2.0	1.0	1.0	1.5	1.75	2.25
PS6	1.0	1.25	1.25	2.0	2.0	0.2	0.2	0.9	1.7	2.0	1.0	1.0	1.5	1.5	1.75
70% etanol	5.75	5.75	5.75	5.75	5.75	4.5	4.5	4.5	4.5	4.5	4.75	4.75	4.74	4.74	4.74

Table 6. Virucidal activity of non-enveloped viruses.

Propolis Sample	Δlg								
	HAdV-5					HRV-14			
	15 min	30 min	60 min	90 min	120 min	15 min	30 min	60 min	90 min	120 min
PS1	0.0	0.0	0.5	1.0	1.0	0.0	0.0	0.1	1.4	1.6
PS2	0.0	0.0	0.7	1.0	1.0	0.0	0.0	0.0	1.2	1.5
PS3	0.0	0.0	0.5	1.0	1.0	0.0	0.0	0.3	1.2	1.6
PS4	0.0	0.0	0.8	0.8	1.2	0.0	0.0	0.2	1.3	1.5
PS5	0.0	0.0	0.4	0.8	0.8	0.0	0.0	0.2	1.3	1.3
PS6	0.0	0.0	0.6	0.8	1.0	0.0	0.0	0.0	1.5	1.8
70% etanol	5.0	5.0	5.0	5.0	5.0	3.2	3.2	3.2	3.2	3.2

Having determined the effect of propolis extracts on extracellular virions and virus replication in the cell, the next step in our research was to follow the effect of the extracts on the adsorption step of the virus to the cell. The experiments were carried out with the two viruses for which we obtained the most distinct activity so far—HSV-1 and HCoV OC-43. The influence was again followed at different time intervals depending on the duration of viral adsorption (up to 60 min for the herpes virus and up to 120 min for the coronavirus). All extracts demonstrated varying degrees of influence on this stage of viral reproduction. The inhibition of the process was more pronounced in HSV-1 compared to HCoV OC-43. The strongest effect on HSV-1 was shown by PS5 and PS6 (Δlg = 2.25) 30 min after exposure, and the influence of PS6 was significant as early as 15 min (Δlg = 1.75). The effect of PS2, PS3 and PS4 (Δlg = 2.0) was also significant at 30 min and remained unchanged until the last investigated time interval of 60 min. PS1, although to a less extent, also affected the adsorption stage of the virus with a decrease in the viral titer with Δlg = 1.75.

When tracking the adsorption of HCoV OC-43 to the HCT-8 cells, a significantly weaker influence of the extracts was observed, and at a slightly longer contact period—at 90 or 120 min. Here, the influence of PS4 is most significant per 120 min (Δlg = 2.25). Additionally, a distinct effect at 120 min of Δlg = 1.75 was demonstrated by PS1, PS2 and PS5. PS3 and PS6 showed weak activity towards the adsorption of HCoV OC-43 (Table 7).

Table 7. Influence of the extracts on the stage of adsorption of HSV-1 and HCoV OC-43 to sensitive cells.

Propolis Sample	Δlg								
	HSV-1				HCoV OC-43				
	15 min	30 min	45 min	60 min	15 min	30 min	60 min	90 min	120 min
PS1	1.5	1.75	1.75	1.75	1.0	1.0	1.0	1.5	1.75
PS2	1.5	2.0	2.0	2.0	1.0	1.0	1.0	1.25	1.75
PS3	1.5	1.75	2.0	2.0	0.75	1.0	1.0	1.25	1.5
PS4	1.5	2.0	2.0	2.0	0.75	0.75	1.0	1.75	2.25
PS5	1.25	2.25	2.25	2.25	1.5	1.5	1.5	1.75	1.75
PS6	1.75	2.25	2.25	2.25	1.5	1.5	1.5	1.5	1.5

After assessing the influence of propolis extracts on various stages of viral reproduction and vitality, we investigated whether these substances provided a protective effect on healthy cells, guarding them against subsequent viral infections. The experiments were once again conducted using the MDBK and HCT-8 cell lines, along with the viral strains HSV-1 and HCoV OC-43. The results revealed that none of the extracts significantly protected HCT-8 cells from HCoV OC-43 infection, with the maximum decrease in viral titer being Δlg = 1.0. However, in the case of MDBK cells, the effect was notably stronger. The most pronounced protection was observed with PS3 after 15 min of treatment (Δlg = 2.0), and this effect further increased to Δlg = 2.25 after 120 min. Similar protection was shown by PS4, but the effect was significant at 30 min. PS2 also has a strong influence, which maintains the same activity during all monitored time intervals (Δlg = 2.0). The other three propolis extracts: PS1, PS5 and PS6, showed weak protective effects on sensitive healthy cells (Table 8).

Table 8. Protective effect of pre-treatment of extracts on healthy cells and subsequent virus infection.

Propolis Sample	Δlg									
	HSV-1					HCoV OC-43				
	15 min	30 min	60 min	90 min	120 min	15 min	30 min	60 min	90 min	120 min
PS1	1.25	1.25	1.25	1.25	1.25	0.5	0.5	0.5	1.0	1.0
PS2	2.0	2.0	2.0	2.0	2.0	0.5	0.5	0.5	1.0	1.0
PS3	2.0	2.0	2.0	2.0	2.25	0.5	0.5	0.5	1.0	1.0
PS4	1.0	2.0	2.0	2.0	2.25	0.5	0.5	0.5	1.0	1.0
PS5	1.0	1.0	1.0	1.0	1.0	1.0	1.0	1.0	1.0	1.0
PS6	1.0	1.0	1.0	1.0	1.25	0.5	0.5	0.5	0.5	0.5

4. Discussion

In some viruses that cause respiratory, intestinal, skin, and sexually transmitted infections in humans, antiviral agents have been developed that significantly reduce the severity of symptoms and shorten the recovery period. However, therapy failure is increasingly observed due to the selection of therapy-resistant mutants [21]. This is a major reason for the intensive search for new, unconventional antiviral agents closer to the cell components, are low toxic, cause fewer side effects, and can serve as an alternative to the currently used antiviral therapeutics.

In recent decades, more and more data have been accumulated from the study of the various biological activities of propolis. One of its benefits is the impact on developing viral infections [22,23]. The exact mechanisms by which its influence is carried out are still not sufficiently studied. Two main directions of its action have been established: (1) direct interaction with the virus or on stages of its replication [13,23–27] and (2) stimulation of the immune system to overcome infection [28–30].

The mechanism and strength of the antiviral effect of propolis are determined by the multitude of substances included in its composition, which is determined by the geographical location and the season of the year in which it was obtained [30]. In the temperate zones of Europe, North America and Asia, where the predominant source is poplar tree species, most often black poplar (*Populus nigra*), propolis contains mainly flavanones and flavones and smaller amounts of phenolic acids and their esters [31]. Propolis flavonoids show antiviral activity against DNA and RNA viruses [32] and immunomodulatory action [30]. Over 150 types of flavonoids have been found in different types of propolis [33]. Propolis from tropical countries contains mainly complex phenolic compounds such as prenylated para-coumaric acids, prenylated flavonoids, caffeoylquinic acid derivatives and lignans [34].

Some studies show that propolis can affect virus replication [28] by reducing the synthesis of viral RNA transcripts in cells and thus reducing the number of coronavirus particles [13] or by inhibiting Varicella zoster virus DNA polymerase [27]. Another potential mechanism of inhibition of viral replication is the proven inhibitory activity of Sulawesi propolis compounds against the enzymatic activity of SARS-CoV-2 main protease [25].

Much research has shown that in contact with the viral particle, propolis destroys the ability of the pathogen to enter the cell [28,35,36]. Virus particles with altered morphology were observed, suggesting possible damage to viral envelope proteins. Virions were also found in an electrodense layer formed around the cell membrane. This has been suggested to affect the entry of the virus into the host cell and disrupt its replication cycle [37].

Basically, viruses are divided into two groups—enveloped and non-enveloped. Non-enveloped viruses are covered with specific viral proteins (capsomeres) (forming the viral capsid), which can be the target of the action of various substances outside the cell. In most cases, these proteins are stable and difficult to influence. Among the viruses we use with such a structure are human rhinovirus type 14 (HRV-14) and human adenovirus type 5 (HadV-5). In enveloped viruses, there is another shell on top of the capsid, which comprises lipids and proteins. This envelope is much more sensitive to the action of different substances, so it has been experimentally shown that when the same substance acts on an enveloped and a non-enveloped virus, the effect on the enveloped virus is significantly stronger. Among the enveloped viruses used in our study are: human coronavirus strain OC-43 (HCoV OC-43), human respiratory syncytial virus type 2 (HRSV-2), and human herpes simplex virus type 1 (HSV-1). Our studies on the virucidal activity of propolis extracts prove the observation described above.

After entering the host cell, each type of virus has its specific viral enzymes, thanks to which it manages to build many of its structural components and assemble them into new virus particles that leave the host cell. These specific viral enzymes differ in a number of characteristics from cellular ones and are putative targets for attack by structural components contained in propolis extracts. As a result of such an interaction, the processes of transcription, translation and replication of viral components and/or their assembly and exit from the host cell are disrupted.

Our results were in agreement with other scientific publications on this topic. We used different experimental setups, each adding the propolis extracts at different stages of the viral infection cycle. In a similar way, the data obtained by other researchers were close to ours, proving a significant influence of propolis on extracellular virions, especially in enveloped viruses, as well as on the stage of viral adsorption on susceptible cells [35,36,38]. A more detailed study of the specific mechanism of action will take place after a more detailed study of the chemical composition of the samples.

We obtained similar results in our previous study of Canadian propolis [39], where we demonstrated an effect on virions and the adsorption stage of Herpes simplex virus types 1 and 2 to MDBK cells. In the present study with the Bulgarian propolis, an influence on the viral replicative cycle was also found in some of the samples, which is clearly due to the differences in the composition of the propolis due to their different geographical origin.

The main belief among human society is that propolis acts on the immune system and thus helps the body overcome various infectious diseases. Our research proves that propolis also has an inhibitory effect on various viruses, directly damaging their structure or affecting stages of their replication.

We also introduced a new research setup where we treated the still healthy cells with the propolis extracts and determined the degree of protection that propolis shows on the cell membrane from subsequent viral infection. Our results reconfirm the data obtained by other teams who used a similar experimental methodology and found that the application of 0.5 mg/mL EEP two hours before infection in MDBK cells caused a reduction in the number of Aujeszky's disease virus formed plaque compared to the other treatments used or to the infected and untreated culture [37].

This proves that if propolis is used prophylactically, it could significantly protect healthy cells in our body from viral invasion. In combination with its already proven antiviral and immunomodulating activities, its application, especially in periods of epidemics, would reduce morbidity and shorten the recovery period.

5. Conclusions

The present study once again confirms our previously reported data on the antiviral activity of propolis. The study of the activity against the replication of structurally distinct viruses showed a different degree of inhibition in individual propolis samples. In direct contact of the propolis extracts with the virus particles, it was found that the effect was stronger for enveloped viruses, which is most likely the result of the interaction of the components included in the composition of propolis with the viral proteins of the envelope, necessary for attachment and entry into the cell, resulting in inactivation of the virus. The data accumulated so far on the activity of propolis present it as a promising candidate for inclusion in the prevention and treatment of many infectious diseases. Still, it is necessary to expand the knowledge of its mechanism of action for its more complete application.

Author Contributions: Conceptualization, N.M.V.-I., I.N.N. and Y.D.T.; propolis samples providing, Y.D.T.; methodology, N.M.V.-I., I.N.N., N.Y.N., Z.D.P., I.G.I., M.M.T. and M.G.T.; formal analysis, N.M.V.-I., I.G.I., Y.D.T. and I.N.N.; investigation, N.M.V.-I., I.N.N., N.Y.N., Z.D.P., M.S.T., D.I.H., I.G.I., Y.D.T., M.M.T. and M.G.T.; resources, N.M.V.-I., I.G.I., Y.D.T. and I.N.N.; data curation, N.M.V.-I., I.G.I., Y.D.T. and I.N.N.; writing—original draft preparation, N.M.V.-I., Y.D.T., M.S.T. and Z.D.P.; writing—review and editing, N.M.V.-I., I.N.N. and Y.D.T.; software, N.M.V.-I. and I.N.N.; visualization, N.M.V.-I., I.G.I., Y.D.T. and I.N.N.; project administration, N.M.V.-I.; funding acquisition, N.M.V.-I.; supervision, N.M.V.-I. All authors have read and agreed to the published version of the manuscript.

Funding: This research was funded by the National Science Fund at the Ministry of Education and Science, Bulgaria, approved by Research Grant No КП-06- K1/3 "Biopolymer-based functional platforms for advanced in vitro target and co-delivery of therapeutic payloads for the treatment of coronavirus infection".

Institutional Review Board Statement: Not applicable.

Informed Consent Statement: Not applicable.

Data Availability Statement: Not applicable.

Conflicts of Interest: The authors declare no conflict of interest. The sponsors had no role in the design, execution, interpretation, or writing of the study.

References

1. Sforcin, J.M. Propolis and the immune system: A review. *J. Ethnopharmacol.* **2007**, *113*, 1–14. [CrossRef]
2. Wagh, V.D. Propolis: A wonder bees product and its pharmacological potentials. *Adv. Pharmacol. Sci.* **2013**, *2013*, 308249. [CrossRef]

3. Bouchelaghem, S. Propolis characterization and antimicrobial activities against Staphylococcus aureus and Candida albicans: A review. *Saudi J. Biol. Sci.* **2022**, *29*, 1936–1946. [CrossRef] [PubMed]
4. Bankova, V.; Popova, M.; Trusheva, B. New emerging fields of application of propolis. *Maced. J. Chem. Chem. Eng.* **2016**, *35*, 1–11. [CrossRef]
5. Salomao, K.; Dantas, A.P.; Borba, C.M.; Campos, L.C.; Machado, D.G.; Aquino Neto, F.R.; de Castro, S.L. Chemical composition and microbicidal activity of extracts from Brazilian and Bulgarian propolis. *Lett. Appl. Microbiol.* **2004**, *38*, 87–92. [CrossRef] [PubMed]
6. Cauich-Kumul, R.; Campos, M.R.S. Bee propolis: Properties, chemical composition, applications, and potential health effects. In *Bioactive Compounds*; Elsevier: Amsterdam, The Netherlands, 2019; pp. 227–243. [CrossRef]
7. Ripari, N.; Sartori, A.A.; da Silva Honorio, M.; Conte, F.L.; Tasca, K.I.; Santiago, K.B.; Sforcin, J.M. Propolis antiviral and immunomodulatory activity: A review and perspectives for COVID-19 treatment. *J. Pharm. Pharmacol.* **2021**, *73*, 281–299. [CrossRef]
8. Amoros, M.; Sauvager, F.; Girre, L.; Cormier, M. In vitro antiviral activity of propolis. *Apidologie* **1992**, *23*, 231–240. [CrossRef]
9. Serkedjieva, J.; Manolova, N.; Bankova, V. Anti-influenza virus effect of some propolis constituents and their analogues (esters of substituted cinnamic acids). *J. Nat. Prod.* **1992**, *55*, 294–297. [CrossRef] [PubMed]
10. Kujumgiev, A.; Tsvetkova, I.; Serkedjieva, Y.; Bankova, V.; Christov, R.; Popov, S. Antibacterial, antifungal and antiviral activity of propolis of different geographic origin. *J. Ethnopharmacol.* **1999**, *64*, 235–240. [CrossRef]
11. Demir, S.; Atayoglu, A.T.; Galeotti, F.; Garzarella, E.U.; Zaccaria, V.; Volpi, N.; Karagoz, A.; Sahin, F. Antiviral activity of different extracts of standardized propolis preparations against HSV. *Antivir. Ther.* **2020**, *25*, 353–363. [CrossRef]
12. Fiorini, A.C.; Scorza, C.A.; de Almeida, A.C.G.; Fonseca, M.C.M.; Finsterer, J.; Fonseca, F.L.A.; Scorza, C.A. Antiviral activity of Brazilian Green Propolis extract against SARSCoV-2 (Severe Acute Respiratory Syndrome—Coronavirus 2) infection: Case report and review. *Clinics* **2021**, *76*, e2357. [CrossRef]
13. Labská, K.; Plodková, H.; Pumannová, J.; Sensch, K.H. Antiviral activity of propolis special extract GH 2002 against Varicella zoster virus in vitro. *Die Pharm.-Int. J. Pharm. Sci.* **2018**, *73*, 733–736.
14. El Hady, F.K.A.; Hegazi, A.G. Egyptian Propolis: 2. Chemical Composition, Antiviral and Antimicrobial Activities of East Nile Delta Propolis. *Z. Für Naturforschung C* **2002**, *57*, 386–394. [CrossRef] [PubMed]
15. González-Búrquez, M.J.; González-Díaz, F.R.; García-Tovar, C.G.; Carrillo-Miranda, L.; Soto-Zárate, C.I.; Canales-Martínez, M.M.; Penieres-Carrillo, J.G.; Crúz-Sánchez, T.A.; Fonseca-Coronado, S. Comparison between In Vitro Antiviral Effect of Mexican Propolis and Three Commercial Flavonoids against Canine Distemper Virus. *Evid. -Based Complement. Altern. Med.* **2018**, *2018*, 7092416. [CrossRef]
16. Nikolova, I.; Paunova-Krasteva, T.; Petrova, Z.; Grozdanov, P.; Nikolova, N.; Tsonev, G.; Triantafyllidis, A.; Andreev, S.; Trepechova, M.; Milkova, V.; et al. Bulgarian medicinal extracts as natural inhibitors with antiviral and antibacterial activity. *Plants* **2022**, *11*, 1666. [CrossRef] [PubMed]
17. Tumbarski, Y.D.; Todorova, M.M.; Topuzova, M.G.; Georgieva, P.I.; Ganeva, Z.A.; Mihov, R.B.; Yanakieva, V.B. Antifungal Activity of Carboxymethyl Cellulose Edible Films Enriched with Propolis Extracts and Their Role in Improvement of the Storage Life of Kashkaval Cheese. *Curr. Res. Nutr. Food Sci.* **2021**, *9*, 487–499. [CrossRef]
18. Ivanov, I.G.; Vrancheva, R.Z.; Marchev, A.S.; Petkova, N.T.; Aneva, I.Y.; Denev, P.P.; Georgiev, V.G.; Pavlov, A.I. Antioxidant activities and phenolic compounds in Bulgarian Fumaria species. *Int. J. Curr. Microbiol. Appl. Sci.* **2014**, *3*, 296–306.
19. Reed, L.J.; Muench, H. A simple method of estimating fifty percent endpoints. *Am. J. Hyg.* **1938**, *27*, 493–497.
20. Tumbarski, Y.; Ilieva, V.; Yanakieva, V.; Topuzova, M.; Todorova, M.; Merdzhanov, P. Application of propolis extracts as biopreservatives in a cosmetic product (hand cream). *J. Hyg. Eng. Des.* **2023**, *43*. in press.
21. Sargsyan, K.; Mazmanian, K.; Lim, C. A strategy for evaluating potential antiviral resistance to small molecule drugs and application to SARS-CoV-2. *Sci. Rep.* **2023**, *13*, 502. [CrossRef]
22. Lima, W.G.; Brito, J.C.M.; da Cruz Nizer, W.S. Bee products as a source of promising therapeutic and chemoprophylaxis strategies against COVID-19 (SARS-CoV-2). *Phyther. Res.* **2020**, *35*, 743–750. [CrossRef]
23. Refaat, H.; Mady, F.M.; Sarhan, H.A.; Rateb, H.S.; Alaaeldin, E. Optimization and evaluation of propolis liposomes as a promising therapeutic approach for COVID-19. *Int. J. Pharm.* **2020**, *592*, 120028. [CrossRef] [PubMed]
24. Alvarez-Suarez, M. (Ed.) *Bee Products—Chemical and Biological Properties*; Springer International Publishing AG: Berlin/Heidelberg, Germany, 2017.
25. Silva-Beltrán, N.P.; Balderrama-Carmona, A.P.; Umsza-Guez, M.A.; Souza Machado, B.A. Antiviral effects of Brazilian green and red propolis extracts on Enterovirus surrogates. *Environ. Sci. Pollut. Res.* **2020**, *27*, 28510–28517. [CrossRef]
26. Sahlan, M.; Irdiani, R.; Flamandita, D.; Aditama, R.; Alfarraj, S.; Javed Ansari, M.; Cahya Khayrani, A.; Kartika Pratami, D.; Lischer, K. Molecular interaction analysis of Sulawesi propolis compounds with SARS-CoV-2 main protease as preliminary study for COVID-19 drug discovery. *J. King Saud Univ. Sci.* **2020**, *33*, 101234. [CrossRef]
27. Kumar, V.; Dhanjal, J.K.; Bhargava, P.; Kaul, A.; Wang, J.; Zhang, H.; Kaul, S.C.; Wadhwa, R.; Sundar, D. Withanone and Withaferin-A are predicted to interact with transmembrane protease serine 2 (TMPRSS2) and block entry of SARS-CoV-2 into cells. *J. Biomol. Struct. Dyn.* **2020**, *40*, 1–13. [CrossRef]
28. Ożarowski, M.; Karpiński, T.M. The Effects of Propolis on Viral Respiratory Diseases. *Molecules* **2023**, *28*, 359. [CrossRef] [PubMed]

29. Zulhendri, F.; Chandrasekaran, K.; Kowacz, M.; Ravalia, M.; Kripal, K.; Fearnley, J.; Perera, C.O. Antiviral, Antibacterial, Antifungal, and Antiparasitic Properties of Propolis: A Review. *Foods* **2021**, *10*, 1360. [CrossRef] [PubMed]
30. Ma, X.; Guo, Z.H.; Li, Y.A.; Yang, K.; Li, X.H.; Liu, Y.L.; Shen, Z.Q.; Zhao, L.; Zhang, Z.Q. Phytochemical Constituents of Propolis Flavonoid, Immunological Enhancement, and Anti-porcine Parvovirus Activities Isolated From Propolis. *Front. Vet. Sci.* **2022**, *9*, 7183. [CrossRef]
31. Mendonca, R.Z.; Nascimento, R.M.; Fernandes, A.C.O.; Silva, P.L., Jr. Antiviral action of aqueous extracts of propolis from Scaptotrigona aff. Postica against Zica, Chikungunya, and Mayaro virus. *bioRxiv* **2022**. [CrossRef]
32. Bankova, V. Recent trends and important developments in propolis research. *Evid. Based Complement. Altern. Med.* **2005**, *2*, 29–32. [CrossRef]
33. Debiaggi, M.; Tateo, F.; Pagani, L.; Luini, M.; Romero, E. Effects of propolis flavonoids on virus infectivity and replication. *Microbiologica* **1990**, *13*, 207–213.
34. Huang, S.; Zhang, C.P.; Wang, K.; Li, G.Q.; Hu, F.L. Recent Advances in the Chemical Composition of Propolis. *Molecules* **2014**, *19*, 19610–19632. [CrossRef]
35. Schnitzler, P.; Neuner, A.; Nolkemper, S.; Zundel, C.; Nowack, H.; Sensch, K.H.; Reichling, J. Antiviral activity and mode of action of propolis extracts and selected compounds. *Phyther. Res.* **2010**, *24*, S20–S28. [CrossRef] [PubMed]
36. Nolkemper, S.; Reichling, J.; Sensch, K.H.; Schnitzler, P. Mechanism of herpes simplex virus type 2 suppression by propolis extracts. *Phytomedicine* **2010**, *17*, 132–138. [CrossRef]
37. Búrquez, M.d.J.G.; Mosqueda, M.d.L.J.; Mendoza, H.R.; Zárate, C.I.S.; Miranda, L.C.; Sánchez, T.A.C. Protective effect of a Mexican propolis on MDBK cells exposed to aujeszky's disease virus (pseudorabies virus). *Afr. J. Tradit. Complement. Altern. Med.* **2015**, *12*, 106–111. [CrossRef]
38. Búfalo, M.C.; Figueiredo, A.S.; De Sousa, J.P.B.; Candeias, J.M.G.; Bastos, J.K.; Sforcin, J.M. Anti-poliovirus activity of Baccharis dracunculifolia and propolis by cell viability determination and real-time PCR. *J. Appl. Microbiol.* **2009**, *107*, 1669–1680. [CrossRef] [PubMed]
39. Bankova, V.; Galabov, A.S.; Antonova, D.; Vilhelmova, N.; Di Perri, B. Chemical composition of propolis extract ACF® and activity against herpes simplex virus. *Phytomedicine* **2014**, *21*, 1432–1438. [CrossRef] [PubMed]

Disclaimer/Publisher's Note: The statements, opinions and data contained in all publications are solely those of the individual author(s) and contributor(s) and not of MDPI and/or the editor(s). MDPI and/or the editor(s) disclaim responsibility for any injury to people or property resulting from any ideas, methods, instructions or products referred to in the content.

Article

Anti-*Helicobacter pylori*, Antioxidant, Antidiabetic, and Anti-Alzheimer's Activities of Laurel Leaf Extract Treated by Moist Heat and Molecular Docking of Its Flavonoid Constituent, Naringenin, against Acetylcholinesterase and Butyrylcholinesterase

Aisha M. H. Al-Rajhi [1], Husam Qanash [2], Majed N. Almashjary [3,4], Mohannad S. Hazzazi [3,4], Hashim R. Felemban [3,5] and Tarek M. Abdelghany [6,*]

[1] Department of Biology, College of Science, Princess Nourah bint Abdulrahman University, P.O. Box 84428, Riyadh 11671, Saudi Arabia; amoalrajhi@pnu.edu.sa
[2] Department of Medical Laboratory Science, College of Applied Medical Sciences, University of Ha'il, Hail 55476, Saudi Arabia; h.qanash@uoh.edu.sa
[3] Department of Medical Laboratory Sciences, Faculty of Applied Medical Sciences, King Abdulaziz University, Jeddah 22254, Saudi Arabia; malmashjary@kau.edu.sa (M.N.A.); mshazzazi@kau.edu.sa (M.S.H.); hrfelemban@kau.edu.sa (H.R.F.)
[4] Hematology Research Unit, King Fahd Medical Research Center, King Abdulaziz University, Jeddah 22254, Saudi Arabia
[5] Special Infectious Agents Unit-BSL3, King Fahd Medical Research Center, King Abdulaziz University, Jeddah 21362, Saudi Arabia
[6] Botany and Microbiology Department, Faculty of Science, Al-Azhar University, Cairo 11725, Egypt
* Correspondence: tabdelghany.201@azhar.edu.eg

Citation: Al-Rajhi, A.M.H.; Qanash, H.; Almashjary, M.N.; Hazzazi, M.S.; Felemban, H.R.; Abdelghany, T.M. Anti-*Helicobacter pylori*, Antioxidant, Antidiabetic, and Anti-Alzheimer's Activities of Laurel Leaf Extract Treated by Moist Heat and Molecular Docking of Its Flavonoid Constituent, Naringenin, against Acetylcholinesterase and Butyrylcholinesterase. *Life* **2023**, *13*, 1512. https://doi.org/10.3390/life13071512

Academic Editors: Stefania Lamponi and Seung Ho Lee

Received: 31 May 2023
Revised: 23 June 2023
Accepted: 3 July 2023
Published: 5 July 2023

Copyright: © 2023 by the authors. Licensee MDPI, Basel, Switzerland. This article is an open access article distributed under the terms and conditions of the Creative Commons Attribution (CC BY) license (https://creativecommons.org/licenses/by/4.0/).

Abstract: It is worth noting that laurel (*Laurus nobilis* L.) contains several pharmacologically and nutritionally active compounds that may differ according to the pretreatment process. The current study is designed to clarify the effect of moist heat on the phenolic and flavonoid constituents and anti-*Helicobacter pylori*, antioxidant, antidiabetic, and anti-alzheimer's activities of laurel leaf extract (LLE). Unmoist-heated (UMH) and moist-heated (MH) LLEs showed the presence of numerous flavonoid and phenolic constituents, although at different levels of concentration. MH significantly induced ($p < 0.05$) the occurrence of most compounds at high concentrations of 5655.89 µg/mL, 3967.65 µg/mL, 224.80 µg/mL, 887.83 µg/mL, 2979.14 µg/mL, 203.02 µg/mL, 284.65 µg/mL, 1893.66 µg/mL, and 187.88 µg/mL, unlike the detection at low concentrations of 3461.19 µg/mL, 196.96 µg/mL, 664.12 µg/mL, 2835.09 µg/mL, 153.26 µg/mL, 254.43 µg/mL, 1605.00 µg/mL, 4486.02 µg/mL, and 195.60 µg/mL using UMH, for naringenin, methyl gallate, caffeic acid, rutin, ellagic acid, coumaric acid, vanillin, ferulic acid, and hesperetin, respectively. Chlorogenic acid, syringic acid, and daidzein were detected in the UMH LLE but not in the MH LLE, unlike pyrocatechol. The anti-*H. pylori* activity of the UMH LLE was lower (23.67 ± 0.58 mm of inhibition zone) than that of the MH LLE (26.00 ± 0.0 mm of inhibition zone). Moreover, the values of MIC and MBC associated with the MH LLE were very low compared to those of the UMH LLE. Via MBC/MIC index calculation, the UMH and MH LLEs showed cidal activity. The MH LLE exhibited higher anti-biofilm activity (93.73%) compared to the anti-biofilm activity (87.75%) of the MH LLE against *H. pylori*. The urease inhibition percentage was more affected in the UMH LLE compared to the MH LLE, with significant ($p < 0.05$) IC$_{50}$ values of 34.17 µg/mL and 91.11 µg/mL, respectively. Promising antioxidant activity was documented with a very low value of IC$_{50}$ (3.45 µg/mL) for the MH LLE compared to the IC$_{50}$ value of 4.69 µg/mL for the UMH LLE and the IC$_{50}$ value of 4.43 µg/mL for ascorbic acid. The MH LLE showed significantly higher ($p < 0.05$) inhibition of α-glucosidase and butyrylcholinesterase activities, with IC$_{50}$ values of 9.9 µg/mL and 17.3 µg/mL, respectively, compared to those of the UMH LLE at 18.36 µg/mL and 28.92 µg/mL. The molecular docking of naringenin showed good docking scores against acetylcholinesterase 1E66 and butyrylcholinesterase 6EMI, indicating that naringenin is an intriguing candidate for additional research as a possible medication for Alzheimer's disease.

Keywords: *Laurus nobilis* L.; anti-*Helicobacter pylori*; antioxidant; antidiabetic; anti-Alzheimer; naringenin

1. Introduction

There is still debate regarding the biological activities of many natural extracts and how to use them in the right way, as well as the methods of extracting them to release large quantities of active substances. In the current investigation, the *Laurus nobilis* plant was used due to the importance of this plant in human life. *L. nobilis*, belonging to the lauraceae family, is a plant commonly dispersed in the Mediterranean Region and some other countries, such as Algeria, Turkey, Greece, Portugal, Morocco, Spain, Mexico, Italy, Greece, Morocco, U.S.A., and Belgium [1].

This plant, also known as sweet bay, bay, Roman laurel, bay laurel, and daphne, is aromatic and evergreen; is utilized as a cooking spice in several countries, including Western and Asian countries; and is applied in traditional medicine as a stimulant, an antiseptic, a stomachic digestive, and a sudorific asset. Some studies reported the biological activities of *L. nobilis* extracts as well as *L. nobilis* essential oils (EOs); for instance, Sakran et al. [2] revealed the antimicrobial activities of an ethanol extract of *L. nobilis* against *Escherichia coli*, *Salmonellae typhi*, and *Staphylococcus aureus*, with different levels of inhibition zones and minimum inhibitory concentrations (MICs). The antifungal and antioxidant activities of *L. nobilis* flower EOs were documented [3,4].

L. nobilis leaf EOs were applied for food preservation to prevent microbial contamination, and da Silveira et al. [5] evaluated the effect of *Yersinia enterocolitica* and *E. coli* on the quality of fresh Tuscan sausage stored at 7 °C for 14 days and supplemented with *L. nobilis* leaf EOs. Moreover, Marques et al. [6] preserved seafood, meat, and some agricultural products via the addition of *L. nobilis* leaf EOs because of their antimicrobial and antioxidant properties. The antagonistic potential of *L. nobilis* leaf extract was documented against *Proteus vulgaris* and *Staphylococcus saprophyticus* in fresh lamb meat [7]. Therefore, lamb meat's shelf life increased from 1 to 3 days at room temperature and from 6 to 13 days at refrigeration temperature as a result of spraying fresh lamb meat with 10% (v/v) *L. nobilis* leaf extract.

Other previous studies reported that *L. nobilis* extract was utilized as an additive in cosmetic and food products due to the existence of aromatic and flavor constituents [8,9]. Pharmacological effects such as immune-modulating and cytotoxic effects were also attributed to *L. nobilis* extract [3]. The activity of *L. nobilis* extract may depend on several factors, such as the extracted organ of the plant and the used solvent. Fidan et al. [10] reported that *L. nobilis* leaf EO exhibited antimicrobial activity against nearly all tested bacteria as well as fungi, while only *S. aureus* from the tested microorganisms was inhibited when using twig EOs.

In the present study, the impact of moist heat on the phenolic and flavonoid constituents as well as some of the biological activities of *L. nobilis* leaf extract was investigated. The effects of other processes on *L. nobilis* extract had been studied; for example, drying via air and a microwave was effective regarding the chemical contents and antimicrobial activity of *L. nobilis* EOs [11]. Similarly, Khodja et al. [12] studied the effects of two kinds of drying approaches, microwave-assisted drying ranging from 180 to 900 W and drying via air and an oven at temperatures ranging from 40 °C to 120 °C, on the total phenolic content and antioxidant activity of *L. nobilis* leaf extract. Extraction methods, such as hydro-distillation, hydro-steam distillation, ohmic-assisted hydro-distillation, and microwave-assisted hydro-distillation, had been experimented with and evaluated to study their effects on the yield, antioxidant activity, and chemical composition of *L. nobilis*

EOs [13], which indicated that microwave-assisted hydro-distillation gave higher quantities of oxygenated monoterpenes.

Helicobacter pylori is one of the most common chronic bacterial pathogens of human beings at all age stages. Numerous illnesses of the upper gastrointestinal tract, such as gastric inflammation and gastric and duodenal ulcers, have been associated with *H pylori* [14,15]. Gastritis development and peptic ulceration are caused by ammonia release from *H. pylori* urease that acts as virulence and colonization factors. However, while the antimicrobial activity of *L. nobilis* leaf extract has been studied extensively on several microorganisms, there is limited research on *H. pylori*. As mentioned previously, *H. pylori* was inhibited by *L. nobilis* leaf extract with an inhibition zone of 24 mm at 200 mg/mL [16]. Secondary metabolites of the plant have been recommended as possible alternatives for the treatment of *H. pylori* as well as other microbes [17–22]. Alzheimer's disease (AD) is a chronic neurodegenerative disease that is considered one of the most common health issues in the world, and its prevalence is increasing among the elderly population [23]. Investigators have continued to develop and discover bioactive substances of a natural origin in the creation of novel drugs for efficient treatment as a result of the continuous rise in the incidence of neurological disorders in humanity. AD is associated with acetylcholine depletion in the brain, which is due to acetylcholinesterase (AChE) and butyrylcholinesterase (BChE) catalytic activities [24], thereby making AChE and BChE potential targets for drug design in the management of AD. Most published reports have explained the biological activities of different fresh or dried organs of *L. nobilis* extract, but without considering the effect of moist heat on plant constituents. There is scarce information about the impact of moist heat on the biological activities of *L. nobilis* extract. Therefore, in this investigation, we present the effects of moist heat on the chemical constituents of *L.nobilis*, and its anti-*H. pylori*, antioxidant, antidiabetic, and anti-Alzheimer's activities

2. Materials and Methods

2.1. Chemical Used

Acetonitrile (molar mass: 41.053 g·mol^{-1}, density: 0.786 g/cm^3 at 25 °C, solubility in water: miscible, and purity: 99.5%), dimethylsulfoxide (DMSO) (molar mass: 78.13 g./mol, density: 1.1004 g/cm, solubility in water: miscible, and purity: 99%), 2,2-diphenyl-1-picrylhydrazyl (DPPH) (molar mass: 394.32 g/mol, density: 1.4 g/cm^3, solubility in water: insoluble, and purity: 99.13%), methanol (molar mass: 32.04 g/mol, density: 0.792 g/cm^3, solubility in water: miscible, and purity: 99.85%), and Mueller–Hinton agar were purchased from Sigma-Aldrich (Steinheim, Germany).

2.2. Sample Collection and Effect of Moist Heat

Air-dried leaves of laurel (*Laurus nobilis* L.) were collected from a supermarket in Saudi Arabia. The identification of the plant was achieved by the taxonomist Prof. Marei A. Hamed from the Botany and Microbiology Department, Faculty of Science, Al-Azhar University, Egypt. The laurel leaves were washed with distilled water to remove any dust particles and then re-dried in air. The re-dried air leaves were divided into two parts: the first part was kept at it was for 15 min at room temperature and then used for unmoistheat (UMH) treatment, while the second part, at the same time, was autoclaved at 80 °C for 15 min and used for the moist heat (MH) treatment. This was followed by the extraction of each part using methanol. The UMH and MH laurel leaf extracts (LLEs) (200 g for each) were grinded and then mixed with 500 mL of methanol with a magnetic stirrer for 1 day. The mixture was centrifuged for 10 min at 5000 rpm. The obtained supernatant was concentrated utilizing a rotary evaporator, completely concentrated to obtain 5 g of crude extract, re-dissolved in DMSO, and then kept at 4 °C for further investigations.

2.3. HPLC Analysis of Phenolic and Flavonoid Contents of UMH and MH LLEs

Phenolic and flavonoid compound identification of the UMH and MH LLEs was carried out utilizing an Agilent Series 1260 device (Agilent, Santa Clara, CA, USA) for High-

Performance Liquid Chromatography (HPLC). The HPLC device consisted of a quaternary HP pump (series 1200), an auto-sampling injector, a solvent degasser, the 1100 ChemStation software (DFAR 252.227-7014), an Eclipse C18 column (4.6 mm, 250 mm i.d., 5 μm, and 40 °C) and an ultraviolet (UV) detector adjusted to 280 nm for phenolic acids and to 330 nm for flavonoids, correspondingly. Separation of flavonoid was performed by utilizing 50 mM of H_3PO_4 (pH of 2.5) (solution A) and acetonitrile/acetic acid (40/60, v/v) (solution B) as the mobile phases in the gradient, in the following subsequent stages: isocratic elution 95%/5% (A/B) for 0–5 min; in the gradient linear from 95%/5% (A/B) to 50%/50% (A/B) for 5–55 min; isocratic elution 50%/50% (A/B) for 55–65 min; and in the gradient linear from 50%/50% (A/B) to 95%/5% (A/B) for 65–67 min, with 0.7 mL/min of the mobile phase as the flow rate. The following solvent system was utilized for the separation of phenolic acids, consisting of 2.5% aqueous acetic acid (A), 8% aqueous acetic acid (B), and acetonitrile (C) in the following subsequent stages: 5% B at 0 min, 10% B at 20 min, 30% B at 50 min, 50% B at 55 min, 50% B and 50% C at 60 min, 100% B at 100 min, and 100% C from 110 min to 120 min. The solvent flow rate was 1 mL/min of solvent. A total of 5 μL of the separated compounds was injected into the HPLC device. Quantitative detection of the separated compounds was calculated according to the data of standard compounds. The solution of standard stock of flavonoid and phenolic compounds was prepared at different dilutions (10–80 μg/mL) in methanol solvent, followed by injection into the HPLC device [25].

2.4. Total Phenolic and Flavonoid Content Detection

The UMH or MH LLE was added to Folin–Ciocalteau's reagent (10% v/v), and then Na_2CO_3 (7.5%) was added to the reaction mixture, followed by incubation for 40 min at 45 °C. At 765 nm, the absorbance was measured to determine the total phenol content, which was expressed as mg of gallic acid equivalent (GAE)/g of dry weight of the extract. On the other hand, the total flavonoid content was determined via the addition of the extract to 0.5 mL of methanol, 500 μL of 10% aluminum chloride, 50 μL of 1 M potassium acetate, and 1.4 mL of distilled water. This was followed by incubation for 30 min at 25 °C. At 415 nm, absorbance was measured to determine the total flavonoid content, which was expressed as mg of quercetin equivalent (QE)/g of dry weight of the extract.

2.5. Anti-H. pylori Activity of UMH and MH Extracts of L. nobilis Leaves

The agar-well diffusion method using Mueller–Hinton (MH) medium was used to test the UMH and MH LLEs against *H. pylori* based on standard 2 of McFarland's turbidity. An inoculum of *H. pylori* was seeded onto the surface of the M-H-containing agar using a sterile cotton swab. The agar wells were then drilled using a sterile 6 mm drill and filled with 100 μL of each extract at 100 mg/mL. Some wells were filled with 10% dimethylsulfoxide (DMSO) as a negative control and others with 0.05 μg/mL clarithromycin as a standard control. All plates were refrigerated at 4 °C for 30 min in order to allow the samples to spread before the growth of *H. pylori* began. The plates containing the inoculums were incubated in a GasPak™ anaerobic system "Oxoid" under suitable conditions, including a microaerobic environment at 37 °C, and an incubation time of 72 h. The zones of inhibition that appeared were then measured in millimeters [14].

2.6. Assay of Minimal Inhibitory Concentration (MIC) and Minimal Bactericidal Concentration (MBC)

Mueller–Hinton broth was used as the micro-dilution broth to determine the MICs of the UMH and MH LLEs against *H. pylori*. Various serial dilutions of the UMH and MH LLEs were prepared, ranging from 0.98 to 1000 μg/mL. A 96-well polystyrene microtiter plate was used, and 200 μL of each dilution of the UMH and MH LLEs was dispensed per well. Fresh *H. pylori* inoculum was prepared in sterile NaCl (0.85%) to achieve the required McFarland turbidity of 1.0. A total of 2 μL of tested organisms was inoculated into each well to give a final dose of 5×10^4 colony-forming units/mL. The plates were incubated at 37 °C for 72 h. The MIC was then visually assessed to determine how effectively the overall growth of the tested microorganisms was inhibited. Each micro-plate containing an inoculum of *H. pylori* without the UMH and MH LLEs was utilized as a positive control, while the micro-plates containing the UMH or MH LLE without *H. pylori* were used as a negative control [26]. The MBC was detected from the micro-dilution plates utilized for the MIC test. Aliquots (10 µL) of each well without visible growth were transferred to Mueller–Hinton agar plates and incubated at 37 °C for 72 h. The lowest concentration of the UMH and MH LLEs that fully inhibited *H. pylori* growth on the plates is known as the MBC. To characterize whether the UMH and MH LLEs had bacteriostatic or bactericidal abilities, the MBC/MIC ratio was recorded. If the MBC/MIC ratio of any extract is no more than four times the value of the MIC, the extract is called a bactericidal agent [27].

2.7. Anti-Biofilm Activity of UMH and MH LLEs

In 96-well polystyrene flatbottom plates, the impact of the UMH and MH LLEs on the development of *H. pylori* biofilms was assessed. In brief, 300 µL of freshly inoculated trypticase soy yeast broth (TSY) with a final concentration of 10^6 CFU/mL was aliquoted into each well of a microplate and cultivated in the presence of previously calculated sub-lethal doses of MBC (75, 50, and 25%). As the controls, wells with the medium and those with methanol and no extracts were employed. For 48 h, the plates were incubated at 37 °C. After incubation, the supernatant was removed, and free-floating *H. pylori* cells were completely cleaned from each well using sterile distilled H_2O. The plates were then allowed to air dry for 30 min, and the biofilm that had developed was stained for 15 min at room temperature using a 0.1% crystal violet aqueous solution. Following incubation, the extra stain was washed away with sterile distilled H_2O. Finally, the attached dye to *H. pylori* cells was solubilized via the addition of 250 µL of ethanol (95%) to each well, and then absorbance was measured using a microplate reader at a wavelength of 570 nm after 15 min of incubation. The biofilm inhibition % was recorded using the following formula:

$$\text{Biofilm inhibition \%} = 1 - \left(\frac{\text{Absorbance of extract} - \text{absorb. Blank}}{\text{Absorbance of control} - \text{bsorb. Blank}}\right) \times 100$$

Absorbance of the media only represents the blank sample, absorbance of *H. pylori* from the treatment represents the extract, and absorbance of *H. pylori* without any treatment represents the control [28].

2.8. Urease Inhibition by UMH or MH LLE

The reaction mixture solution was composed of 850 µL of urea, the UMH or MH LLE (up to 1000 µg/mL), and 100 mM of phosphate buffer (pH of 7.4) to reach a total value of 985 µL. The reaction mixture started via the addition of urease enzyme (15 µL) and was then measured by determining the concentration of ammonia at 60 min, utilizing 500 µL of solution A (composed of 0.5 g of phenol and 2.5 mg of sodium nitroprusside dissolved in 50 mL of distilled H_2O) and 500 µL of solution B (composed of 250 mg of NaCl and 820 µL of 5% NaOCl hypochlorite dissolved in 50 mL of H_2O) for 30 min at 37 °C. Uninhibited urease activity was taken as 100% of the control activity. According to the Berthelot spectrophotometric method, absorbance was measured at 625 nm, and then the urease inhibition percentage was recorded using the following formula:

$$\text{Urease inhibition \%} = 1 - \left(\frac{\text{Absorbance of extract or positive control}}{\text{Absorbance of enzyme as control}}\right) \times 100$$

The concentration that stimulates an inhibition value halfway between the minimum and maximum responses of the extract (IC_{50}) was detected by checking the inhibition influence of different concentrations of the extract in the test. Urease inhibition was recorded using hydroxyurea as the standard compound [29].

2.9. DPPH Radical Scavenging Activity for Antioxidant Activity Assessment of UMH and MH LLEs

The capacity of the LLEs to scavenge free radicals was evaluated using the 1,1-diphenyl-2-picrylhydrazyl (DPPH) assay. Briefly, 50 µL of the extract solution in water (1.95 to 1000 µg/mL) was combined with 2950 µL of a 60 M solution of DPPH in methanol. After giving the mixture a good shake, it was left to stand at room temperature for 30 min. Then, using a Varian 50 Bio spectrophotometer, absorbance was determined at 517 nm. A higher free radical scavenging capacity of the reaction mixture was indicated by a lower absorbance value. A standard antioxidant (ascorbic acid) was applied in this experiment. The following equation was used to determine the ability of the extracts to scavenge DPPH radicals:

$$\text{DPPH scavenging }(\%) = \frac{C_0 - LE1}{C_0} \times 100$$

where C_0 means the absorbance of the control reaction, while LE1 means the absorbance of the LLEs. The concentration of the LLEs required to inhibit 50% (IC_{50}) of DPPH free radicals was recorded using a log dose–inhibition curve [30].

2.10. Assay of α-Glucosidase Inhibition by UMH and MH LLEs

The LLEs were evaluated for α-glucosidase inhibitory potential according to the approach presented by Pistia-Brueggeman and Hollingsworth [31] with minor modifications. The reaction mixture containing 50 µL of each extract at different concentrations, ranging from 1.97 to 1000 µg/mL, was mixed with the α-glucosidase (10 µL of 1 U/mL) enzyme solution. Phosphate buffer (125 µL of 0.1 M at a pH of 6.8) was added to the reaction mixture, and the mixture was then incubated at 37 °C for 20 min. At the end of the incubation period, 20 µL of 1 M p-nitrophenyl-α-D-glucopyranoside (pNPG) as a substrate was added to start the reaction, followed by incubation for 30 min. Then, 50 µL of 0.1 N Na_2CO_3 was added to terminate the reaction. Absorbance at 405 nm was measured via a Biosystm 310 plus spectrophotometer. The α-glucosidase inhibition % was calculated using the following formula:

$$\alpha - \text{Glucosidase inhibition \%} = \left(\frac{\text{OD BLANK} - \text{OD extract}}{\text{OD BLANK}}\right) \times 100$$

One unit of enzyme can be expressed as the quantity of α-glucosidase required for the formation of p-Nitrophenol (one µmol) from p-NPG per min. This concentration is requisite to prevent 50% of the enzyme activity (IC_{50}) and is calculated utilizing a regression equation obtained through plotting the concentration (1.97–1000 µg/mL) and inhibition (%) for different concentrations.

2.11. Butyrylcholinesterase Inhibition Assay

Butyrylcholinesterase (BTchI) inhibition was assayed according to the Ellman method with some modifications [32,33]. The buffer and solutions for BChE were freshly made. This involved preparing 0.022 M S-butyrylthiocholine iodide (BTchI) solution (7.0 mg of BTchI was dissolved in 1 mL of water) and 0.44 U/mL of BChE solution (2.9762 mg of BChE enzyme was dissolved in 6.746 mL of buffer at a pH of 8.0). To achieve a final concentration of 1000 µg/mL, each LLE was first dissolved in DMSO and then in distilled water to obtain a concentration of 44 mg/mL. The BChE inhibition assay was determined via measuring absorbance using a microplate reader; 200 µL of the buffer, 5 µL of BChE enzyme, 5 µL of Ellman's reagent 5,5'-dithiobis-2-nitrobenzoic acid (DTNB), and 5 µL of the LLE at a concentration of 40 mg/mL were combined and kept in a solution for 15 min at 30 degrees Celsius in a temperature-controlled water bath. The enzymatic reaction was then started by adding 5 µL of the BTchI substrate solution to the mixture. Absorbances were measured at 410 nm utilizing the microplate reader at 45 s intervals 13 times at a controlled

30 °C. Using the following formula, the measured absorbance was used to determine the enzymatic inhibition:

$$\text{Butyrylcholinesterase inhibition } (\%) = 100 - \left[\left(\frac{\text{R extract}}{\text{R max}}\right)\right] \times 100$$

where R extract means the change rate in the absorbance of the test containing the LLE (Δabs/Δtime), while R max means the maximum change rate in the absorbance of the blank sample without any inhibitor.

2.12. Molecular Docking Investigation

The crystal structures of acetylcholinesterase (PDB code 1E66) and butyrylcholinesterase (PDB code 6EMI) proteins were downloaded from the protein data bank (https://www.rcsb.org/, accessed on 30 March 2021). Naringenin was docked via the MOE 2019 software. Both receptor and compound creation and optimization were carried out using the software's default method for structural optimization. Hydrogen atoms were added after the elimination of all water molecules around the proteins. Then, naringenin optimized its shape with the lowest binding energy while using the MMFF94x force field to achieve this optimization. Alpha-site spheres were created using the site finder in the MOE module. The five best binding configurations with flexible molecular rotation were created using the DFT-optimized structure of naringenin. The free binding energy (S, in kcal/mol), which represents the binding affinity, was ranked using hydrogen bonds that developed between the proteins and naringenin. Validation was performed by re-docking naringenin into the binding pocket and measuring the root-mean-square deviation (RMSD) between positions.

2.13. Statistical Analysis

The outcomes are presented as mean ± SD (standard deviation), and the results were taken as the average of three replicates. The statistical analysis was performed using the computer programs Microsoft Excel version 365 and SPSS v.25 (Statistical Package for the Social Sciences version 25.00). Quantitative data with parametric distribution between the different treatments were analyzed utilizing one-way analysis of variance (ANOVA) and Tukey's post hoc test, at 0.05 probability level.

3. Results and Discussion

3.1. Phenolic and Flavonoid Characterization of LLEs

This study examined the influence of moist heat on the phenolic and flavonoid contents of LLE compared to air-dried LLE, as well as anti-*H. pylori*, antioxidant, antidiabetic, and anti-Alzheimer's activities (Figure 1). The LLE, either unmoist-heated (UMH) or moist-heated (MH), was enriched with several flavonoid and phenolic compounds (Table 1 and Figures 2 and 3). Naringenin, methyl gallate, catechin, and ellagic acid were the most detected compounds with the highest concentrations, while apigenin and cinnamic acid were detected with the lowest concentrations in either the UMH or MH LLE. Surprisingly, significantly high ($p < 0.05$) concentrations of 3967.65 µg/mL, 224.80 µg/mL, 887.83 µg/mL, 2979.14 µg/mL, 203.02 µg/mL, 284.65 µg/mL, 1893.66 µg/mL, 5655.89 µg/mL, 51.22 µg/mL, 95.03 µg/mL, 547.19 µg/mL, and 187.88 µg/mL were detected for methyl gallate, caffeic acid, rutin, ellagic acid, coumaric acid, vanillin, ferulic acid, naringenin, cinnamic acid, apigenin, kaempferol, and hesperetin in the MH LLE, compared to their concentrations of 3461.19 µg/mL, 196.96 µg/mL, 664.12 µg/mL, 2835.09 µg/mL, 153.26 µg/mL, 254.43 µg/mL, 1605.00 µg/mL, 4486.02 µg/mL, 43.25 µg/mL, 88.55 µg/mL, 206.95 µg/mL, and 195.60 µg/mL, respectively, in the UMH LLE. These results indicated that moist heat promoted the release of phenolic and flavonoid compounds. Juániz et al. [34] reported that heat treatments of vegetables increased the concentration of phenolic constituents and proposed that thermal damage of cell walls and sub-cellular compartments throughout the cooking process stimulates the discharge of constituents. However, chlorogenic acid was detected at a high concentration of 1157.51 µg/mL, besides syringic acid and daidzein, in the UMH LLE, but it was not detected in the MH LLE. These disappeared compounds might be unstable under heat treatment or might have transformed into other compounds. Therefore, pyrocatechol was detected at a concentration of 868.77 µg/mL only in the MH LLE but not in the UMH LLE. Table 1 shows the total contents of phenolic compounds and flavonoids in the UMH and MH LLEs. The present findings showed that the UMH LLE contained less total phenolic content (1.87 ± 0.33 mg GAE/g) and total flavonoid content (0.68 ± 0.10 mg QE/g) than the MH LLE (2.65 ± 0.17 mg GAE/g and 1.05 ± 0.10 mg QE/g, respectively).

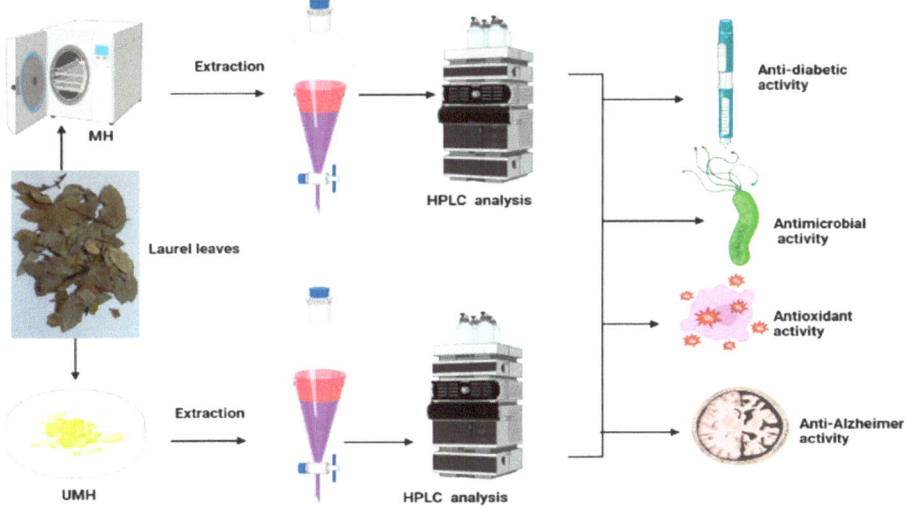

Figure 1. Planned experiments for moist-heated (MH) and unmoist-heated laurel leaves.

Previously, Routray and Orsat [35] mentioned that exposure of natural extracts to higher microwave power resulted in the degradation of some phenolic constituents. Dobroslavić et al. [36] found the presence of variable amounts of 29 phenolic and flavonoid compounds in LLE, with the most abundance being quercetin glycosides and kaempferol. Moreover, they noticed that temperature increases from 40 to 80 °C resulted in higher total phenolic content. Higher total phenolic content was obtained as a result of exposure of *L. nobilis* to boiling water for 3 h, compared to the amount obtained at room temperature after 72 h [37]. In another report, Bulut Kocabas et al. [38] noticed that the total phenolic content of *L. nobilis* leaves exposed to heated water at 80 °C for 45 min was 10-fold more than that obtained at room temperature for 24 h, with apigenin and its glycosides representing the most detected flavones in the LLE [39]. But in our study, apigenin was detected at a low concentration. Four phenolic compounds, including gallic acid, coumaric acid, pyrogallol, and resorcinol, were detected in the *L. nobilis* extract via HPLC [40]. Alejo-Armijo et al. [39] mentioned that various phenolic compounds, such as flavonoids, tannins (proanthocyanidins), phenolic acids, and lignans, were associated with LLE, particularly flavonoids, which represent the most abundant constituents with a variety of identified compounds. Fidan et al. [10] mentioned that the differences in chemical constituents in *L. nobilis* are perhaps due to the dissimilar genotypes, the planted and climatic conditions where the plants were cultivated, and the extraction process [10].

3.2. Anti-H. pylori Activity

The UMH LLE exhibited anti-*H. pylori* activity but less than the MH LLE, with inhibition zones of 23.67 ± 0.58 mm and 26.00 ± 0.0 mm, respectively (Table 2 and Figure 4). The inhibitory potential of both UMH and MH LLEs was good compared to the positive control that showed an inhibition zone of 20.33 ± 0.58 mm against *H. pylori*. Our results were consistent with previous findings [16], where acetone and hexane extracts of *L. nobilis* inhibited

H. *pylori* with inhibition zones of 24 mm and 26 mm, respectively. Dobroslavić et al. [41] demonstrated the antimicrobial activity of LLE. The presence of numerous flavonoid and phenolic compounds in the LLE is one of the main reasons for the anti-*H. pylori* activity. For instance, ellagic acid [42], naringenin [43], and ferulic acid [15] exhibited anti-*H. pylori* activity. Several investigations have shown that flavonoids play a critical role in inhibiting the spread of *H. pylori* [15,44,45]. The MH LLE showed greater inhibitory activity than the MH LLE, which might be related to the fact that the most analyzed compounds via HPLC were detected at the highest concentration compared to the fresh extract. It is very clear that the MIC and MBC values of the MH LLE were significantly ($p < 0.05$) lower (1.9 µg/mL) than the MIC and MBC values of 7.8 µg/mL of the UMH LLE (Table 2). Via the calculation of the MBC/MIC index, it is obvious that both UMH and MH LLEs possess cidal activity due to the MBC/MIC index being one (Table 2). The biofilm formation of *H. pylori* was more affected by the MH LLE than the UMH LLE. The anti-biofilm activity increment with an increase in MBC concentrations of the MH and UMH LLEs. In the same line, it was observed that the anti-biofilm activity of the MH LLE was higher than the UMH LLE; for instance, the anti-biofilm activity was 93.73 and 87.75% at 75% for the MH LLE's MBC and the UMH LLE's MBC, respectively (Figure 5A). The anti-biofilm activity of the MH and UMH LLEs was also indicated by the change in the color of stained *H. pylori* biofilm in the microtiter plate, which was dependent on anti-biofilm activity (Figure 6B).

3.3. Urease Inhibition

In the stomach, *H. pylori* uses urease enzyme to neutralize and protect itself from acidic conditions via ammonia production, and the activity of urease is necessary for *H. pylori* colonization. Urease inhibition represents one mechanism that determines the anti-*H. pylori* activity of antimicrobial compounds. In the present study, urease inhibition was documented using both the MH and UMH LLEs with different inhibition percentages depending on the concentration of the extracts. Unlike the results of anti-*H. pylori* activity, MIC, MBC, and anti-biofilm activity, which were more affected by the MH LLE, urease inhibition was more affected by the UMH LLE compared to the MH LLE (Figure 6). For instance, at 1.95, 62.5, 250, and 500 µg/mL, urease inhibition was 20.2, 56.0, 71.0, and 85.9% in the UMH LLE, while it was 3.1, 43.6, 62.5, and 83.0%, respectively, in the MH LLE. A lower significant ($p < 0.05$) value of IC_{50} (34.17 µg/mL) was recorded in the UMH LLE compared to the IC_{50} (91.11 µg/mL) of the MH LLE. These results can be explained by the possibility of the presence of another mechanism of anti-*H. pylori* activity, such as inhibition of DNA gyrase, N-acetyltransferase, and dihydrofolate reductase associated with the presence of compounds in the extract, or that the mechanism of the compounds in the UMH LLE may differ from the mechanism of the compounds in the MH LLE. Hydroxyurea was applied as a standard compound for urease inhibition, which showed an IC_{50} of 37.5 µg/mL. Surprisingly, the obtained IC_{50} value of the MH LLE was less (34.17 µg/mL) than that of hydroxyurea. This outcome suggests the capability of this extract to be applied as a promising extract to manage illnesses of the gastrointestinal tract alone or in combination with authorized drugs for anti-*H. pylori* treatment. As mentioned in the results of the HPLC analysis, naringenin was the main detected compound in both the MH and UMH LLEs, and this compound at a concentration of 300 µg/mL showed urease inhibition (34%) of *H. pylori* [44]. According to Biglar et al. [46], *L. nobilis* extract was tested against *H. pylori* growth and exhibited an attractive value of IC_{50} (48.69 µg/mL) for urease inhibition.

3.4. Antioxidant Activity

The presence of many flavonoid and phenolic compounds in the LLEs made us expect antioxidant activity, but not to the extent noted in Table 3, suggesting that the flavonoid and phenolic constituents present in the LLEs have a high level of hydroxylation, which is manifested in their high ability to provide protons and, therefore, stabilize DPPH. The antioxidant activity measured via DPPH scavenging % increased with an increase in extract concentration up to 1000 µg/mL of the UMH and MH LLEs, giving values of 95.5% and

97.1%, respectively (Table 3). Papageorgiou et al. [47] reported that the extracted phenolic acids of LLE reflected antioxidant activity. The present results showed that the antioxidant capacity of the MH LLE was higher than that of the UMH LLE, but with a slight difference. Not surprisingly, the IC_{50} (3.45 µg/mL) of the MH LLE was less than the IC_{50} value (4.69 µg/mL) of the UMH LLE, but the most surprising thing is that it is less than the IC_{50} value (4.43 µg/mL) of ascorbic acid. LLE, because of its wide presence of bioactive molecules, is an excellent source of antioxidant potential for pharmaceutical and cosmetic applications [41]. Our investigation on the antioxidant activity of LLE was consistent with many studies using UMH LLE [36,40,48], but there were no reported studies about MH LLE. The efficacy of conventional heat-reflux extraction (CRE), ultrasound-assisted extraction (UAE), and microwave-assisted extraction (MAE) was studied to evaluate the antioxidant activity of LLE, and the highest antioxidant capacity was obtained via CRE [41].

3.5. Antidiabetic Activity

α-glucosidase activity inhibition is considered a successful strategy for controlling the level of blood sugar and regulating food-related hyperglycemia. In our investigation, both the MH and UMH LLEs were effective inhibitors of glucosidase in a dose-dependent manner (Figure 7). However, the MH LLE showed the highest inhibitory activity toward this enzyme compared to the UMH LLE, particularly at low concentrations of up to 31.25 µg/mL; for instance, at 1.95, 7.81, and 31.25 µg/mL using the MH LLE, glucosidase inhibition was 26.9, 48.4, and 55.9%, while using the UMH LLE, α-glucosidase inhibition was 16.6, 38.4, and 58.7%, respectively. The values of IC_{50} of the two extracts were promising, but from the present findings, the IC_{50} of the MH LLE was significantly ($p < 0.05$) more promising (9.9 µg/mL) compared to the IC_{50} of the UMH LLE (18.36 µg/mL). These results may be because of the presence of some compounds in the UMH LLE. According to a study performed by Khan et al. [49], the lipid and glucose profiles of diabetic patients were improved because of treatment with powdered *L. nobilis* leaves in capsule form. In another study, *L. nobilis* (bay laurel) extracted with different solvents inhibited α-glucosidase metabolic activity [50]. Our finding endorses the utilization of LLE, particularly MH, for further investigations to identify their potential for regulating type II diabetes. Earlier investigations have shown that several natural plants contain rich constituents with valuable inhibitory potential toward α-glucosidase. According to a previous study [51], quercetin and rutin could inhibit glucosidase via binding with alpha-glucosidase to form a new complex. A recent report showed that catechin and quercetin displayed significant inhibitory potential toward α-glucosidase with different IC_{50} values of 12.78 and 92.87 µg/mL, respectively, while some compounds such as gallic acid and ellagic acid displayed moderate α-glucosidase inhibitory capacity, with IC_{50} values of 102.35 and 222.80 µg/mL, respectively [52].

3.6. Anti-Alzheimer's Activity

In the current study, inhibition of butyrylcholinesterase (BChE) was recorded as a result of exposure to the LLEs (Figure 8). This enzyme is used as a marker of Alzheimer's disease (AD) development. Recently, Falade et al. [53] reported that *L. nobilis* may be considered a promising agent for the nutraceutical management of AD. Figure 8 shows that the inhibition of BChE activity increases with an increase in both the UMH and MH LLEs up to 100 µg/mL, with inhibition of 82.3 and 88.5%, respectively. The MH LLE has the best inhibitory potential for BChE, with an IC_{50} of 17.3 µg/mL, compared to the UMH LLE, with an IC_{50} of 28.92 µg/mL, at all concentrations used; these results may be related to the high concentrations of the most detected compounds in the MH LLE as recorded in the HPLC analysis. Inhibition of two enzymes, namely BChE and acetylcholinesterase (AChE), was observed (100% inhibition) as a result of exposure to 30 µg/mL of *L. nobilis* extract, with IC_{50} values of 4.76 ± 0.36 µg/mL and 4.21 ± 0.50 µg/mL, respectively [53]. In vivo, phytoconstituents of *L. nobilis* had inhibitory action against BChE and AChE in scopolamine-induced rats [54].

Table 1. Identified phenolic and flavonoid compounds in moist-heated (MH) and unmoist-heated (UMH) laurel leaf extracts, with the total phenolic and flavonoid contents.

Compound	UMH Laurel Leaf Extract				MH Laurel Leaf Extract			
	Retention Time	Area	Area (%)	Conc. (µg/mL)	Retention Time	Area	Area (%)	Conc. (µg/mL)
Gallic acid	3.34	332.50	7.68	1435.58	3.339	325.41	6.79	1405.00
Chlorogenic acid	3.97	169.05	3.90	1157.51	4.169	0.00	0.00	0.00
Catechin	4.53	266.21	6.15	3296.24	4.516	260.61	5.44	3226.94
Methyl gallate	5.44	1268.23	29.30	3461.19	5.427	1453.81	30.34	3967.65
Caffeic acid	6.22	51.22	1.18	196.96	6.199	58.46	1.22	224.80
Syringic acid	6.57	83.73	1.93	283.90	6.528	0.00	0.00	0.00
Pyrocatechol	6.72	0.00	0.00	0.00	6.660	120.75	2.52	868.77
Rutin	8.02	114.38	2.64	664.12	8.002	152.91	3.19	887.83
Ellagic acid	8.53	305.87	7.06	2835.09	8.517	321.42	6.71	2979.14
Coumaric acid	9.04	97.16	2.24	153.26	9.003	128.70	2.69	203.02
Vanillin	9.77	116.26	2.69	254.43	9.750	130.07	2.71	284.65
Ferulic acid	10.31	469.84	10.85	1605.00	10.307	554.34	11.57	1893.66
Naringenin	10.68	744.04	17.19	4486.02	10.673	938.07	19.58	5655.89
Daidzein	12.24	10.24	0.24	31.64	12.312	0.00	0.00	0.00
Quercetin	12.80	110.03	2.54	757.79	12.782	60.99	1.27	420.02
Cinnamic acid	14.12	46.96	1.08	43.25	14.102	55.61	1.16	51.22
Apigenin	14.68	23.27	0.54	88.55	14.655	24.97	0.52	95.03
Kaempferol	15.12	53.38	1.23	206.95	15.298	141.13	2.95	547.19
Hesperetin	15.75	67.11	1.55	195.60	15.731	64.46	1.35	187.88
Total phenol		1.87 ± 0.33 mg GAE/g				2.65 ± 0.17 mg GAE/g		
Total flavonoid		0.68 ± 0.10 mg QE/g				1.05 ± 0.10 mg QE/g		

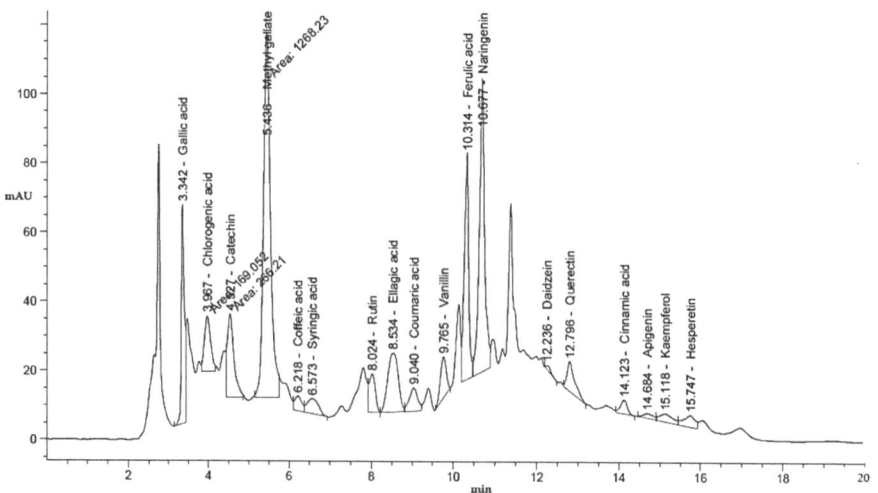

Figure 2. HPLC chromatogram of identified phenolic and flavonoid constituents in unmoist-heated laurel leaf extract.

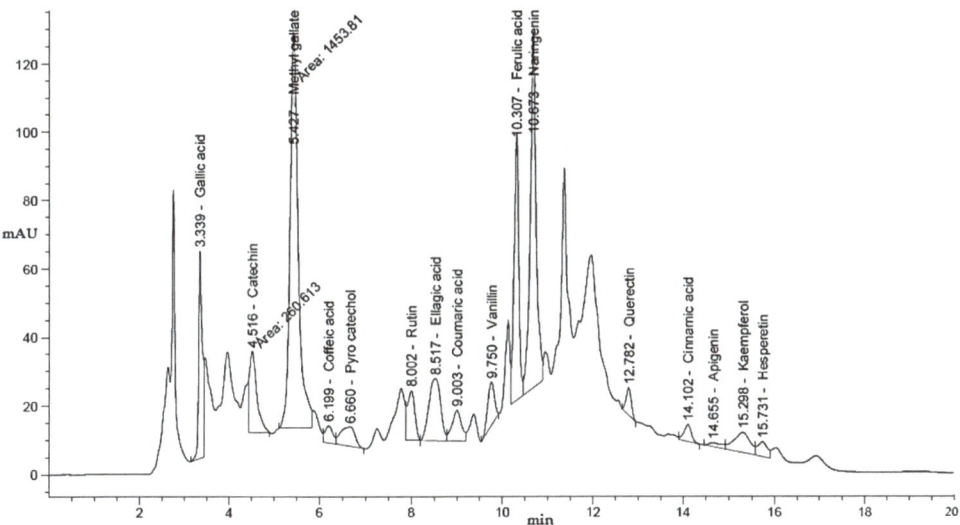

Figure 3. HPLC chromatogram of identified phenolic and flavonoid constituents in moist-heated laurel leaf extract.

Table 2. Inhibitory activity, MIC, and MBC of MH and UMH LLEs against *H. pylori*.

Mean Inhibition Zone (mm)				MIC (µg/mL)		MBC (µg/mL)		MBC/MIC Index	
UMH	MH	Control	Negative	UMH	MH	UMH	MH	UMH	MH
23.67 ± 0.58	26.00 ± 0.0	20.33 ± 0.58	0.0 ± 0.0	7.8 ± 0.1	1.9 ± 0.17	7.8 ± 0.35	1.9 ± 0.1	1.0	1.0

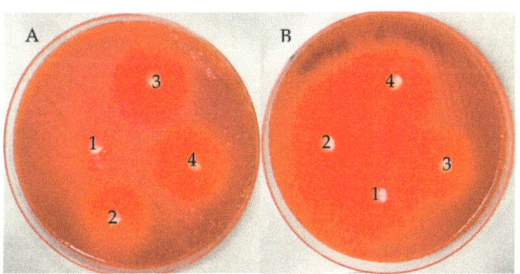

Figure 4. Antimicrobial activity of UMH (**A**) and MH (**B**) laurel leaf extracts (1, negative control; 2, positive control; 3 and 4, two wells of the plant extract).

3.7. Molecular Docking of Naringenin with Acetylcholinesterase and Butyrylcholinesterase

The technology of molecular docking is an effective approach to imagine the interaction among ligands of micromolecular size and receptors of macromolecular size, allowing the detection of probable binding sites and the ligands' steric conformations. Thus, in the current investigation, a molecular docking technique was employed to predict a potent ligand (naringenin) that could inhibit acetylcholinesterase 1E66 and butyrylcholinesterase 6EMI. The investigation also aimed to gain a better understanding of how the inhibitor interacts with the protein binding sites. The investigation was conducted using naringenin, and the RMSD values (Figures 9 and 10) were found to be 0.978201 Å and 0.733427 Å for docking with acetylcholinesterase 1E66 and butyrylcholinesterase 6EMI, respectively, demonstrating that the MOE-Dock approach is trustworthy for docking this inhibitor and that our docking method is valid for the investigated inhibitor. The binding interactions with (1E66) revealed that the ligand forms strong hydrogen bonding interactions through O 29 and O 30 with TYR 70 and HIS 440 residues at a distance of 2.81 Å and 2.82 Å, respectively. Moreover, it is observed that naringenin binds well with butyrylcholinesterase 6EMI via O 29 through the TRP 430 residue in the active pocket of the protein. An analysis of the binding interactions (Tables 4 and 5) illustrated the binding of naringenin with 1E66 and 6EMI proteins and the formation of several hydrogen bonds. Naringenin achieves favorable binding to both target proteins with negative scores of −6.78716 Kcal/mole and −6.14549 Kcal/mole. All docking scores and energies are presented in Tables 6 and 7. Molecular docking interactions were performed in other studies to support the biological activities of several natural constituents [30,55,56]. Recently, several phytoconstituents were docked against AChE and BChE [54].

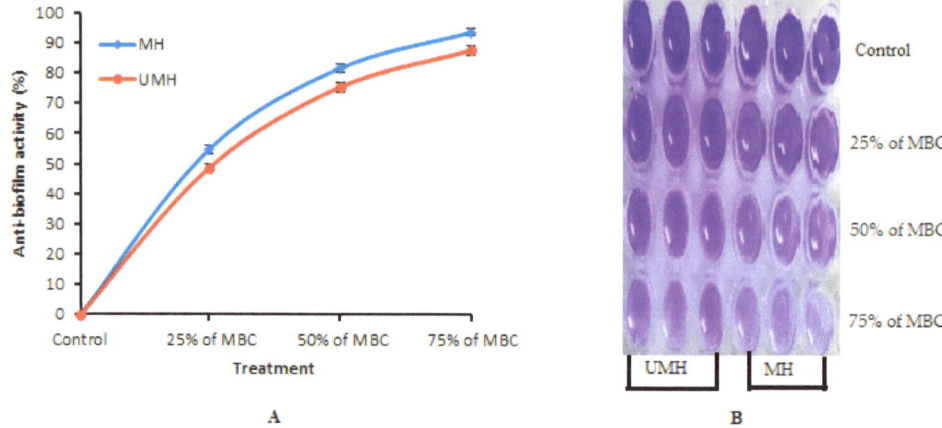

Figure 5. Anti-biofilm activity of MU LLE and UMH LLE against *H. pylori*: (**A**) microtiter plate reveals changes in stain color as a pointer of icreased anti-biofilm formation of *H. Pylori* (**B**) under different treatments of media + *H. Pylori* (Cont.); 25% of MBC, 50% of MBC, and 75% of MBC.

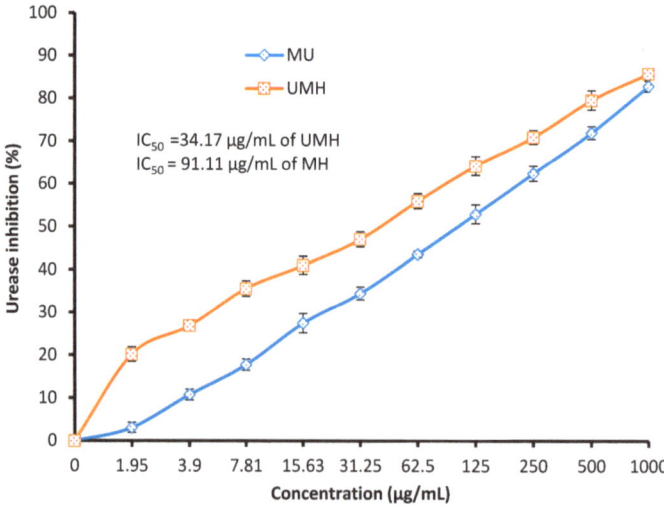

Figure 6. Urease inhibition at different concentrations of MU LLE and UMH LLE.

Table 3. DPPH scavenging % of ascorbic acid, MH LLE, and UMH LLE.

Concentration (µg/mL)	Ascorbic Acid DPPH Scavenging %	UMH LLE DPPH Scavenging %	MH LLE DPPH Scavenging %
1000	97.0 ±0.004	95.5 ± 0.003	97.1 ± 0.006
500	94.2 ± 0.001	93.3 ± 0.001	94.7 ± 0.001
250	90.0 ± 0.004	87.6 ± 0.005	90.3 ± 0.003
125	83.1 ± 0.004	81.7 ± 0.002	85.2 ± 0.005
62.50	76.4 ± 0.004	75.6 ± 0.003	77.7 ± 0.003
31.25	69.3 ± 0.002	68.9 ± 0.003	71.0 ± 0.006
15.63	62.5 ± 0.003	61.80.003	64.5 ± 0.004
7.81	55.2 ± 0.002	54.6 ± 0.004	57.8 ± 0.004
3.90	48.3 ± 0.002	47.6 ± 0.003	50.5 ± 0.007
1.95	40.2 ± 0.002	39.9 ± 0.004	42.1 ± 0.002
0	0.0 ± 0.000	0.0 ± 0.000	0.0 ± 0.000
IC_{50}	4.43 µg/mL	4.69 µg/mL	3.45 µg/mL

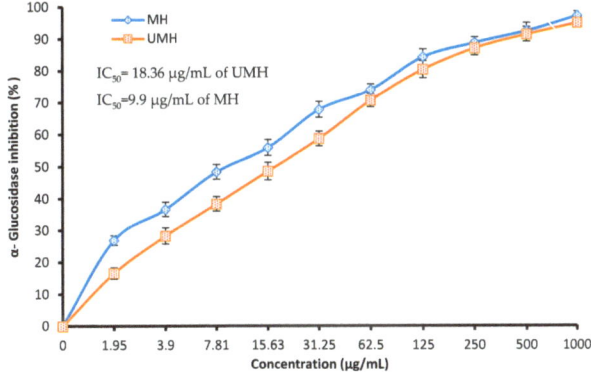

Figure 7. Antidiabetic activity at different concentrations of MU LLE and UMH LLE.

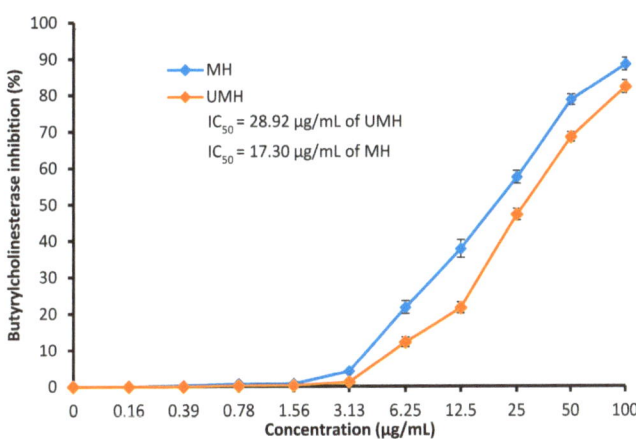

Figure 8. Anti-Alzheimer's activity at different concentrations of MU LLE and UMH LLE.

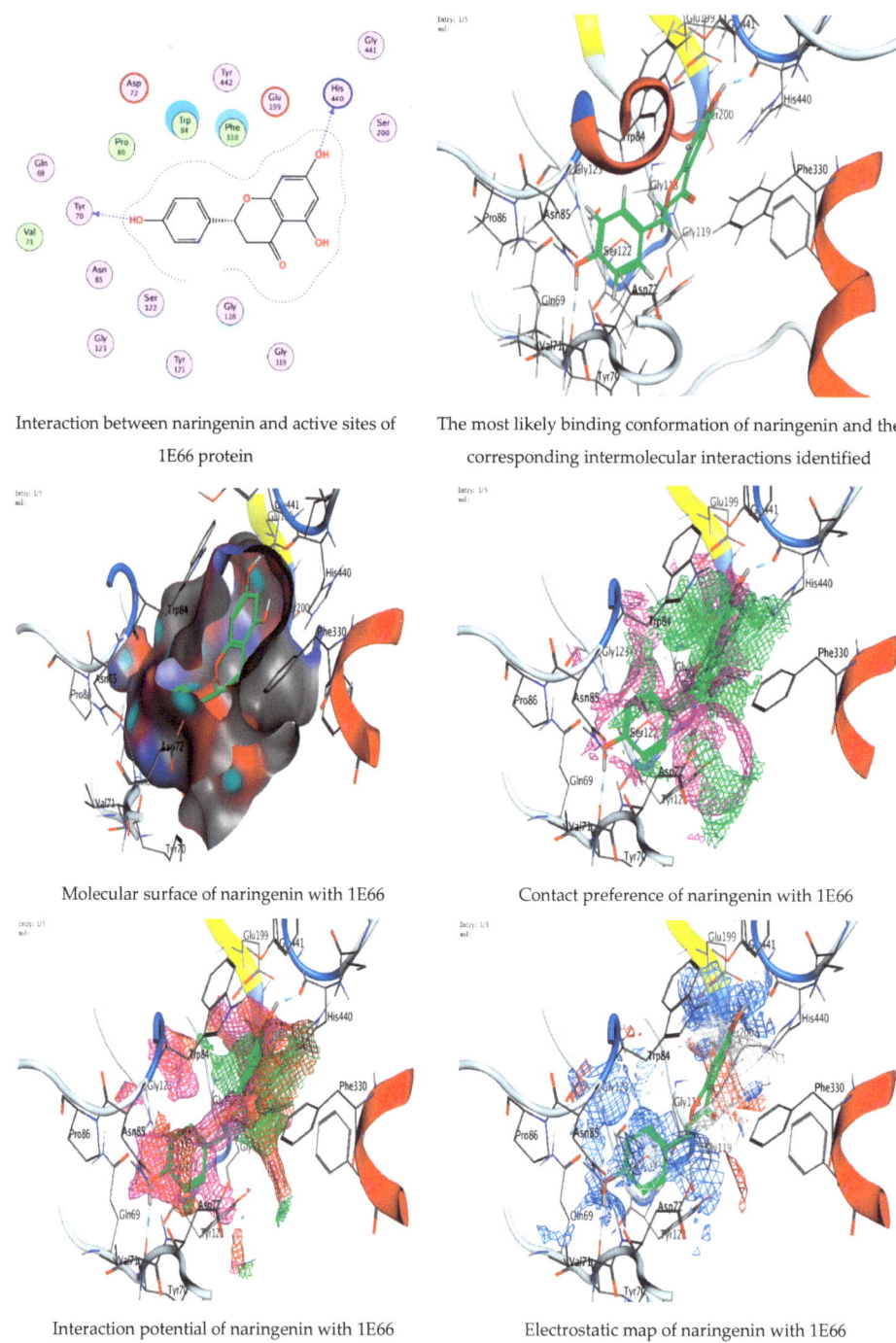

Figure 9. Molecular docking process of naringenin with 1E66.

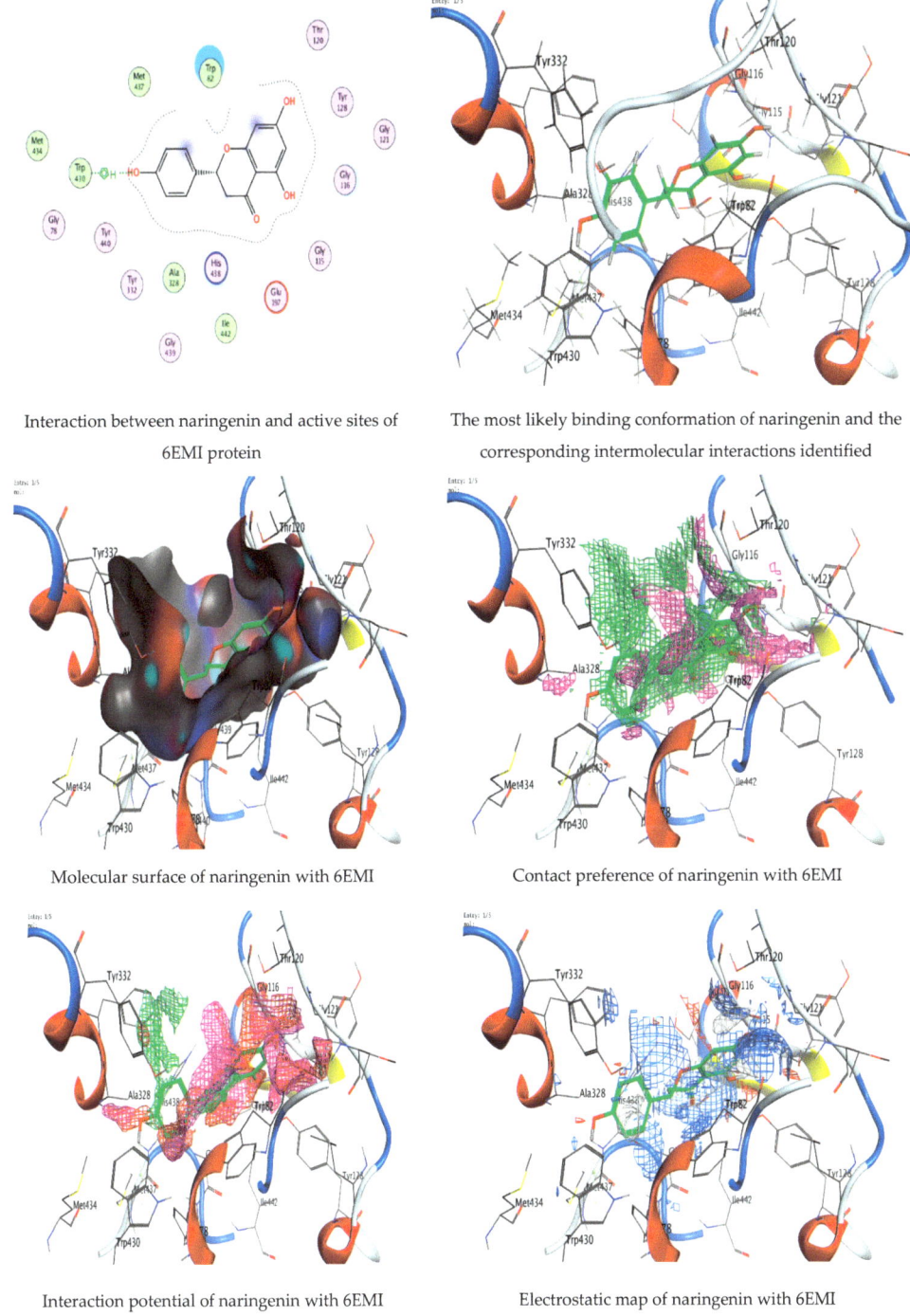

Interaction between naringenin and active sites of 6EMI protein

The most likely binding conformation of naringenin and the corresponding intermolecular interactions identified

Molecular surface of naringenin with 6EMI

Contact preference of naringenin with 6EMI

Interaction potential of naringenin with 6EMI

Electrostatic map of naringenin with 6EMI

Figure 10. Molecular docking process of naringenin with 6EMI.

Table 4. Docking scores and energies of naringenin with the crystal structure of acetylcholinesterase 1E66.

Mol	rseq	mseq	S	rmsd_refine	E_conf	E_place	E_score1	E_refine	E_score2
Naringenin	1	1	−6.78716	0.978201	−39.4713	−76.5186	−13.1741	−36.5393	−6.78716
Naringenin	1	1	−6.67799	1.277979	−34.1638	−68.3821	−12.2916	−36.1206	−6.67799
Naringenin	1	1	−6.62979	1.202798	−40.401	−90.799	−12.9332	−38.2054	−6.62979
Naringenin	1	1	−6.58196	0.838779	−33.5025	−72.8311	−12.7582	−35.0069	−6.58196
Naringenin	1	1	−6.55275	1.554429	−39.5703	−71.615	−12.3801	−32.3347	−6.55275

Table 5. Docking scores and energies of naringenin with the crystal structure of butyrylcholinesterase 6EMI.

Mol	rseq	mseq	S	rmsd_refine	E_conf	E_place	E_score1	E_refine	E_score2
Naringenin	1	1	−6.14549	0.733427	−39.739	−86.8222	−11.8804	−26.7900	−6.14549
Naringenin	1	1	−6.02164	1.344088	−38.5364	−65.6819	−11.5795	−26.8067	−6.02164
Naringenin	1	1	−5.88513	0.995142	−39.3925	−65.6398	−11.7741	−26.4780	−5.88513
Naringenin	1	1	−5.81087	1.068273	−37.007	−80.6394	−12.5956	−27.5993	−5.81087
Naringenin	1	1	−5.74699	1.258355	−37.1225	−74.7382	−12.0082	−24.9304	−5.74699

Table 6. Interaction of naringenin with the crystal structure of acetylcholinesterase 1E66.

Mol	Ligand	Receptor	Interaction	Distance	E (kcal/mol)
Naringenin	O 29	O TYR 70 (A)	H-donor	2.81	−2.7
	O 31	O HIS 440 (A)	H-donor	2.82	−3.3

Table 7. Interaction of naringenin with the crystal structure of butyrylcholinesterase 6EMI.

Mol	Ligand	Receptor	Interaction	Distance	E (kcal/mol)
Naringenin	O 29	5-ring TRP 430 (A)	H-Pi	4.03	−0.6

4. Conclusions

To the best of our knowledge, there are no previous reports on the relation between the effects of moist heat and the contents of flavonoid and phenolic compounds, anti-*H. pylori*, antioxidant, antidiabetic, and anti-Alzheimer's activities of LLE. In the present investigation, moist heat was effective in improving the release and the yield of flavonoid and phenolic compounds, which was accompanied by increases in their biological activities. The molecular docking study led to the discovery of significant ligand interactions with regard to the target protein's binding site. The main detected compound (naringenin) in the LLEs was docked molecularly with acetylcholinesterase and butyrylcholinesterase as indicators of Alzheimer's disease development. As a result of our investigation, we concluded that naringenin, which was theoretically examined here, exhibited good docking scores and binding interactions. Therefore, naringenin may be an effective therapeutic candidate, and their effectiveness against acetylcholinesterase (1E66) and butyrylcholinesterase (6EMI) may be boosted. LLE under the effect of moist heat is an excellent base for releasing more valuable compounds with potential applications in the pharmacological and food industries. Although natural extracts have been documented as brilliant substitutes for synthetic drugs, pharmacological utilization is challenging due to the diversity and complication of bioactive constituents present in such extracts. Therefore, the special influences of natural extracts at the level of cell therapy need to be better and more deeply investigated to complement the deficiencies in in vitro investigations in future studies.

Author Contributions: Conceptualization, Writing—review and editing A.M.H.A.-R.; formal analysis, investigation, H.Q.; investigation, Writing—review and editing, M.N.A.; methodology, investigation, M.S.H.; resources, formal analysis, H.R.F.; supervision, writing—original draft preparation, T.M.A.; All authors have read and agreed to the published version of the manuscript.

Funding: This research was funded by the Princess Nourah bint Abdulrahman University Researchers Supporting Project number (PNURSP2023R217), Princess Nourah bint Abdulrahman University, Riyadh, Saudi Arabia.

Institutional Review Board Statement: Not applicable.

Informed Consent Statement: Not applicable.

Data Availability Statement: Not applicable.

Acknowledgments: All authors thank the Princess Nourah bint Abdulrahman University Researchers Supporting Project number (PNURSP2023R217), Princess Nourah bint Abdulrahman University, Riyadh, Saudi Arabia.

Conflicts of Interest: The authors declare no conflict of interest.

References

1. Derwich, E.; Benziane, Z.; Boukir, A.; Mohamed, S.; Abdellah, B. Chemical composition and antibacterial activity of leaves essential oil of *Laurus nobilis* from Morocco. *Aust. J. Basic Appl. Sci.* **2009**, *3*, 3818–3824.
2. Sakran, K.A.; Raharjo, D.; Mertaniasih, N.M. Antimicrobial Activities of *Laurus nobilis* Leaves Ethanol Extract on *Staphylococcus aureus*, *Salmonellae typhi*, and *Escherichia coli*. *Indones. J. Trop. Infect. Dis.* **2021**, *9*, 120–125. [CrossRef]
3. Sırıken, B.; Yavuz, C.; Güler, A. Antibacterial Activity of *Laurus nobilis*: A review of literature. *Med. Sci. Discov.* **2018**, *5*, 374–379. [CrossRef]
4. Mssillou, I.; Agour, A.; El Ghouizi, A.; Hamamouch, N.; Lyoussi, B.; Derwich, E. Chemical Composition, Antioxidant Activity, and Antifungal Effects of Essential Oil from *Laurus nobilis* L. Flowers Growing in Morocco. *J. Food Qual.* **2020**, *2020*, 8819311. [CrossRef]
5. da Silveira, S.M.; Luciano, F.B.; Fronza, N.; Cunha, A., Jr.; Scheuermann, G.N.; Vieira, C.R.W. Chemical composition and antibacterial activity of *Laurus nobilis* essential oil towards foodborne pathogens and its application in fresh Tuscan sausage stored at 7 C. *LWT-Food Sci. Technol.* **2014**, *59*, 86–93. [CrossRef]
6. Marques, A.; Teixeira, B.; Nunes, M.L. Bay laurel (*Laurus nobilis*) oils. In *Essential Oils in Food Preservation, Flavor and Safety*; Academic Press: Cambridge, MA, USA, 2016; pp. 239–246. [CrossRef]
7. Mahmoud, S.H.; Wafa, M.M. The antibacterial activity of *Laurus nobilis* leaf extract and its potential use as a preservative for fresh lamb meat. *Afr. J. Microbiol. Res.* **2020**, *14*, 617–624. [CrossRef]
8. De Corato, U.; Maccioni, O.; Trupo, M.; Di Sanzo, G. Use of essential oil of *Laurus nobilis* obtained by means of a supercritical carbon dioxide technique against post harvest spoilage fungi. *Crop Prot.* **2010**, *29*, 142–147. [CrossRef]
9. Jemâa, J.M.B.; Tersim, N.; Toudert, K.T.; Khouja, M.L. Insecticidal activities of essential oils from leaves of *Laurus nobilis* L. from Tunisia, Algeria and Morocco, and comparative chemical composition. *J. Stored Prod. Res.* **2012**, *48*, 97–104. [CrossRef]
10. Fidan, H.; Stefanova, G.; Kostova, I.; Stankov, S.; Damyanova, S.; Stoyanova, A.; Zheljazkov, V.D. Chemical Composition and Antimicrobial Activity of *Laurus nobilis* L. Essential Oils from Bulgaria. *Molecules* **2019**, *24*, 804. [CrossRef]
11. Dahak, K.; Bouamama, H.; Benkhalti, F.; Taourirte, M. Drying methods and their implication on quality, quantity and antimicrobial activity of the essential oil of *Laurus nobilis* L. from Morocco. *Online J. Biol. Sci.* **2014**, *14*, 94–101. [CrossRef]
12. Khodja, Y.K.; Dahmoune, F.; Madani, K.; Khettal, B. Conventional method and microwave drying kinetics of *Laurus nobilis* leaves: Effects on phenolic compounds and antioxidant activity. *Braz. J. Food Technol.* **2020**, *23*, e2019214. [CrossRef]
13. Taban, A.; Saharkhiz, M.J.; Niakousari, M. Sweet bay (*Laurus nobilis* L.) essential oil and its chemical composition, antioxidant activity and leaf micromorphology under different extraction methods. *Sustain. Chem. Pharm.* **2018**, *9*, 12–18. [CrossRef]
14. Yahya, R.; Al-Rajhi, A.M.H.; Alzaid, S.Z.; Al Abboud, M.A.; Almuhayawi, M.S.; Al Jaouni, S.K.; Selim, S.; Ismail, K.S.; Abdelghany, T.M. Molecular Docking and Efficacy of *Aloe vera* Gel Based on Chitosan Nanoparticles against *Helicobacter pylori* and Its Antioxidant and Anti-Inflammatory Activities. *Polymers* **2022**, *14*, 2994. [CrossRef] [PubMed]
15. Al-Rajhi, A.M.H.; Qanash, H.; Bazaid, A.S.; Binsaleh, N.K.; Abdelghany, T.M. Pharmacological Evaluation of *Acacia nilotica* Flower Extract against *Helicobacter pylori* and Human Hepatocellular Carcinoma In Vitro and In Silico. *J. Funct. Biomater.* **2023**, *14*, 237. [CrossRef]
16. Guzeldag, G.; Kadioglu, L.; Mercimek, A.; Matyar, F. Preliminary examination of herbal extracts on the inhibition of *Helicobacter pylori*. *Afr. J. Tradit. Complement. Altern. Med.* **2013**, *11*, 93–96.
17. Qanash, H.; Bazaid, A.S.; Binsaleh, N.K.; Alharbi, B.; Alshammari, N.; Qahl, S.H.; Alhuthali, H.M.; Bagher, A.A. Phytochemical Characterization of Saudi Mint and Its Mediating Effect on the Production of Silver Nanoparticles and Its Antimicrobial and Antioxidant Activities. *Plants* **2023**, *12*, 2177. [CrossRef]
18. Mashraqi, A.; Modafer, Y.; Al Abboud, M.A.; Salama, H.M.; Abada, E. HPLC Analysis and Molecular Docking Study of *Myoporum serratum* Seeds Extract with Its Bioactivity against Pathogenic Microorganisms and Cancer Cell Lines. *Molecules* **2023**, *28*, 4041. [CrossRef]

19. Aldayel, T.S.; MBadran, M.; HAlomrani, A.; AlFaris, N.A.; ZAltamimi, J.; SAlqahtani, A.; ANasr, F.; Ghaffar, S.; Orfali, R. Chitosan-Coated Solid Lipid Nanoparticles as an Efficient Avenue for Boosted Biological Activities of *Aloe perryi*: Antioxidant, Antibacterial, and Anticancer Potential. *Molecules* 2023, 28, 3569. [CrossRef]
20. Al-Rajhi, A.M.H.; Yahya, R.; Bakri, M.M.; Yahya, R.; Abdelghany, T.M. In situ green synthesis of Cu-doped ZnO based polymers nanocomposite with studying antimicrobial, antioxidant and anti-inflammatory activities. *Appl. Biol. Chem.* 2022, 65, 35. [CrossRef]
21. Al-Rajhi, A.M.; Salem, S.S.; Alharbi, A.A.; Abdelghany, T.M. Ecofriendly synthesis of silver nanoparticles using Kei-apple (*Dovyalis caffra*) fruit and their efficacy against cancer cells and clinical pathogenic microorganisms. *Arab. J. Chem.* 2022, 15, 103927. [CrossRef]
22. Qanash, H.; Yahya, R.; Bakri, M.M.; Bazaid, A.S.; Qanash, S.; Shater, A.F.; Abdelghany, T.M. Anticancer, antioxidant, antiviral and antimicrobial activities of Kei Apple (*Dovyalis caffra*) fruit. *Sci. Rep.* 2022, 12, 5914. [CrossRef]
23. Baig, M.H.; Ahmad, K.; Rabbani, G.; Choi, I. Use of Peptides for the Management of Alzheimer's Disease: Diagnosis and Inhibition. *Front. Aging Neurosci.* 2018, 10, 21. [CrossRef] [PubMed]
24. Adedayo, B.C.; Oyeleye, S.I.; Okeke, B.M.; Oboh, G. Anti-cholinesterase and antioxidant properties of alkaloid and phenolic-rich extracts from pawpaw (*Carica papaya*) leaf: A comparative study. *Flavour. Fragr. J.* 2020, 36, 47–54. [CrossRef]
25. Al-Rajhi, A.M.H.; Abdel Ghany, T.M. Nanoemulsions of some edible oils and their antimicrobial, antioxidant, and anti-hemolytic activities. *BioResources* 2023, 18, 1465–1481. [CrossRef]
26. Andrews, J.M. Determination of minimum inhibitory concentrations. *J. Antimicrob. Chemother.* 2001, 48 (Suppl. S1), 5–16. [CrossRef]
27. French, G.L. Bactericidal Agents in the Treatment of MRSA Infections—The Potential Role of Daptomycin. *J. Antimicrob. Chemother.* 2006, 58, 1107. [CrossRef]
28. Antunes, A.L.S.; Trentin, D.S.; Bonfanti, J.W.; Pinto, C.C.F.; Perez, L.R.R.; Macedo, A.J.; Barth, A.L. Application of a feasible method for determination of biofilm antimicrobial susceptibility in staphylococci. *Acta Patol. Microbiol. Immunol. Scand.* 2010, 118, 873–877. [CrossRef]
29. Mahernia, S.; Bagherzadeh, K.; Mojab, F.; Amanlou, M. Urease Inhibitory Activities of some Commonly Consumed Herbal Medicines. *Iran. J. Pharm. Res.* 2015, 14, 943–947.
30. Al-Rajhi, A.M.H.; Yahya, R.; Abdelghany, T.M.; Fareid, M.A.; Mohamed, A.M.; Amin, B.H.; Masrahi, A.S. Anticancer, Anticoagulant, Antioxidant and Antimicrobial Activities of *Thevetia peruviana* Latex with Molecular Docking of Antimicrobial and Anticancer Activities. *Molecules* 2022, 27, 3165. [CrossRef] [PubMed]
31. Pistia-Brueggeman, G.; Hollingsworth, R.I. A preparation and screening strategy for glycosidase inhibitors. *Tetrahedron* 2001, 57, 8773–8778. [CrossRef]
32. Ellman, G.L.; Courtney, K.D.; Andres, V., Jr.; Feather-Stone, R.M. A new and rapid colorimetric determination of acetylcholinesterase activity. *Biochem. Pharmacol.* 1961, 7, 88–95. [CrossRef] [PubMed]
33. Sezgin, Z.; Biberoglu, K.; Chupakhin, V.; Makhaeva, G.F.; Tacal, O. Determination of binding points of methylene blue and cationic phenoxazine dyes on human butyrylcholinesterase. *Arch. Biochem. Biophys.* 2013, 532, 32–38. [CrossRef] [PubMed]
34. Juániz, I.; Ludwig, I.A.; Huarte, E.; Pereira-Caro, G.; Moreno-Rojas, J.M.; Cid, C.; De Peña, M.P. Influence of heat treatment on antioxidant capacity and (poly)phenolic compounds of selected vegetables. *Food Chem.* 2016, 197 Pt A, 66–73. [CrossRef]
35. Routray, W.; Orsat, V. Microwave-Assisted Extraction of Flavonoids: A Review. *Food Bioprocess Technol.* 2012, 5, 409–424. [CrossRef]
36. Dobroslavić, E.; Elez Garofulić, I.; Zorić, Z.; Pedisić, S.; Dragović-Uzelac, V. Polyphenolic Characterization and Antioxidant Capacity of *Laurus nobilis* L. Leaf Extracts Obtained by Green and Conventional Extraction Techniques. *Processes* 2021, 9, 1840. [CrossRef]
37. Ramos, C.; Teixeira, B.; Batista, I.; Matos, O.; Serrano, C.; Neng, N.R.; Nogueira, J.M.F.; Nunes, M.L.; Marques, A. Antioxidant and antibacterial activity of essential oil and extracts of bay laurel *Laurus nobilis* Linnaeus (Lauraceae) from Portugal. *Nat. Prod. Res.* 2012, 26, 518–529. [CrossRef]
38. Bulut Kocabas, B.; Attar, A.; Peksel, A.; Altikatoglu Yapaoz, M. Phytosynthesis of CuONPs via *Laurus nobilis*: Determination of antioxidant content, antibacterial activity, and dye decolorization potential. *Biotechnol. Appl. Biochem.* 2021, 68, 889–895. [CrossRef]
39. Alejo-Armijo, A.; Altarejos, J.; Salido, S. Phytochemicals and biological activities of laurel tree (*Laurus nobilis*). *Nat. Prod. Commun.* 2017, 12, 743–757. [CrossRef]
40. Muñiz-Márquez, D.B.; Rodríguez, R.; Balagurusamy, N.; Carrillo, M.L.; Belmares, R.; Contreras, J.C.; Nevárez, G.V.; Aguilar, C.N. Phenolic content and antioxidant capacity of extracts of *Laurus nobilis* L., *Coriandrum sativum* L. and *Amaranthus hybridus* L. *CyTA-J. Food* 2014, 12, 271–276. [CrossRef]
41. Dobroslavić, E.; Repajić, M.; Dragović-Uzelac, V.; Elez Garofulić, I. Isolation of *Laurus nobilis* Leaf Polyphenols: A Review on Current Techniques and Future Perspectives. *Foods* 2022, 11, 235. [CrossRef]
42. Martini, S.; D'Addario, C.; Colacevich, A.; Focardi, S.; Borghini, F.; Santucci, A.; Figura, N.; Rossi, C. Antimicrobial activity against *Helicobacter pylori* strains and antioxidant properties of blackberry leaves (*Rubus ulmifolius*) and isolated compounds. *Int. J. Antimicrob. Agents* 2009, 34, 50–59. [CrossRef] [PubMed]

43. Duda-Madej, A.; Stecko, J.; Sobieraj, J.; Szymańska, N.; Kozłowska, J. Naringenin and Its Derivatives—Health-Promoting Phytobiotic against Resistant Bacteria and Fungi in Humans. *Antibiotics* **2022**, *11*, 1628. [CrossRef] [PubMed]
44. Bae, E.A.; Han, M.J.; Kim, D.H. In vitro anti-*Helicobacter pylori* activity of some flavonoids and their metabolites. *Planta Med.* **1999**, *65*, 442–443. [CrossRef] [PubMed]
45. Widelski, J.; Okińczyc, P.; Suśniak, K.; Malm, A.; Bozhadze, A.; Jokhadze, M.; Korona-Głowniak, I. Correlation between Chemical Profile of Georgian Propolis Extracts and Their Activity against *Helicobacter pylori*. *Molecules* **2023**, *28*, 1374. [CrossRef] [PubMed]
46. Biglar, M.; Sufi, H.; Bagherzadeh, K.; Amanlou, M.; Mojab, F. Screening of 20 commonly used Iranian traditional medicinal plants against urease. *Iran. J. Pharm. Res.* **2014**, *13*, 195–198.
47. Papageorgiou, V.; Mallouchos, A.; Komaitis, M. Investigation of the antioxidant behavior of air- and freeze-dried aromatic plant materials in relation to their phenolic content and vegetative cycle. *J. Agric. Food Chem.* **2008**, *56*, 5743–5752. [CrossRef]
48. Mohammed, R.R.; Omer, A.K.; Yener, Z.; Uyar, A.; Ahmed, A.K. Biomedical effects of *Laurus nobilis* L. leaf extract on vital organs in streptozotocin-induced diabetic rats: Experimental research. *Ann. Med. Surg.* **2020**, *61*, 188–197. [CrossRef]
49. Khan, I.; Shah, S.; Ahmad, J.; Abdullah, A.; Johnson, S.K. Effect of incorporating Bay leaves in cookies on postprandial glycemia, appetite, palatability, and gastrointestinal well-being. *J. Am. Coll. Nutr.* **2017**, *36*, 514–519. [CrossRef]
50. Duletić-Laušević, S.; Oalđe, M.; Alimpić-Aradski, A. In vitro evaluation of antioxidant, antineurodegenerative and antidiabetic activities of *Ocimum basilicum* L., *Laurus nobilis* L. leaves and *Citrus reticulata* Blanco peel extracts. *Lek. Sirovine* **2019**, *39*, 60–68. [CrossRef]
51. Li, Y.Q.; Zhou, F.C.; Gao, F.; Bian, J.S.; Shan, F. Comparative evaluation of quercetin, isoquercetin and rutin as inhibitors of alpha-glucosidase. *J. Agric. Food Chem.* **2009**, *57*, 11463–11468. [CrossRef]
52. Shen, H.; Wang, J.; Ao, J.; Cai, Y.; Xi, M.; Hou, Y.; Luo, A. Inhibitory kinetics and mechanism of active compounds in green walnut husk against α-glucosidase: Spectroscopy and molecular docking analyses. *LWT* **2022**, *172*, 114179. [CrossRef]
53. Falade, A.O.; Omolaiye, G.I.; Adewole, K.E.; Agunloye, O.M.; Ishola, A.A.; Okaiyeto, K.; Oboh, G.; Oguntibeju, O.O. Aqueous Extracts of Bay Leaf (*Laurus nobilis*) and Rosemary (Rosmarinus officinalis) Inhibit Iron-Induced Lipid Peroxidation and Key-Enzymes Implicated in Alzheimer's Disease in Rat Brain-in Vitro. *Am. J. Biochem. Biotechnol.* **2022**, *18*, 9–22. [CrossRef]
54. Banpure, S.G.; Chopade, V.V.; Chaudhari, P.D. Anti-Alzhimer Activity of Bay Leaves in Scopolamineinduced Rat Model. *J. Drug Deliv. Technol.* **2023**, *13*, 17–22. [CrossRef]
55. Dehghan, M.; Fathinejad, F.; Farzaei, M.H.; Barzegari, E. In silico unraveling of molecular anti-neurodegenerative profile of *Citrus medica* flavonoids against novel pharmaceutical targets. *Chem. Pap.* **2023**, *77*, 595–610. [CrossRef]
56. Qanash, H.; Alotaibi, K.; Aldarhami, A.; Bazaid, A.S.; Ganash, M.; Saeedi, N.H.; Ghany, T.A. Effectiveness of oil-based nanoemulsions with molecular docking of its antimicrobial potential. *BioResources* **2023**, *18*, 1554–1576. [CrossRef]

Disclaimer/Publisher's Note: The statements, opinions and data contained in all publications are solely those of the individual author(s) and contributor(s) and not of MDPI and/or the editor(s). MDPI and/or the editor(s) disclaim responsibility for any injury to people or property resulting from any ideas, methods, instructions or products referred to in the content.

Article

Protection of Liver Functions and Improvement of Kidney Functions by Twelve Weeks Consumption of Cuban Policosanol (Raydel®) with a Decrease of Glycated Hemoglobin and Blood Pressure from a Randomized, Placebo-Controlled, and Double-Blinded Study with Healthy and Middle-Aged Japanese Participants

Kyung-Hyun Cho [1,*], Ji-Eun Kim [1], Tomohiro Komatsu [2,3] and Yoshinari Uehara [2,3]

[1] Raydel Research Institute, Medical Innovation Complex, Daegu 41061, Republic of Korea; ths01035@raydel.co.kr

[2] Center for Preventive, Anti-Aging and Regenerative Medicine, Fukuoka University Hospital, 8-19-1 Nanakuma, Johnan-ku, Fukuoka 814-0180, Japan; komatsu@fukuoka-u.ac.jp (T.K.); ueharay@fukuoka-u.ac.jp (Y.U.)

[3] Faculty of Sports and Health Science, Fukuoka University, 8-19-1 Nanakuma, Johnan-ku, Fukuoka 814-0180, Japan

* Correspondence: chok@raydel.co.kr; Tel.: +82-53-964-1990; Fax: +82-53-965-1992

Citation: Cho, K.-H.; Kim, J.-E.; Komatsu, T.; Uehara, Y. Protection of Liver Functions and Improvement of Kidney Functions by Twelve Weeks Consumption of Cuban Policosanol (Raydel®) with a Decrease of Glycated Hemoglobin and Blood Pressure from a Randomized, Placebo-Controlled, and Double-Blinded Study with Healthy and Middle-Aged Japanese Participants. *Life* **2023**, *13*, 1319. https://doi.org/10.3390/life13061319

Academic Editors: Stefania Lamponi and Seung Ho Lee

Received: 17 April 2023
Revised: 30 May 2023
Accepted: 2 June 2023
Published: 4 June 2023

Copyright: © 2023 by the authors. Licensee MDPI, Basel, Switzerland. This article is an open access article distributed under the terms and conditions of the Creative Commons Attribution (CC BY) license (https://creativecommons.org/licenses/by/4.0/).

Abstract: Policosanol consumption has been associated with treating blood pressure and dyslipidemia by increasing the level of high-density lipoproteins-cholesterol (HDL-C) and HDL functionality. Although policosanol supplementation also ameliorated liver function in animal models, it has not been reported in a human clinical study, particularly with a 20 mg doage of policosanol. In the current study, twelve-week consumption of Cuban policosanol (Raydel®) significantly enhanced the hepatic functions, showing remarkable decreases in hepatic enzymes, blood urea nitrogen, and glycated hemoglobin. From the human trial with Japanese participants, the policosanol group (n = 26, male 13/female 13) showed a remarkable decrease in alanine aminotransferase (ALT) and aspartate aminotransferase (AST) from baseline up to 21% ($p = 0.041$) and 8.7% ($p = 0.017$), respectively. In contrast, the placebo group (n = 26, male 13/female 13) showed almost no change or slight elevation. The policosanol group showed a 16% decrease in γ-glutamyl transferase (γ-GTP) at week 12 from the baseline ($p = 0.015$), while the placebo group showed a 1.2% increase. The policosanol group exhibited significantly lower serum alkaline phosphatase (ALP) levels at week 8 ($p = 0.012$), week 12 ($p = 0.012$), and after 4-weeks ($p = 0.006$) compared to those of the placebo group. After 12 weeks of policosanol consumption, the ferric ion reduction ability and paraoxonase of serum were elevated by 37% ($p < 0.001$) and 29% ($p = 0.004$) higher than week 0, while placebo consumption showed no notable changes. Interestingly, glycated hemoglobin (HbA$_{1c}$) in serum was lowered significantly in the policosanol group 4 weeks after consumption, which was approximately 2.1% ($p = 0.004$) lower than the placebo group. In addition, blood urea nitrogen (BUN) and uric acid levels were significantly lower in the policosanol group after 4 weeks: 14% lower ($p = 0.002$) and 4% lower ($p = 0.048$) than those of the placebo group, respectively. Repeated measures of ANOVA showed that the policosanol group had remarkable decreases in AST ($p = 0.041$), ALT ($p = 0.008$), γ-GTP ($p = 0.016$), ALP ($p = 0.003$), HbA$_{1c}$ ($p = 0.010$), BUN ($p = 0.030$), and SBP ($p = 0.011$) from the changes in the placebo group in point of time and group interaction. In conclusion, 12 weeks of 20 mg consumption of policosanol significantly enhanced hepatic protection by lowering the serum AST, ALT, ALP, and γ-GTP via a decrease in glycated hemoglobin, uric acid, and BUN with an elevation of serum antioxidant abilities. These results suggest that improvements in blood pressure by consumption of 20 mg of policosanol (Raydel®) were accompanied by protection of liver function and enhanced kidney function.

Keywords: policosanol; liver function; antioxidant; anti-glycation; glycated hemoglobin; blood urea nitrogen; blood pressure

1. Introduction

Liver damage is associated with many diseases with metabolic disorders or acute and chronic infections, which can be linked with life-threatening events [1,2]. Although the warning signs of liver diseases are jaundice, vomiting blood, and abdomen swelling, but most liver diseases often show no symptoms until they have progressed to significant damage [3]. Many blood biomarkers, such as liver aspartate aminotransferase (AST), alanine aminotransferase (ALT), γ-glutamyl transferase (γ-GTP), and alkaline phosphatase (ALP) have been used clinically [4,5] to monitor liver function for regular check-ups. Many liver diseases are caused by viral infections, inherited genetic factors, obesity, exposure to xenobiotics, and misuse of alcohol [6]. On the other hand, except for infections and genetic factors, liver damage, including fatty liver changes, is intimately associated with lifestyle, such as lack of exercise [7], alcohol consumption [8], and frequent use of drugs [9]. In particular, underlying diseases, such as diabetes, hypertension, and coronary heart disease, are risk factors for liver damage in middle-aged populations [10,11].

Many functional foods, such as milk thistle, ginseng, licorice, and turmeric, have been developed and marketed to protect against liver damage and enhance hepatic functions, such as lowering the serum AST and ALT levels [12]. In the Republic of Korea, 13 kinds of herbal and fermented extracts have been registered in the Ministry of Food and Drug Safety (MFDS) for hepatic health, https://www.foodsafetykorea.go.kr/portal/healthyfoodlife/functionalityView.do?viewNo=13 (accessed on 22 February 2023). On the other hand, their recommended dosage is too high, approximately 300 mg (Pinitol)–3150 mg (*Rubus coreanus* extract powder) per day, suggesting low efficacy. The common concerns of herbal medicine or folk medicine were associated with causing hepatotoxicity due to the hidden or unidentified ingredients and adulterants in the extracts, as summarized previously [13,14]. Therefore, to avoid herb-induced liver injury (HILI) and drug-induced liver injury (DILI), it is necessary to develop new agents to improve the hepatic functions without liver and kidney toxicity.

Because liver damage is closely associated with the progression of metabolic syndrome [15], including hypertension, diabetes, and dyslipidemia, improvement of the diabetic parameters and kidney functions should be accompanied by improvements in hepatic functions. Many studies to develop nutraceuticals to protect against liver damage could not focus only on lowering the AST, ALT, and γ-GTP levels without comparing the glycated hemoglobin and kidney function parameters. In a Chinese and Japanese study, dyslipidemia and nephrolithiasis were correlated to have a co-incidental risk factor of being overweight, hypertension [16], and glycated hemoglobin [17].

Cuban policosanol (Raydel®) consumption (10–20 mg/day) for 8–12 weeks was associated with the treatment of prehypertension [18,19] and dyslipidemia [20] via raising HDL-C and enhancing HDL functionality in randomized human studies in healthy Korean participants. Long-term clinical studies for 24 weeks also showed that policosanol consumption exerted an anti-glycation activity and antioxidant activity to protect HDL and LDL [21], which made a good agreement with the protection of apoA-I and apo-B from proteolytic degradation from in vitro studies [22,23]. Furthermore, in vitro and in vivo studies showed that reconstituted HDL containing Cuban policosanol exerted potent antioxidant, anti-glycation, and anti-inflammatory activities [18–23] with cholesterol efflux ability [19]. Recently, the reconstituted HDL-containing Cuban policosanol displayed a larger particle size with potent anti-glycation activity to protect apoA-I and antioxidant activity to protect LDL, while three reconstituted HDLs, each containing Chinese policosanol, did not [24]. These in vitro potentials of policosanol to enhance the HDL functionality are linked with the in vivo efficacy in a human clinical study to improve blood pressure and dyslipidemia in healthy Korean and Japanese participants [18–21,25].

A recent clinical study with healthy middle-aged Japanese participants showed that 12 weeks of policosanol (Raydel®) consumption significantly improved blood pressure, hepatic parameters, BUN, and HbA$_{1c}$ with enhanced HDL functionalities [25]. These results suggest that the efficacy of policosanol consumption in hypertension and dyslipidemia might be associated with the protection of liver function by improving HDL functionalities and kidney function parameters.

A desirable therapeutic agent should simultaneously protect against liver and kidney damage by improving fatty liver disease and hypertension without adverse effects. Based on the previous study, this study analyzed the effects of policosanol consumption on hepatic parameters, such as AST, ALT, γ-GTP, and ALP, as well as kidney parameters, HbA$_{1c}$, uric acid, and blood urea nitrogen during 16 weeks, including 12 weeks of consumption and four weeks post-consumption.

2. Materials and Methods

2.1. Policosanol

Raydel® policosanol tablet (10 mg/tablet, two tablets for total 20 mg per day), which was manufactured with Cuban policosanol at Raydel Australia (Thornleigh, Sydney, Australia), was obtained from Raydel Japan (Tokyo, Japan). Cuban policosanol was defined as genuine policosanol with a specific ratio of each ingredient [26]: 1-tetracosanol ($C_{24}H_{49}OH$, 0.1–20 mg/g); 1-hexacosanol ($C_{26}H_{53}OH$, 30.0–100.0 mg/g); 1-heptacosanol ($C_{27}H_{55}OH$, 1.0–30.0 mg/g); 1-octacosanol ($C_{28}H_{57}OH$, 600.0–700.0 mg/g); 1-nonacosanol ($C_{29}H_{59}OH$, 1.0–20.0 mg/g); 1-triacontanol ($C_{30}H_{61}OH$, 100.0–150.0); 1-dotriacontanol ($C_{32}H_{65}OH$, 50.0–100.0 mg/g); and 1-tetratriacontanol ($C_{34}H_{69}OH$, 1.0–50.0 mg/g).

2.2. Participants, Study Design, and Analysis

Healthy male and female volunteers with normal lipid levels and normal blood pressure were recruited nationwide in Japan via newspaper and internet advertisements between September 2021 and May 2022, as described in the preceding paper [25]. The inclusion criteria were LDL-C levels in the normal range (120–159 mg/dL) and age between 20 and 65 years old. The exclusion criteria were as follows: (1) maintenance treatment for metabolic disorder, including dyslipidemia, hypertension, and diabetes; (2) severe hepatic, renal, cardiac, respiratory, endocrinological, and metabolic disorder disease; (3) allergies; (4) heavy drinkers (more than 30 g of alcohol per day); (5) taking medicine or functional food products that may affect the lipid metabolism, including raising HDL-C or lowering LDL-C concentration, and lowering triglyceride concentration; (6) current or past smoker; (7) women in pregnancy, lactation, or planning to become pregnant during the study period; (8) person who had more than 200 mL of blood donation within one month or 400 mL of blood within three months before starting this clinical trial; (9) a person who participated in another clinical trial within the last three months or currently is participating in another clinical trial; (10) those who consumed more 2000 kcal per day; and (11) others considered unsuitable for the study at the discretion of the principal investigator. The study was approved by the Koseikai Fukuda Internal Medicine Clinic (Osaka, Japan, IRB approval number 15000074, approval date on 18 September 2021).

As shown in Figure 1A, this study was a double-blinded, randomized, and placebo-controlled trial with a 12-week treatment period. The selected participants were healthy male and female volunteers (n = 52, average age 52.1 ± 1.3 years old) with a sedentary lifestyle and without hypertension or any complaint of endocrinological disorder. All participants had unremarkable medical records without illicit drug use or a past history of chronic diseases. All participants received advice to avoid excess food (1800 kcal and 1500 kcal for men and women, respectively, per day), cholesterol (600 mg per day), alcohol drinking (<30 g and <15 g of ethanol for men and women, respectively, per day), and smoking, both direct and indirect, which can interfere with liver and kidney metabolism.

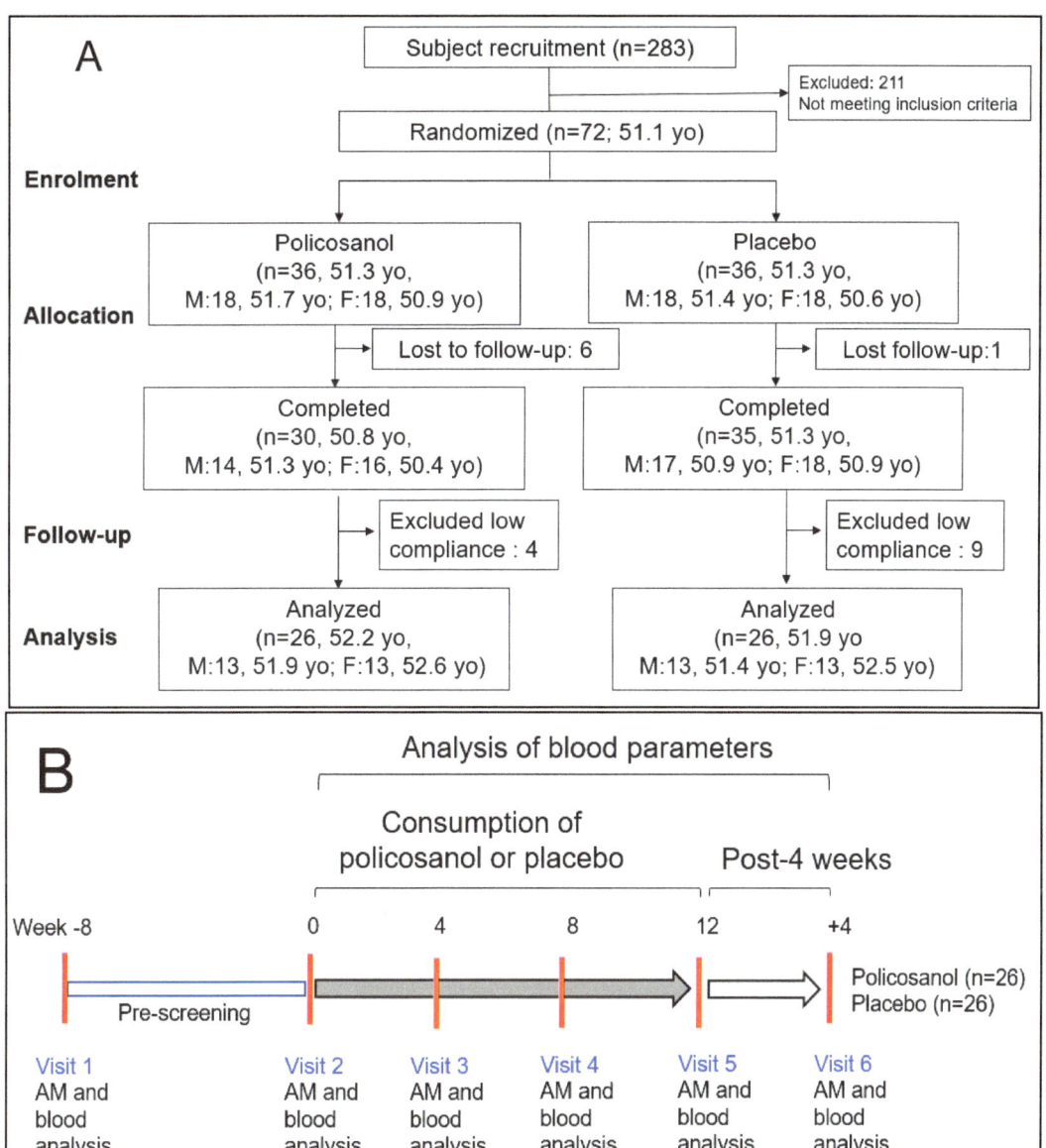

Figure 1. Study design and participant allocation for analysis (**A**) and visiting schedule of participants (**B**). AM, anthropometric measurements; yo, years old.

After allocating the participants into two groups, they were directed to take two tablets per day containing policosanol 10 mg/tablet (Raydel®) or a placebo. The tablet for the policosanol group included policosanol (10 mg/tablet), hydroxypropyl cellulose, carboxymethyl cellulose, maltodextrin, lactose, and crystalline cellulose. The tablet for the placebo group contained maltodextrin (10 mg/tablet) instead of policosanol. The blood parameters of all participants who completed the program were analyzed after 12 weeks of consumption and four weeks post-consumption (Figure 1B).

2.3. Anthropometric Analysis and Blood Analysis

The blood pressure was measured using an Omron HEM-907 (Kyoto, Japan) with three measurements with the average recorded. The height, bodyweight, and body mass index (BMI) were measured individually using a DST-210N (Muratec KDS Co., Ltd., Kyoto, Japan).

After fasting overnight, blood samples were collected in ethylenediaminetetraacetic acid (EDTA)-coated tubes and centrifuged at $3000 \times g$ for 15 min at 4 °C for the plasma assays. The samples were subjected to 19 blood biochemical assays by BML Inc. (Tokyo, Japan): total protein, albumin, aspartate transferase (AST), alanine aminotransferase (ALT), gamma-glutamyl transpeptidase (γ-GTP), creatinine, glucose, uric acid, blood urea nitrogen (BUN), lactate dehydrogenase (LDH), total bilirubin, glycated hemoglobin (hemoglobin A_{1c}, HbA_{1c}), and high sensitivity C-reactive protein (hsCRP). The protocol of human blood donation was conducted according to the guidelines of the Declaration of Helsinki and approved by the Koseikai Fukuda Internal Medicine Clinic (Osaka, Japan), with the IRB approval number 15000074.

2.4. Ferric Ion Reducing Ability Assay

The ferric ion-reducing ability (FRA) was determined using the method reported by Benzie and Strain [27]. Briefly, the FRA reagents were prepared freshly by mixing 20 mL of 0.2 M acetate buffer (pH 3.6), 2.5 mL of 10 mM 2,4,6-tripyridyl-S-triazine (Fluka Chemicals, Buchs, Switzerland), and 2.5 mL of 20 mM $FeCl_3 \cdot 6H_2O$. The antioxidant activities of serum were estimated by measuring the increase in absorbance induced by the ferrous ions generated. Freshly prepared FRA reagent (300 µL) was mixed with serum as an antioxidant source. The FRA was determined by measuring the absorbance at 593 nm every two min over a 60 min period at 25 °C using a UV-2600i spectrophotometer.

2.5. Paraoxonase Assay

The paraoxonase-1 (PON-1) activity in serum toward paraoxon was determined by evaluating the hydrolysis of paraoxon into p-nitrophenol and diethylphosphate, which was catalyzed by the enzyme [28]. The PON-1 activity was determined by measuring the initial velocity of p-nitrophenol production at 37 °C, as determined by the absorbance at 415 nm (microplate reader, Bio-Rad model 680; Bio-Rad, Hercules, CA, USA).

2.6. Data Analysis

All analyses in the Tables were normalized using a homogeneity test of the variances through Levene's statistics. Nonparametric statistics were performed using a Kruskal–Wallis test if not normalized. For Table 1, repeated measure ANOVA was used to compare the score changes in the hepatic parameters, renal parameters, and SBP between the two groups during the same period. The differences in the placebo or policosanol groups over the follow-up time were analyzed to compare in point of time and group interaction. Significant changes between the baseline and follow-up values within the groups were assessed using a paired t-test.

For Supplemental Tables S1–S5, comparisons between the policosanol and placebo with respect to BP, anthropological assessments, hematological data in blood, protein data in serum, and inflammatory assessments were analyzed using an analysis of covariance (ANCOVA) with the independent variable as the baseline and treatment. As a post hoc analysis, the Bonferroni test was used to determine the significance of the differences in the continuous variables to identify the differences between the two groups. Spearman rank correlation analysis was carried out to find a positive or negative association. The statistical power was estimated using the program G*Power 3.1.9.7 (G*Power from the University of Düsseldorf, Düsseldorf, Germany). All tests were two-tailed, and the statistical significance was $p < 0.05$. The data were analyzed using the SPSS software version 29.0 (IBM, Chicago, IL, USA).

3. Results

3.1. Improvements in Hepatic Functions and Kideny Functions in the Policosanol Group

At baseline (week 0), all subjects showed a normal range of each parameter in blood and serum data without any difference between the policosanol group and the placebo group, as summarized in the Supplemental Table S1. However, after 12 weeks of consumption, the policosanol group showed significantly lower AST, ALT, γ-GTP, and ALP than those of the placebo group in a time-dependent manner, particularly at week 12 and after 4 weeks, as shown in Table 1.

Table 1. Repeated measures ANOVA of blood parameters with hepatic functions, biliary systems, and kidney functions between placebo group and policosanol (PCO) group during 16 weeks [¶].

Variables	Groups	Week 0	Week 4	Week 8	Week 12	Post 4 Weeks	Sources	F	$p^{\ddagger}$
		Mean ± SEM	Mean ± SEM	Mean ± SEM	Mean ± SEM	Mean ± SEM			
AST	placebo	20.1 ± 1.0	20.4 ± 1.1	20.3 ± 1.2	20.9 ± 1.2	22.3 ± 1.5	Time × Group	2.715	0.041
	PCO 20 mg	20.8 ± 1.6	19.8 ± 0.8	19.1 ± 0.7	18.6 ± 0.8	18.6 ± 0.8 *			
	$p^{\dagger}$	0.724	0.312	0.196	0.017	0.008			
ALT	placebo	18.2 ± 1.7	19.1 ± 1.9	19.7 ± 2.5	18.8 ± 1.8	20.4 ± 2.1	Time × Group	3.954	0.008
	PCO 20 mg	21.5 ± 3.3	18.7 ± 1.2	17.4 ± 0.9	17.1 ± 1.6	15.8 ± 1.0			
	$p^{\dagger}$	0.367	0.176	0.124	0.041	0.001			
ALT/AST (ratio)	placebo	0.88 ± 0.06	0.92 ± 0.07	0.91 ± 0.06	0.88 ± 0.06	0.90 ± 0.06	Time × Group	3.404	0.010
	PCO 20 mg	0.98 ± 0.06	0.94 ± 0.04	0.91 ± 0.04	0.89 ± 0.05	0.84 ± 0.04			
	$p^{\dagger}$	0.258	0.256	0.112	0.145	0.001			
γ-GTP	placebo	26.2 ± 3.6	28 ± 3.9	26.9 ± 4.1	27.8 ± 4.3	30.0 ± 5.2	Time × Group	3.312	0.016
	PCO 20 mg	28.7 ± 4.3	27.4 ± 3.3	24.2 ± 2.7	24.1 ± 2.7	23.7 ± 2.5			
	$p^{\dagger}$	0.666	0.128	0.039	0.015	0.011			
ALP	placebo	64.8 ± 3.9	66.0 ± 3.5	66.7 ± 4.2	66.1 ± 3.8	67.0 ± 3.8	Time × Group	4.240	0.003
	PCO 20 mg	69.0 ± 2.6	69.3 ± 2.3	66.6 ± 2.3	65.9 ± 2.5	65.7 ± 2.1			
	$p^{\dagger}$	0.414	0.958	0.012	0.012	0.006			
HbA$_{1c}$	placebo	5.48 ± 0.05	5.48 ± 0.06	5.35 ± 0.05	5.45 ± 0.07	5.46 ± 0.06	Time × Group	7.129	0.010
	PCO 20 mg	5.46 ± 0.05	5.42 ± 0.05	5.30 ± 0.04	5.42 ± 0.05	5.34 ± 0.04			
	$p^{\dagger}$	0.791	0.230	0.250	0.599	0.004			
BUN	placebo	13.6 ± 0.6	14.5 ± 0.7	14.0 ± 0.7	14.2 ± 0.7	14.8 ± 0.7	Time × Group	2.944	0.030
	PCO 20 mg	13.4 ± 0.4	13.4 ± 0.6	13.3 ± 0.5	12.8 ± 0.6	12.6 ± 0.6 *			
	$p^{\dagger}$	0.767	0.193	0.411	0.053	0.002			
Uric acid (mg/dL)	placebo	5.1 ± 0.3	5.0 ± 0.3	4.9 ± 0.3	5.1 ± 0.3	5.2 ± 0.3	Time × Group	1.437	0.241
	PCO 20 mg	5.2 ± 0.2	5.0 ± 0.3	5.0 ± 0.3	5.1 ± 0.2	5.0 ± 0.2 *			
	p	0.859	0.528	0.708	0.418	0.048			
SBP (mmHg)	placebo	112.0 ± 2.1	114.7 ± 2.5	105.9 ± 3.1	110.4 ± 2.2	115.7 ± 2.2	Time × Group	3.359	0.011
	PCO 20 mg	114.0 ± 3.1	111.0 ± 3.3	104.0 ± 2.8	104.7 ± 2.9	107.4 ± 2.6 *			
	$p^{\dagger}$	0.586	0.045	0.231	0.004	0.001			

[¶] Data are expressed as the mean ± SEM. Estimated statistical power was 99.8% from the selected participants in both group (n = 52) based on calculations using the program G*Power 3.1.9.7 (G*Power from the University of Düsseldorf, Düsseldorf, Germany). The participants meet the inclusion criteria and were instructed to avoid alcohol drinking (<30 g and <15 g of ethanol for men and women, respectively, per day), and smoking, both direct and indirect. $p^{\dagger}$ indicates whether the ANCOVA is statistically significant. $p^{\ddagger}$ indicates whether the repeated measures ANOVA is statistically significant. *, Statistically significantly different mean value by independent t-test between the placebo group and policosanol 20 mg group. AST, alanine transaminase; ALT, alanine aminotransferase; γ-GTP, gamma-glutamyl transferase; ALP, alkaline phosphatase; HbA$_{1c}$, glycated hemoglobin; BUN, blood urea nitrogen; SBP, systolic blood pressure; PCO, policosanol.

Repeated measures ANOVA revealed that there were significant differences in serum AST, ALT, ALT/AST (ratio), γ-GTP, and ALP between the placebo and policosanol groups in a time-dependent manner. The policosanol group significantly lowered the hepatic function parameters, especially in week 12 and after 4 weeks of consumption in terms of time and group interaction. The diabetic marker (HbA$_{1c}$) and kidney function parameters (BUN and uric acid) were also significantly decreased in the policosanol group in a time-dependent manner (Table 1). Repeated measures ANOVA revealed that HbA$_{1c}$ and BUN showed significant differences between the placebo and policosanol groups in terms of time and group interaction.

As shown in Table 1, serum AST levels were 11% ($p = 0.017$) and 17% ($p = 0.008$) lower in the policosanol group than the placebo group at week 12 and after 4 weeks, respectively. The policosanol group also exhibited significantly lower AST in a time-dependent manner: up to 17% decrease at week 16 ($p^† = 0.008$) compared to week 0, as shown in Table 1 and Supplemental Figure S1. Serum ALT levels were 9% ($p^† = 0.041$) and 23% ($p^† < 0.001$) lower in the policosanol group than the placebo group at week 12 and after 4 weeks, respectively. Repeated measures ANOVA revealed that policosanol group showed significantly lower AST ($p^‡ = 0.041$) and ALT ($p^‡ = 0.008$) than the placebo group, as shown in Table 1.

Interestingly, although the two groups had similar alcohol consumption of 9–10 g of ethanol/day during the 16 weeks, γ-GTP was 10% ($p^† = 0.039$), 14% ($p^† = 0.015$), and 23% ($p^† = 0.011$) lower in the policosanol group than those of placebo group at week 8, week 12, and after 4 weeks, respectively, in a time-dependent manner, as shown in Table 1 and Supplemental Figure S2. The serum alkaline phosphatase (ALP) level was decreased in the policosanol group in a time-dependent manner during the 16 weeks: approximately 4.8% lower than week 0 ($p^† = 0.006$), while the placebo group showed a 3.3% increase from week 0 to week 16 (Table 1). The policosanol group showed significantly lower ALP at week 8 ($p^† = 0.012$), week 12 ($p^† = 0.012$), and after 4 weeks ($p^† = 0.006$) compared with those of the placebo group, although both groups showed a normal range of ALP during the 16 weeks. At week 16, the policosanol group showed a 2.1% lower ALP level ($p^† = 0.006$) than that of the placebo group, but policosanol group showed a 6.5% higher level than the placebo group at week 0. Repeated measures ANOVA revealed that the policosanol group showed significantly lower γ-GTP ($p^‡ = 0.016$) and ALP ($p^‡ = 0.003$) than placebo group as shown in Table 1.

These results suggest that liver function is protected by policosanol consumption for 12 weeks: lowered serum levels of hepatic enzymes, AST, ALT, γ-GTP, and ALP, in the policosanol group. Interestingly, the hepatic protection effects of policosanol were maintained at post-4 week consumption.

3.2. Improvements in Kidney Functions in the Policosanol Group

As a kidney function parameter, the blood urea nitrogen (BUN) decreased in the policosanol group in a time-dependent manner with up to a 6% decrease ($p^† = 0.002$) from weeks 0 to 16, while the placebo group showed a 9% increase from week 0 to 16 (Table 1 and Supplemental Figure S3A). The policosanol group also exhibited a 15% lower BUN ($p^† = 0.002$) than the placebo group at week 16. Repeated measures ANOVA of BUN showed that the policosanol group showed a significant difference ($p^‡ = 0.030$) from the placebo group in the point of time and group interaction (Table 1). On the other hand, uric acid was also decreased in the policosanol group in a time-dependent manner: 4% lower than week 0 and the placebo group at week 16 ($p^† = 0.048$), as shown in Table 1 and Supplemental Figure S3B. Although repeated measures ANOVA showed no significant difference ($p^‡ = 0.241$) in the group and time interactions during the 16 weeks between the two groups, the policosanol group showed a significant difference between week 0 and 16 (after 4 weeks of consumption) compared to the placebo group. The other parameters for kidney functions, such as electrolytes, inorganic phosphorus (P), calcium (Ca), sodium (Na), potassium (K), and chloride (Cl), were similar in both groups, which fell in the normal range, as listed in Supplemental Table S2. These results suggested that the policosanol consumption induced enhancement of kidney function via a decrease of BUN and uric acid without impairment of electrolyte metabolism in kidney.

3.3. Decrease of ALT/AST Ratio and SBP in the Policosanol Group

The ALT/AST ratio decreased gradually and significantly in the policosanol group from 0.98 ± 0.06 at week 0 to 0.86 ± 0.05 at week 16 and after four weeks of consumption, while the placebo group did not show a change around 0.88~0.90 during the 16 weeks, as shown in Table 1 and Supplemental Figure S4A. Repeated measures ANOVA showed that there was a significant difference ($p = 0.010$) in the ALT/AST ratio between the two groups in the group and the time interactions during the 16 weeks (Table 1). The SBP decreased gradually and significantly to an 8.2% reduction in the policosanol group from 114.0 ± 3.1 mmHg at week 0 and to 104.7 ± 2.9 mmHg at week 12, as shown in Table 1 and Supplemental Figure S4B. On the other hand, the placebo group did not show a notable change in the SBP during 16 weeks—approximately 112–115 mmHg. Repeated measures ANOVA showed a significant difference ($p = 0.011$) in the SBP between the two groups in the group and time interactions during the 16 weeks (Table 1). These results suggest positive correlations between the decrease in the ALT/AST ratio and SBP.

There was no age difference (approximately 52.1 ± 1.3 years old) between the policosanol group (n = 26, M13/F13) and the placebo group (n = 26, M13/F13) during the consumption period. Although there was no difference in SBP between the groups at week 0, the policosanol group showed a 6.2% ($p = 0.005$) and 7.2% ($p = 0.004$) lower SBP than that of the placebo group at week 12 and week 16 (after 4 weeks), respectively, as shown in Supplemental Table S3. The policosanol group also exhibited a significantly lower SBP time-dependent manner: up to a 5.2% ($p = 0.004$) and 7.2% ($p < 0.001$) decrease in the SBP at week 12 and week 16, respectively, compared with week 0 (Supplemental Table S1). Although the policosanol group showed a significant decrease in the SBP at week 12 and 16, after 12 weeks of consumption, the SBP remained in the normal range without an abrupt decrease below hypotension (<90 mmHg).

The DBP and pulse rate were similar in both groups at approximately 63–69 mmHg and 68–73 bpm, respectively, in the normal ranges during the 16 weeks. Interestingly, the policosanol group showed 2% lower body weight ($p = 0.031$) and BMI ($p = 0.022$) than the placebo group only at week 4. The two groups showed no difference with normal ranges of total serum protein, albumin, creatinine, glucose, and lactate dehydrogenase (LDH) levels, which were similar to the group during the 16 weeks (Supplemental Tables S3–S5). These results suggest that policosanol consumption caused an improvement in the BP without significant impairment of the protein and carbohydrate metabolism in the liver and kidney, as shown in Table 1 and Supplemental Tables S2–S5. Except for the decrease in SBP in the policosanol group, there was no difference in DBP, pulse rate, body weight, or BMI between the placebo and policosanol groups during the 16 weeks.

Overall, these results suggest that the consumption of policosanol caused several beneficial effects to protect liver functions (lowering AST, ALT, and γ-GTP), hepatobiliary systems (lowering ALP), and kidney functions (lowering BUN, uric acid, and HbA_{1c})0. These beneficial activities contributed to the enhancement of the liver function and kidney function, which are connected to the decrease in SBP, without impairment of electrolyte metabolism. These results suggest that policosanol consumption may prevent or attenuate the incidence of liver disease, kidney disease, and diabetes, which can explain why blood pressure improves.

3.4. Enhancement of the Antioxidant Abilities of Serum

As shown in Figure 2A, at week 12, the policosanol group showed a 37% increase in the ferric ion reduction ability (FRA) around 118 μM of ferrous equivalents than that of week 0 ($p < 0.001$). In contrast, the placebo group did not change (63–70 μM of ferrous equivalents). The paraoxonase (PON) activity was elevated 29% in the policosanol group to approximately 116 μU/L/min at week 12 compared with week 0 ($p = 0.004$). In contrast, the placebo group did not change: it was approximately 88–90 μU/L/min (Figure 2B). These results suggest that the consumption of policosanol for 12 weeks was linked with the enhancement of the serum antioxidant abilities, such as FRA and PON.

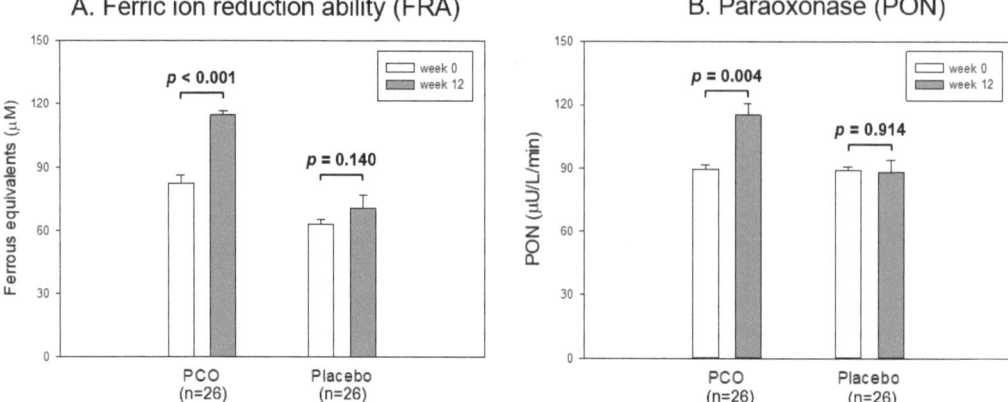

Figure 2. Antioxidant abilities of serum from each group between week 0 and 12. The data are expressed as the mean ± SD from three independent experiments with duplicate samples. FRA and PON activity in each group between week 0 and week 12 were compared using a paired *t*-test. (**A**) Comparison of the ferric ion reduction ability (FRA). FRA was expressed as the concentration of vitamin C (mM), equivalent to reducing the amount of ferric ion (μM) per hour. (**B**) Comparison of the paraoxonase (PON) activity. PON activity was expressed as the initial velocity of *p*-nitrophenol production per min (μU/L/min) at 37 °C during 60 min incubation.

3.5. Changes of Hematological Data and Serum Protein Data

As shown in Supplemental Table S4, the policosanol and placebo groups showed a normal range of numbers in white blood cells (WBC), hematocrit (Hct), and platelets (Plt) between during the 16 weeks. These results suggest that there was no difference in complete blood count between the two groups without indication of leukemia, anemia, or thrombocytosis. Although the mechanism is unclear, a mild increase in RBC number and Hb number was associated with an enhancement of oxygen carrying ability. Other hematologic parameters, mean corpuscular volume (MCV), mean corpuscular hemoglobin (MCH), and mean corpuscular hemoglobin concentration (MCHC) were similar in both groups within normal range during 16 weeks (Supplemental Table S4). These results suggest that policosanol consumption did not impair the red blood cell size and volume parameters; all values were in the normal range without risk of anemia.

Interestingly, the glycated hemoglobin (HbA_{1c}) level was significantly lower in the policosanol group, even though the policosanol group showed a higher hemoglobin (Hb) level than the placebo group during 16 weeks (Supplemental Figure S5A). At week 16, the policosanol group showed a 2.2% lower HbA_{1c} level ($p = 0.004$) than the placebo group, as shown in Table 1 and Supplemental Figure S5B. During the 16 weeks, repeated measures ANOVA of HbA_{1c} showed that the policosanol group showed a significant difference ($p = 0.010$) from the placebo group in point of time and group interaction, as shown in Table 1. These results suggest that policosanol consumption inhibited glycation in blood proteins, particularly hemoglobin, in a time-dependent manner at week 16 after four weeks of consumption. These decreases in glycated hemoglobin were closely associated with the decrease in SBP ($r = 0.197$, $p = 0.161$) by policosanol consumption, but the mechanism is unclear.

As shown in Supplemental Table S5, the placebo group and policosanol group showed normal levels of total protein and albumin in the blood, around 7.0–7.1 g/dL and 4.3–4.4 g/dL, respectively, without any difference between the groups during the 16 weeks. Lactate dehydrogenase (LDH) and creatinine showed normal ranges around 155–168 mg/dL and 0.71–0.77 mg/dL, respectively, in both groups without any difference between the groups. The acute inflammatory parameter, hsCRP, was no different between the groups within the normal range around 0.03–0.22 mg/dL during the 16 weeks, suggesting that there were no remarkable infections or autoimmune inflammatory responses such as rheumatoid arthritis. Overall, these results suggest that policosanol consumption did not impair protein synthesis and nutritional metabolism in liver and kidney functions.

4. Discussion

Patients with dyslipidemia are frequently associated with nonalcoholic fatty liver disease (NAFLD) accompanied by impairment of hepatic functions: elevation of hepatic parameters including AST, ALT, ALP, and γ-GTP [29,30]. Therefore, lowering the hepatic parameters has been recognized as a potent efficacy of nutraceuticals to protect against liver damage in NAFLD [31]. The blood biomarkers level of liver health should be checked regularly to diagnose the progression of liver damage or diseases, such as fatty liver disease, hepatitis, and cirrhosis, particularly in middle-aged populations. Globally, NAFLD and metabolic syndrome have increased rapidly [32], particularly due to the westernized transition, and are more prevalent in middle-aged populations (45–65 years old) to exhibit hypertension, dyslipidemia (low HDL-C and high triglyceride), diabetes, and chronic kidney disease to impair the quality of life [33]. More importantly, improving the serum lipid profile by increasing HDL-C and decreasing the TG and glucose levels in middle-aged adults was associated with an enhancement of cognition and lower Alzheimer's disease risk [34].

In the middle-aged Japanese participants, the consumption of Cuban policosanol lowered blood pressure and glycated hemoglobin by raising HDL-C with an improvement in the HDL quality and functionality [25]. The current results also revealed the policosanol group to have a 2.2% lower HbA_{1c} level than the placebo group during the 16 weeks in point of group and time interaction ($p^{\ddagger} = 0.010$), and a 2.2% decrease in HbA_{1c} from baseline (Supplemental Figure S5 and Supplemental Tables S1 and S4). Glycated hemoglobin is a risk factor for cardiovascular diseases, all-cause mortality [35], and hypertension [36]. Therefore, lowering HbA_{1c} by policosanol consumption might help alleviate cardiovascular mortality because HbA_{1c} is a reliable risk factor for all-cause mortality and cardiovascular mortality in nondiabetic and diabetic populations [37]. Indeed, in vitro tests and human clinical studies showed that Cuban policosanol exhibited potent anti-glycation activity against fructation [22–24] and less glycation extent of apoA-I in HDL [21] and apo-B in LDL/VLDL [25]. A comparison study with various origins of policosanol showed that Cuban policosanol inhibited glycation by protecting apoA-I from proteolytic degradation. In contrast, three Chinese policosanols did not inhibit fructose-mediated glycation [24]. The in vitro anti-glycation activities of Cuban policosanol agreed with in vivo protection of apoA-I with less multimerization of apoA-I from human studies with Korean [20,21] and Japanese participants [25].

Although a few of results look similar and overlap with the preceding report [25], however, there are many different points between the preceding paper and current paper. In the previous paper, change in blood pressure, glycated hemoglobin, AST, ALT, and γ-GTP were reported via analysis of covariance between week 0 and week 12 without data from repeated measures ANOVA of a 4 week interval. The preceding paper [25] focused mainly on improvement of lipid profile, lipoprotein properties, HDL and LDL quality, and HDL functionality.

The current study is very different from the preceding paper [25] in terms of its longer analysis period and different parameters of repeated measures ANOVA. In the current study, 16 weeks of total data were analyzed by repeated measures ANOVA with blood pressure, glycated hemoglobin, liver function, and kidney function parameters. However, the preceding paper [25] showed total 12-week data, which was analyzed by ANCOVA with the independent variables as baseline (week 0) and treatment (week 12) in each group.

Many efficacy studies with functional foods have focused only on the lowering effects of AST and ALT and γ-GTP, without comparing ALP, glycated hemoglobin, and kidney function parameters, such as blood urea nitrogen (BUN) and uric acid. ALP is a diagnostic marker of cholestatic hepatitis and hepatic fibrosis in patients with nonalcoholic steatohepatitis [37] and an independent predictor for hepatic disease-related death [38]. BUN is elevated in patients with nonalcoholic fatty liver disease [39] and is positively associated with the elevation of HbA_{1c} in the older population [40]. In addition to protecting against liver function, enhancing kidney function and blood pressure is more desirable to prevent NAFLD and metabolic syndrome in middle-aged populations.

High serum ALT and AST were associated with a high risk of hypertension and increased BP in young Chinese populations [41]. A clinical study with Korean subjects suggested that AST, ALT, and γ-GTP were positively associated with SBP and DBP from correlation analysis of the liver enzyme and cardiovascular factors [42]. As shown in Supplemental Figures S1 and S4 a decrease in AST, ALT, and AST/ALT ratio all correlate well with a decrease in BP in the policosanol group. Interestingly, among four liver enzymes (AST, ALT, γ-GTP, and ALP), only ALT was negatively associated with the serum apoA-I level ($r = -0.028$, $p = 0.010$) and HDL-C ($r = -0.224$, $p < 0.001$) and positively associated with the apo-B level ($r = 0.114$, $p = 0.022$) and SBP ($r = 0.148$, $p = 0.002$) from the clinical observations [42]. A recent randomized and double-blinded clinical study with Japanese participants also showed that 12 weeks of policosanol consumption resulted in an increase in apoA-I, a decrease in ALT, and a decrease in SBP [25]. Overall, these reports strongly suggest that enhancing the HDL quantity and quality were related to protecting against liver function and preventing hypertension.

The current results agree well with previous reports, which showed the hepatoprotective activity of policosanol with antioxidant activity. Policosanol (25 and 100 mg/kg) protected against acute liver injury, carbon tetrachloride (CCl_4)-induced hepatic injury in rats, and a model of hepatotoxicity in which the process of lipid peroxidation [43]. Oral supplementation (100 mg/kg) of policosanol alleviates CCl_4-induced liver fibrosis by lowering AST, ALT, ALP, and γ-GTP [41]. The hepatoprotection activity was also linked with the reduction of serum levels of interleukin (IL-6), tumor necrosis factor (TNF-α), and malondialdehyde (MDA) [44]. Without fatty liver change, interestingly, serum AST and ALT were decreased significantly by policosanol supplementation for eight weeks in hyperlipidemic zebrafish [45]. In spontaneously hypertensive rats (SHR), eight weeks of feeding the policosanol resulted in a remarkable lowering of the BP with a lowering of the serum CRP [46]. The zebrafish and SHR also significantly decreased fatty streak lesions, inflammatory cell infiltration, and reactive oxygen species [45,46]. These ameliorations of the fatty liver changes agreed with another report that showed policosanol alleviated hepatic lipid accumulation in mice models by regulating bile acids metabolism [47]. In the same context, policosanol attenuated cholesterol synthesis via AMP-activated protein kinase (AMPK) in hepatoma cells [48] and hypercholesterolemic rats [49]. Overall, the policosanol consumption ameliorated the fatty liver change and lowered the AST and ALT levels in human and various animal models.

A new therapeutic agent with identified active ingredients was developed to protect against liver and kidney damage with a relatively lower dosage (<20–100 mg/day) without adverse effects. On the other hand, the almost registered ingredients of functional foods in MFDS of Korea have a higher dosage (e.g., 3150 mg/day of *Rubus coreanus* Extract) and many unknown and unidentified ingredients, which can help induce HILI and DILI [50,51]. Moreover, heavy metal contamination of registered herbal supplements and synthetic drugs as common adulterants in herbal products are frequently associated with high morbidity and mortality from HILI [52,53].

As far as we know, this paper is the first report to show that short-term (12 week) consumption of 20 mg of policosanol resulted in improved liver functions and kidney functions simultaneously by lowering the serum AST, ALT, γ-GTP, ALP, HbA_{1c}, BUN, and SBP in a time- and group-dependent manner. These enhancements can explain why policosanol supplementation ameliorated fatty liver change, inhibited the inflammatory response and ROS production in hepatic tissue, and lowered BP in animal and human studies, as reported previously [21–25].

In conclusion, 12 weeks of 20 mg consumption of policosanol (Raydel®) protected liver function and enhanced kidney functions, and improved blood pressure (SBP) from randomized, placebo-controlled, and double-blinded trials with healthy Japanese participants.

Supplementary Materials: The following supporting information can be downloaded at: https://www.mdpi.com/article/10.3390/life13061319/s1, Figure S1. Graphical expression of change in the parameters in hepatic function, serum AST (A), and ALT (B), during the 16 weeks between the policosanol and placebo group. AST, Aspartate transaminase; ALT, Alanine aminotransferase. *, $p < 0.05$ versus placebo; **, $p < 0.01$ versus placebo; ***, $p < 0.001$ versus placebo from the analysis of covariance (ANCOVA) model with the independent variable as the baseline and treatment. p ‡ value in blue font indicates the significance of time and group interaction during 16 weeks from repeated measurement ANOVA. Figure S2. Graphical expression of change in the parameters in hepatic function, serum γ-GTP (A) and ALP (B), during the 16 weeks between the policosanol and placebo group. γ-GTP, Gamma-glutamyl transferase; ALP, Alkaline phosphatase. *, $p < 0.05$ versus placebo from the analysis of covariance (ANCOVA) model with the independent variable as the baseline and treatment. p ‡ value in blue font indicates the significance of time and group interaction during 16 weeks from repeated measurement ANOVA. Figure S3. Graphical expression of change in the parameters of the kidney functions, blood urea nitrogen (A), and uric acid (B), during the 16 weeks between the policosanol and placebo group. BUN, blood urea nitrogen. *, $p < 0.05$ versus placebo; **, $p < 0.01$ versus placebo from the analysis of covariance (ANCOVA) model with the independent variable as the baseline and treatment. p ‡ value in blue font indicates the significance of time and group interaction during 16 weeks from repeated measurement ANOVA. Figure S4. Graphical expression of change in the hepatic parameters, ALT/AST ratio, (A) and SBP (B) during the 16 weeks between the policosanol and placebo group. *, $p < 0.05$ versus placebo; **, $p < 0.01$ versus placebo; ***, $p < 0.001$ versus placebo from the analysis of covariance (ANCOVA) model with the independent variable as the baseline and treatment. p ‡ value in blue font indicates the significance of time and group interaction during 16 weeks from repeated measurement ANOVA. Figure S5. Graphical expression of change in the blood hemoglobin (A) and glycated hemoglobin (B) contents during the 16 weeks between the policosanol and placebo groups. Hb, hemoglobin; HbA_{1c}, glycated hemoglobin. *, $p < 0.05$ versus the placebo from the analysis of covariance (ANCOVA) model with the independent variable as the baseline and treatment. p ‡ value in blue font indicates the significance of the time and group interaction during 16 weeks from repeated measurement ANOVA. Table S1. Comparison of baseline (week 0) data between placebo and policosanol 20 mg group ¶. Table S2. Parameters for kidney function test. Table S3. Change in the blood pressure and anthropological data between placebo group and policosanol (PCO) group during 16 weeks ¶. Table S4. Changes in the hematologic data in blood between placebo group and policosanol (PCO) group during 16 weeks ¶. Table S5. Serum proteins, glucose, and inflammatory parameters between placebo group and policosanol (PCO) group during 16 weeks ¶.

Author Contributions: Conceptualization, K.-H.C. and Y.U.; methodology, J.-E.K. and T.K.; writing—original draft preparation, K.-H.C.; supervision, K.-H.C.; data curation and investigation, Y.U. All authors have read and agreed to the published version of the manuscript.

Funding: This research received no external funding.

Institutional Review Board Statement: The protocol of human blood donation was conducted according to the guidelines of the Declaration of Helsinki and was approved by the Koseikai Fukuda Internal Medicine Clinic (Osaka, Japan), with the IRB approval number 15000074, approval date on 18 September 2021.

Informed Consent Statement: Not applicable.

Data Availability Statement: The data used to support the findings of this study are available from the corresponding author upon reasonable request.

Conflicts of Interest: The authors declare no conflict of interest.

References

1. Rosselli, M.; Lotersztajn, S.; Vizzutti, F.; Arena, U.; Pinzani, M.; Marra, F. The metabolic syndrome and chronic liver disease. *Curr. Pharm. Des.* **2014**, *20*, 5010–5024. [CrossRef] [PubMed]
2. Talwani, R.; Gilliam, B.L.; Howell, C. Infectious diseases and the liver. *Clin. Liver Dis.* **2011**, *15*, 111–130. [CrossRef] [PubMed]
3. Anstee, Q.M.; Castera, L.; Loomba, R. Impact of non-invasive biomarkers on hepatology practice: Past, present and future. *J. Hepatol.* **2022**, *76*, 1362–1378. [CrossRef]
4. Limdi, J.; Hyde, G. Evaluation of abnormal liver function tests. *Postgrad. Med. J.* **2003**, *79*, 307–312. [CrossRef] [PubMed]
5. McGill, M.R. The past and present of serum aminotransferases and the future of liver injury biomarkers. *EXCLI J.* **2016**, *15*, 817. [CrossRef]
6. Beier, J.I.; Arteel, G.E. Environmental exposure as a risk-modifying factor in liver diseases: Knowns and unknowns. *Acta Pharm. Sin. B* **2021**, *11*, 3768–3778. [CrossRef]
7. Harrison, S.A.; Day, C.P. Benefits of lifestyle modification in NAFLD. *Gut* **2007**, *56*, 1760–1769. [CrossRef]
8. Osna, N.A.; Donohue, T.M., Jr.; Kharbanda, K.K. Alcoholic liver disease: Pathogenesis and current management. *Alcohol Res. Curr. Rev.* **2017**, *38*, 147.
9. Imani, F.; Motavaf, M.; Safari, S.; Alavian, S.M. The therapeutic use of analgesics in patients with liver cirrhosis: A literature review and evidence-based recommendations. *Hepat. Mon.* **2014**, *14*, e23539. [CrossRef]
10. Chang, W.H.; Mueller, S.H.; Chung, S.-C.; Foster, G.R.; Lai, A.G. Increased burden of cardiovascular disease in people with liver disease: Unequal geographical variations, risk factors and excess years of life lost. *J. Transl. Med.* **2022**, *20*, 2. [CrossRef]
11. Bonora, E.; Targher, G. Increased risk of cardiovascular disease and chronic kidney disease in NAFLD. *Nat. Rev. Gastroenterol. Hepatol.* **2012**, *9*, 372–381. [CrossRef] [PubMed]
12. Mega, A.; Marzi, L.; Kob, M.; Piccin, A.; Floreani, A. Food and nutrition in the pathogenesis of liver damage. *Nutrients* **2021**, *13*, 1326. [CrossRef] [PubMed]
13. Teschke, R.; Eickhoff, A. Herbal hepatotoxicity in traditional and modern medicine: Actual key issues and new encouraging steps. *Front. Pharmacol.* **2015**, *6*, 72. [CrossRef] [PubMed]
14. Cho, J.-H.; Oh, D.-S.; Hong, S.-H.; Ko, H.; Lee, N.-H.; Park, S.-E.; Han, C.-W.; Kim, S.-M.; Kim, Y.-C.; Kim, K.-S. A nationwide study of the incidence rate of herb-induced liver injury in Korea. *Arch. Toxicol.* **2017**, *91*, 4009–4015. [CrossRef] [PubMed]
15. Marchesini, G.; Forlani, G.; Bugianesi, E. Is liver disease a threat to patients with metabolic disorders? *Ann. Med.* **2005**, *37*, 333–346. [CrossRef]
16. Ding, Q.; Ouyang, J.; Fan, B.; Cao, C.; Fan, Z.; Ding, L.; Li, F.; Tu, W.; Jin, X.; Wang, J. Association between dyslipidemia and nephrolithiasis risk in a Chinese population. *Urol. Int.* **2019**, *103*, 156–165. [CrossRef]
17. Kabeya, Y.; Kato, K.; Tomita, M.; Katsuki, T.; Oikawa, Y.; Shimada, A.; Atsumi, Y. Associations of insulin resistance and glycemic control with the risk of kidney stones. *Intern. Med.* **2012**, *51*, 699–705. [CrossRef]
18. Park, H.-J.; Yadav, D.; Jeong, D.-J.; Kim, S.-J.; Bae, M.-A.; Kim, J.-R.; Cho, K.-H. Short-term consumption of Cuban policosanol lowers aortic and peripheral blood pressure and ameliorates serum lipid parameters in healthy Korean participants: Randomized, double-blinded, and placebo-controlled study. *Int. J. Environ. Res. Public Health* **2019**, *16*, 809. [CrossRef]
19. Cho, K.-H.; Kim, S.-J.; Yadav, D.; Kim, J.-Y.; Kim, J.-R. Consumption of cuban policosanol improves blood pressure and lipid profile via enhancement of HDL functionality in healthy women subjects: Randomized, double-blinded, and placebo-controlled study. *Oxid. Med. Cell. Longev.* **2018**, *2018*, 4809525. [CrossRef]
20. Kim, J.-Y.; Kim, S.-M.; Kim, S.-J.; Lee, E.-Y.; Kim, J.-R.; Cho, K.-H. Consumption of policosanol enhances HDL functionality via CETP inhibition and reduces blood pressure and visceral fat in young and middle-aged subjects. *Int. J. Mol. Med.* **2017**, *39*, 889. [CrossRef]
21. Kim, S.-J.; Yadav, D.; Park, H.-J.; Kim, J.-R.; Cho, K.-H. Long-term consumption of cuban policosanol lowers central and brachial blood pressure and improves lipid profile with enhancement of lipoprotein properties in healthy korean participants. *Front. Physiol.* **2018**, *9*, 412. [CrossRef] [PubMed]
22. Lim, S.-M.; Yoo, J.-A.; Lee, E.-Y.; Cho, K.-H. Enhancement of high-density lipoprotein cholesterol functions by encapsulation of policosanol exerts anti-senescence and tissue regeneration effects via improvement of anti-glycation, anti-apoptosis, and cholesteryl ester transfer inhibition. *Rejuvenation Res.* **2016**, *19*, 59–70. [CrossRef] [PubMed]
23. Cho, K.-H.; Bae, M.; Kim, J.-R. Cuban sugar cane wax acid and policosanol showed similar atheroprotective effects with inhibition of LDL oxidation and cholesteryl ester transfer via enhancement of high-density lipoproteins functionality. *Cardiovasc. Ther.* **2019**, *2019*, 8496409. [CrossRef] [PubMed]
24. Cho, K.-H.; Baek, S.H.; Nam, H.-S.; Kim, J.-E.; Kang, D.-J.; Na, H.; Zee, S. Cuban Sugar Cane Wax Alcohol Exhibited Enhanced Antioxidant, Anti-Glycation and Anti-Inflammatory Activity in Reconstituted High-Density Lipoprotein (rHDL) with Improved Structural and Functional Correlations: Comparison of Various Policosanols. *Int. J. Mol. Sci.* **2023**, *24*, 3186. [CrossRef]
25. Cho, K.-H.; Nam, H.-S.; Baek, S.-H.; Kang, D.-J.; Na, H.; Komatsu, T.; Uehara, Y. Beneficial Effect of Cuban Policosanol on Blood Pressure and Serum Lipoproteins Accompanied with Lowered Glycated Hemoglobin and Enhanced High-Density Lipoprotein Functionalities in a Randomized, Placebo-Controlled, and Double-Blinded Trial with Healthy Japanese. *Int. J. Mol. Sci.* **2023**, *24*, 5185. [CrossRef]
26. Canavaciolo, V.L.G.; Gómez, C.V. "Copycat-policosanols" versus genuine policosanol. *Rev. CENIC Cienc. Quím.* **2007**, *38*, 207–213.

27. Benzie, I.F.; Strain, J. [2] Ferric reducing/antioxidant power assay: Direct measure of total antioxidant activity of biological fluids and modified version for simultaneous measurement of total antioxidant power and ascorbic acid concentration. In *Methods in Enzymology*; Elsevier: Amsterdam, The Netherlands, 1999; Volume 299, pp. 15–27. [CrossRef]
28. Blatter Garin, M.-C.; Moren, X.; James, R.W. Paraoxonase-1 and serum concentrations of HDL-cholesterol and apoA-I. *J. Lipid Res.* 2006, *47*, 515–520. [CrossRef]
29. Speliotes, E.K.; Balakrishnan, M.; Friedman, L.S.; Corey, K.E. treatment of Dyslipidemia in Common liver Diseases. *Clin. Liver Dis.* 2019, *14*, 161–162. [CrossRef]
30. Hadizadeh, F.; Faghihimani, E.; Adibi, P. Nonalcoholic fatty liver disease: Diagnostic biomarkers. *World J. Gastrointest. Pathophysiol.* 2017, *8*, 11. [CrossRef]
31. Cicero, A.F.; Colletti, A.; Bellentani, S. Nutraceutical approach to non-alcoholic fatty liver disease (NAFLD): The available clinical evidence. *Nutrients* 2018, *10*, 1153. [CrossRef]
32. Zhou, F.; Zhou, J.; Wang, W.; Zhang, X.J.; Ji, Y.X.; Zhang, P.; She, Z.G.; Zhu, L.; Cai, J.; Li, H. Unexpected rapid increase in the burden of NAFLD in China from 2008 to 2018: A systematic review and meta-analysis. *Hepatology* 2019, *70*, 1119–1133. [CrossRef] [PubMed]
33. Kuma, A.; Kato, A. Lifestyle-Related Risk Factors for the Incidence and Progression of Chronic Kidney Disease in the Healthy Young and Middle-Aged Population. *Nutrients* 2022, *14*, 3787. [CrossRef] [PubMed]
34. Zhang, X.; Tong, T.; Chang, A.; Ang, T.F.A.; Tao, Q.; Auerbach, S.; Devine, S.; Qiu, W.Q.; Mez, J.; Massaro, J. Midlife lipid and glucose levels are associated with Alzheimer's disease. *Alzheimer's Dement.* 2023, *19*, 181–193. [CrossRef] [PubMed]
35. Cavero-Redondo, I.; Peleteiro, B.; Álvarez-Bueno, C.; Rodriguez-Artalejo, F.; Martínez-Vizcaíno, V. Glycated haemoglobin A1c as a risk factor of cardiovascular outcomes and all-cause mortality in diabetic and non-diabetic populations: A systematic review and meta-analysis. *BMJ Open* 2017, *7*, e015949. [CrossRef]
36. Bower, J.K.; Appel, L.J.; Matsushita, K.; Young, J.H.; Alonso, A.; Brancati, F.L.; Selvin, E. Glycated hemoglobin and risk of hypertension in the atherosclerosis risk in communities study. *Diabetes Care* 2012, *35*, 1031–1037. [CrossRef]
37. Hu, J.; Zhang, X.; Gu, J.; Yang, M.; Zhang, X.; Zhao, H.; Li, L. Serum alkaline phosphatase levels as a simple and useful test in screening for significant fibrosis in treatment-naive patients with hepatitis B e-antigen negative chronic hepatitis B. *Eur. J. Gastroenterol. Hepatol.* 2019, *31*, 817–823. [CrossRef]
38. Rafiq, N.; Bai, C.; Fang, Y.; Srishord, M.; McCullough, A.; Gramlich, T.; Younossi, Z.M. Long-term follow-up of patients with nonalcoholic fatty liver. *Clin. Gastroenterol. Hepatol.* 2009, *7*, 234–238. [CrossRef]
39. Liu, X.; Zhang, H.; Liang, J. Blood urea nitrogen is elevated in patients with non-alcoholic fatty liver disease. *Hepato-Gastroenterol.* 2013, *60*, 343–345.
40. Lan, Q.; Zheng, L.; Zhou, X.; Wu, H.; Buys, N.; Liu, Z.; Sun, J.; Fan, H. The value of blood urea nitrogen in the prediction of risks of cardiovascular disease in an older population. *Front. Cardiovasc. Med.* 2021, *8*, 614117. [CrossRef]
41. Zhu, L.; Fang, Z.; Jin, Y.; Chang, W.; Huang, M.; He, L.; Chen, Y.; Yao, Y. Association between serum alanine and aspartate aminotransferase and blood pressure: A cross-sectional study of Chinese freshmen. *BMC Cardiovasc. Disord.* 2021, *21*, 472. [CrossRef]
42. Park, E.-O.; Bae, E.J.; Park, B.-H.; Chae, S.-W. The associations between liver enzymes and cardiovascular risk factors in adults with mild dyslipidemia. *J. Clin. Med.* 2020, *9*, 1147. [CrossRef] [PubMed]
43. Noa, M.; Mendoza, S.; Mas, R.; Mendoza, N. Effect of policosanol on carbon tetrachloride-induced acute liver damage in Sprague-Dawley rats. *Drugs R&D* 2003, *4*, 29–35. [CrossRef]
44. Zein, N.; Yassin, F.; Makled, S.; Alotaibi, S.S.; Albogami, S.M.; Mostafa-Hedeab, G.; Batiha, G.E.-S.; Elewa, Y.H.A. Oral supplementation of policosanol alleviates carbon tetrachloride-induced liver fibrosis in rats. *Biomed. Pharmacother.* 2022, *150*, 113020. [CrossRef] [PubMed]
45. Lee, E.-Y.; Yoo, J.-A.; Lim, S.-M.; Cho, K.-H. Anti-aging and tissue regeneration ability of policosanol along with lipid-lowering effect in hyperlipidemic zebrafish via enhancement of high-density lipoprotein functionality. *Rejuvenation Res.* 2016, *19*, 149–158. [CrossRef] [PubMed]
46. Cho, K.-H.; Yadav, D.; Kim, S.-J.; Kim, J.-R. Blood pressure lowering effect of cuban policosanol is accompanied by improvement of hepatic inflammation, lipoprotein profile, and HDL quality in spontaneously hypertensive rats. *Molecules* 2018, *23*, 1080. [CrossRef]
47. Zhai, Z.; Niu, K.M.; Liu, H.; Lin, C.; Tu, Y.; Liu, Y.; Cai, L.; Ouyang, K.; Liu, J. Policosanol alleviates hepatic lipid accumulation by regulating bile acids metabolism in C57BL6/mice through AMPK–FXR–TGR5 cross-talk. *J. Food Sci.* 2021, *86*, 5466–5478. [CrossRef]
48. Singh, D.K.; Li, L.; Porter, T.D. Policosanol inhibits cholesterol synthesis in hepatoma cells by activation of AMP-kinase. *J. Pharmacol. Exp. Ther.* 2006, *318*, 1020–1026. [CrossRef]
49. Nam, D.-E.; Yun, J.-M.; Kim, D.; Kim, O.-K. Policosanol attenuates cholesterol synthesis via AMPK activation in hypercholesterolemic rats. *J. Med. Food* 2019, *22*, 1110–1117. [CrossRef]
50. Amadi, C.N.; Orisakwe, O.E. Herb-induced liver injuries in developing nations: An update. *Toxics* 2018, *6*, 24. [CrossRef]
51. Suk, K.T.; Kim, D.J. Drug-induced liver injury: Present and future. *Clin. Mol. Hepatol.* 2012, *18*, 249. [CrossRef]

52. Obi, E.; Akunyili, D.N.; Ekpo, B.; Orisakwe, O.E. Heavy metal hazards of Nigerian herbal remedies. *Sci. Total Environ.* **2006**, *369*, 35–41. [CrossRef] [PubMed]
53. Ernst, E. Adulteration of Chinese herbal medicines with synthetic drugs: A systematic review. *J. Intern. Med.* **2002**, *252*, 107–113. [CrossRef] [PubMed]

Disclaimer/Publisher's Note: The statements, opinions and data contained in all publications are solely those of the individual author(s) and contributor(s) and not of MDPI and/or the editor(s). MDPI and/or the editor(s) disclaim responsibility for any injury to people or property resulting from any ideas, methods, instructions or products referred to in the content.

MDPI AG
Grosspeteranlage 5
4052 Basel
Switzerland
Tel.: +41 61 683 77 34

Life Editorial Office
E-mail: life@mdpi.com
www.mdpi.com/journal/life

Disclaimer/Publisher's Note: The statements, opinions and data contained in all publications are solely those of the individual author(s) and contributor(s) and not of MDPI and/or the editor(s). MDPI and/or the editor(s) disclaim responsibility for any injury to people or property resulting from any ideas, methods, instructions or products referred to in the content.

www.ingramcontent.com/pod-product-compliance
Lightning Source LLC
LaVergne TN
LVHW072328090526
838202LV00019B/2371